# Coastal Lagoons

*Ecosystem Processes and Modeling for Sustainable Use and Development*

# Coastal Lagoons

## Ecosystem Processes and Modeling for Sustainable Use and Development

*edited by*
**I. Ethem Gönenç**
IGEM Research & Consulting Co.
Turkey

**John P. Wolflin**
US Fish and Wildlife Service
USA

## CRC PRESS

Boca Raton   London   New York   Washington, D.C.

Cover Art: Koycegiz–Dalyan Lagoon, Turkey. (Photograph by Cem Gazioglu. Photo processing by Ali Ertürk.)

## Library of Congress Cataloging-in-Publication Data

Coastal lagoons : ecosystem processes and modeling for sustainable use and development/
edited by I. Ethem Gönenç and John P. Wolflin.
   p. cm.
   Includes bibliographical references and index.
   ISBN 1-56670-686-6 (alk. paper)
   1. Coastal ecology. 2. Sustainable development. I. Gönenç, I. Ethem. II. Wolflin, John.

QH541.5.C65C5915 2004
577.7'8--dc22                                                                           2004051926

**Visit the CRC Press Web site at www.crcpress.com**

© 2005 by CRC Press

No claim to original U.S. Government works
International Standard Book Number 1-56670-686-6
Library of Congress Card Number 2004051926
Printed in the United States of America  1  2  3  4  5  6  7  8  9  0
Printed on acid-free paper

# Preface

Coastal lagoons are the most valuable components of coastal areas. In most NATO coastal countries, the majority of the population lives within a 50-km coastal band. Increasing human use and development pressures in coastal areas and, in fact, coastal lagoons, make these dynamic and productive areas very sensitive and vulnerable to deterioration. This has resulted in direct and indirect impacts that have considerably reduced the ability of these areas to meet an ever-increasing demand for their use and development. For this reason, the NATO Committee on the Challenges of Modern Society (CCMS) Pilot Study on Ecosystem Modeling of Coastal Lagoons for Sustainable Management was initiated in 1995. The purpose of the pilot study was to develop strategies for sustained use of these areas. Turkey was the Pilot Study Director, and the United States was the Co-Pilot Study Director. Canada, Italy, Poland, Portugal, Romania, the Russian Federation, and Spain were the main participants. Azerbaijan, Denmark, Kazakhstan, Kyrgyzstan, Lithuania, Turkmenistan, the Ukraine, and the United Kingdom attended as observer and contributing countries.

The pilot study investigated the existing knowledge on hydrodynamics and ecology of lagoons. This book, compiled as a product of the pilot study, summarizes the role of modeling as a support system for decision making to provide sustained use and development of lagoon ecosystems. Sustainable use management is a conscious social decision that provides for the long-term health of both the ecological and economic systems of a lagoon. Lagoon ecosystem components, which describe the "supply" side or natural capital of the lagoon system, and are available to the socio-economic or "demand" side, are defined in detail. The hierarchical interrelationships of these ecosystem components are considered to be the basic conceptual and methodological elements used to define and manage sustainability. As reflected in the case studies and other investigations by the scientists who have authored chapters of this book and contributed to the pilot study, the finite capacity of the natural capital of lagoons cannot meet the growing demands of the socio-economic system without a strategy of sustainable use and development.

As discussed, decision makers must balance the cost and benefits of alternative uses of the natural capital of the environment. This highlights that the use of models as tools in the decision-making process provides awareness of the interrelationships between input and output variables within the ecosystem and related environments. The authors present the body of knowledge in the field and the experience gained from the pilot study.

The use of a model to predict the outputs of the ecological system provides the basis for estimating both the ecological and economic changes expected. As a model predicts outputs, changes in the natural capital need to be recorded as assets or liabilities for the affected economy. Changes in input variables result in corresponding output change in the natural capital that needs to be documented in economic terms. Assigning monetary values to goods and services is critical to decision makers.

Modeling further enhances the accuracy of predictions for and awareness of the consequences of human actions and decisions concerning the use and development of lagoon systems as a whole. It allows the consequences of decisions on use of natural capital and the future benefit, or loss to public or private financial interests. This book emphasizes the essential role of modeling in modern coastal lagoon management.

# Acknowledgments

This book is a product of the Ecosystem Modeling of Coastal Lagoons for Sustainable Management Pilot Study supported by the NATO Committee on the Challenges of Modern Society (CCMS), which was initiated in 1995. This book would not exist without the support of the NATO-CCMS managers in Brussels as well as the national coordinators of participating countries. We particularly want to thank Dr. Deniz Beten, CCMS Program Director, and Prof. Dr. Nejat Ince, National CCMS Coordinator for Turkey, for their dedication and support.

This book would not have been possible without the dedication of many individuals and organizations. We are deeply indebted to our colleagues, the scientists who wrote the various chapters. Their chapters reflect their dedication to improving the management of coastal lagoons and to scientific accuracy and thoroughness. Less evident, but equally essential to the success of this project, was their receptiveness to editorial suggestions, their patience during the extensive review process, and their persistence in seeing the manuscript through to a successful conclusion.

Many workshops were organized in the participating countries during the pilot study. We deeply appreciate all the national and local institutions of the participating countries, which honored the contributors of this book with their support and contributions to the workshops. Many scientists provided valuable counsel.

On behalf of the contributors, we especially want to thank Dr. Richard Wetzel, David Fausel, and Christine Andreasen for editorial guidance. We were fortunate to have several graduate students—Bade Cebeci, Nusret Karakaya, Ali Ertürk, Constantin Cazacu, Flavio Martins, Kiziltan Yüceil, Alpaslan Ekdal, and many others—who unselfishly contributed to the development of the manuscript. We appreciate the support and assistance of Nanci Henningsen of the Center of Marine Biotechnology, University of Maryland, and Dr. Winston Lung of the University of Virginia. We recognize Ms. Lynn Schoolfield of the U.S. EPA as part of the pilot study family and one of our most ardent supporters.

Finally, we want to express our appreciation to the pilot study family, which includes not only the chapter authors but also their families. The support and understanding of these families were instrumental in ensuring the completion of this project. As the pilot study family has developed over time, we have identified ourselves as the LEMSM (Lagoon Ecosystem Modeling for Sustainable Management) family.

We, the editors, dedicate this book to those responsible for lagoon ecosystem management worldwide in the hope that these valuable areas are conserved for future generations. This conservation can only be achieved through the implementation of sustainable management practices such as those describe herein. We are also particularly grateful for the summer offshore lagoon breezes, called *meltem* in the Mediterranean countries, which sustained us during our studies and the development of the manuscript.

**I. Ethem Gönenç and John P. Wolflin**

# Biographies

## THE EDITORS

**I. Ethem Gönenç, Ph.D.,** is an internationally recognized scientist who is considered a visionary. He is currently CEO of IGEM Research & Consulting Co. in Istanbul, Turkey, and Chair of the Sustainable Ecosytem Society (SES). He received his Ph.D. from Istanbul Technical University (ITU), where he worked for 28 years. He has been the recipient of NATO and World Health Organization fellowships. At Denmark Technical University he developed a new model for biofilm kinetics with Professor Paul Harremoes for which he received the Danish Royalty Award.

Dr. Gönenç is the director of the NATO-CCMS Pilot Study on Ecosystem Modeling of Coastal Lagoons for Sustainable Management, which was initiated in 1995. He brought together expert scientists from 17 countries and was the catalyst behind the development of the manuscript for this book. Dr. Gönenç has done pioneering work in the field of environmental science and technology. He has worked on master plans and feasibility studies in his home country of Turkey, including those for Istanbul, a mega-metropolis of 12 million people, and Ankara, the capital. He was responsible for the planning and development of the innovative approach to wastewater management in Istanbul.

Dr. Gönenç has published extensively in international journals and has made presentations at numerous conferences and congresses throughout the world on a wide range of environmental management topics including sustainable watershed management, environmental policy, environmental impact assessment, ecological modeling, diffusion in environment, biofilm kinetics, and nitrification/denitrification. He is currently involved in several multidisciplinary research projects. Although recently retired from ITC, he continues to be dedicated to implementation of science-based environmental management practices within the context of sustained use and development of natural resources on an international basis.

IGEM Research & Consulting Co. and SES
Ormankent Sitesi 1
Villa No. 713
Sirapinar Koyu
tel/fax: 90 216 4355484
e-mail: iegonenc@igemconsulting.comd35@isnet.net.tr

**John P. Wolflin, M.S.,** teamed with I. Ethem Gönenç in 1995 as the co-director of the NATO-CCMS Pilot Study on Ecosystem Modeling of Coastal Lagoons for Sustainable Managemen. He is one of the principal editors of the manuscript developed by the scientific experts affiliated with the pilot study. Mr. Wolflin has a B.S. in biology and behavioral science from Western Michigan University, Kalamazoo, and an M.S. in biology with a special emphasis on coastal ecology from Southern Connecticut State University, New Haven.

Mr. Wolflin is a tenured scientist with more than 30 years of experience in natural resource management and education. Currently, he is the Supervisor of the U.S. Fish and Wildlife Service Chesapeake Bay Office, Annapolis, Maryland, where he oversees conservation programs on wetlands regulation and restoration, Federal navigation and highway projects, environmental contaminant assessment and remediation, endangered and invasive species, and application of conservation provisions of other Federal environmental laws and international treaties that focus on migratory birds and interjurisdictional fisheries.

Mr. Wolflin began his career as a field scientist. He is recognized for his work on complex and controversial land and water resource developments. He chaired an interdisciplinary team that developed a comprehensive management strategy for fish and wildlife, commercial navigation, and recreational use of the upper Mississippi River. He has completed the Department of the Interior's Executive Management Development Program. He supervised the Service's Idaho Office and has worked with the Philippine Bureau of Fisheries and Aquatic Resources as a U.S. Peace Corps volunteer.

Mr. Wolflin has lectured on sustained use and development of natural resources at universities in the U.S. and abroad. He taught at the Johns Hopkins University from 1990 to 1995. He has written extensively on resource management and made presentations at numerous conferences including the International Wetlands Research Bureau, Wetlands International, and the European Union/U.S. Information Agency World Net Broadcast.

U.S. Fish and Wildlife Service
Chesapeake Bay Office, Annapolis, MD 21617, U.S.A.
tel: 410 573 4573; fax: 410 266 9127
e-mail: john_wolflin@fws.gov

## BIOGRAPHIES OF THE PRINCIPAL CHAPTER AUTHORS

**Eugeniusz Andrulewicz, Ph.D.,** is a senior scientist and project leader at the Sea Fisheries Institute, Gdynia, Poland. He received his M.Sc. in chemistry from Gdansk University and his Ph.D. in natural sciences from the Agriculture Academy in Szczecin. For 20 years he worked as a marine chemist at the Institute of Meteorology and Water Management Maritime Branch in Gdynia, where he conducted field and laboratory measurements on salinity, oxygen, nutrients, trace metals, chlorinated hydrocarbons, and petroleum hydrocarbons. His research areas include marine chemistry, marine ecology, and underwater exploration. He is currently involved in research and assessments of the biological effects of anthropogenic activities, ecosystem health studies, and ecosystem-based management.

Dr. Andrulewicz is the author of numerous papers in peer-reviewed journals and of more than 100 scientific, technical, and popular publications. He has been involved in various international governmental organizations, such as ICES, HELCOM,

SCOR, IMO-MEPC, and GIWA, and in nongovernmental organizations, including WWF, CBO (Conference of the Baltic Oceanographers), and BMB (Baltic Marine Biologists). He was a member of the ICES Advisory Committee on Marine Pollution, the vice chairman of the HELCOM Environment Committee and the HELCOM Working Group on Monitoring and Assessments, the co-chairman of the HELCOM Working Group on Management Plans for Coastal Lagoons and Wetlands, and a member of the Editorial Board of the German *Journal of Hydrography*. He is now co-chairman of GIWA Baltic Sea Sub-Region No. 17, and a member of the Advisory Committee on the Marine Environment, the ICES Advisory Committee on Ecosystems, and the ICES Marine Habitat Committee.

Department of Fisheries Oceanography and Marine Ecology
Sea Fisheries Institute
ul. Kollataja 1, 81-332 Gdynia, Poland
tel: 48 58 620 17 28; fax: 48 58 620 28 31
e-mail: Eugene@mir.gdynia.pl

**Boris Chubarenko, Ph.D.,** received his M.Sc. in mechanical engineering in 1994 from the Moscow Institute for Physics and Technology. He defended his Ph.D. thesis on numerical methods of wave dynamics in nonlinear media in 1987. He studied marine environmental pollution prevention in Sweden in 1994 and environmental water management, monitoring, and modeling in Denmark in 1997. Since 1987 he has worked at Atlantic Branch of P.P. Shirshov Institute of Oceanology of the Russian Academy of Sciences, headed by Dr. V. Paka. He is currently a Laboratory Head for Coastal Systems Study at the P.P. Shirishov Institute.

From 1990 to 1992 Dr. Chubarenko was a head of the Ecological Laboratory at Kaliningrad State University. He was a guest scientist at the Department of Environmental Flows, headed by Professor Kolumban Hutter, at Darmstadt Technical University, Germany, where he participated in theoretical and observational projects on physical limnology that focused on Lake Constance. Currently, he lectures on modeling of coastal water systems at Kaliningrad State University. His research interests focus on physical coastal oceanography, integrative studies of lagoons, estuaries and sandy coasts, and application of numerical models in environmental impact assessment, environmental monitoring, and operational oceanography. He is currently concentrating on integrative studies of the Baltic lagoons (the Vistula, Curonian, and Darss–Zingst Bodden Chain lagoons) and of the Baltic coastal zone, including monitoring and environmental impact assessment of port developments. Since 1994, he has prepared and coordinated international projects supported by European funds. He has published extensively in international journals.

Atlantic Branch of P.P. Shirshov Institute of Oceanology
Russian Academy of Sciences, Prospect Mira 1
Kaliningrad 236000, Russia
tel: 7 0112 451574; fax: 7 0112 516970
e-mail: chuboris@ioran.baltnet.ru

**Sofia Gamito, Ph.D.,** is an assistant professor at the Faculdade de Ciências do Mar e do Ambiente (Department of Marine and Environmental Sciences) at the University of Algarve, Portugal. She received her degree in biology in 1983 from the University of Lisbon. In 1990 she completed a degree equivalent to an M.Sc. in marine ecology and in 1994 she obtained her Ph.D. in biological sciences, with a speciality in marine ecology. Her main research interests are in marine ecology, particularly biological processes, ecological modeling, trophic food webs, experimental ecology, and environmental management, focusing on lagoonal ecology and management of protected areas. She has published extensively and has participated in several national and international projects related to ecology and management of coastal areas.

Faculty of Marine and Environmental Sciences
Campus de Gambelas
University of Algarve, 8005-139 Faro, Portugal
tel: 35 1 289 800976; fax: 35 1 289 818353
e-mail: sgamito@ualg.pt

**Javier Gilabert, Ph.D.,** is an associate professor at the Technical University of Cartagena, Spain. He obtained his M.Sc. degree in marine biology in 1985 from the University of the Balearic Islands and his Ph.D. in ecology in 1992 from the University of Murcia. He worked for the Remote Sensing Department of Indra Space, a Sapin-based international IT company, and he was an EU Marie Curie postdoctoral fellow for 2 years at the Plymouth Marine Laboratory in the U.K. He has published extensively and been a reviewer for several aquatic ecology journals. His current research interest is plankton dynamics from an interdisciplinary point of view, from physical (hydrodynamics) to chemical (nutrient diffusion) to biological processes (uptake of nutrients by phytoplankton and grazing impact of zooplankton). He has been studying the dynamics of the planktonic size structure in different environments in order to incorporate the mechanisms controlling the planktonic size structure into the ecosystem models. Projects concerning these aspects have been carried out at several spatial scales ranging from coastal lagoons such as the Mar Menor in Spain to the Mediterranean Sea to the Atlantic Ocean (90° latitudinal transects). He is also involved in the study of harmful regional algae blooms.

Department of Chemical and Environmental Engineering
Technical University of Cartagena
Alfonso XIII, 44. 30202 Cartagena, Spain
tel: 34 968 325 669; fax: 34 968 325 435
e-mail: javier.gilabert@upct.es

**Melike Gürel, Ph.D.,** is an assistant professor in the Environmental Engineering Department of Istanbul Technical University (ITU), Turkey. She obtained her B.Sc. in 1992, her M.Sc. in 1994, and her Ph.D. in 2000, all in environmental engineering, from ITU. She has worked in the Environmental Engineering Department of ITU since 1994. From 2001 to 2002, she was a visiting scientist in the field of water quality modeling in the Ecosystems Research Division of the U.S. Environmental

Protection Agency, Athens, Georgia. Her current areas of interest are eutrophication, water quality modeling, and integrated watershed management.

Dr. Gürel is a member of the International Water Association (IWA), the National Committee of Turkey on Water Pollution Control (SKATMK), and the Chamber of Environmental Engineers, Turkey (CMO). She has published in international SCI journals and in international proceedings. She has taken part in two NATO-CCMS projects, one in NATO CR, and many national projects supported by ITU or the Scientific and Technical Research Council of Turkey.

Department of Civil Engineering
Istanbul Technical University
34469 Maslak, Istanbul, Turkey
tel: 90 212 285 65 79; fax: 90 212 285 65 87
e-mail: mgurel@ins.itu.edu.tr

**Vladimir G. Koutitonsky, Ph.D.,** is a coastal physical oceanographer engaged in fundamental and applied research at the Institut des Sciences de la Mer de Rimouski (ISMER), Université du Québec à Rimouski, Québec, Canada. He obtained his B.Sc. in physics and chemistry in 1969 from the American University in Cairo, Egypt; his M.Sc. in physical oceanography in 1973 from McGill University, Montréal, Canada; and his Ph.D. in coastal oceanography in 1985 from the Marine Sciences Research Centre, State University of New York at Stony Brook, U.S.A.

Dr. Koutitonsky's research focuses on coastal environmental hydraulics, modeling and field observations as described in numerous publications in international journals and research and environmental impact reports. Most of his fundamental research deals with mesoscale processes occurring at inertial, tidal, and synoptic frequencies in stratified coastal waters, primarily in the St. Lawrence estuary. His practical expertise includes planning, supervising, and conducting environmental impact studies involving field measurements and numerical models (the MIKE21 and MIKE3 systems, www.dhi.dk) for studying water circulation and quality, eutrophication, aquaculture site selection, transport of particles (sediments, larvae, oil spills, and pollutants), and refraction of waves in the coastal zone, estuaries, lagoons, and bays. Dr. Koutitonsky teaches coastal dynamics courses worldwide and graduate courses at ISMER.

Institut des Sciences de la Mer de Rimouski (ISMER)
310 Allee des Ursulines, Rimouski, Quebec, Canada, G5L-3A1
tel: 418 723-1986, ext. 1763; fax: 418 724-1842
e-mail: VGK@uqar.qc.ca

**Concepción Marcos Diego, Ph.D.,** did graduate work in biology at the University of La Laguna, Canary Islands, Spain and received her Ph.D. from the University of Murcia on ecological planning in the coastal zone. She is Professor Titular of Ecology and a researcher for the Ecology and Management of Coastal Marine Ecosystems group in the Department of Ecology and Hydrology at the University of Murcia. She has published on management of coastal ecosystems, land capability, biological indicators, marine protected areas, coastal lagoons ecology (mainly in the Mar Menor lagoon), and the population dynamics and ecology of echinoderms and benthic fishes. Her research was conducted in the Galapagos Islands, the Atlantic archipelagos (Azores, Madeira, Canary Islands, and Cabo Verde), the Antarctic, and the Mediterranean. She collaborates as a scientific assessor for several environmental management and research projects with local administrations such as the Consejería de Agrigultura and the Agua y Medio Ambiente of the Region de Murcia Government. She is a subdirector of the Institute of the Water and the Environment at the University of Murcia.

Department of Ecology and Hydrology
Campus de Espinardo
University of Murcia, 30100 Murcia, Spain
tel: 34 968 364978; fax: 34 968 363963
e-mail: cmarcos@um.es

**Ramiro Neves, Ph.D.,** is an associate professor at the Instituto Superior Técnico (IST) of the Technical University of Lisbon, Portugal. He received his degree in mechanical engineering in 1979 from IST and his Ph.D. in applied sciences in 1986 from the University of Liège, Belgium. His main interests are hydrodynamic modeling, circulation, and sediment transport. His team has developed several hydrodynamic models for European estuaries and coastal seas to access marine environmental problems. One of their goals is to achieve exploitable products to solve marine environmental issues such as eutrophication, dredging, circulation, and water contaminants dispersal.

MARETEC, Instituto Superior Técnico (IST)
Technical University of Lisbon, Pavilhão de Turbomáquinas
Av. Rovisco Pais, 1049-001 Lisbon, Portugal
tel: 35 1 21 841 7397; fax: 35 1 21 841 7398
e-mail: ramiro.neves@ist.utl.pt

**Angel Pérez-Ruzafa, Ph.D.,** did graduate work in biology at the University of Murcia, Spain and the University of La Laguna, Canary Islands. He received his Ph.D. from the University of Murcia with a thesis on the ecology of the benthic assemblages of the Mar Menor Lagoon. He is Professor Titular of Ecology and Head of the Research Group on Ecology and Management of Coastal Marine Ecosystems of the Department of Ecology and Hydrology in the Department of Biology at the University of Murcia. He has published extensively on coastal lagoon ecology, mainly on the Mar Menor Lagoon, environmental impact, biological indicators, marine protected areas, and the dynamics of populations and ecology of echinoderms and benthic fishes. He has conducted national and international research projects on the Galapagos Islands, the Atlantic archipelagos (the Azores, Madeira, Canary Islands and Cabo Verde), the Antarctic, and the Mediterranean. He has been Vice-Chancellor of University Extension and International Relationships at the University of Murcia. He is a scientific advisor to several environmental management and research projects conducted by national and local administrations such as the Consejería de Agricultura, Agua y Medio Ambiente of the Murcia government and the Dirección General de Puertos y Costas of the Spanish Ministerio de Fomento.

Department of Ecology and Hydrology
Campus de Espinardo
University of Murcia, 30100 Murcia, Spain
tel: 34 968 364998; fax: 34 968 363963
e-mail: angelpr@um.es

**Rosemarie C. Russo, Ph.D.,** is the director of the Ecosystems Research Division of the National Exposure Research Laboratory, Office of Research and Development, U.S. Environmental Protection Agency, Athens, Georgia, U.S.A. She received her B.S. in chemistry in 1964 from the University of Minnesota, Duluth and her Ph.D. in inorganic chemistry in 1972 from the University of New Hampshire. Her research interests are aquatic toxicology, ammonia and nitrite toxicity in aquatic organisms, and water quality criteria. Research conducted in the EPA laboratory in Athens focuses on identification and quantification of transformation processes affecting behavior of contaminants in environmental systems and the development of mathematical models to assess the response of aquatic systems to stresses from natural and anthropogenic sources. Field and laboratory studies support process research, model development, testing and validation, and characterize variability and prediction uncertainty. She has written extensively on toxicity of ammonia to fishes and national water quality criteria documents. She has worked on international environmental research projects in Russia, the Ukraine, Poland, Estonia, Lithuania, and China.

Ecosystems Research Division
U.S. EPA, 960 College Station Road
Athens, Georgia 30605-2700 U.S.A.
tel: 1-706-355-8001
e-mail: russo.rosemarie@epa.gov

**Aysegül Tanik, Ph.D.,** is a full professor at Istanbul Technical University (ITU) in Turkey. She obtained her B.Sc. in chemical engineering in 1981 and her M.Sc. in environmental engineering in 1984 from Bogazici University, Istanbul. She received her Ph.D in environmental engineering in 1991 from ITU, where she has been a member of the teaching staff of the Department of Environmental Engineering since 1992 and she became a full professor in 2002.

From 1984 to 1992 Dr. Tanik worked as a project and research engineer for various contracting firms dealing with water and wastewater treatment. Her current fields of interest are determination and management of diffuse sources of pollutants, water quality management, water quality modeling, and watershed management. She has lectured recently to undergraduate environmental engineering students on environmental impact assessment, marine pollution and its control, and environmental law, and to graduate students on diffuse pollution, integrated watershed management, and fate and transport of pollutants in the environment.

Dr. Tanik is a member of the International Water Association (IWA), the Turkish National Committee on Water Pollution Control (SKATMK), and the Chamber of Civil Engineers. She has published extensively. She has participated in two NATO-CCMS projects, one in NATO CR and many national projects supported by ITU and the Scientific and Technical Research Council of Turkey.

Department of Environmental Engineering
Istanbul Technical University
34469 Maslak, Istanbul, Turkey
tel: 90 212 2856884; fax: 90 212 2853781
e-mail:tanika@itu.edu.tr

**Karen Terwilliger, M.S.,** is a nature conservation and communication consultant who has worked in both the private and public sectors with local, state, federal, and international government and nongovernmental organizations. Ms. Terwilliger received her B.S. from Purdue University, Layfayette, Indiana, in 1976 and her M.S. from Old Dominion University, Norfolk, Virginia, in 1981. She directed Virginia's Wildlife Diversity and Endangered Species Program for 15 years, and worked with the U.S. Forest Service, U.S. Fish and Wildlife Service, U.S. Army Corps of Engineers before starting her own company. Based on her 30 years of diverse expertise, Terwilliger Consulting Inc. provides natural resource conservation and communication services to private and public landowners and organizations. Her international work includes work with the International Association of Fish and Wildlife Agencies and NATO. She is a member of several federal, regional, and state technical committees and endangered species recovery teams. She has held state and regional positions in professional societies and has received numerous awards for her outstanding contributions.

Terwilliger Consulting, Inc.
28295 Burton Shore Rd.
Locustville, VA 23404, U.S.A.
tel: 757-787-2637, ext. 11; fax: 757-787-2411
e-mail: natural@visi.net

**Georg Umgiesser, Ph.D.,** is a senior scientist at the ISMAR-CNR, the Institute of Marine Sciences of the National Research Council in Venice, Italy. He graduated in 1986 from the Institut für Meereskunde in Hamburg, Germany, after completing his thesis on the modeling of the Venice Lagoon. His main interests are hydrodynamic numerical modeling, circulation, and sediment transport. He developed a framework of finite element models for the application to shallow water bodies, SHYFEM, which resolves hydrodynamic processes, as well as processes of water quality and sediment transport. This model has been applied extensively to the Venice Lagoon and other Italian and Mediterranean lagoons. He has participated in various European Union projects dealing with the North Sea, the Mediterranean Sea, turbulence, and the application of three-dimensional hydrodynamic models. He teaches at the University of Venice and he has been a visiting professor at Kyushu University in Japan.

Dr. Umgiesser is the author of numerous publications in peer-reviewed journals. He is currently researching the sea–lagoon exchange, the influence of the hydrological cycle on the lagoon environment, and sediment transport in very shallow water bodies.

ISMAR-CNR (Institute of Marine Sciences)
S. Polo 1364, 30125 Venice, Italy
tel: 39 041 5216875; fax.: 39 041 2602340
e-mail: georg.umgiesser@ismar.cnr.it

**Angheluta Vadineanu, Ph.D.,** graduated in biology in 1972 and received his Ph.D. in ecology in 1980 from the University of Bucharest, Romania. He is a full professor and head of the Department of Systems Ecology and Sustainability (UNIBUC-ECO), and he holds the UNESCO-Cousteau Chair in Ecotechnie at the University of Bucharest. He has a broad and varied experience in teaching and training of human resources in the field of systems ecology, sustainability, knowledge development, and use of natural capital and the socio-economic system. For more than 25 years his research and management activities have focused on the dynamics of the structure and productivity of the Lower Danube River Wetlands System, including the Razim-Sinoe Lagoon system, the identification and management of the Romanian ecological network, the structure and functioning of benthic communities, population dynamics and the energy budget, the assessment of energy flow and biogeochemical cycles in wetlands and terrestrial ecosystems; and more recently, the social and economic valuation of natural capital and assessment of the ecological footprint for local or national socio-economic systems. He is author or co-author of more than 120 scientific papers and eight books. He was a member of the Environmental Advisory Committee for the European Bank for Reconstruction and Development from 1992 to 1994. He is a member of several international scientific and advisory bodies, including the Scientific Committee of the European Center for Nature Conservation and IUCN Commission on Ecosystem Management. He also coordinates the National Network for Long Term Ecological Research.

Department of Systems Ecology and Sustainability
University of Bucharest SPL
Independentei 91-95
76201 Bucharest, Romania
tel: 40 1 411 23 10; fax: 40 1 411 23 10
e-mail: anvadi@bio.bio.unibuc.ro

# Other Contributors

**Bilsen Beler Baykal, Ph.D.**
Department of Civil Engineering
Istanbul Technical University
34469 Maslak
Istanbul, Turkey

**Irina Chubarenko, Ph.D.**
Laboratory for Coastal Systems Study
Atlantic Branch, P.P. Shirshov Institute
  of Oceanology
Russian Academy of Sciences
Prospect Mira 1
236000 Kaliningrad
Russia

**Ertugrul Dogan, Ph.D.**
Institute of Marine Sciences, Vefa
Istanbul University
Istanbul, Turkey

**Alpaslan Ekdal, M.Sc.**
Department of Civil Engineering
Istanbul Technical University
34469 Maslak
Istanbul, Turkey

**Ali Ertürk, M.Sc.**
Department of Civil Engineering
Istanbul Technical University
34469 Maslak
Istanbul, Turkey

**Nusret Karakaya, M.Sc.**
Department of Civil Engineering
Istanbul Technical University
34469 Maslak
Istanbul, Turkey

**Erdogan Okus, Ph.D.**
Institute of Marine Sciences, Vefa
Istanbul University
Istanbul, Turkey

**Dursun Z. Seker, Ph.D.**
Department of Civil Engineering
Istanbul Technical University
34469 Maslak
Istanbul, Turkey

**Aylin Bederli Tümay, Ph.D.**
Department of Civil Engineering
Istanbul Technical University
34469 Maslak
Istanbul, Turkey

**Kiziltan Yüceil, Ph.D.**
Department of Civil Engineering
Istanbul Technical University
34469 Maslak
Istanbul, Turkey

# Contents

**Chapter 9**

*I. Ethem Gönenç, Vladimir G. Koutitonsky, Angel Pérez-Ruzafa,*
*Concepción Marcos Diego, Javier Gilabert, Eugeniusz Andrulewicz,*
*Boris Chubarenko, Irina Chubarenko, Melike Gürel, Aysegül Tanik,*
*Ali Ertürk, Ertugrul Dogan, Erdogan Okus, Dursun Z. Seker, Alpaslan Ekdal,*
*Kiziltan Yüceil, Aylin Bederli Tümay, Nusret Karakaya,*
*and Bilsen Beler Baykal*

# 1  Introduction

## I. Ethem Gönenç and John P. Wolflin

## CONTENTS

## 1.1  BACKGROUND: ISSUES AND APPROACH

Lagoons are the most valuable components of coastal areas in terms of both the ecosystem and natural capital. Surrounding areas of lagoons provide excellent opportunities for agriculture and tourism sectors on the one hand and for fishery and aquatic products sectors on the other hand. Sustainable use management is a conscious social decision that provides for the long-term health of both the ecological and economic systems of a lagoon and surrounding areas. However, the concept of sustainable management of lagoons is often either not clearly understood or not applied.

The NATO Committee on the Challenges of Modern Society (CCMS) study was initiated in 1995 to define and promote sustainable use management in lagoons. The focus was on integrating management decisions with current modeling methodologies. This book, *Coastal Lagoons: Ecosystem Processes and Modeling for Sustainable Use and Development,* is a product of this NATO-CCMS study.

Coastal lagoons are shallow aquatic ecosystems that develop at the interface between coastal terrestrial and marine ecosystems. They are driven to a major extent by the high density of noncommercial auxiliary energy and mass exchanged with the surrounding ecosystems. The rate of structural and functional change of hydrogeomorphological units and biological communities is particularly dependent on the exchanges of auxiliary energy and mass. Although lagoons are intricately connected to surrounding environments, they develop mechanisms for structural and functional regulation, which result in specific biological productivity and carrying capacities.

Continental and marine environments influence coastal lagoons by definition of location. Historically, coastal regions have been areas prone to human habitation. The resulting rural and urban landscapes reflect human orientation toward the use of the natural capital of lagoons. Lagoons are sensitive areas that play an important role among the coastal zone ecosystems as they provide suitable breeding areas for many species. Today many lagoons are deteriorating because of overuse of their natural capital. Fisheries and aquaculture, tourism, urban, industrial, and agricultural developments are typical uses that are not only uncontrolled but also competing. The result is that the existing quality and future ability to sustain the productivity of natural capital is being compromised. The environmental deterioration can be characterized by dissolved oxygen deficits, aquatic toxicity, variation in organism structure, disappearance of benthic animals, turbidity and odors, fish mortality, sedimentation, and clogging of channels. These problems hinder future use of the lagoons and surrounding environments and lead to loss of agriculture, fisheries, and aquatic production as well as hinder tourism.

The loss and deterioration of coastal environments are being recognized. More than 30% of the special protection areas designated under the European Union directives for conservation are coastal and many NATO and EAPC countries have developed a considerable body of protective legislation in recognition of the value of coastal environments. Therefore, special emphasis must be given to the concept of sustainable use management in decision making on the use and development of the natural capital of coastal areas.

## 1.2  PURPOSE OF THE BOOK

This book suggests a basic framework for making informed decisions and taking positive actions for the sustainable management of lagoon systems. The individual chapters present the current status of available information on lagoon systems and models that describe the processes and mechanisms of the interrelationships and energy flow within a lagoons systems. The data and models are useful for demonstrating the cause and effect relationship of changing input variables to predict the alternative future outputs for a lagoon ecosystem. They form the basis for decision-making. It is suggested that a decision support system should be established and maintained on a continuing basis in order for sustainable management decisions to be effectively integrated into the socio-economic system influencing the natural lagoon system.

It must be recognized that many decisions will be made temporally (over a time scale) and spatially (across a wide geographic area and diverse societal infrastructure units and levels) that affect each lagoon. It is critically important to provide the best available knowledge and information in a coordinated way to result in decisions that foster the sustainable management of these threatened coastal systems. It is the task of decision makers to make choices that affect the lagoon system using the best available information and tools. These decisions inevitably center on finding the balance between the finite capacity of the lagoon system and the many demands being placed upon it by the socio-economic system that depends on it. Further, it is imperative to establish a process or plan by which informed decisions can be made over time, and which provides consistency and ensures coordination by the multitude of "users" of the lagoon system about the future of the lagoon.

This book focuses on the issue of sustainable use and promotes modeling as a tool for decision making in the context of the following:

1. Developing and improving the concept of hierarchical organization of physical, chemical, and biological environment, which promotes understanding of a holistic or ecosystem approach to management for sustainability of natural capital in the coastal lagoon management
2. Positioning of the lagoon systems within the hierarchy
3. Establishing the structural dynamic models as a crucial step for research, monitoring, and integrated management programs of coastal lagoons
4. Selecting scientific achievements relevant for policy makers, planners, and practitioners working in the field of sustainable management of the lagoons and land/seascapes to whom they belong
5. Unifying a conceptual and operational framework for ecosystem modeling of coastal lagoons
6. Integrating sustainable use management principles into the decision-making process temporally and spatially

## 1.3  OVERVIEW OF TOPICS

The following guidance is provided to inform readers from various disciplines and professional work areas about the various topics discussed in the individual chapters of this book. As noted above, Chapter 1 briefly identifies the focus of the book.

Chapter 2 provides the overall framework for the rest of the chapters. This chapter defines the organization and functional structural specifications of lagoons. A philosophy of sustainable management is related. The processes of transport, biogeochemical cycle, and ecology are described, and the challenges of modeling in lagoon environments are discussed. Chapter 2 is highly recommended for decision makers and managers because it is an overview of the key issues that should be considered in management.

Chapter 3 reviews the physical processes that drive transport in lagoons. Equations for defining mass, momentum, and energy transfer are presented along with equations for determining temperature, salinity, and sediment transport. Finally, required input and boundary conditions and the boundary processes are discussed for lagoon ecosystems. This material is intended for physical scientists. It presents the reader with an excellent background on physical modeling of lagoons.

Chapter 4 consists of detailed reviews on biogeochemical cycles (nutrients and organic chemicals) in lagoons and related processes and mechanisms, special conditions that impact these processes and mechanisms, and the equations used to formulate these concepts. This information is a sound tool for nutrient and toxic modeling studies. This chapter is strongly recommended to readers with a particular interest in understanding the internal dynamics of the lagoons, the impacts of inputs from the socio-economic system into a lagoon, and the evaluation of lagoon carrying capacities.

Chapter 5 outlines the changes in lagoons under different eutrophication states and morphologic conditions regarding main biological features and processes. The chapter provides a framework for ecological modeling studies. The targeted readers for this chapter are those who have a specific interest in assessing the trophic state in lagoons by a structural analysis of producers and consumers, as well as changes that result from the impacts of socio-economic inputs into the lagoon environment.

Chapter 6 presents valuable information on premodeling analysis and model selection, critical considerations for model implementation, stability and accuracy problems of numerical modeling, and model analysis. This chapter is recommended as a reference guide for experienced modelers.

Chapter 7 reviews the principles of developing a lagoon monitoring system as the first step toward modeling and management. The relationships between monitoring and modeling and guidance on evaluation of the monitoring results are discussed. This chapter is intended as a common reference for all readers.

Chapter 8 discusses decision-making processes. A thorough assessment of how modeling and other tools should be employed in integrated sustainable use management for a lagoon is presented. This chapter should be reviewed in the context of the information presented in the book prior to this chapter. The mission of this book and the pilot study is accomplished only if the information and knowledge presented in this chapter are used. This chapter is required reading for elected officials, managers, and decision makers.

Chapter 9 consists of selected case studies from different areas of the world. They provide detailed information and knowledge on how to apply the methodologies and approaches given in the book and how to use tools for sustainable use management of lagoons.

Naturally, readers from various disciplines involved with different aspects of lagoon assessment and management may not need to absorb information provided in every chapter in full detail. Readers who study lagoon hydrodynamics should focus on Chapters 3 and 6. Those interested in lagoon ecology will find Chapters 2, 4, and 5 of value. Finally, readers interested in lagoon management will find relevant information in Chapters 2 and 8, and other chapters, as necessary. Chapters 1 and 7 are intended for all readers.

## 1.4  THE FUTURE

Backed by NATO-CCMS, scientists, operators, managers, and students from a wide geographic range encompassing Central Asia, Europe, and North America have contributed to this book. They reunited annually for workshops in contributing countries to study lagoons for a 5-year period. These workshops enabled the participants to test the reliability of the information provided in the chapters of this book for the targeted purposes outlined above. Contributors from different disciplines and with different levels of background shared information, made arguments, investigated lagoons, and together developed the common vision and understanding reflected in this book.

At the beginning of the new millennium, countries at all stages of development are aware that environmental pollution must be managed, particularly in coastal areas. The international team of expert scientists who contributed to this book appreciate your attention and wish you well as you continue on the path of designing and implementing sustainable use management strategies.

# 2 Identification of the Lagoon Ecosystems

*Angheluta Vadineanu*

## CONTENTS

## 2.1 INTRODUCTION

Lagoon ecosystems are ecotones, or transition units of landscapes and sea/waterscapes. A key aspect of lagoons is highly sensitive areas known as wetlands, the interface areas between the land and the water.

According to the definition accepted by the Ramsar Convention, wetlands exist in a wide range of local ecosystems and landscapes or waterscapes distributed over continents and at the land/sea interface. They are natural, seminatural, and human-dominated ecological systems that altogether cover an average of 6% of the Earth's land surface.[1]

Wetlands are diverse in nature. They include or are part of areas such as beaches, tidal flats, lagoons, mangroves, swamps, estuaries, floodplains, marshes, fens, and bogs.[1,2] The world's wetlands consist of about three quarters inland wetlands and one quarter coastal wetlands. Palustrine and estuarine wetlands, which include lagoons, account for most of them.[1]

Exponential increase in human population and the corresponding demand for food and energy resources as well as for space and transport have in the last century stimulated the promotion of economic growth driven by the principles of neoclassical economy. Current philosophy has promoted, and unfortunately

1-56670-686-6/05/$0.00+$1.50
© 2005 by CRC Press

still promotes today, the extensive substitution of natural and seminatural eco-logical systems, or the self-maintained components of Natural Capital (NC), into human-dominated components. Consequently, most of the natural and semi-natural components, particularly wetlands, have been seen until recently as "wastelands." These areas are being extensively replaced by intensive crop farms, tree plantations, commercial fish culture, harbors, and industrial complexes or human settlements.[2-6]

The lack of scientific background for understanding and estimating the multi-functional role of wetlands associated with the sectoral approach has resulted in lack of appreciation by policy and decision makers of the resources and services that these types of systems have produced.

However, these are some of the most productive units in the ecosphere. They provide a wide range of self-maintained resources and services, from the viewpoint of energy and raw materials. They replace such self-regulated systems totally or, to a very great extent, they depend on the input of fossil auxiliary energy and inorganic matter (e.g., chemical fertilizers) as well as on human control mechani-zation (e.g., high-tech equipment for agriculture). Thus the ecological footprint (EF) of many local and national socio-economic systems (SESs) themselves become highly dependent on fossil fuels and underperform in providing services. The EF basically tries to assess how much biologically productive area is needed to supply resources and services, to absorb wastes, and to host the built-up infrastructure of any particular SES.[7]

There has been an increase in scientific understanding and awareness among a growing number of policy and decision makers, especially in recent years. They now recognize that the structure and metabolism of any sustainable SES should be well rooted in a diverse, self-maintained, and productive EF. This has launched a new philosophy, derived from the theory of systems ecology and ecological eco-nomics, dealing with "sustainable market and sustainable socio-economic develop-ment." This is an ecosystem approach, and new managerial patterns have emerged, consisting of ecosystem rehabilitation or reconstruction for the improvement of the EF and conservation through adaptive management of spatio-temporal relationships among SES and the components of NC.

In recent years much work has been done to promote these new concepts. Objectives and patterns now focus on reconstruction and management of natural or seminatural ecological components (e.g., wetlands) as major initiatives in the EF of many SESs. However, principally we are still in the process of conceptual clarifi-cation, strategy, and policy development as well as designing and developing the operational infrastructure or smaller scale of projects implementation.

This chapter presents a comprehensive analysis of the existing concepts, knowl-edge, and practical achievements in the integrated or ecosystem approach for sus-tainability or adaptive management of the relationships among SES and the compo-nents of NC. It is an attempt to improve the conceptual framework and provide an operational infrastructure for modeling and sustainable use and development of lagoons, one of the components of the coastal landscape most sensitive and vulnerable to human impact. This chapter thus provides the overall framework for developments discussed in the following chapters of the book.

## 2.2 CONCEPTUAL FRAMEWORK OF SUSTAINABLE USE AND DEVELOPMENT

Since the Brundtland Report (1987 WCED) considerable effort has been directed toward the development of a general definition of sustainability in order to implement the vision of sustainability in practical policy decisions. There has been worldwide recognition of the global "ecological crisis" faced by human civilization especially after the UNCED Conference/Rio 1992. This has prompted those responsible for formulating and implementing strategies and policies for economic development to balance the spatio-temporal structure and metabolism of SES with the spatio-temporal organization of the "environment" or with biophysical structures, the NC, and their production and carrying capacity.

In this respect, this is an attempt to assess and integrate a wide range of operational definitions that have been developed and checked in recent years.[3–6,8–24] The following were identified as the basic requirements that must be met in order to put into practice the concept of sustainability.

1. Assessment of the conceptual and methodological development of sustainability that ensures establishment of state-of-the-art definition and identification of main gaps and shortcomings and, therefore, the need for further development and improvement.
2. Formulation of the basic elements of a dynamic model for co-development of SES and NC or for sustainable use and development to serve as the basis for promoting local, regional, and global transition.
3. Identification of the advantages and opportunities that each country and region may have as well as the limits or constraints with which they may be faced in the designing and implementing of long-term "co-development" strategies and action plans.
4. Identification of existing shortages and gaps in the policy and decision-making process dealing with sustainability and formulation of a comprehensive and dynamic model for the "decision support systems (DDSs)." This will serve as the interface, or the operational infrastructure, and thus enable us to balance the spatio-temporal relationships and the mass and energy exchanges between the NC structure, serving as the footprint, and the SES.

What follows is a brief description of the basic conceptual and methodological elements to be relied upon in the co-development of SES $\Leftrightarrow$ NC vision of sustainability as well as the structure of the dynamic DSS that can put sustainability into practice.

The concepts and methods dealing with the "environment" have changed and improved as ecological theory usually described as "biological ecology" has developed from its early stage. The current ecological theory is more often and more appropriately defined as "systems ecology" (Figure 2.1). The identification and description of the natural, seminatural, human-dominated, and human-created environment has changed as well. This change was from a former conceptual model that defined the environment as an assemblage of factors—air, water, soil, biota, and human settlements—to the

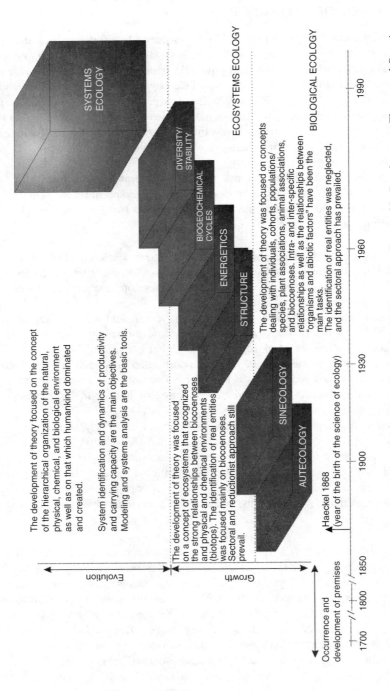

**FIGURE 2.1** Growth and evolution of the science of ecology. (After Vadineanu, A., *Sustainable Development: Theory and Practice*, Bucharest University Press, Bucharest, 1998. With permission.)

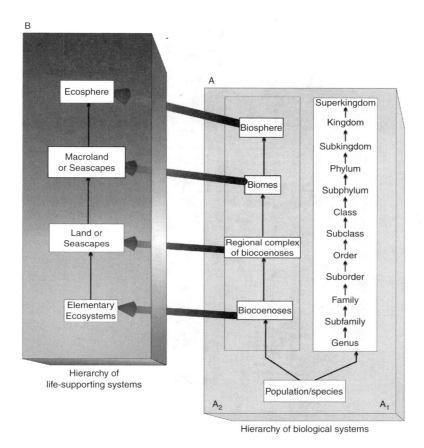

**FIGURE 2.2** Relationships between taxonomic and organizational hierarchies of the living systems (A) and their integration within the hierarchy of life-supporting systems or ecological systems (B) $A_1$ = diversity of living organisms and hierarchical order of the taxa established based on the similarity between ordered entities. $A_2$ = hierarchical organization of living organisms in large and complex biological systems. B = hierarchy of spatio-temporal organization of the upper layer of lithosphere, hydrosphere, troposphere, and biosphere. (After Vadineanu, A., *Sustainable Development: Theory and Practice*, Bucharest University Press, Bucharest, 1998. With permission.)

most recent thinking that considers the environment as a "hierarchical spatio-temporal organization."[6,25–27] (Figure 2.2 and Figure 2.3). Ecological systems, as organized units and components of the hierarchy, are described as self-organizing and self-maintaining systems, or as "life-supporting systems." They have been described as nonlinear dynamic and adaptive systems with evolving production and carrying capacity. These nonlinear systems go through successive phases of adaptive cycles: growth (R); accumulation or maturization (K); release or "creative destruction" (Ω); and restructuring and reorganization (α).[28,29]

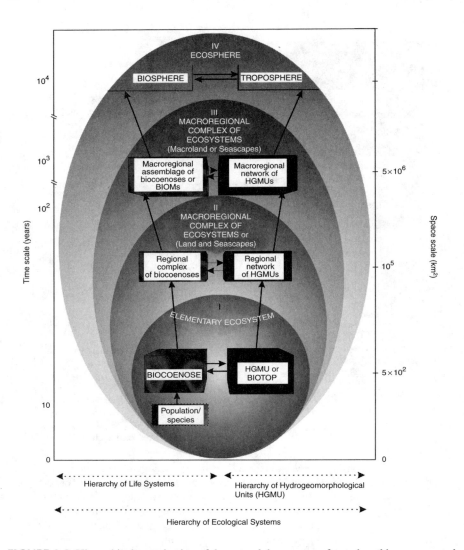

**FIGURE 2.3** Hierarchical organization of the natural, human-transformed, and human-created physical, chemical, and biological environment. According to existing knowledge concerning the organization of life, we can distinguish five hierarchical levels above biological individuals and four spatio-temporal levels within the ecological hierarchy. It must be noted that three-dimensional space of the hierarchical organization integrates upper lithosphere, ocean basins, and troposphere, and the time constants of the ecological systems are in years, decades, centuries, or millennia. (After Vadineanu, A., *Sustainable Development: Theory and Practice*, Bucharest University Press, Bucharest, 1998. With permission.)

The ecological hierarchy comprises two main hierarchical chains of ecological systems that show a marked and evolving dichotomy in spatio-temporal development:

1. Self-maintained natural and seminatural ecological systems that provide a wide range of natural resources and services
2. Human-dominated ecological systems that depend to varying degrees on commercial auxiliary energy and material inflow (e.g., agriculture, aquaculture) and human-made systems (e.g., urban ecosystems, industrial complexes), which are totally dependent on commercial energy and material inflow.[6,25,27]

The divergent dynamics of these systems is the core of the so-called "ecological crisis." Thus, the ecological hierarchy integrates both the components of the NC and those of the SES.

Accordingly, the term *biodiversity* in its broad meaning covers, on the one hand, the components of NC together with their taxonomic and genetic diversity and, on the other hand, human social organization, and ethnic, linguistic, and cultural diversity.

Biodiversity consists of NC and social and cultural capital. It provides both the EF that supports the SES with resources and services and the interface between NC and the structure and metabolism of the "economic subsystem" (Figure 2.4).

It must be noted that, in order to make the transition from the current status of a strong dichotomy between SES⇔NC to that of co-development, there is a need to establish an internal balance between the economic subsystem and social and cultural capital.

In the last decade a rapid shift has been observed from the sectoral, reductionistic, and inappropriate temporal (months and years) and spatial scale approach toward a holistic, adaptive, and long-term approach (decades and centuries). Systems analysis and modeling are used more extensively for the identification and description of the ecological systems (including SES) as large, complex, dissipative, and dynamic systems.

However, the relationship between humans and nature more recently referred to as a "development and environmental" relationship or "economy and ecology" should be further reformulated. It should be recast as the mediated and dynamic relationship at local, regional, and global scales between the structure and metabolism of SES on one side, and the structure, productivity, and carrying capacity of the natural, seminatural, and human-dominated systems (NC) on the other (see Chapter 8 for details).

The following conclusions are set forth:

1. Sustainability deals with co-development or balancing the dynamics of the spatio-temporal relationship between SES and NC.
2. The principles of free market economy, which negatively limit NC from contributing to SES, should be replaced by principles of "sustainable market economy." This will require identification of the overall dynamic framework for co-development, according to the structure, productivity,

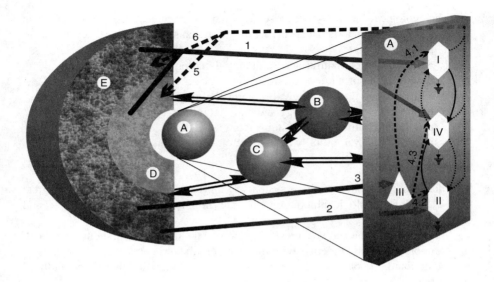

**FIGURE 2.4** The general physical model of the socio-economic system and its relationships with Natural Capital (NC).

A = the human-made physical capital: I = the infrastructure of the economic subsystem dependent on the renewable resources provided by the components of the NC; II = the industrial infrastructure of the economic subsystem dependent on "non-renewable" resources; III = systems for commercial energy production using fossil and nuclear fuels and hydro-power potential as primary resources; and IV = the human settlements infrastructure. 4.1, 4.2, and 4.3 identify the energy flow pathways; B = social capital; C = cultural capital; D = human-dominated components of the NC; E = natural and semi-natural components of the NC: 1 = flow of renewable resources; 2 = flow of raw materials; 3 = flow of fossil and nuclear fuels; 4 = flow of electrical energy; 5 = material and energy inputs (fertilization, agrotechnical works, irrigation, selection, etc.) to support the management of human-dominated systems; 6 = dispersion of heat and of secondary products (wastes) in the troposphere and in the HGMU components. (After Vadineanu, A., *Sustainable Development: Theory and Practice,* Bucharest University Press, Bucharest, 1998. With permission.)

and carrying capacities of the local, regional, and global NC. In addition, ethical and moral criteria for sharing of resources and services within and among generations and among jurisdictions must be considered.

3. There is a need to establish thresholds for the constituent units of the NC and for the spatial relationship between NC and SES. Specifically, self-maintained and self-regulated ecological systems should represent more than 50% of the total NC of a country or region. So, the structure and metabolism of a particular SES should have a high degree of complementarity with the structure, productivity, and carrying capacity of the domestic NC.

4. Although we refer to the NC as the EF for a particular SES, wetlands, and in particular lagoons, are a major component in the EF of any SES

(for example, they provide more than 24% of global net primary production and 60–65% of the world's fish and shellfish production).[2]

5. Finally, the need for a holistic or ecosystem approach to all our economic, social, and engineering activities is not merely a sustainable development strategic paper as often described by politicians, decision makers, and the public.

It might be easier to use terms such as ecological crisis, integrated or interdisciplinary approach to the environment, or carrying capacity. However, it is very difficult to conceptualize the link between the ecological crisis and the dichotomy in the development of NC components and SES. The integrated or systemic approach also requires an understanding that the physical, chemical, and biological environment has a hierarchical organization that integrates the SES as human-dominated and human-created ecological systems dependent on mass and energy transfer with the other components of the hierarchy. It also must be understood that the carrying capacity of NC is linked to stability in a broad sense as well as to the dynamic capacity of the ecological systems to provide goods and services and to assimilate the wastes of SES.[5,6,11,16]

## 2.3  SPATIO-TEMPORAL ORGANIZATION OF LAGOON ECOSYSTEMS

The basic structural and functional units of the "environment," widely known as ecosystems, and those from the next hierarchical level (Figure 2.3), known as land or sea/waterscapes, are the entities on which both scientific investigation and integrated or sustainable management are focused. Coastal lagoon ecosystems, and in particular the associated wetlands, are components of mixed land/sea/waterscapes that are complex dynamic systems.

To approach and understand how these systems work and how they can be managed as NC, resources and service providers as well as spatio-temporal organization and structure must be identified. This structural model that represents the real world environment by depicting the dynamic components and their relationships in time and space is called a *homomorph model*.[30,31] Homomorph models are necessary for most scientists and managers to operate in the real world. Development and understanding of homomorph models are necessary for integrated management and for sustained use of NC that provide support for the SES. There have been, and still are, users of basic theoretical principles of the science of systems ecology who cannot associate these concepts with any real counterpart. Or, even if they do, such a structural model is either very superficial (with inappropriate space scales or oversimplification) or has no true agreement with the real system.

One of the major targets in the field of applied systems ecology is the development of a specific methodology for ecosystem identification and landscape or sea/waterscape identification.[26,31–38]

Identification of the lagoons and the land/sea/waterscapes to which they belong is a step-by-step process that involves:

1. Development and implementation of extensive and intensive research and monitoring programs, at appropriate time and space scales, consisting of field observations, measurements, and sampling, combined with air photography and remote sensing
2. Analysis of historical information and data
3. Identification of fauna and flora taxa and estimation of biomass, abundance, distribution, and dominance, as well as the trophic niche, relationship (food webs), production, and demographic structure
4. Assessment and description of the three-dimensional space distribution of major components of the hydrogeomorphic unit (HGMU) and variability of the lagoons (e.g., water volume, water movement, water retention time, stratification, and water-level oscillation, bottom nature, and chemistry)
5. Identification of lagoon ecotones, boundary conditions, and external driving forces

In summary, all these steps are described in detail in various chapters of this book. This chapter identifies the crucial need for information systems dealing with the functioning and dynamics of lagoons in order to carry out sustainable use or adaptative management of lagoon resources and services. The remainder of this section provides a brief summary of information relative to lagoon function, dynamics, and management for sustained use and development.

## 2.3.1 LAGOON ECOTONE

The ecotones, or transition zones, are the border areas between the local ecosystems. They are elementary structural and functional units in various types of landscapes and sea/waterscapes. The physical, chemical, and biological components of ecotones have a linear development of tens of kilometers and usually a narrow transversal development of a few meters or, only very rarely, of hundreds of meters. In ecotones the joint HGMUs exhibit a marked discontinuity in at least one constituent (see Chapter 3).

There is a very extensive literature dealing with the role of ecotone components of lagoons.[39–53] Useful conclusions that support managerial purposes are:

- A spatio-temporal organization for biological components allows for the understanding of mass and energy exchanges between lagoon systems and surrounding ecosystems (e.g., agricultural, forests, urban, or marine shelf ecosystems). In fact, lagoon ecotones modulate and establish boundary conditions that are driving forces for the lagoon's inner structural and functional dynamics.
- As buffers, wetlands are more sensitive to the antropogenic forces as well as regional and global climate changes. Wetlands and, in fact, lagoon ecotones are habitats for many vulnerable species, a space for microevolution, or a space for longitudinal migration.
- Due to their structural and functional features, lagoon ecotones should receive special consideration in any strategy and management program

for both natural and human-oriented landscape management actions and should be key sites in the monitoring for climate changes. The ecotone component protects lagoons from anthropogenic activities in surrounding areas. For example, where farming is intensively practiced or urban development has occurred within the catchments area, wetlands provide an effective control of nutrients and other chemical compounds or mineral particles from agricultural or urban areas. Wetlands are also the habitats of many efficient predators or pests that affect crop production. Therefore, preservation, restoration, or development of these lagoon ecotones, particularly the wetland components, is important for dealing with pollution abatement and/or integrated pest control. Likewise, these ecotones are the major habitats of many vulnerable plant and animal species. The management of ecotones should be designed to reach biodiversity conservation objectives, since biodiversity is clearly indicative of conservation of NC and likely of SES benefits (see Chapter 8 for details).

### 2.3.2  HGMU Spatio-Temporal Organization

Chapter 6 of this book provides strong arguments to support the need for an accurate description of the structure and spatio-temporal organization of a lagoon's HGMU. Such arguments are needed in order to select or develop appropriate hydrodynamic and transport models as a part of the modeling package used for the overall description of the structure and function dynamics (productivity and carrying capacity) of the lagoon system. Based on data concerning water exchange between lagoons and adjacent seas, the Kjerfve classification system of choked, restricted, and leaky lagoons is proposed (see Chapter 5). Ten morphohydrometric parameters are introduced that, when quantified for a particular lagoon, provide an in-depth understanding of inner physical and hydrochemical heterogeneity (see Chapter 6, Section 6.3, for details).

Time series measurements of morphometric parameters as described in Chapter 7 provide information on changes in the shape and bottom relief of a lagoon HGMU under the pressure of many external driving forces (e.g., tides, waves, floods, erosion, or deposition). The complementary relationship between the spatio-temporal organization of a lagoon HGMU and biocoenoses must be stressed again. On the one hand, changes in many physical and chemical variables (e.g., salinity, temperature, dissolved oxygen, nutrient availability, depth, water renewal rate, turbulence, light availability, and bottom structure) describe the dynamic state of HGMUs as inner driving forces for component populations and the entire community (e.g., species composition, population size, distribution, cost of maintenance, and primary and secondary productivity). On the other hand, the ranges of fluctuations of those variables are usually modulated by the activity of biological components (see Chapter 5 for details).

The physical, chemical, and biochemical processes widely involved in biogeochemical cycling of nutrients and chemical compounds in lagoon systems are described in Chapters 3 and 4. These chapters focus on some of the main features of the processes of energy and mass transfer, which are of great importance for sustainable management.

### 2.3.3  Biocoenose's Spatio-Temporal Organization

To identify the spatio-temporal organization of a lagoon biocoenose, it is extremely important to give special attention to the intensive and extensive field investigation procedures employed (see Chapter 7 for details). This includes the sample size, which needs to consider estimation of the component species/population size at an error rate below 20%; sampling frequency, which should reflect the specific features of the life cycles of the component species; sampling methods (e.g., transects, random and systematic sampling, mark–recapture); and equipment, in order to identify the heterogeneity inside HGMUs and the mobility and dispersion of individuals and cohorts from different populations.[54] The bulk of the data gained by field sample analysis should be used to estimate population size, spatial distribution, abundance, biomass, and dispersion through immigration and emigration. Populations whose combined abundance and biomass account for up to 80% and 90%, respectively, are those populations that play a significant role in the spatio-temporal organization of the biocoenose. In addition, specific sampling analysis should allow the identification of trophic spectrum or functional niche, age structure, estimation of production, and biomass turnover time for each dominant population.

If reliable data for the above parameters are available, then the next steps are aggregating the dominant populations into modules according to some effective criteria and establishing the network of mass and energy transfer among dominant populations from different trophic modules.

A critical step in any attempt to identify the biocoenose of a lagoon system is dealing with the identification of the network of trophodynamic modules, which preserve structural and functional attributes of the respective biocoenose. The solution should be a homomorph model. This avoids structural oversimplification (e.g., traditional trophic levels and linear pathways connecting them) but accepts loss of structural information (e.g., retain only the dominant populations or cohorts after the first phase of empirical data analysis and then establish the trophodynamic modules through aggregation of the dominant populations by applying effective procedures). It is expected that, similar to the real world, the network of trophodynamic modules through which we identify a dynamic and complex biocoenose resembles food webs more than food chains.[26,37]

The internal structure of a lagoon system (components and the direct and indirect relationships among them, or so-called network patterns) should be identified in order to be able to find out reasonable answers to questions such as:

1. What is the relationship between stability, used here with a broader meaning than resilience[55] of a lagoon ecosystem, and the spatio-temporal organization and dynamic diversity among trophodynamic modules?
2. To what extent does the nature and heterogeneity of an HGMU determine the internal organization of the biocoenose?

The identity of the network compartments and the nature of spatio-temporal interaction patterns are crucial for the construction of an adequate homomorph model and are dependent on the quality of research and monitoring programs (see below and Chapter 7 for details).

Typical temporal and spatial scales on which a trophic unit operates and the relationship to the network is the meaning of spatio-temporal organization of the biocoenose underlying a lagoon HGMU. The biocoenose is subject to continuous change because any unit, like a population or a group of populations, can have only a finite lifetime and a finite spatial distribution, as described in Chapter 5.

The basic unit for the network describing the spatio-temporal organization of a lagoon biocoenose is a trophic dynamic module (TDM), first introduced by Pahl-Wostl.[26] The modules are defined for both dynamic and trophic function. According to Pahl-Wostl,[26] a trophic dynamic module comprises all populations that have the same dynamic characteristics and the same functional niche in the trophic web and that coexist over the same finite period in time and space. Thus, having reliable data and estimates concerning the variables listed earlier for a given biocoenose, we can identify the homomorph or structural model by applying the following stepwise procedures.

1. Based on the dynamic characteristics of the identified dominant populations, or cohorts ($\tau$), the dynamic classes $C_i$ are established.
2. By locating these Classes ($C_i$) in time and space, the dynamic modules ($MD_{ikr}$) are established.
3. By taking into consideration the functional characteristics of populations or cohorts, as expressed by their functional niches ($C_{ik}$), the trophic-dynamic modules are established ($MTD_{ikra}$).

The following definitions apply:
- Cohort—Embodies a group of individuals that are part of the structure of a given population that have the same or similar age and share the same functional niche.
- Dynamic Class ($C_i$)—Includes all populations or cohorts whose biomass turnover time ($\tau$) is in a range where $\tau$ remains a constant value on a logarithmic scale.
- Dynamic Module ($MD_{ikr}$)—Comprises all dominant populations and cohorts belonging to the same dynamic class ($C_i$) that co-exist within the same time and space interval.
- Trophic–Dynamic Module ($MTD_{ikra}$)—Comprises all dominant populations and cohorts belonging to a dynamic module ($MD_{ikr}$) that have the same functional niche ($C_{ia}$).
- Functional Niche ($C_{ia}$)—Indicates the position occupied by dominant populations and cohorts within the trophic connection network, or their position within the network as matter and energy carriers. Primary producers (Figure 2.5) are taken as the reference position for establishing the position within the network of the other functional niches.
- $i$ = dynamic class ($i = 1, \ldots, n$)
  $k$ = time interval by units $\tau_i$
  $r$ = space indicated by units $S_i$
  $a$ = functional niche
  $\tau$ = biomass turnover time (B/P)

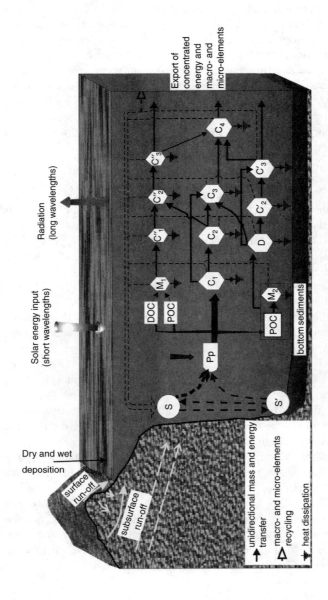

**FIGURE 2.5** A potential homomorph model for the identification of lagoon ecosystems. DOC = dissolved organic carbon; POC = particulate organic carbon; Pp = primary producers; $M_1$ = bacterioplankton; $M_2$ = benthic micro-organisms; $C_1$ = herbivores; $C_1''$ = microfiltrators (e.g., rotifers, small cladocera); D = detritus-feeding populations; $C_2$ and $C_2''$ = zooplanktonic carnivores; $C_2'$ = carnivore invertebrate species; $C_3'$ = benthos-feeding fish species; $C_3$ = plankton-feeding fish species; $C_4$ = predator fish species; S and S′ = available stock of chemical elements or compounds.

## 2.3.4 GENERAL HOMOMORPH MODEL FOR LAGOONS

When the identification process of a given lagoon system is completed, the result should be a structural and functional model that preserves the basic structural and functional attributes of the lagoon and its spatio-temporal relations to other ecosystems from the land–seascape complex to which it belongs (Figures 2.5 and 2.6).

Thus, the structural and functional or biophysical model through which a given lagoon system is identified comprises:

1. The major components in the structure of HGMU and the trophic–dynamic modules in the structure of biocoenose
2. The most active relationships for mass, energy, and information transfer within the lagoon itself and between it and surrounding ecosystems (including lower troposphere)

The network and coupled processes are continuously fueled by solar, chemical, or auxiliary energy (e.g., wind and tide energy; concentrated energy in

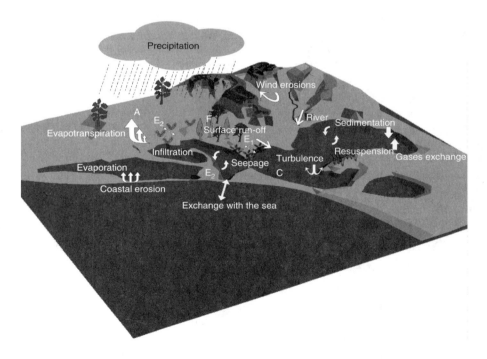

**FIGURE 2.6** HGMU's identification of the lagoon system and its spatial relations with ecosystems from the land–seascape complex. A = agricultural system; E, $E_1$, $E_2$ = Ecotones; $E_1$ = hedgerow and $E_2$ = riparian vegetation as ecotones; C = lagoon; F = Forest.

nonliving organic matter or fossil energy), and makes the lagoon function as a productive, self-regulating, and self-maintaining system. This involves a permanent inner transfer of mass, energy, and information that consists of three overall processes:

1. Energy flow
2. Biogeochemical cycling of chemical elements which ends in production of natural resources and services as well as an entropy dissipation, and
3. Information flows which develop multiple self-regulating mechanisms

The identification process also requires defining the structural and functional parameters and corresponding vectors by which the state of the lagoon at a given time $t_i$ as well as the set of external driving forces and boundary conditions could be described (see Chapter 3 and Chapter 4 for details). In other words, we may consider that the homomorph model of a lagoon system is a simplified copy of the real system, developed by ignoring some elementary components and aggregating others. A very important fact is that the model should preserve the characteristics of spatio-temporal organization of the lagoon and its connectivity to the upper hierarchical system, that is, the land/sea/waterscape.

The homomorph model provides the only operational and effective way to cope with the complexity of lagoons when designing and implementing appropriate research and monitoring programs. Further, a homomorph model is the most powerful tool for designing and implementing sustainable management plans.

Productivity and self-regulation of a particular lagoon system define its carrying capacity or potential role in the EF for an SES. Lagoons are also dynamic, nonlinear systems driven by both natural and anthropogenic external forces as well as internal ones. Usually, at large time scales they exhibit structural and functional changes that, in turn, lead to changes in their production and carrying capacity. Particular reference is made to Chapter 5 for information concerning structural and functional changes in lagoon systems in relation to eutrophication, renewal rates, and practices for extensive and intensive fisheries management.

## 2.4  SCIENTIFIC ACHIEVEMENTS RELEVANT FOR SUSTAINABLE MANAGEMENT OF LAGOONS AND LAND/SEASCAPES

The last decades of the 20th century were very productive in terms of significant achievements in energetics, biogeochemistry, and ecotoxicology within a wide range of ecological systems. Critical analysis and integration of the results that have been carried out by many ecologists[6,32,55–64] prove that natural and seminatural ecological systems are resources and service providers to the SES, usually at their own expense. Among the components of the NC, lagoon ecosystems have proved to be the most productive. This is the reason for trying to discriminate and bring to the forefront some fundamental achievements, which may help in formulating and implementing strategies and action plans for integrated and sustainable management of complex land/seascapes, where lagoons are major components.

- On average, only 0.25% of the solar energy reaching the land and ocean surface and 0.5% of the solar energy absorbed by the primary producers are concentrated in biomass[26,54,58] (Figure 2.7a).
- The greatest part of the planet, about 75% of the surface, including oceans and land, has the lowest efficiency in absorption and concentration of solar energy. The density of the energy flow concentrated in these types of ecological systems does not exceed 1000 kcal·$m^{-2}$·$yr^{-1}$, and this provides only 35% of the ecosphere's gross primary production (35% of the energy is concentrated by the primary producers of the ecosphere).
- The density of the energy flow is concentrated at the level of the primary producers from the following natural and semi-natural ecological systems: estuaries and lagoons, coral reefs, wet forests, floodplains, and agriculture. These ecological systems represent only 10% of the total of the ecosphere, but nonetheless the density of energy flow is of the greatest values (10,000–40,000 kcal·$m^{-2}$·$yr^{-1}$). The primary producers of these categories of ecological systems provide almost 38% of the ecosphere's gross primary production.
- Agriculture covers about 0.8–1% of the total surface of the planet and provides 5% of the gross primary production of the ecosphere. In order to maintain a concentrated solar energy flow, equal to that of the estuaries, lagoons, wet tropical, and subtropical forests, etc., agriculture requires a large amount of fossil fuel (up to 1000–2000 kcal·$m^{-2}$·$yr^{-1}$).
- The quantity of energy absorbed and concentrated by the dominant populations or cohorts in the tropho-dynamic modules, represented by the first-order (herbivorous) and by the second- and third-order consumers, decreases from one tropho-dynamic module to another. In general, the energy assimilated (absorbed and concentrated) by a tropho-dynamic module made up of heterotrophic species represents only 5–20% of the energy concentrated by the tropho-dynamic module, which is the energy source (Figures 2.7b and c).

This rule explains why the sequence of tropho-dynamic modules directly or indirectly using the energy concentrated by the primary producers (Figure 2.7a) includes, in any ecological system, only three to four modules. Note also that organic matter (and the potential energy it stores) that is not consumed or digested by the components of a trophic module, or that is represented by intermediary metabolites (containing considerable quantities of concentrated energy) is transferred into two main reservoirs. These are dissolved organic carbon (DOC) and particulate organic carbon (POC). They become, in turn, sources of concentrated energy that support the other two series of tropho-dynamic modules which are complementary to the first one. Note that natural and seminatural ecological systems have and develop structures which use "waste" having a concentrated energy content. They

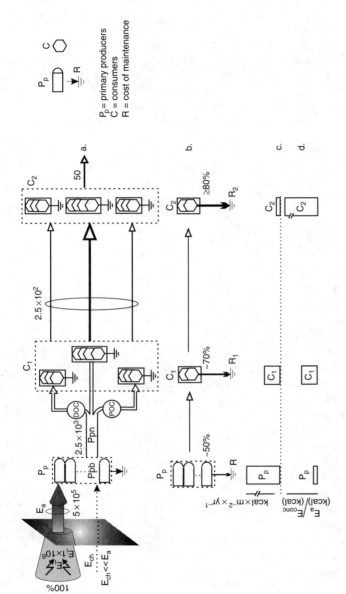

**FIGURE 2.7** Energy flow through ecosystems. a. A general view of the trophodynamic structure of the ecosystems showing the three types of food chains and the average density of energy flow. Values represent the average density flows in kcal·m⁻²·y⁻¹. $E_i$ = incident solar energy at the land and ocean surface; $E_a$ = absorbed solar energy by primary producers (Pp); $E_{conc}$ = concentrated or assimilated energy, Ppb = gross primary production; $P_{pn}$ = net primary production; $C_1$ = herbivores; $C_2$ = carnivores; DOC = dissolved organic matter; POC = particulate organic matter; $E_{ch}$ = energy released by oxidation of certain inorganic compounds. b. Increasing cost of maintenance in food chains. c. Decreasing amount of concentrated energy in food chains. d. Increasing energy concentration (quality) in food chains. (Compiled after Botnariuc, N. and Vadineanu, A., 1982; Odum, E., 1993; and Pahl-Wostl, C., 1995.)

have the possibility of recycling the raw material necessary to photosynthesis and chemosynthesis and maximizing the efficiency of using the concentrated energy by the primary producers and thus, of the density of the concentrated energy flow. In the land/seascapes, recycling of raw materials (especially micro- and macronutrients) and maximizing the concentrated energy flow are made possible by the functional differentiation of certain ecosystems that are partially or totally "heterotrophic."

- In the sequence of mechanisms for the transfer and conversion of the solar energy that has as its support the sequence of tropho-dynamic modules and the trophic network of any ecosystem, there is a process of concentrating the energy from more diluted forms (low quality) into more concentrated forms (high quality and with higher information content) (Figure 2.7d).

Therefore, the energy flow in an ecological system is unidirectional from the permanent solar energy source toward the higher tropho-dynamic modules. Along this sequence, the quantity of energy decreases but its quality increases. The same quantity of energy, but of a different form and of a different quality, ensures different mechanical work. In this respect, from the viewpoint of quality, the very concentrated energy existing at the level of the third-order consumers or the energy of the fossil fuels (oil, natural gas) has a much higher quality than diluted solar energy.[58]

The concentration or quality of energy can be expressed in terms of the amount of energy of a certain type and quality required to develop energy of higher quality (Figure 2.7d). Odum[58] estimated that, on average, for one kcal of energy concentrated in the biomass of primary producers, almost 200 kcal of solar energy are required; in the biomass of herbivores 2000 kcal of solar energy are required; and in the biomass of carnivores 10,000 kcal of diluted solar energy are required. In addition, it can be estimated that 50–100 kcal of primary producers and 5–10 kcal of herbivores are required for one kcal of energy concentrated by first-order carnivores.

- In any given ecological system the concentrated energy flow is maintained with important energy expenses. For energy absorption and concentration, each population in the structure of tropho-dynamic modules spends very large quantities of the concentrated energy for its own maintenance. In this respect, biological systems convert large quantities (>40%) of the energy that is concentrated by the process of assimilation and biosynthesis into heat, through respiration. The energy needed to concentrate energy and for maintenance (or development) of spatio-temporal organization increases from the primary producers (40–85% of gross primary production with an average of 50%) to the third-order consumers (an average of 80% of the assimilated energy). It is very important to emphasize that, when dealing with herbivores and carnivores, only 20–40% of the energy assimilated is concentrated in their biomass. The rest of the energy is spent on self-maintenance of the spatio-temporal organization and recycling the macro- and microelements.

- Natural and seminatural ecological systems are able to produce the biological resources (concentrated energy) and recycle the raw material necessary for biological productivity, by gradually concentrating solar energy. This complex process is very expensive from an energy point of view. Solar radiation is a permanent source of energy that helps cover these expenses, while biological systems are the most efficient converters and dissipative structures.
- Cycling of chemical elements and compounds within the ecological systems involves transfer, transformation, accumulation, and concentration phenomena. The behavior of chemical elements in an ecological system may be associated with toxic effects for biological components. The toxic effects may result in reduction of biological diversity and deterioration of genetic heritage (including the human population's genetic structure). The disturbance of the distribution of certain chemical elements or compounds or the introduction of new human-made compounds in key compartments, or reservoirs, of an ecological systems may bring about structural and functional changes in the network of tropho-dynamic modules supporting both the energy and information flows and biogeochemical cycles (i.e., ecotoxicological effects). These effects are propagated and amplified in time and space due to the accumulation, concentration, and transfer of chemical elements and compounds as well as their derivatives.

  Deterioration of quality and health of ecological systems (including human populations, particularly through deterioration of their genetic structures) is possible. Management of transformed or created ecosystems that neglected the phenomena related to the transformation, concentration, and remote transfer of chemical elements and compounds in general and those having a high risk for the biological systems, in particular, are highly subject to deterioration.

  In order to substantiate the above statement, it is necessary to select a few of the key concepts that could provide a basis for identifying and understanding the elements of an extremely complex process specific to ecological systems. This knowledge should improve and strengthen management of both the SES and the NC and address sustainable use of goals. At the same time, this knowledge could become a major pillar of an information system that could develop on the basis of consultation of the literature and continuous improvement through implementing extensive and long-term research programs.

- Biogeochemical cycles are based on the ecological system structure (ecosystems—micro- and macro-landscapes and seascapes—ecosphere) at which level we can very clearly differentiate the following systems: (1) the primary cycling system comprised of the dynamic components of the hydrogeomorphological units and of the troposphere and (2) the secondary cycling system comprised of the network of tropho-dynamic modules, which in fact carries out the most active phenomena and processes that the biogeochemical cycles depend on (Figure 2.8). The compartments of the HGMU, represented by soil, parental rock, and interstitial water in the terrestrial systems, and bodies of water and the

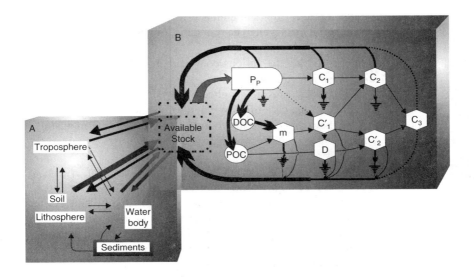

**FIGURE 2.8** A general model that may be used to identify large ecological systems and to describe biogeochemical cycles.

A = Primary cycling system that comprises the major components of hydrogeomorphological units (HGMU): soil/lithosphere; water body; sediment; and the troposphere. Low rate physical, chemical, and geological processes mainly support exchanges among the components.

B = Secondary cycling system that is very dynamic and built on the trophic structure of biocoenoses, biomes, and biosphere. Recycling rate of macro- and micro-elements by dominant populations of each trophodynamic module is strongly correlated with the level of energy expenses. Exchanges between the fast recyclable stock and less dynamic pools in the primary cycling system have worked as very efficient buffering mechanisms for biogeochemical cycles of all chemical elements. Human impact mainly affects the balance of such exchanges and finally leads to the increase of available stock.

DOC = dissolved organic carbon; POC = particulate organic carbon; $P_p$ = primary producers; C = consumers; m = microorganisms; D = detritus feeders.

sediment in the aquatic systems togeher with the troposphere have a certain loading of chemical elements and compounds, as they function as reservoirs, or pools of the biogeochemical cycles (Figure 2.8A).

Direct exchanges occur among the reservoirs due to physical and chemical phenomena and processes (adsorption, absorption, hydrolysis, oxidation-reduction, precipitation, flocculation, ion exchange, sedimentation, wet and dry deposition, turbulence, currents, etc.) or they are mediated by certain components or by the entire recycling network of trophodynamic modules. Through these exchanges and the transport systems of water and air masses, the biogeochemical cycles occurring at the ecosystem levels (local cycles) and at the micro- and macro-landscapes and seascapes are closely interconnected, and in fact integrated in the global biogeochemical cycles. (See Chapter 4 for details.)

- It is well defined that a chemical element is preferentially stored in one of the pools and is, at this level, to be found in more or less stable chemical combinations.

  The exchanges among the various compartments of the hydrogeomorphological units and the troposphere are made very slowly (at a time scale of geological processes) and are supplied by physical, chemical, and geological processes. In the case of most chemical elements, there is at least one pool where they can be partially found in more unstable combinations and are consequently available to the primary producers in a network of tropho-dynamic modules.

  Thus, the primary cycling process, which involves physical, chemical, and slowly evolving geological phenomena and mechanisms, overlaps the secondary cycling process. And, as a result, it is supported by much faster physical, chemical, and biological mechanisms and phenomena. In between these cycling processes, which take place on different levels and at various rates, are established contacts and exchanges that are identifiable in the more dynamic compartments, reservoirs, or pool. These usually lead to either an increase of the recyclable stock or a decrease due to the leaks from and to the primary cycling system.

  The following section refers to the fast cycling system that involves the intervention of the biological components, which strictly conditions the ecological systems' functioning as productive units.
- Consider the complexity of homomorphic models through which each ecological system is identified by relating the variety of coupled mechanisms ensuring the cycling of the elements, as well as their dynamics, according to the key driving forces. From this we can understand the requirements and standards in researching and monitoring biogeochemical cycles and also the reason for the underlying long distance and time delay between the place and moment when a significant change of a driving force occurs and when the effects are recorded. Further, the same structure represents the cycling support for all the chemical elements (including the heavy metals and radionuclides) and for man-made chemical compounds. This fact accounts for the accumulation of the effects. The complexity of biogeochemical cycles as being the fundamental processes of the ecological systems has valuable significance. This is especially important for managers, decision makers, and politicians who want to apply programs for sustainable use of NC and to prevent deterioration of the quality of resources (e.g., water, air, soil, biological) and human health.

In addition to the above statement, we may add a complementary series of interpretations that are intended to be taken as important coordinates of the decision-making process and management sustainable use and development goals. (See Chapter 8 for details.)

- Biogeochemical cycles of chemical elements and compounds are associated with the energy flow, thus supporting and strictly controlling the quality and amount of resources and services provided by ecological systems.

- Recyclable stocks of many chemical elements and compounds are usually distributed in that compartment of HGMU that performs the most active exchanges with the cycling compartment (Figure 2.8), in relatively low concentrations. However, the density of flow between the dominant components of the tropho-dynamic modules is increased, and more often multiplied by hundreds and thousands of times, through the phenomena of accumulation and concentration.
- In many cases, the concentration of certain heavy metals, radionuclides, and chemical compounds in the main pool of the cycling system is maintained below safe threshold levels established by toxicity tests. However, most of the components in the terminal tropho-dynamic modules raise the concentration far beyond safe threshold levels. Thus, the trophic chains directly open to human populations more often transfer significantly higher amounts of "toxics" and expose individuals and human populations to mutagenic effects and chronic diseases to a greater extent than expected.

  Therefore, the real level of exposure and the potential toxic effects cannot be estimated to an acceptably accurate degree unless we combine the classical methodology based on toxicity tests with the evaluation of the density of mass transfer along the trophic chains. Under such circumstances, the common procedure of discharging gaseous or liquid wastes into the troposphere and large surface water bodies (freshwater or marine) assuming high dilution coefficients must be seriously reconsidered.
- The recycling of the "raw material" or the macro- and microelements is one of the key driving forces for gross primary productivity. It is carried out by the dominant populations of the tropho-dynamic modules of an ecological system, not only in a single functional category.

  Figure 2.8B shows the main recycling mechanisms of the macro- and microelements occurring in the aquatic and terrestrial ecological systems. These mechanisms are provided by the dominant populations of the tropho-dynamic modules network needing high-energy expenses. The energy requirement may comprise up to 50 to 80% of the total energy concentrated in the form of gross primary production or energy assimilated by the autotrophic and heterotrophic populations. These mechanisms that are coupled by mass and energy transfer chains make up the cycling system functioning in any natural or semi-natural ecological system. They consist of a sequence of coupled natural "technologies" effective in processing the organic by-products ("wastes") and recycling the raw materials that are required for the maintenance or maximization of the concentrated energy flow, as well as for the production of natural goods and services.

  In contrast, in ecological systems created or dominated by the human population, processing of the organic by-products (dissolved and particulate organic matter) and the solid organic waste is performed by technologies that are bound to a considerable input of concentrated energy, usually in fossil fuels.
- Recent statistical data processed by various specialized institutions have shown that more than 100,000 chemical compounds (e.g., pesticides, plastics,

medicines, additives, food preserving agents, fertilizers) are produced and used daily. Among these compounds more than 7000 are produced and commercialized in large quantities.[64–66]

In addition, and possibly more important, the diversity of synthesized chemical compounds increases annually by 500 to 1000 new products. Most of these new chemical compounds are now being traded internationally on a large scale. Also important is that for approximately 80% of the total number of chemical products produced there are no data regarding their potential toxic effects, for 18% there are incomplete and uncertain data, and for only 2% do we have a complete database of toxic effects.

This picture is even more complicated if we correlate correct toxicity data. Much of the waste from the technological processes has a significantly high risk for human health and that of ecological systems. In industrialized countries 70% of waste comes from the chemical and petro-chemical industries. And the risk is even higher in the case of radioactive waste.

Finally, there has been a great amount of data collected (in fact exceeding a critical mass necessary for an empirical understanding). Although chemical compounds may not have immediate identifiable effects, there are resounding long-term effects as a result of accumulation over time. These effects may endanger the ecological systems as well as the human population.

- It has been recognized that all chemical compounds produced, or resulting as by-products in technological processes, when released into the "environment" follow similar pathways for mass movement and cycling similar to natural chemical elements and compounds. Thus, management of harmful chemical compounds needs to be promoted with new conceptual and methodological developments as shown in Figures 2.9, 2.10, and 2.11.

## 2.5  CHALLENGES FOR ECOSYSTEM MODELING

The following has logically emerged from the foregoing design and analysis: This has a broad formulation instead of being focused only on lagoon systems, to underlie their applicability to all types of ecological systems (including socio-economic systems) to again stress existing opportunities and gaps in the operational infrastructure for sustainability. Some of these are further discussed and applied to lagoon systems (see following chapters).

According to Jorgensen[67] more than 4000 ecological models have been developed and used in the last three decades as tools in the research of complex and dynamic ecological systems or in environmental management. A few years ago, Jorgensen[68] published an excellent book that integrated the best available experience in the field of ecological modeling, following review of more than 400 models. However, the power of existing models to describe the complexity and dynamic behavior of different categories of ecological systems or to explain and give reliable prognoses for specific environmental problems is still very limited due to the following constraints:

- Poor or very poor identification of the particular ecological systems at a spatial and temporal scale before developing the models. This results in limited real world assessment.

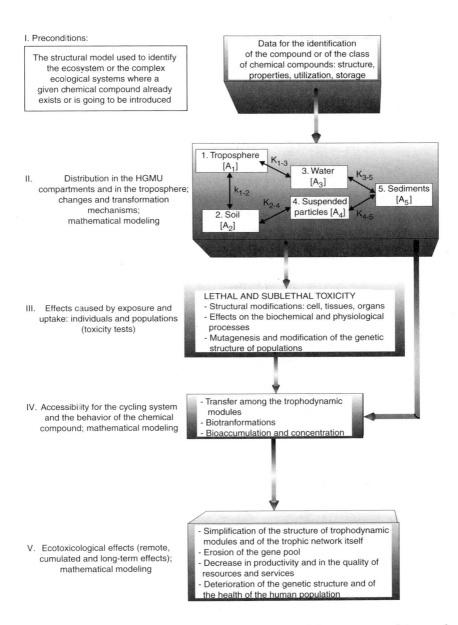

I. Preconditions:

The structural model used to identify the ecosystem or the complex ecological systems where a given chemical compound already exists or is going to be introduced

Data for the identification of the compound or of the class of chemical compounds: structure, properties, utilization, storage

II. Distribution in the HGMU compartments and in the troposphere; changes and transformation mechanisms; mathematical modeling

1. Troposphere [A₁]
$K_{1-3}$
3. Water [A₃]
$K_{3-5}$
5. Sediments [A₅]
$k_{1-2}$
$K_{2-4}$
4. Suspended particles [A₄]
$K_{4-5}$
2. Soil [A₂]

III. Effects caused by exposure and uptake: individuals and populations (toxicity tests)

LETHAL AND SUBLETHAL TOXICITY
- Structural modifications: cell, tissues, organs
- Effects on the biochemical and physiological processes
- Mutagenesis and modification of the genetic structure of populations

IV. Accessibility for the cycling system and the behavior of the chemical compound; mathematical modeling

- Transfer among the trophodynamic modules
- Biotranformations
- Bioaccumulation and concentration

V. Ecotoxicological effects (remote, cumulated and long-term effects); mathematical modeling

- Simplification of the structure of trophodynamic modules and of the trophic network itself
- Erosion of the gene pool
- Decrease in productivity and in the quality of resources and services
- Deterioration of the genetic structure and of the health of the human population

**FIGURE 2.9** The structure of the model of integrated approach for assessment of the transfer pathways, of the changes and potential harmful effects of various classes of chemical compounds.

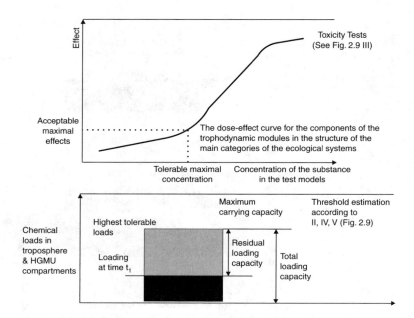

**FIGURE 2.10** Tools for the evaluation of the impact of toxic chemical substances and for the drawing up of low risk solutions (depending on the quality of data knowledge base, continuously consolidated according to the model illustrated in Figure 2.9).

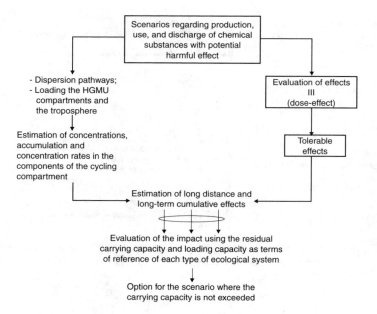

**FIGURE 2.11** Diagram for drawing up and checking the scenarios regarding management of toxic chemical compounds.

- The model development based on the unrealistic assumption that ecological systems maintain rigid structures, and on fixed and incomplete parameters
- Weak quality data sets available for model development and especially for parameter estimation
- In most cases, the structural and functional diversity of ecological systems has been neglected and human society only recently has started to be identified and modeled as a dominant component of the socio-ecological complexes

In order to overcome the above constraints, different research initiatives have been launched in the last decade for development and improvement of ecological modeling concepts and techniques:

    i. Identification of ecological systems, both components of NC and SES by specific and dynamic structural and functional models (dynamic homomorph models), which preserve their specific dynamic properties at the most appropriate spatio-temporal dimensions (see Section 2.3).

Such structural dynamic models integrate: the network of major components in the structure of HGMUs and troposphere; the network of trophodynamic modules describing the spatio-temporal organization of biocoenoses or the network of modules in the economic subsystem; the patterns for inner mass, energy, and information transfer and the boundary conditions or the pathways of the so-called "metabolism" of SES as shown in Figure 2.4.

To each structural dynamic model used for the identification of a particular category of ecological system there must be a set of external driving forces, a set of structural and functional parameters, and the corresponding sets of state variables.

    ii. In order to improve the knowledge and data quality concerning the structural and functional dynamics of ecological systems, the development of the conceptual framework and methodology to assess the NC and SES has started. The diversity of ecological structure at the national and macroregional scale, is viewed as the networks of different types of ecological systems (especially belonging to the hierarchical level of microlandscapes or seascapes). This includes components of socio-economic systems where long term ecological research and integrated monitoring programs are carried out.

The land/seascapes and corresponding lagoon systems are, or should be, represented in such networks. In this respect, the current NATO-CCMS pilot study is a very promising starting point as demonstrated in the following chapters.

    iii. Development and improvement of mathematical modeling techniques in order to cope with: (a) poor databases, (b) specific decision making and management issues, and (c) to take the complexity, adaptability and structural and functional dynamics of the ecological systems into account (Figure 2.12).

       a. To manage the constraints linked to poor databases in the case of many specific ecological systems or specific environmental issues, both "fuzzy models" and modeling techniques based on chaos and fractal

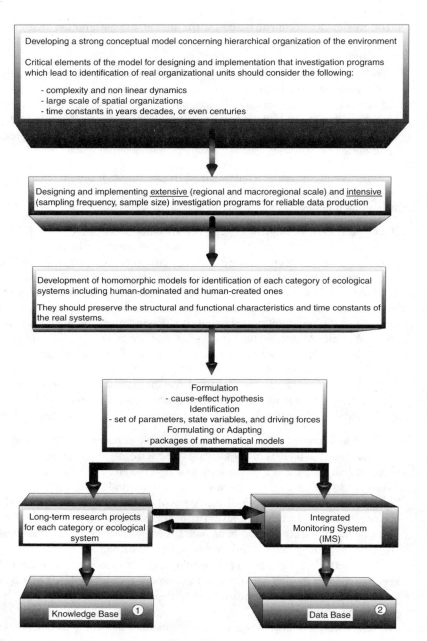

**FIGURE 2.12** Main steps in the development of information systems concerning the organization, dynamics, productivity, and carrying capacity of ecological systems. (1) In the first instance it is supplied with information produced by the implementation of extensive and intensive research projects for system identification and that gained during critical analysis and integration of the historical data. Each phase in a long-term research project provides new information as long as the hypotheses are validated. The information concerns the structural and functional mechanisms, phenomena, and processes in a given category of ecological systems. They help describe the dynamics, productivity, and carrying capacity of those systems. (2) Data base is developed following the structure of the integrated monitoring system. The data concerning the dynamics of key state variables make possible the assessment of the system status at any particular time.

theory have been developed and used in order to improve the parameter estimation.[67,69,70]

b. Artificial neural networks (ANNs) methods and especially the "multi-layer feed-forward neural network" (BPN) and the "Kohonen self-organizing mapping" (SOM) have been used extensively for ecological modeling[71] and in particular for performing specific tasks in different fields of applied ecology. These fields include soil hydrology;[72] modeling the greenhouse effect;[73] modeling water and carbon fluxes above European coniferous forests;[74] modeling phytoplankton primary production;[75,76] applying ANNs tools to ocean color remote sensing;[77] predicting P/B ratio of animal populations;[78] predicting collembolan diversity and abundance in riparian habitats; and predicting the response of zooplankton biomass to climatic and oceanic changes.[79]

The technique of stochastic dynamic programming, previously used in agricultural economics[80] and in commercial fisheries,[81] has been improved. This has been used to obtain solutions for maintaining a minimum viable population size with minimum economic loss, and it suggests that this approach can have a "universal applicability in conservation biology."[82]

The numerical method of analysis and input-output models also has been used for the assessment of "ecological sustainability" of a regional economy[83] or a national economy.[84]

c. Despite many and persisting constraints, a great effort has been made in recent years to develop and apply techniques for modeling structural changes in ecological systems, based on the catastrophe theory[67,70] and for mathematical modeling of the structural dynamic homomorph models describing ecological systems (Figure 2.13). Also promising are the models consisting of linear differential equations with time varying parameters[85] and especially the dynamic mathematical models developed by using the exergy as a goal function.[67]

Above is a brief presentation of the main findings of a much more comprehensive critical analysis of recent approaches and developments in the field of ecological modeling. It can be concluded that there are strong concepts available enabling a systemic approach to environment and an almost complete range of methods, modeling tools and logistics for:

- Designing and implementing study programs for spatio-temporal identification of NC and SES[s] as well as quality assessments of historical data and knowledge
- Designing and developing the initial structure for the information system of knowledge and database for each category of identified ecological systems (Figure 2.14)
- Developing or adapting the most appropriate package of mathematical models, integrating all types of a, b, and c models, in order to describe specific phenomena, processes, or structural changes, and dynamics of the whole system by using existing knowledge and data as well as the set of hypotheses dealing with uncertainties and gaps in knowledge and data

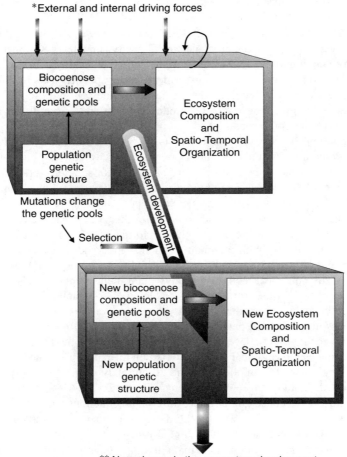

**FIGURE 2.13** Ecosystem development: conceptual frame.

*At a time scale of years, decades, and centuries changes occur in their number and nature or in their range of fluctuations, intensity, and frequency. Their impact on populations and biocoenose composition lead to time-delayed and long-term effects.

**Ecosystem development is a long-term process that involves a long series of successive phases of growth and evolution.

(After Vadineanu, A., *Sustainable Development: Theory and Practice*, Bucharest University Press, Bucharest, 1998. With permission.)

- Building the interface and guidelines for information system management
- Establishing the network of microlandscapes or seascapes according to the diversity of the identified ecological network
- Launching long-term ecological research and integrated monitoring programs in order to feed and improve knowledge and data bases

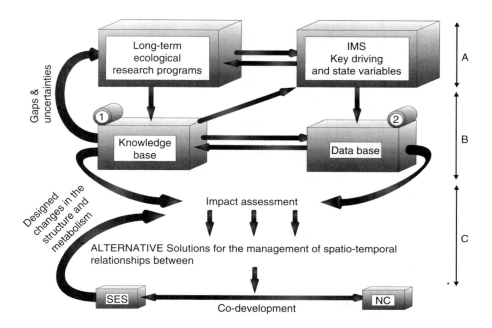

**FIGURE 2.14** A general diagram showing the complementary relationships between research and integrated monitoring programs. (A) The structure and use of the information system (B) and the significance of balancing the relationships between SES and NC or their co-development. (C) Strategic and project impact assessment requires the cost-benefit analysis in a large socio-economic context and at a large time scale and provides alternative solutions either for rehabilitation and reconstruction of the damaged system or for maintenance of the dynamics of healthy ecological systems within the limits of their production and carrying capacities.

# REFERENCES

1. Williams, M., *Wetlands – A Threatened Landscape*, Basil Blackwell, Oxford, U.K., 1991.
2. Schot, P.P., Wetlands, in *Environment Management in Practice*, Vol. 3, Narth, B., Hens, L., Compton, P., and Devuyst, D., Eds., Routledge, New York, 1999.
3. Arrow, K., Bolin, B., Costanza, R., Dasgupta, P., Folke, C., Holling, S.C., Jansson, O.J., Levin, S., Maler, G.K., Perrings, C., and Pimentel, D., Economic growth, carrying capacity and the environment, *Ecological Economics*, 15(2), 91, 1995.
4. Costanza, R., Economic growth, carrying capacity and the environment, *Ecological Economics*, 15(2), 89, 1995.
5. Musters, M.J.C., De Graaf, J.H., and Keurs, J.W., Defining socio-environmetnal systems for sustainable development, *Ecological Economics,* 26(3), 243, 1998.
6. Vadineanu, A., *Sustainable Development, Theory and Practice*, Bucharest University Press, Bucharest, 1998.
7. Wackernagel, M. and Ress, W., *Our Ecological Footprint, Reducing Human Impact on the Earth,* New Society Publisher, Gabriola Island, Philadelphia, 1996.

8. Daly, H.E., The economic growth debate: what some economists have learned, but many have not? *J. Environ. Ecol. Manag.,* 14, 323, 1987.

9. Pearce, D.W. and Turner, E.K., *Economics of Natural Resources and the Environment,* John Hopkins University Press, Baltimore, 1990.

10. Costanza, R. and Daly, H., Natural capital and sustainable development, *Conserv. Biol.,* 6, 37, 1992.

11. Costanza, R., Segura, O., and Martinez–Alier, J., Eds., *Getting Down to Earth: Practical Applications of Ecological Economics,* Island Press, Washington, D.C., 1996.

12. Rennings, K. and Wiggeringi, H., Steps towards indicators of sustainable development: linking economic and ecological aspects, *Ecological Economics,* 20(1), 25, 1997.

13. Hinterberger, K., Luks, F., and Bleek-Schmidt, F., Material flows vs. "Natural Capital." What makes an economy sustainable? *Ecological Economics,* 23(1), 1, 1997.

14. Proops, R.C.J., Faber, M., Manstetten, R., and Jost, F., Achieving a sustainable world, *Ecological Economics,* 17(3), 133, 1996.

15. Walpole, C.S. and Sinden, A.J., BCA & GIS: Integration of economic and environmental indicators to aid land management decisions, *Ecological Economics,* 23(1), 25, 1997.

16. Ring, I., Evolutionary strategies in environmental policy, *Ecological Economics,* 23(3), 237, 1997.

17. Ruijgrok, E., Vellinga, P., and Goosen, H., Dealing with nature, *Ecological Economics* 28, 347, 1999.

18. Berkes, R. and Folke, C., Investing in cultural capital for sustainable use of natural capital, in *Investing in Natural Capital: The Ecological Economics Approach to Sustainability,* Jansson, A.M, Hammer, Folke, C., and Costanza, R., Eds., Island Press, Washington, D.C., 1994, 128.

19. Cairns, J., Jr., Sustainability, ecosystems services and health, *Int. J. Sustain. Dev. World. Ecol.,* 4(3), 153, 1997.

20. De Groot, R., Environmental functions and the economic value of natural systems, in *Investing in Natural Capital: The Ecological Economics Approach to Sustainability,* Jansson, A.M., Hammer, M., Folke, C., and Costanza, R., Eds., Island Press, Washington, D.C., 1994, p. 151.

21. De Groot, R., Van der Perk, J., Chiesura, A., and van Vliet, A., Importance and threat as determining factors for criticality of natural capital, *Ecological Economics,* 44(2–3), 187, 2003.

22. Wackernagel, M., Onisto, L., Bello, P., Callejas, L., Falfan L. S., Garcia, M.J., Isabel, S., and Guerrero, S.M., National natural capital accounting with the ecological footprint concept, *Ecological Economics,* 29(3), 375, 1999.

23. Ayres, R.U. and Simonis, U.E., Eds., *Industrial Metabolism: Restructuring for Sustainable Development,* United Nations University Press, Tokyo, 1994.

24. Oglethorpe, R.D. and Sanderson, A.R., An ecological economic model for agri-environmental policy analysis, *Ecological Economics,* 28(2), 245, 1999.

25. Odum, E.P., *Ecology: A Bridge between Science and Society,* Sinauer Associates, Sunderland, MA, 1997.

26. Pahl-Wostl, C., *The Dynamic Nature of Ecosystems,* John Wiley & Sons, New York, 1995.

27. Vadineanu, A., *Sustainable Development Theory and Practice Regarding the Transition of Socio-Economic Systems Towards Sustainability,* UNESCO-CEPES, Bucharest, 2001.

28. Holling, C.S., Understanding the complexity of economic, ecological and social systems, *Ecosystems,* 4, 390, 2001.

29. Holling, C.S. and Gunderson, H.L., Resilience and adaptative cycles, in *Panarcy: Understanding Transformations in Human and Natural Systems*, Gunderson, H.L. and Holling, C.S., Eds., Island Press, Washington, D.C., 2001, p. 26.

30. Patten, C.B., *Systems Analysis and Simulation in Ecology*, Academy Press, London, 1971.

31. Ziegler, P.B., Multilevel multiformalism modeling: an ecosystem example, in *Theoretical Systems Ecology*, Halfon, E., Ed., Academic Press, New York, 1979, p. 18.

32. Vadineanu, A. and Teodorescu, I., Controlul pe baze ecologice al populatiilor de insecte din agrosisteme, *Ocrot. Nat. Med. Inconj.*, 31(1), 28, 1987.

33. Naveh, Z. and Lieberman, A., *Landscape Ecology: Theory and Application*, Springer-Verlag, Heidelberg, 1994.

34. Wiegert, R., Holism and reductionism in ecology: hypotheses, scale and systems models, *Oikos*, 53, 267, 1988.

35. West, B., *An Essay on the Importance of Being Nonlinear*, Springer-Verlag, Heidelberg, 1985.

36. Lauenroth, K.W., Skogerboe, V.G., and Flug, M., *Analysis of Ecological Systems: State-of-the-Art in Ecological Modeling*, Elsevier, Amsterdam, 1993.

37. Ulanowicz, R.R., *Growth and Development: Ecosystems Phenomenology*, Springer-Verlag, New York, 1986.

38. Salthe, S.N., *Evolving Hierarchical Systems: Their Structure and Representation*, Columbia University Press, New York, 1985.

39. Décamp, H. and Naiman, R.J., L'ecologie des fleuves, *La Recherche*, 20, 310, 1989.

40. Décamp, H. and Tabacchi, E., Species richness in vegetation along river margins, in *Aquatic Ecology: Scale, Pattern and Processes*, Giller, P.S., Hildrew, A.G., and Raffaelli, D.G., Eds., Blackwell, Oxford, 1994, p. 1.

41. Décamp, H., The renewal of floodplain forests along rivers: a landscape perspective, *Verh. Internat. Verein Limnol.*, 26, 35, 1996.

42. Naiman, R.J. and Décamp, H., Eds., *The Ecology and Management of Aquatic-Terrestrial Ecotones*, UNESCO, Paris and The Parthenon Publishing Group, Carnforth, U.K., 1990.

43. Mitsch, W.J., Mitsch, R.H., and Turner, R.E., Wetlands of the old and new worlds: ecology and management, in *Global Wetlands: Old World and New*, Mitsch, W.J., Ed., Elsevier, Amsterdam, 1994, p. 3.

44. Maltby, E., Hogan, V.D., and McInnes, J.R., Functional Analysis of European Wetlands Ecosystems (FAEWE) Phase I, Ecosystem Research Report No. 18, Brussels, 1996.

45. Pinay, G., Décamp, H., Chauvet, E., and Fustec, E., Functions of ecotones in fluvial systems, in *The Ecology and Management of Aquatic—Terrestrial Ecotones*, Naiman, R. and Décamp, H., Eds., Parthenon Press, London, 1990.

46. Pinay, G. and Décamp, H., The role of riparian woods in regulating nitrogen fluxes between the alluvial aquifer and surface water: a conceptual model, *Regulated Rivers*, 2, 507, 1988.

47. Merot, P. and Durand, P., Modeling the interaction between buffer zones and the catchment, in *Buffer Zones: Their Processes and Potential in Water Protection*, Haycock, N., Burt, T., Goulding, K., and Pinay, G., Eds., White Crescent Press, London, 1997, p. 208.

48. Gold, J.A. and Kellogg, Q.D., Modeling internal processes of riparian buffer zones, in *Buffer Zones: Their Processes and Potential in Water Protection*, Haycock, N., Burt, T., Goulding, K., and Pinay, G., Eds., White Crescent Press, London, 1997, p. 192.

49. Haycock, N.E. and Pinay, G., Nitrate retention in grass and poplar vegetated riparian buffer strips during the winter, *J. Environm. Qual.*, 27, 273, 1993.

50. Vadineanu, A., Cristofor, S., Ignat, G., Iordache, V., Sarbu, A., Ciubuc, C., Romanca, G., Teodorescu, I., Postolache, C., Adamescu, M., and Florescu, C., Functional assessment of the wetlands ecosystems in the Lower Danube River System, *Limnologische Berichte (Donau, 1997)*, WIEN, 63, 1997.
51. Cristofor, S., Sarbu, A., Vadineanu, A., Ignat, G., Iordache, V., Postolache, C., Dinu, C., and Ciubuc, C., Effects of hydrological regimes on riparian vegetation in the Lower Danube floodplain, *Limnologische Berichte*, WIEN, 233, 1997.
52. Postolache, C., Iordache, V., Vadineanu, A., Ignat, G., Cristofor, S., Neagoe, A., and Bodescu, F., Effects of hydrological conditions on the dynamic of nutrients in the Lower Danube floodplain, *Limnologische Berichte (Donau 1997)*, WIEN, 15, 1997.
53. Sarbu, A., Cristofor, S., and Vadineanu, A., Effects of hydrological regime on submerged macrophytes in the Lower Danube floodplain and delta, *Limnologische Berichte (Donau, 1997)*, WIEN, 237, 1997.
54. Botnariuc, N. and Vadineanu, A., *Ecologie Generala*, Edit. Did. Pedag, Bucharest, 1982.
55. Pimm, S.L., *Food Webs*, Chapman & Hall, London, 1982.
56. Whittaker, R.H. and Likens, E.G., Primary production of the biosphere, *Hum. Biol.*, 1, 301, 1971.
57. Margalef, R., Diversity, stability and maturity in natural ecosystems, in *Unifying Concepts in Ecology*, Junk Publ., The Hague, 1975, p. 151.
58. Odum, E.P., *Ecology and Our Endangered Life-Support Systems*, Sinauer Associates, Sunderland, MA, 1993.
59. Odum, E.P. and Biever, J.L., Resource quality, mutualism and energy partitioning in food chains, *Am. Nat.*, 124, 360, 1984.
60. Odum, H.T., *Environmental Accounting: Energy and Environmental Decision Making*, John Wiley & Sons, New York, 1996.
61. Hillbricht-Ilowska, A., Trophic relations and energy flow in plankton, *Pol. Ecol. Studies*, 3(1), 1, 1977.
62. Vitousek, P.M., Ehrlich, R.P., Ehrlich K.A., and Matson, A.P., Human appropriation of the products of photosynthesis, *Bioscience*, 36, 368, 1986.
63. Norberg, J., Linking nature's services to ecosystems: some general ecological concepts, *Ecological Economics*, 29(2), 183, 1999.
64. Connell, W.D., Ecotoxicology—a framework for investigations of hazardous chemicals in the environment, *Ambio*, 16(1), 47, 1987.
65. Novotny, V. and Somlyody, L., Eds., *Remediation and Management of Degraded River Basins: With Emphasis on Central and Eastern Europe*, Springer-Verlag, Berlin, 1995.
66. Sigel, H., Metal ions in biological systems, in *Circulation of Metals in the Environment*, Marcel Dekker, New York, 1987, Vol. 18.
67. Jorgensen, S.E., State of the art of ecological modeling with emphasis on development of structural dynamic models, *Ecol. Model.*, 120, 75, 1999.
68. Jorgensen, S.E., *Fundamentals of Ecological Modeling*, 2nd ed., Elsevier, Amsterdam, 1994.
69. Jorgensen, S.E., The growth rate of zooplankton at the edge of chaos, *J. Theor. Bio.*, 175, 13, 1995.
70. Jorgensen, S.E., *Integration of Ecosystem Theories: A Pattern*, 2nd ed., Kluwer, Dordrecht, 1997.
71. Lek, S. and Guegan, F.J., Artificial neural networks as a tool in ecological modeling, an introduction, *Ecol. Model.*, 120, 65, 1999.

72. Vila, P.J., Wagner, V., Neven, P., Voltz, M., and Lagacherie, P., Neural network architecture selection: New Bayesian perspectives in predictive modeling: application to a soil hydrology problem, *Ecol. Model.*, 120, 119, 1999.

73. Seginer, I., Boulard, T., and Bailey, B.J., Neural network models of the green house climate, *J. Agric. Eng. Res.,* 59, 203, 1994.

74. Van Wijk, T.M. and Bouten, W., Water and carbon fluxes above European coniferous forest modeled with artificial neural networks, *Ecol. Model.*, 121, 181, 1999.

75. Scardi, M. and Harding, W.I., Developing an empirical model of phytoplankton primary production: a neural network case study, *Ecol. Model.*, 120, 213, 1999.

76. Recknagel, F., ANNA—Artificial neural networks model predicting blooms and succession of blue-green algae, *Hydrobiology*, 349, 47, 1997.

77. Gross, L., Thiria, S., and Frovin, R., Applying artificial neural networks methodology to ocean color remote sensing, *Ecol. Model.*, 120, 237, 1999.

78. Brey, T., Teichmann J., and Barlick, O., Artificial neural networks vs. multiple linear regression predicting P/B ratios from empirical data, *Mar. Ecol. Prog. Ser.*, 140, 251, 1996.

79. Aoki, I., Komatsu, T., and Hwang, K., Prediction of response of zooplankton biomass to climatic and the oceanic changes, *Ecol. Model.*, 120, 261, 1999.

80. Kennedy, J.O.S., *Dynamic Programming Applications to Agricultural and Natural Resources*, Elsevier, London, 1986.

81. Gillis, D.M., Pikitch, E.E., and Peterman, R.M., Dynamic discarding decisions: foraging theory for high grading in a trawl fishery, *Behav. Ecol.*, 6, 146, 1995.

82. Doherty, F.P., Jr., Marschall, A.E., and Grubb, C.T., Jr., Balancing conservation and economic gain: a dynamic programing approach, *Ecological Economics,* 29(3), 349, 1999.

83. Eder, P. and Narodoslawsky, M., What environmental pressures are a region's industries responsible for? A method of analysis with descriptive indices and input-output models, *Ecological Economics,* 29(3), 359, 1999.

84. Lawn, A.P. and Sanders, D.R., Has Australia surpassed its optimal macroeconomic scale? Finding out with the aid of benefit and cost account and sustainable benefit index, *Ecological Economics*, 28(2), 213, 1999.

85. Patten, C.B., Bear model for Adirondack National Park, *Ecol. Model.*, 100, 11, 1997.

# 3 Physical Processes

*Georg Umgiesser and Ramiro Neves*

## CONTENTS

1-56670-686-6/05/$0.00+$1.50

## 3.1 INTRODUCTION TO TRANSPORT PHENOMENA

Chapter 2 concluded that the calculation of spatio-temporal distribution of major components of a lagoon's hydrogeomorphological unit and biocoenose is important for the description of the structure and function dynamics (productivity and carrying capacity) of the lagoon system and, consequently, for sustainable management. This concept is required to understand the transport phenomena that describe the evolution of properties due to fluid motion (advection) and/or molecular and turbulent dynamics (diffusion). In the case of turbulent flows the small-scale motion of the fluid particles is actually random, and this nonresolved advection is also treated as diffusion (eddy diffusion).

A mathematical description of the transport phenomena (transport equations) is based on the concept of conservation principle, which is valid in any application. Conservation principle can be stated as

{The rate of accumulation of a property inside a control volume}

   = {what flows in minus what flows out}+{production minus consumption}

Using this conservation principle, transport equations for any property inside a control volume can be derived if production and consumption mechanisms are known and if the control volume and transport processes are quantified. The control volume is presented as the largest volume for which one can consider the interior properties as uniformly distributed as well as fluxes across the surface.

In previous coastal lagoon studies the control volume was often implicitly defined as the whole lagoon. Concepts of residence and flushing time were derived from this global approach (see Chapter 5 for details). In that case, only fluxes at the boundaries were required. This type of integral approach cannot describe gradients and is consequently not sufficient to support process-oriented research or management. For these purposes the system has to be divided into homogeneous parts (control volumes) and fluxes between them have to be calculated. This approach requires a numerical model. The number of space dimensions required to describe the control volumes equals the number of dimensions of the model. The resolution of the transport equations in practical situations has been made possible using numerical methods and computers (see Chapter 6 for details). Before the advent of computers transport processes had to be studied using empirical formulations (derived from experiments) or analytical solutions in simple geometries or boundary conditions.

This chapter presents a general transport equation (also called an evolution equation) based on the concepts of (1) control volume, (2) advective flux, and (3) diffusive flux. Based on this generic equation, equations for hydrodynamics, temperature, salinity, and suspended sediments are also introduced. Special flows and simplification of dimensionality and boundary processes and conditions, particularly for coastal lagoons, are described in detail for use in lagoon modeling studies.

## 3.2 FLUXES AND TRANSPORT EQUATION

### 3.2.1 VELOCITY AND DIFFUSIVITY IN LAMINAR AND TURBULENT FLOWS AND IN A NUMERICAL MODEL

For transport purposes, fluids are considered a continuum system. Velocity is defined in a macroscopic way based on the concept of continuum system. Because fluids are not a real continuum, system velocity cannot describe transport processes at the molecular scale. The nonrepresented processes are represented by diffusion. Although the concept of velocity is well known, it is reconsidered for modeling purposes.

Diffusion in laminar flows occurs from movements at a molecular scale not represented by the velocity. In turbulent flows, velocity, as defined for laminar flows, becomes time dependent, changing at a frequency that is too high to be represented analytically. As a consequence time average values must be considered, following the Reynolds approach (see Section 3.4.2.4 for details). Transport processes, not described by this average velocity, are represented by turbulent diffusion (using an eddy diffusivity, which is several orders of magnitude higher than molecular diffusivity). More information on this topic is given in Section 3.4.

Most numerical models use grids with spatial and time steps larger than those associated with turbulent eddies. Again, processes not resolved by velocity computed by models have to be accounted for by diffusion (subgrid diffusion).

The box represented in Figure 3.1 is commonly used to illustrate the concept of diffusion in laminar flows. The same box could be used to illustrate the concept of eddy diffusion in turbulent flows or subgrid diffusion in numerical models. In molecular diffusion white and black dots represent molecules, while in other cases they represent eddies. In the initial conditions (stage (a) in Figure 3.1) two different fluids are kept apart by a diaphragm. Molecules inside each half-box move randomly, with velocities not described by our model (Brownian or eddy). When the diaphragm is removed, particles from each side keep moving, resulting in the possibility of

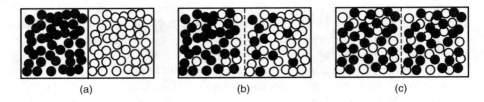

**FIGURE 3.1** Distribution of molecules of two different substances, inside a box at three different moments: initially kept apart by a diaphragm (a), during mixing (b), and when spatial gradient has disappeared (c).

mixing (stage (b) in Figure 3.1). After some time the proportion of white and dark fluid is the same in both the box halves. At this stage, the probability of a white molecule moving from the right side of the box to the left side is equal to the probability of another white molecule moving the opposite way and, therefore, there is no net exchange (stage (c) in Figure 3.1).

A macroscopic view of the mixing is represented in Figure 3.2. In the initial condition (stage (a) in Figure 3.2), a black and a white side is observed (stage (b) in Figure 3.2); during the mixing process a growing gray area occurs, corresponding to the mixing zone; and after complete mixing, a homogeneous gray fluid is observed (stage (c) in Figure 3.2).

The velocity of each elementary portion of fluid, like the velocity of any other material point, is defined using two consecutive locations, as shown in Figure 3.3:

$$\vec{u} = \lim_{\Delta t \to 0} \left( \frac{d\vec{x}^{t+\Delta t} - \vec{x}^t}{\Delta t} \right) = \frac{d\vec{x}}{dt} = \frac{dx_i}{dt} = u_i \tag{3.1}$$

Knowing the velocity of each individual molecule, it will be possible to fully characterize transport. But, according to Heisenberg's uncertainty principle, it will never be possible to know the place and the velocity of each molecule simultaneously. Consequently, the fluid has to be considered a continuum system, for which a velocity is defined.

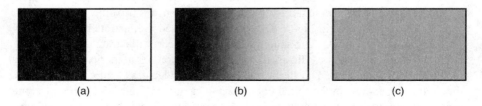

**FIGURE 3.2** Macroscopic view of the fluid composed by the molecules represented in Figure 3.1.

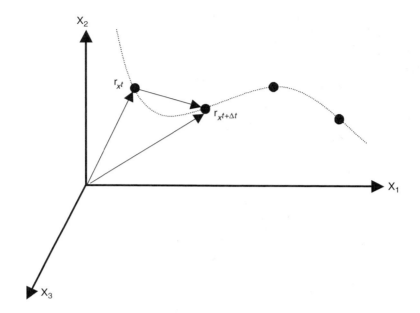

**FIGURE 3.3** Trajectory of an elementary portion of fluid showing consecutive locations apart in time of $\Delta t$.

Considering a fluid as a continuum system, an elementary volume of fluid is a portion of fluid so small that its properties (including velocity) can be considered as uniform, but it is much bigger than the size of a molecule in laminar flow and than an eddy in turbulent flow. In the case of a numerical model, an elementary volume is the volume included inside a grid cell. As a consequence, diffusivity increases with grid size in numerical models.

### 3.2.2 ADVECTIVE FLUX

The advective flux accounts for the amount of a property transported per unit of time due to fluid velocity across a surface perpendicular to the motion. Its dimensions are $[BT^{-1}]$ and can be expressed as[†]

$$\Phi_{adv} = \frac{B}{V}\frac{V}{t} \tag{3.2}$$

where $V$ is the volume. The quantity $\beta = B/V$ has the dimensions of a specific quantity (amount per unit of volume) and is called the concentration. The ratio between the volume and the time $[L^3T^{-1}]$ is the flow rate that can be calculated as the product

---

[†] $[B]$ means "dimensions of $B$"; $T$ means time.

of the velocity $[LT^{-1}]$ and the area $[L^2]$ of the surface. The transport produced by the velocity per unit of area is

$$\phi_{adv} = \beta \cdot \vec{u}$$

where $\vec{u}$ is the velocity relative to the velocity of the surface. This flux is a vector parallel to the velocity. The flux across an elementary area $dA$ not perpendicular to the velocity is given by

$$d(\Phi_{adv}) = \beta(\vec{u} \cdot \vec{n})\, dA$$

where $\vec{n}$ is the external normal to the elementary surface $dA$. In the case of a finite surface $A$, the total flux is the summation of the elementary fluxes across the elementary areas composing it. This summation can be represented by the integral over that surface:

$$\Phi_{adv} = \int_A d(\Phi_{adv}) = \int_A \beta(\vec{u} \cdot \vec{n})\, dA \qquad (3.3)$$

This is the generic definition of the advective flux of a property $B$ across a surface $A$.

### 3.2.3  DIFFUSIVE FLUX

Diffusive flux is the net transport associated with the Brownian movement of molecules in the case of a laminar flow or due to both molecular and turbulent movement in the case of a turbulent flow, depending on the property gradient. Diffusion transports the property down the direction of the gradient, as determined by Fick, who stated that the diffusive flux per unit of area is given by

$$\vec{\phi}_{dif} = -\varphi(\vec{\nabla}\beta) \qquad (3.4)$$

where $\varphi$ is the diffusivity and the quantity inside the parentheses is the gradient of $\beta$, the specific value of B (concentration in case of mass).

Both $\beta$ gradient and diffusivity can vary spatially, implying the calculation of the flux on elementary surfaces:

$$d(\phi_{dif}) = -\varphi(\vec{\nabla}\beta)\vec{n}\, dA$$

and its integration along the overall surface:

$$\Phi_{dif} = \int_A d(\phi_{dif}) = \int_A -\varphi(\vec{\nabla}\beta)\vec{n}\, dA \qquad (3.5)$$

where $\vec{n}$ is the normal to the elementary surface.

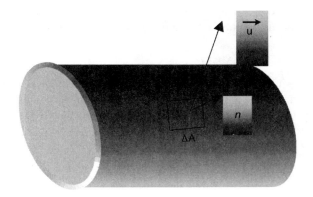

**FIGURE 3.4** Representation of a generic transparent surface, an elementary area on that surface, the respective exterior normal, and the fluid velocity.

### 3.2.4 ELEMENTARY AREA AND ELEMENTARY VOLUME

The transport across a finite surface $A$ is obtained as the integration of the transport across elementary areas where properties can be assumed to be uniform. Mathematically, elementary areas are infinitesimal, but physically they are just small enough to allow the assumption that properties assume constant values on their surfaces, allowing for the substitution of the integrals by summations. Fluxes across an elementary surface are obtained by multiplying fluxes per unit of area with the area of the elementary surface. This is a basic assumption of modeling.

An elementary volume is a volume limited by elementary areas, inside which properties can be considered as having uniform values. The total amount of a property contained inside an elementary volume is given by the product of its specific value with its elementary volume.

### 3.2.5 NET FLUX ACROSS A CLOSED SURFACE

Let us consider a closed surface as represented in Figure 3.4. In the figure an elementary area $\Delta A$, local normal $n$, and velocity $u$, which can vary from point to point, are represented. In regions where normal and velocity have opposite senses the internal product is negative, meaning that property $B$ is being advected (transported) into the interior of the volume limited by the surface. Where the internal product is positive, the property is being transported outward. Thus, the integral of the flux (advective or diffusive) over a closed surface gives the difference between the amount of property being transported outward and inward.

## 3.3  TRANSPORT AND EVOLUTION

An evolution equation describes the transformations suffered by a property as time progresses. The properties of a fluid limited by solid surface can be modified only by production (sources) or destruction (sinks) processes. If the boundary of the volume is permeable, allowing for advective and/or diffusive fluxes, transport also will contribute for the evolution of the property inside the volume.

### 3.3.1  RATE OF ACCUMULATION

The rate of accumulation of a property $B$ inside an elementary volume is the ratio between the variation of the total amount contained in the volume and the time interval during which accumulation happened. This concept can be translated by the algebraic equation:

$$\frac{(\beta V)^{t+\Delta t} - (\beta V)^t}{\Delta t}$$

In this equation, the total amount of $B$ inside an elementary volume $V$ in each moment is given by the product of the specific value $\beta$ multiplied with the volume.[†] If an infinitesimal time interval is considered, the previous equation can be written in a differential form:

$$\frac{d(\beta V)}{dt}$$

This equation puts into evidence the physical meaning of the time derivative, showing that it describes the rate of accumulation.

### 3.3.2  LAGRANGIAN FORM OF THE EVOLUTION EQUATION

If there is no flux across the boundary of the control volume, the rate of accumulation accounts for the sources ($S_0$) minus the sinks ($S_i$):

$$\frac{d(\beta V)}{dt} = V(S_o - S_i)$$

This can happen in the case of a volume limited by a solid boundary or in the case of a volume moving at the same velocity as the flow where diffusivity can be neglected.

Faecal bacteria are traditionally assumed to be unable to grow in saline water and have a first-order decay rate. For such a variable in a volume where advective and diffusive fluxes are null and the sink is the mortality of bacteria, the evolution equation would be written as

$$\frac{d(\beta V)}{dt} = -m\beta V$$

where $m$ is the rate of mortality.

---

[†] In a general case, the total amount inside a volume $V$ of a property with a specific value $\beta$ is the integral of $\beta$ inside the volume. In the case that $\beta$ is constant, the integral is the product of $\beta$ times the volume.

If the surface limiting the volume is permeable, a diffusive flux exists proportional to the gradient and the evolution equation becomes

$$\frac{d(\beta V)}{dt} = -\int_A -\varphi(\vec{\nabla}\beta)\vec{n}dA + V(S_0 - S_i) \tag{3.6}$$

Please note that a positive flux is directed out of the elementary volume and is, therefore, lowering the mass of the property inside the control volume. The negative sign in front of the integral accounts for this fact.

Equation (3.6) describes the evolution of a generic property $B$, with a specific value $\beta$, if there is no advective exchange between the elementary volume of fluid and the surrounding volume. That is the case when the elementary volume moves at the same velocity as the fluid (null relative velocity), corresponding to the Lagrangian formulation of the problem. Measuring instruments moving freely, transported by the flow (e.g., attached to a buoy), would carry out this type of measurement. That is not, however, the common way of measuring.

### 3.3.3 Eulerian Form of the Evolution Equation

In general, field measurements are carried out in fixed stations (see Chapter 7 for details). If such a monitoring strategy is adopted, the elementary volume to be considered in the evolution equation is fixed in space, and the velocity relative to the surface of the volume becomes the flow velocity. In that case the evolution equation becomes:

$$\frac{d(\beta V)}{dt} = -\int_A [\beta \vec{u}\vec{n} - \varphi(\vec{\nabla}\beta\vec{n})]\, dA + V(S_0 - S_i) \tag{3.7}$$

The partial derivative states that the control volume does not move and, consequently, the velocity to be considered in the advective flux is the flow velocity. This equation holds for situations where the volume is time dependent. If we consider the case of a rigid volume, we can take it out of the time derivative.

### 3.3.4 Differential Form of the Transport Equation

Let us consider a Cartesian reference and an elementary cubic volume, as represented in Figure 3.5. Being an elementary volume, it is small enough to assume that properties have uniform values on the surface and that the value of the property can be considered uniform inside the volume. Considering this approach and the geometric properties indicated in the figure, we can write an equation for the volume and for the fluxes. The volume is given by

$$\Delta V = \Delta x_1\, \Delta x_2\, \Delta x_3$$

The integral of fluxes becomes the summation of their specific values across each elementary area (control volume surfaces) multiplied by the area of the

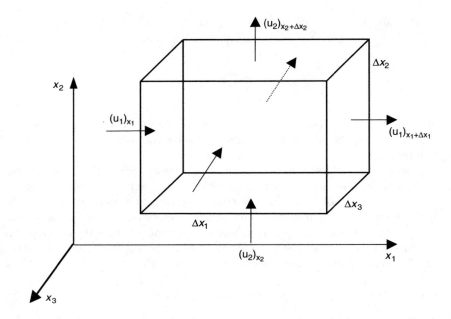

**FIGURE 3.5** Cubic type elementary control volume. Velocity components are represented on every surface of the volume.

corresponding surface. Keeping the orientation of the normal in mind, we can write the fluxes for surfaces perpendicular to the $x_1$ axis:

$$\left[\left(\beta u_1 - \varphi\frac{\partial\phi}{\partial x_1}\right)_{x_1+\Delta x_1} - \left(\beta u_1 - \varphi\frac{\partial\phi}{\partial x_1}\right)_{x_1}\right]\Delta x_2\Delta x_3$$

Doing a similar calculation for other directions, dividing the whole equation by $\Delta V$ and letting $\Delta x_i$ converge to zero, we obtain Equation (3.8):

$$\frac{\partial\beta}{\partial t} + \frac{\partial\beta u_i}{\partial x_i} = \frac{\partial}{\partial x_j}\left(\varphi\frac{\partial\phi}{\partial x_j}\right) + (S_o - S_i) \tag{3.8}$$

In this equation the Einstein convention[†] for summation is used and sources and sinks are calculated per unit of volume. Equation (3.7) and Equation (3.8) are general and hold for any property. They can also be used to derive continuity and momentum equations in the next sections.

---

[†] In the Einstein convention, a doubled index represents a summation of the terms obtained replacing that index by each dimension of the physical space (3 in a three-dimensional space).

Following a similar procedure and assuming an incompressible fluid—an equation equivalent to Equation (3.8) on a Lagrangian reference can be obtained from Equation (3.6):

$$\frac{d\beta}{dt} = \frac{\partial}{\partial x_j}\left(\varphi \frac{\partial \beta}{\partial x_j}\right) + (S_o - S_i) \tag{3.9}$$

Comparing Equation (3.8) and Equation (3.9) and using the fact that for an incompressible fluid $\partial u_i/\partial x_i = 0$ (see Section 3.4.1.1), the relation between total and partial time derivatives is obtained:[†]

$$\frac{d\beta}{dt} = \frac{\partial \beta}{\partial t} + u_i \frac{\partial \beta}{\partial x_i} \tag{3.10}$$

This equation simply says that the property of a fixed elementary volume in a moving fluid element can change through local changes in the fluid and advective changes transported by the fluid into the elementary volume.

In the field of mathematical modeling, especially physical processes, there is a strong tradition of obtaining the discretized equations starting from their differential form using, for example, the Taylor series. In this text discretized equations will be obtained using both a finite-volume approach and the Taylor series.

### 3.3.5 BOUNDARY AND INITIAL CONDITIONS

Partial differential equations relate values of properties to time and space derivatives. Spatial derivatives relate values of the property in neighboring points, and consequently the evaluation at boundaries requires the knowledge of information from outside the study area called boundary conditions. Similarly, time derivative relates values of the property in sequential instants of time and their evaluation requires knowledge of the solution in the first instant of time called initial conditions.

Initial conditions have to be specified in terms of property values. In contrast, spatial boundary conditions can be specified in terms of values or their derivatives. Comparing differential and integral forms of evolution equations (Equation (3.7) and Equation (3.8)) shows that imposing boundary conditions in terms of spatial derivatives in differential equations is, in fact, equivalent to imposing fluxes across the boundary in integral equations.

Physically boundaries can be divided into two main groups: solid boundaries and open boundaries. Solid boundaries are impermeable and, consequently, there is neither water flux nor advective transport across them. Diffusive flux across solid boundaries can be neglected for most dissolved substances, but not for momentum, where it is represented by bottom shear stress. Solid matter can be

---

[†] In fact, the condition of null velocity divergence is not required, only the continuity equation. What happens is that the continuity equation becomes the condition of zero velocity divergence in the case of incompressible flows.

deposited or eroded from the bottom, resulting in apparent fluxes across that solid boundary.

In tidal flows some regions are covered and some are uncovered depending on the level of water. The boundary between covered and uncovered regions is a moving, solid boundary, and its location has to be calculated by the model at each time step.

Open boundaries are artificial boundaries and usually require elaborate formulations. At open boundaries, conditions have to be specified using measurements or estimated using the solution calculated inside the domain (radiative boundaries).

The most complex situation at open boundaries is the case of active properties properties that modify the flow. This is the case with baroclinic flows, where gradients of water density generate density flows. In this case, any boundary condition error generates a force that modifies the flow itself, originating a quick propagation of the error to inner points. This can be extremely important for coastal lagoons.

As a general rule, the diffusive flux is much smaller than the advective flux and can be neglected at the open boundary. Doing so, the boundary values of properties have to be specified only when the flow enters into the computation area, as in the case of coastal lagoons. The net advective transport is proportional to local gradients and to local transport. As a consequence, in the case of passive tracers, open boundaries must be located in regions with small velocities and small gradients.

Major problems are found with properties acting directly on flow. That is the case of the sea surface level, which is transported at the speed of gravity waves (proportional to the square root of the water depth), and its local value is the result of waves propagating in many directions. At the boundary, the sea level depends on incoming and outgoing waves.

Incoming waves have to be specified and outgoing waves have to be estimated using the solution inside the domain and an approach for their propagation speed. This is the basis of a radiation condition. The distinction between contributions of incoming and outgoing waves is a major difficulty, especially in tidal systems where measured tide level itself is the result of waves propagating in different directions.

Free surface can be treated as a solid boundary in the sense that there are no advective fluxes. Across this boundary one can exchange momentum, heat, and mass. Mass of any substance can enter the system through this boundary (e.g., precipitation), but only gaseous matter can leave it (e.g., evaporation). Fluxes across this boundary depend mainly on atmospheric conditions and are proportional to its surface, being much more important in the ocean than in coastal lagoons.

In coastal lagoons the relevance of momentum exchange across free surface decreases as tidal amplitude increases. In contrast, the exchange of heat across this boundary tends to be more important than the exchange across the open boundary because in coastal lagoons the ratio between surface and volume is much smaller than that in the sea.

In the next sections additional information on boundary conditions will be supplied for each specific property. It will be seen that the most adequate conditions depend on the availability of the measuring devices used for obtaining the information.

## 3.4 HYDRODYNAMICS

The hydrodynamic behavior is governed by mathematical equations that have been known for more than 100 years. It is believed that these equations have universal character, and the only reason that we are not able to exactly predict the dynamical evolution of a water body is that we do not possess analytical solutions to these equations and our knowledge of boundary and initial conditions is incomplete. These points will be discussed later in the sections that follow.

The basic hydrodynamic equation can be derived in its full form through the application of conservation laws to the basic variables of the system. More specifically, the general conservation equation derived in the last chapter can be directly applied to the variables under consideration.

In the oceanic environment, including coastal lagoons, there are seven variables that completely define the state of the fluid: the water density $\rho$; the three velocity components $u_i, i = 1,3$, in the direction of $x_i, i = 1,3$; the pressure $p$; the temperature $T$; and the salinity $S$. If only freshwater systems are considered, salinity is not a variable, reducing the number of state variables to six.

Remarkably, the same set of variables is used also for the description of the atmosphere. However, in this case, humidity (water vapor content) replaces salinity as a state variable. Moreover, the equations applicable to atmospheric motion differ only slightly from the ones used in the oceans.

### 3.4.1 CONSERVATION LAWS IN HYDRODYNAMICS

As explained previously, the basic hydrodynamic equations can be relatively easily deduced from conservation equations of the single state variables. A rigorous deduction of these equations will not always be shown, but the equations in their original form will be presented and their meanings and implications noted. It will also be shown how in all possible cases the equations can be derived using the general conservation equation (Equation (3.8)), noted earlier in this chapter.

#### 3.4.1.1 Conservation of Mass

The first equation can be deduced through the application of the well-known law of mass conservation. In this case for the quantity to be conserved in Equation (3.8), we use the density $\rho$ which is just mass per unit volume. Diffusion can clearly be neglected in this case (there is no diffusion of water dissolved in water) and with the assumption that there is no sink or source in the control volume:

$$\frac{\partial \rho}{\partial t} + \frac{\partial \rho u_i}{\partial x_i} = 0 \qquad (3.11)$$

The conservation equation of mass is also called the continuity equation. It expresses the fact that a mass flux into a fixed volume will lead to an increase of mass in the volume due to an increase in density.

If we substitute the partial derivative with the total one, we can rewrite the continuity equation as follows:

$$\frac{1}{\rho}\frac{d\rho}{dt} + \frac{\partial u_i}{\partial x_i} = 0$$

In this form it is easy to see that for an incompressible fluid (for which $d\rho/dt = 0$ is valid) the continuity equation reduces to:

$$\frac{\partial u_i}{\partial x_i} = 0 \tag{3.12}$$

This means that the flow field $u_i$ is simply nondivergent. In this case the mass flux into a fixed volume is exactly zero.

It is interesting to see that this form of the continuity equation is the form normally used for hydrographical and oceanographic applications. The main effect in neglecting the compressibility effect is that acoustic waves cannot be described in the water. As acoustic waves are of minor importance in oceanographic applications, neglecting these terms is justified.

### 3.4.1.2 Conservation of Momentum

Conservation of momentum is in its simplest form described by Newton's law

$$F_i = a_i m$$

where $F_i$ is the force acting on a fluid volume, $a_i$ is the acceleration, and $m$ is the mass of the fluid particle.

Using the density as usual in fluid dynamics instead of the mass of a particle we can write

$$\rho \frac{du_i}{dt} = f_i$$

where $f_i$ is the force per unit volume acting on the fluid volume. This equation can be deduced from the general conservation equation (Equation (3.8)) when applied to $\rho u_i$ together with the continuity equation. In this case the source and sink terms are given by $f_i$, and the diffusion terms are neglected for now.

### 3.4.1.2.1 The Euler Equations

The forces acting on a fluid body may be divided conveniently into two classes: the volume forces and the interface forces. An example of the first one is gravitational force, and an example of the second one is pressure gradient force or wind stress.

The influence of the pressure gradient forces can be derived in this manner: given an infinitesimal volume $\Delta V = \Delta x_1 \Delta x_2 \Delta x_3$, the force exerted by the pressure on the left side of the cube in $x_1$ direction is given by $p(x) \Delta x_2 \Delta x_3$ (pressure is force per area). Similarly, the force exerted on the right side of the cube is

$$-p(x + \Delta x_1) \Delta x_2 \Delta x_3 \cong \left( -p(x) - \frac{\partial p}{\partial x_1} \Delta x_1 \right) \Delta x_2 \Delta x_3$$

Here a Taylor series expansion has been applied to $p(x + \Delta x_1)$. So the net force on the volume in the $x_1$ direction, after adding the two contributions, is $-\frac{\partial p}{\partial x_1} dV$, and the force per unit volume is just $-\frac{\partial p}{\partial x_1}$. The same analysis can be carried out for the other two directions, giving the total pressure gradient force per unit volume of $-\frac{\partial p}{\partial x_i}$.

This result can now be used to insert the external forces into the general momentum equation. Including the gravitational force in their form per unit volume $\rho g_i$, we have

$$f_i = -\frac{\partial p}{\partial x_i} + \rho g_i$$

and, therefore,

$$\frac{du_i}{dt} = -\frac{1}{\rho} \frac{\partial p}{\partial x_i} + g_i \qquad (3.13)$$

In this form the equations are called *Euler equations,* or more precisely, Euler equations in a nonrotating frame of reference. Note that the gravitational acceleration is a vector with only the vertical component different from zero

$$g_i = (0, 0, -g)$$

(pointing downward) and $g$ is 9.81 m s$^{-2}$.

### 3.4.1.2.2 The Euler Equations in a Rotating Frame of Reference

The above-derived equations are not suitable for their application to meso-scale or basin-wide scale. This is because the Earth is not an inertial frame of reference but is rotating. Although this has no impact on water bodies that are small in size, for the larger applications it is very important to take account of this effect.

The influence of the Earth's rotation can be described with the introduction of a new apparent volume force called the *Coriolis force.* In this case, the new forces on the water parcel are composed not only of the pressure gradient, but also of the additional Coriolis force, and the new equations (not deduced) can be written as

$$\frac{du_1}{dt} - fu_2 = -\frac{1}{\rho}\frac{\partial p}{\partial x_1} \qquad (3.14a)$$

$$\frac{du_2}{dt} + fu_1 = -\frac{1}{\rho}\frac{\partial p}{\partial x_2} \qquad (3.14b)$$

$$\frac{du_3}{dt} = -\frac{1}{\rho}\frac{\partial p}{\partial x_3} - g \qquad (3.14c)$$

where $f$ is the Coriolis parameter with $f = 2\Omega\sin(\varphi)$, $\Omega$ is the angular frequency of rotation of the earth ($2\pi/24h$), and $\varphi$ denotes the geographical latitude. For ease of notation, we have expanded the index notation into the three equations, one for each coordinate direction. In this form the equations are called *Euler equations in a rotating frame of reference.*

Besides the gravitational force, the Coriolis force is the only other important volume force that acts in a fluid body. Because of the vector product, the Coriolis force always acts perpendicular to the current velocity, in the northern hemisphere to the right and in the southern hemisphere to the left of the fluid flow. It is the Coriolis force that is responsible for all the meso-scale structure we can see on the weather charts that contain cyclones and anticyclones. However, the Coriolis force is important only for large-scale circulations and, therefore, may not be important for many coastal lagoons of the world.

### 3.4.1.2.3   The Navier-Stokes Equations

The above-derived Euler equations describe the flow of a fluid without friction. Often this may be a good approximation, especially if no material boundaries (lateral and vertical) are close to the area of investigation. Internal friction in a fluid is normally very small and the Euler equations provide a satisfactory simplification. However, once the fluid is close to a boundary, friction becomes more important and another area force, the stress tensor, has to be introduced. A moving fluid layer exerts a force on the neighboring fluid layers. The strength of this force is directly proportional to the area of the fluid layer and the velocity difference between these layers, and inversely proportional to the distance of the layers:

$$F = \mu A \Delta u_1 / \Delta x_3$$

where $A$ is the area, $\Delta u_1$ is the velocity difference from one layer to the other, and $\Delta x_3$ is the distance of the fluid layers. The parameter $\mu$ is the constant of proportionality and is called the *dynamic viscosity coefficient.* It is a parameter that depends only on the type of fluid and its temperature and salinity contents, but not on the fluid dynamics.

Taking the finite differences to the infinitesimal limit, the stress (force per area) can be written as

$$\tau_{31} = \mu \frac{\partial u_1}{\partial x_3}$$

Here the subscript 31 means the force exerted by the moving fluid layer in the $x_1$ direction due to the velocity gradient in direction $x_3$. Therefore, the stress tensor has nine components, three terms for every spatial direction.

The effect of this stress tensor on the equation of motion can easily be derived. The derivation is almost equivalent to the incorporation of the pressure gradient force into the Euler equations, and the result for the viscous force $F_{31}^r$ per unit mass in the $x_1$ direction due to the shear in the $x_3$ direction is

$$F_{31}^r = \frac{1}{\rho} \frac{\partial \tau_{31}}{\partial x_3} = \frac{1}{\rho} \frac{\partial}{\partial x_3} \left( \mu \frac{\partial u_1}{\partial x_3} \right) = v \frac{\partial^2 u_1}{\partial x_3^2}$$

where we have used the *kinematic viscosity coefficient* $v = \mu/\rho$ to write the equation more compactly.

In the same way, the viscous forces in the $x_1$ direction due to shear in the other directions can be derived, yielding

$$F_1^r = v \left( \frac{\partial^2 u_1}{\partial x_1^2} + \frac{\partial^2 u_1}{\partial x_2^2} + \frac{\partial^2 u_1}{\partial x_3^2} \right) = v \frac{\partial^2 u_1}{\partial x_j^2}$$

and the viscous forces in the other directions read

$$F_2^r = v \frac{\partial^2 u_2}{\partial x_j^2} \qquad F_3^r = v \frac{\partial^2 u_3}{\partial x_j^2}$$

Including the friction term into the Euler equations we end up with the so-called Navier-Stokes equations:

$$\frac{du_1}{dt} - f u_2 = -\frac{1}{\rho} \frac{\partial p}{\partial x_1} + v \frac{\partial^2 u_1}{\partial x_j^2} \tag{3.15a}$$

$$\frac{du_2}{dt} + f u_1 = -\frac{1}{\rho} \frac{\partial p}{\partial x_2} + v \frac{\partial^2 u_2}{\partial x_j^2} \tag{3.15b}$$

$$\frac{du_3}{dt} = -\frac{1}{\rho} \frac{\partial p}{\partial x_3} - g + v \frac{\partial^2 u_3}{\partial x_j^2} \tag{3.15c}$$

The same result could have been derived by starting directly from Equation (3.8) and again using $\rho u_i$ as the quantity to be conserved. When neglecting spatial variations of $\rho$ in the diffusion terms (which is generally always a good approximation), the Navier-Stokes equations directly result with the diffusivity in Equation (3.8) being the viscosity just derived. As in the Euler equations the additional forces are the pressure gradient and the gravitational acceleration.

It may be interesting to note that the physical process responsible for molecular friction is the molecular diffusion of the fluid particles. If a faster fluid layer moves above a slower fluid layer, some of the particles from the slower fluid layer will diffuse into the faster layer, slowing down the upper layer. On the other hand, faster fluid particles diffusing into the fluid layer below will accelerate the slower fluid layer. From both sides of the layer this results in an effective friction, either slowing down or accelerating the other layer. Therefore, the operator $\partial^2/\partial x_j^2$ is also called the diffusion operator.

### 3.4.1.3  Conservation of Energy

Conservation of energy can be formulated as a conservation equation for temperature. Simply put, the total change in temperature of a fluid volume is given by the rate of heating of the volume. The change of temperature $T$ of the fluid with time can be expressed as

$$\frac{1}{\rho c_p} \frac{\partial Q_s}{\partial x_3}$$

where $Q_s$ is the solar radiation [W/m²] and $c_p$ is the specific heat of water [J kg⁻¹ K⁻¹]. The term on the right-hand side represents the source term of Equation (3.8).

The only other process that is changing the heat content of a fluid volume is molecular diffusion. In this case fluid particles that are warmer and that diffuse into colder fluid because of molecular motion will contribute to the warming up of the colder fluid. As explained previously, the diffusion can be described through the Laplace operator $\partial^2/\partial x_j^2$, and the whole conservation energy of temperature can be written as

$$\frac{dT}{dt} = v_T \frac{\partial^2 T}{\partial x_j^2} + \frac{1}{\rho c_p} \frac{\partial Q_s}{\partial x_3} \tag{3.16}$$

where $v_T$ is the molecular diffusivity for temperature, a parameter that depends only on the properties of the fluid. When compared to Equation (3.8) this is the conservation of $T$ with the external source given above.

### 3.4.1.4  Conservation of Salt

In the oceans and coastal lagoons the water has a certain salt content that varies from nearly zero high up in the estuaries and rivers to values of about 10 psu in brackish water and up to more than 30 psu in ocean waters. In lagoons where evaporation is important these values can be even higher. In all cases the constituents of salt are nearly constant in all places of the world. It is, therefore, possible to express the salt content through a concentration value called salinity. The salinity is measured in grams per kilogram or psu (practical salinity unit).

It is, therefore, important to consider the salt content and its variation using a conservation equation for salt. The salinity can be changed only through mass fluxes such as evaporation and river input. These processes can be viewed as boundary processes that will be discussed in Chapter 4. As with temperature, the only other process that changes the salt content of a fluid volume is diffusion. The equation for salinity is therefore

$$\frac{dS}{dt} = v_S \frac{\partial^2 S}{\partial x_j^2} \tag{3.17}$$

where $S$ is the salinity of the fluid and $v_S$ is the molecular diffusivity for salt, a parameter that again depends only on the properties of the fluid. Note that this equation can be derived directly from Equation (3.8), with $S$ the quantity to be conserved and no external sources.

### 3.4.1.5  Equation of State

The equation of state is a thermodynamic equation that relates different quantities in a fluid in equilibrium to each other. An example is the well-known equation of state for an ideal gas

$$p = \rho R T$$

that relates the pressure $p$, the density $\rho$, and the temperature $T$ to one another. Here $R$ is the gas constant per unit mass that still depends on the specific property of the gas. Its value for dry air is 287 J kg$^{-1}$ K$^{-1}$.

In the case of a fluid the situation is complicated by the fact that the salinity $S$ is also influencing the equation of state. In this case the equation of state can be formally written as

$$\rho = f(T, S, p)$$

so that the density can be computed given a set of values for temperature, salinity, and pressure. The equation itself is a complicated polynomial function of these quantities that has been empirically deduced by different organizations (e.g., UNESCO). The formula is not given here in its general formulation. However, for some applications the equation of state can be simplified to

$$\rho = \rho_0 - \alpha_T(T - T_0) + \alpha_S(S - S_0) \tag{3.18}$$

where the dependence on pressure has been neglected and a simple linear dependence on $T$ and $S$ has been used. Here $\rho_0$ is the reference density at the reference temperature

$T_0$ and the reference salinity $S_0$. The parameter $\alpha_T$ is called the thermal expansion coefficient and $\alpha_S$, the expansion coefficient, changes due to salinity. Both coefficients are positive. As can be seen, denser (heavier) water may result by either lowering the temperature or raising the salinity.

### 3.4.2  SIMPLIFICATION AND SCALE ANALYSIS

The equations given above, especially the equation for momentum, are still too complicated to be used for modeling. This is true mainly because of the variability of the spatial and temporal scales to which these equations are applicable. Although we did not make any assumptions, these equations describe the flow for the global ocean circulation as they describe the flow down to scales where molecular effects become important. Therefore, these equations must describe scales from $10^6$ m on the ocean scale down to about $10^{-6}$ m ($\mu$m) where dissipation becomes important.

The same is true concerning the time scales to which the equations apply. These time scales range from years for the general circulation down to microseconds if we deal with dissipation effects due to friction. Because we are interested in describing motion that takes place in coastal lagoons or estuaries, not all processes included in the equations above are equally important and simplifications have to be made.

#### 3.4.2.1  Incompressibility

Compared to air, water is a relatively incompressible medium. That there is a certain compressibility of the water is clearly seen by the fact that acoustic waves can travel in water. These waves depend completely on the compressibility of the medium. However, it can be shown that, excluding acoustic waves, the effect of compressibility on the dynamics of the oceans is negligible.

Therefore, if we assume that water is an incompressible medium, this can be written mathematically as $d\rho/dt = 0$. If we substitute this into the continuity equation, we have the simplified continuity equation:

$$\frac{\partial u_i}{\partial x_i} = 0 \qquad (3.19)$$

In this version the continuity equation states that the divergence of the mass flow is zero. This is the form of the mass conservation normally used in oceanography.

#### 3.4.2.2  The Hydrostatic Approximation

If we consider a basin of water at rest ($u_i = 0$), the Navier-Stokes equations reduce to

$$0 = -\frac{1}{\rho}\frac{\partial p}{\partial x_i} + g_i$$

The vertical component of this equation may be written as

$$\frac{\partial p}{\partial x_3} = -\rho g$$

This equation is called the *hydrostatic equation* because it is exactly valid in a static fluid with no motion. If we integrate this equation from a depth $z$ up to the surface that is supposed to be at $x_3 = 0$, we have:

$$\int_z^0 \frac{\partial p}{\partial x_3}\, dx_3 = p_0 - p(z) = -g \int_z^0 \rho\, dx_3$$

or

$$p(z) = p_0 + g \int_z^0 \rho\, dx_3 \qquad (3.20)$$

where $p_0$ is the surface or atmospheric pressure. In this form the hydrostatic equation states that the pressure at depth $z$ is due to the weight of the water column above it.

It turns out that this equation is also a very good approximation of the vertical component of the momentum equation in case of a situation where the velocity vector $u_i$ is not zero. Only in regions of very strong vertical convection due to strong bathymetric gradients or cooling of surface water may the vertical acceleration have appreciable effects on the total momentum equation.

### 3.4.2.3 The Coriolis Force

As specified above, the Coriolis force is an apparent force that is due to the rotation of the Earth. The complete expression also takes into account the influence of rotation due to and on vertical motion. However, it can be seen that only the horizontal part of the Coriolis force is important. In fact, the vertical momentum equation has been substituted by the hydrostatic assumption.

If the horizontal components are evaluated there is one term that is multiplied with the vertical velocity $u_3$. This term is nearly always much smaller than the other terms due to the smallness of the vertical velocity. If this term is neglected the remaining Coriolis vector can be written as

$$F_i^C = \left( F_1^C, F_2^C, F_3^C \right) = (-fu_2, +fu_1, 0)$$

where $f$ is called the Coriolis parameter and has already been defined as

$$f = 2\Omega \sin(\varphi)$$

with $\varphi$ the latitude of the point where the equation is evaluated. For applications in coastal lagoons, the value of $f$ varies little and therefore often is kept constant. In this case $\varphi$ is the average latitude of the basin.

As can be seen from the equation, the Coriolis force is always perpendicular to the flow. For example, in a flow that has only a component in the $x_1$ direction ($u_2 = 0$), the Coriolis force is only acting in the $x_2$ direction. For the northern hemisphere, $f$ is positive, and therefore the force is always to the right with respect to the direction of flow.

### 3.4.2.4 The Reynolds Equations

As mentioned above, the Navier-Stokes equations describe all possible water motions from scales on the order of the oceans down to scales where molecular friction is active and dissipation is important. However, for modeling purposes this is not acceptable. For example, the fact that the very small scales are not modeled directly (because the computational grid is just too coarse) means that molecular friction will never be important in the Navier-Stokes equations.

However, the fact that molecular friction is not important for the scales we are considering is somehow misleading. As there will be input of energy in our basin through solar heating, tides, and wind, there also must be some way to convert this energy and eventually dissipate it, otherwise the total energy will continue to increase. But the only way energy can be dissipated is through the molecular friction term. All other terms only take part in redistributing the energy inside the basin. Therefore, there must be some place in the basin where friction becomes important and the molecular forces do their work.

This obvious paradox has been resolved by the British physicist O. Reynolds who applied an averaging technique to the full Navier-Stokes equations. He described the flow as one that can be divided into a slowly varying part and a random fluctuation around this. So, for example, the velocity $u$ is represented as $u = \bar{u} + u'$ where $\bar{u}$ is the slowly varying part and $u'$ is the fluctuations. If all variables are represented like this and are introduced into the hydrodynamic equations and these equations are averaged over a suitable time interval, then new equations result for the slowly varying parts of all variables.

The structure of these averaged equations is very similar to that of the original equations. However, due to the nonlinear nature of the advective terms that are contained implicitly in the total derivative, some new terms appear in the conservation equation for momentum, temperature, and salt. These new terms, called *Reynolds fluxes*, are averages of products of fluctuating variables. Without going into detail, these terms can be formally written in the same way as the diffusion terms in the Navier-Stokes equations. The new terms now read

$$F_i^t = v_M^H \left( \frac{\partial^2 u_i}{\partial x_1^2} + \frac{\partial^2 u_i}{\partial x_2^2} \right) + v_M^V \frac{\partial^2 u_i}{\partial x_3^2}$$

where two new parameters $v_M^H, v_M^V$ have been substituted for the molecular viscosity $v$. A distinction has been made between the horizontal and the vertical component and the subscript $M$ indicates the momentum equation.

It is important to note the different physical processes that lead to these equations. In the case of the Navier-Stokes equations, the friction term was due to the molecular (Brownian) motion of the fluid particles. This motion was statistical in nature and the diffusivity was only a function of the fluid static properties.

In the case of the Reynolds equations, the additional term is now due to the fluctuating part of the variable that is under consideration (here, the velocity). This fluctuating part is also called the turbulent part. It is not actually statistical in nature because the Navier-Stokes equations could at least in principle be used to compute all the small-scale motions. However, as in thermodynamics, it would in principle be possible to describe the motion of $10^{23}$ molecules that are in one mole of air; practically, it is not possible and only a statistical representation of the motion is given.

Therefore, the diffusion of the fluid particles is now due to the turbulence that is acting on scales smaller than the ones that have been retained after the averaging procedure. This turbulent motion is similar to the molecular motion of the fluid particles. However, there are two main differences. Because the turbulent motions (fluctuation) depend on the averaging scale, the turbulent viscosities $v_M^H, v_M^V$ will also depend on the scale of averaging. Even worse, because the turbulence depends on the dynamic state of the fluid, the turbulent viscosities also depend on the dynamics of the fluid. Various factors, such as the local buoyancy or the velocity shear, will influence these parameters.

It is clear now that there is a fundamental difference between the molecular and the turbulent diffusion term. The first one depends only on the static properties of the fluid (type of fluid, temperature, salinity), whereas the second one also depends on the dynamic flow field itself. The molecular viscosity can be given a very accurate value that can be measured or computed in statistical mechanics. However, the turbulent parameter varies over several orders of magnitude depending crucially on the flow field itself. Because of the dynamic nature of the turbulent motion, it is expected that the values for the viscosities will be different in the horizontal and the vertical directions. This fact has been accounted for by having the viscosity take different values $v_M^H, v_M^V$ for the horizontal and vertical dimensions.

Comparing the range of values for the two types of parameters, it can be seen that the turbulent diffusion term is some orders of magnitude bigger than its molecular counterpart. Therefore, the molecular friction is almost always neglected and only the turbulent terms are retained. The value for the turbulent viscosity parameter is often set to a constant average, one that best represents the physical processes to be described. In a more general case a turbulence closure model must be used that will actually compute the parameter $v_M$ in every point of the water body. The description of these turbulence closure models (e.g., $k$-$\varepsilon$, Mellor-Yamada) is beyond the scope of this book.

### 3.4.2.5 The Primitive Equations

In this section the simplified three-dimensional equations are given one more time as a reference for the next section. These equations are also called primitive equations, not because they are easy to solve, but because only basic simplifications have been applied to them, and they are presented in their "primitive" structure.

The conservation of mass leads to the continuity equation

$$\frac{\partial u_1}{\partial x_1} + \frac{\partial u_2}{\partial x_2} + \frac{\partial u_3}{\partial x_3} = 0 \tag{3.21}$$

The two horizontal components of the Reynolds equations read

$$\frac{\partial u_1}{\partial t} + u_1 \frac{\partial u_1}{\partial x_1} + u_2 \frac{\partial u_1}{\partial x_2} + u_3 \frac{\partial u_1}{\partial x_3} - f u_2 = -\frac{1}{\rho} \frac{\partial p}{\partial x_1} + v_M^H \left( \frac{\partial^2 u_1}{\partial x_1^2} + \frac{\partial^2 u_1}{\partial x_2^2} \right) + v_M^V \frac{\partial^2 u_1}{\partial x_3^2} \tag{3.22a}$$

and

$$\frac{\partial u_2}{\partial t} + u_1 \frac{\partial u_2}{\partial x_1} + u_2 \frac{\partial u_2}{\partial x_2} + u_3 \frac{\partial u_2}{\partial x_3} + f u_1 = -\frac{1}{\rho} \frac{\partial p}{\partial x_2} + v_M^H \left( \frac{\partial^2 u_2}{\partial x_1^2} + \frac{\partial^2 u_2}{\partial x_2^2} \right) + v_M^V \frac{\partial^2 u_2}{\partial x_3^2} \tag{3.22b}$$

where the total time derivative has been split into the inertial term that describes the local acceleration and a second term that describes the advective acceleration.

For the vertical component the hydrostatic approximation is used

$$\frac{\partial p}{\partial x_3} = -\rho g \tag{3.23}$$

The conservation equations for heat and salt read

$$\frac{\partial T}{\partial t} + u_1 \frac{\partial T}{\partial x_1} + u_2 \frac{\partial T}{\partial x_2} + u_3 \frac{\partial T}{\partial x_3} = v_T^H \left( \frac{\partial^2 T}{\partial x_1^2} + \frac{\partial^2 T}{\partial x_2^2} \right) + v_T^V \frac{\partial^2 T}{\partial x_3^2} + \frac{1}{\rho c_p} \frac{\partial Q_s}{\partial x_3} \tag{3.24a}$$

and

$$\frac{\partial S}{\partial t} + u_1 \frac{\partial S}{\partial x_1} + u_2 \frac{\partial S}{\partial x_2} + u_3 \frac{\partial S}{\partial x_3} = v_S^H \left( \frac{\partial^2 S}{\partial x_1^2} + \frac{\partial^2 S}{\partial x_2^2} \right) + v_S^V \frac{\partial^2 S}{\partial x_3^2} \tag{3.24b}$$

with $V_T$ and $V_S$ the turbulent diffusivities for temperature and salinity, respectively. Finally, the equation of state reads

$$\rho = f(T, S, p)$$

This completes the discussion of the hydrodynamic equations.

### 3.4.3 SPECIAL FLOWS AND SIMPLIFICATIONS IN DIMENSIONALITY

The equations derived above explain the whole spectrum of dynamic behavior that may be expected in the coastal seas and lagoons. Quite often, however, the flow shows some peculiar characteristics that make it possible to simplify the equations even more. In this section some often-made simplifications are presented that allow the reduction of the dimensions of the problem.

An example may be a flow that is constant in one direction. This means that all derivatives in this direction vanish. Therefore, the solution in this equation is trivial and the component of this direction may be eliminated, resulting in a lower dimensional problem that is easier to track.

#### 3.4.3.1 Barotrophic 2D Equations

A very important simplification can be achieved when the key variables can be considered constant along the water column. This is normally the case when no stratification occurs and the water is well mixed.

In this case, the equations can be integrated over the water column and the two-dimensional barotropic (vertically integrated) equations result. These equations are also called *shallow water equations*, even if they are not applicable only to shallow basins and not all shallow lagoons allow for this simplification.

A better-suited continuity equation may be deduced for this vertically integrated flow. This can be done by applying the kinematic boundary conditions (explained below) to the continuity equation, or it can also be easily derived directly. Consider a control volume over the whole water column with rectangular area $\Delta x_1 \Delta x_2$. The total inflow in $x_1$ direction is $Hu_1(x_1,x_2)\Delta x_2 - Hu_1(x_1+\Delta x_1,x_2)\Delta x_2$. A similar equation holds in the $x_2$ direction and the total water depth is called $H$. The total inflow must result in more water in the control volume and because of incompressibility the only way for water to accumulate is through the rise of the water level $\eta$. The rate of increase in the volume can therefore be written as $(\partial \eta / \partial t)\Delta x_1 \Delta x_2$. If we divide this conservation equation by the area of the control volume $\Delta x_1 \Delta x_2$ we have

$$\frac{\partial \eta}{\partial t} + \frac{Hu_1(x_1+\Delta x_1,x_2) - Hu_1(x_1,x_2)}{\Delta x_1} + \frac{Hu_2(x_1,x_2+\Delta x_2) - Hu_2(x_1,x_2)}{\Delta x_2} = 0$$

and in the limit

$$\frac{\partial \eta}{\partial t} + \frac{\partial Hu_1}{\partial x_1} + \frac{\partial Hu_2}{\partial x_2} = 0$$

Because the variables are considered constant throughout the water column the density $\rho$ that is changing mainly in the vertical direction may be considered constant all over and is denoted as $\rho_0$. With this simplification the hydrostatic equation can be integrated from a depth $z$ to the water surface $\eta$

$$p_0 - p(z) = \int_z^\eta \frac{\partial p}{\partial x_3} dx_3 = -\int_z^\eta g\rho_0 dx_3 = -g\rho_0(\eta - z)$$

Assuming the atmospheric pressure $p_0$ to be constant, we can write the horizontal pressure gradient as

$$\frac{1}{\rho_0}\frac{\partial p}{\partial x_1} = g\frac{\partial \eta}{\partial x_1} \qquad \frac{1}{\rho_0}\frac{\partial p}{\partial x_2} = g\frac{\partial \eta}{\partial x_2}$$

In this way the pressure can be completely eliminated from the equations and is substituted by the water level $\eta$.

One more term that has to be dealt with is the turbulent friction term. If this term is integrated over the whole water column we obtain

$$v_M \int_b^s \frac{\partial^2 u_1}{\partial x_3^2} dx_3 = v_M \left(\frac{\partial u_1}{\partial x_3}\right)^s - v_M \left(\frac{\partial u_1}{\partial x_3}\right)^b$$

where the indices $s$ and $b$ stand for surface and bottom. These two terms are exactly the stress per density in the $x_1$ direction that is exerted over the two interfaces of the surface $\tau^s$ and the bottom $\tau^b$:

$$v_M \int_b^s \frac{\partial^2 u_1}{\partial x_3^2} dx_3 = \frac{1}{\rho_0}\left(\tau_1^s - \tau_1^b\right)$$

These values are boundary values that have to be imposed on the moving fluid or, as is the case with the bottom friction, can be computed from the actual flow field. Their exact formulation will be given in Section 3.5 on boundary processes.

If the vertical integrated equations are divided by the total depth $H$, the 2D shallow water equations result:

$$\frac{\partial u_1}{\partial t} + u_1\frac{\partial u_1}{\partial x_1} + u_2\frac{\partial u_1}{\partial x_2} - fu_2 = -g\frac{\partial \eta}{\partial x_1} + \frac{1}{\rho_0 H}\left(\tau_1^s - \tau_1^b\right) + v_M^H\left(\frac{\partial^2 u_1}{\partial x_1^2} + \frac{\partial^2 u_1}{\partial x_2^2}\right) \quad (3.25a)$$

and

$$\frac{\partial u_2}{\partial t} + u_1\frac{\partial u_2}{\partial x_1} + u_2\frac{\partial u_2}{\partial x_2} + fu_1 = -g\frac{\partial \eta}{\partial x_2} + \frac{1}{\rho_0 H}\left(\tau_2^s - \tau_2^b\right) + v_M^H\left(\frac{\partial^2 u_2}{\partial x_1^2} + \frac{\partial^2 u_2}{\partial x_2^2}\right) \quad (3.25b)$$

Together with the continuity equation

$$\frac{\partial \eta}{\partial t} + \frac{\partial Hu_1}{\partial x_1} + \frac{\partial Hu_2}{\partial x_2} = 0 \qquad (3.26)$$

they form a closed set of equations for the variables $\eta, u_1, u_2$, provided that the boundary conditions for the stress $\tau^s, \tau^b$ and the values of $v_M^H$ are specified. The values for the velocities $u_1, u_2$ represent average values over the whole water column.

Equations for the transport and diffusion of the temperature and salinity may also be derived as above. After vertical integration of the conservation equations, we have

$$\frac{\partial T}{\partial t} + u_1 \frac{\partial T}{\partial x_1} + u_2 \frac{\partial T}{\partial x_2} = v_T^H \left( \frac{\partial^2 T}{\partial x_1^2} + \frac{\partial^2 T}{\partial x_2^2} \right) + \frac{1}{\rho c_p} \frac{Q_s}{H} \qquad (3.27)$$

and

$$\frac{\partial S}{\partial t} + u_1 \frac{\partial S}{\partial x_1} + u_2 \frac{\partial S}{\partial x_2} = v_S^H \left( \frac{\partial^2 S}{\partial x_1^2} + \frac{\partial^2 S}{\partial x_2^2} \right) \qquad (3.28)$$

where the fluxes at the surface and bottom have been set to zero. If these fluxes are important they can easily be included in the equations. The values $T$ and $S$ represent average values of the temperature and salinity over the water column.

It should be noted that these equations could be used to compute the transport and diffusion of temperature or salinity or any other conservative dissolved substance $C$ by the velocity field $u_i$; however, they are not necessary for the solution of the hydrodynamic equations. This is due to the fact that the hydrodynamic equations do not depend on the unknown density $\rho$ but on $\rho_0$, which is constant and does not depend on $T$ and $S$.

### 3.4.3.2 1D Equations (Channel Flow)

A further simplification can be adopted for the hydrodynamic equations if the flow is mainly in one direction. This is true for a flow in a channel where flow perpendicular to the channel axis can be neglected or is of no interest. In this case, the 2D equations may be integrated over the width of the channel.

If this integration is done the continuity equation reads

$$B\frac{\partial \eta}{\partial t} + \frac{\partial BHu_1}{\partial x_1} = 0 \qquad (3.29)$$

where $B$ now denotes the (possibly variable) width of the channel. The momentum equation now reads

$$\frac{\partial u_1}{\partial t} + u_1 \frac{\partial u_1}{\partial x_1} = -g \frac{\partial \eta}{\partial x_1} + \frac{1}{\rho_0 H}\left(\tau_1^s - \tau_1^b\right) + v_M^H \frac{\partial^2 u_1}{\partial x_1^2} \tag{3.30}$$

and the other conservation equations for $T$ and $S$ can be deduced in a similar way.

### 3.4.4 Initial and Boundary Conditions

Boundary processes are the driving force for all movements that can be observed in the oceans or the atmosphere. While the equations of hydrodynamics guarantee the inner consistency of the variables and enforce the conservation of the physical properties, only boundary processes make this consistency interesting for us.

An example may suffice: imagine a bath tub that has been filled recently with water. After the tap has been shut off there is still some movement in the water that slowly declines. After a while the water in the bath tub enters a state of rest. It has reached its equilibrium. If no external forces are applied to the water, such as manual stirring, opening of the water tap, or wind blowing through the open window, from then on nothing more will happen. The fluid is internally consistent, but no interesting features will result from this consistency.

In mathematical language these boundary processes enter the equations through boundary conditions that have to be applied to the equations. Different boundary conditions can be used depending on the type of the problem that has to be tackled.

There are two types of boundary conditions. The first type of condition is called the *Dirichlé condition* and fixes the value of a certain variable. The second type of condition, the *von Neuman condition*, specifies a flux of the variable instead of fixing the value.

The temperature equation can be used as an example. The radiation coming from the sun heats up the upper layer of a water body. This can be taken into account by prescribing the sea surface temperature of the water, as it may be known through measurements at a monitoring station. This case corresponds to the first type of boundary condition.

However, the heat flux from the sun can also be prescribed. In this case the value of the temperature is not fixed but varies according to the heat flux specified. This condition is of the second type. Both types of boundary conditions will eventually change the temperature at the sea surface and, through advection and diffusion, also change the interior temperature of the water body.

### 3.4.4.1 Initial Conditions

Initial conditions can be viewed as a special type of boundary condition. Whereas the normal boundary conditions refer to the spatial coordinates, the initial conditions are boundary conditions for the time coordinate.

Every problem that depends on time must have initial conditions for all unknowns; otherwise, the solution is not possible. However, very often we are in a situation where our knowledge of initial conditions is limited. In the example of a coastal lagoon we may still be in the situation to specify the water level (maybe constant to a first approximation), but we surely do not know what the current velocities are in the interior.

Fortunately, there is a way out of this dilemma. Many problems, after a certain time, converge to a solution that is the same for different initial conditions. These types of problems are therefore not sensitive to their initial conditions. An example is a lagoon where a tidal elevation is prescribed on the inlet. In this case the solutions of the hydrodynamic variables tend to converge even if different initial conditions have been imposed. This is due to the fact that through the tidal movement at the inlets, energy is created in the lagoon that is itself destroyed by the frictional forces. After a while all energy that was in the system due to the initial state has been dissipated, and from then on water movement in the basin is due only to the forcing of the tide. The lagoon has "forgotten" its initial state.

The duration of this memory effect depends mostly on the strength of the friction (e.g., how long it takes to remove the initial energy from the system). In the case of numerical simulations, this time is called the *spin up time* of the simulation. After this initial spin up, the model is in equilibrium with the boundary conditions and the effect of the unknown initial conditions has been damped out.

### 3.4.4.2 Conditions on Material Boundaries

Conditions on material (land) boundaries have to be specified for the scalar quantities and the transports. On lateral material boundaries normally no-flux conditions are applied for the scalar quantities. For example, there should be no heat or salinity flux over (or exchange through) these material boundaries. These conditions take into account the fact that there are actually no exchange processes going on through material boundaries.

Two conditions must be considered for the current transports. The first is the condition that the transport through a material boundary must be zero,

$$u_i \cdot n_i = 0$$

where $n_i$ denotes the direction of the normal of the boundary.

For flows with friction there is another boundary condition to be prescribed. On material boundaries the fluid particles actually adhere to the wall, and in the very vicinity of the wall there is no tangential movement of the fluid. This phenomenon is observed only very close to the material boundary, as it is due to molecular friction. However, the scale of the modeling is much larger than the molecular scale, and, therefore, it is inadequate to impose no tangential flow along the boundary.

Although it is feasible to impose some lateral friction on the lateral boundary, very often this lateral friction is set to zero. This is a suitable description if the area-to-boundary ratio of the water body is high, meaning that the influence of the lateral boundary can be neglected. In this case the boundary condition is called a *free slip* condition. However, it must be clear that in doing so the only process that can remove energy from the system is the bottom friction (to be discussed below). For coastal lagoons and estuaries this hypothesis is adequate, but when dealing with deep-water bodies it may not hold true.

### 3.4.4.3  Conditions on Open Boundaries

Open boundaries do not actually exist in nature. Nevertheless, all those boundaries of the artificially created water body that are not made out of material boundaries (e.g., the boundary is in the water body itself) are called open boundaries. In this case the conditions for material boundaries cannot be applied because there is actually a strong flux of all the variables through the open inlets.

Mathematically speaking, fluxes have to be described through the open boundaries. In the case of mass flow the water level or the current velocities as well as river discharges must be provided. For the other variables heat and salt fluxes have to be prescribed.

An alternative, as described previously, is to fix temperature and salinity values. This corresponds to a boundary condition of the first type. In this case, it must be pointed out that the value of the property can be described only in conditions of inflow, when the information is transported into the domain. In the case of outflow, no information is necessary as the value at the boundary is adjusting dynamically to the value that is advected out of the water body.

The Venice Lagoon is an example. The lagoon communicates with the Adriatic Sea through three inlets through which it exchanges water. On these boundaries either the mass flux into the lagoon (velocities) or the tide (water level) must be prescribed. Moreover, properties of the water such as temperature or salinity must be imposed if the water is entering the lagoon.

The same parameters and variables also must be specified on the inner boundary where the lagoon waters combine with the mainland. Here rivers enter the basin, bringing different water masses into the lagoon. In this case it is more convenient to impose fluxes instead of water levels because they are very often better known.

### 3.4.4.4  Conditions on the Sea Surface and the Sea Bottom

In lagoons and estuaries the water surface and the bottom of the water body are the largest surfaces in contact with the water. This is true because, for a typical basin, the horizontal dimensions are always much larger than the vertical ones. Therefore, the fluxes across these two interfaces will be very important for dynamics of the fluid and particular attention must be devoted to these boundaries.

At the sea surface, fluxes of various kinds can be observed. There is momentum input due to the mechanical action of the wind across the air–sea interface. It is this force that drives large parts of the oceanic circulation, and it is also important on a local scale. The wind stress acting across the surface can be parameterized by the bulk formula

$$\tau_1^s = c_D \rho_a |u_i^w| u_1^w \qquad \tau_2^s = c_D \rho_a |u_i^w| u_2^w$$

where $\rho_a$ is the density of air, $u_i^w$ is the components of the wind vector, and $|u_i^w|$ is the modulus of the wind speed. The parameter $c_D$ is called the *drag coefficient* and it has an empirical value between $1.5 \cdot 10^{-3}$ and $3.2 \cdot 10^{-3}$. Several formulations exist where the drag coefficient is parameterized through the wind velocity.

Other fluxes at and through the sea surface concern the mass, salt, and heat flux. Mass and salt fluxes are due to precipitation and evaporation that is occurring at the water interface. Rain contributes to the mass flux, also diluting the surface waters, whereas evaporation represents a loss of water and consequentially an increase in salinity. Solar radiation heats the upper part of the oceans through absorption of short waves, while outgoing long wave radiation and evaporation contribute to a heat loss of water. Sensible heat flux (due to conduction of heat) also plays a role in the heat budget of the oceans.

The bottom of the water body represents a material boundary to the fluid. It is, therefore, clear that the above-mentioned no-flux condition is valid also for this lower boundary. In this case, the role of sediment as a reservoir for nutrients, true water, and other variables has been completely neglected. If these processes have to be included in the description of the dynamics then a different approach has to be taken. However, the no-flux condition is normally a good approximation.

On the other side a very important momentum flux is taking place across the bottom boundary that normally cannot be ignored. The bottom exerts a drag (friction) on the water column that slows down the fluid layers in contact with the bed. This is the same principle as the wind action through the surface that accelerates the upper fluid layers.

The bottom friction stress can be expressed in a similar way as the wind stress. A bulk formula of the form

$$\tau_1^b = c_B \rho_0 \, |u_i| \, u_1 \qquad \tau_2^b = c_B \rho_0 \, |u_i| \, u_2$$

can be used with $\rho_0$ the density of the water, $u_i$ the components of the current velocity in the vicinity of the bottom, and $|u_i|$ the modulus of the current velocity. The parameter $c_B$ is called the *bottom friction coefficient* and has an empirical value of about $2.5 \times 10^{-3}$, similar to the wind drag coefficient. Other forms of bottom friction parameterizations are also possible but will not be discussed further here.

## 3.5 BOUNDARY PROCESSES

In the above paragraphs, generic evolution equations have been given for a generic property and for properties directly involved on water movement. Evolution equations involve the computation of the transport of properties between neighboring points and the computation of the rate of accumulation of those properties. The transport of properties requires computation of fluxes, including computation along the boundary. In the same way, the simulation of the time evolution requires the knowledge of initial conditions. Boundary conditions are usually not well known and their specification has to be based on a good knowledge of boundary processes.

Any marine system has two physical boundaries: the bottom and the free surface. Marine systems also have an artificial boundary: the open boundary. The bottom boundary condition along the coastal line is often considered a lateral solid boundary, which can be fixed or mobile.

### 3.5.1  Bottom Processes

The bottom boundary can be a well-defined solid boundary (e.g., rocky or formed by compact sediment) or a mobile boundary formed by sand or easily erodable cohesive sediment. Even if sediment is mobile, its velocity is much smaller than that of the fluid velocity and, for flow modeling purposes, it can be considered at rest.

Exchanges of water between the bottom and the water column involve mostly interstitial water and are negligible in terms of water balance. In contrast, the exchanges of dissolved matter associated with interstitial water are usually important due to high concentrations in that water.

Exchanges of particulate matter and momentum are the most important exchanges associated with the flow properties. Exchange of momentum is the bottom shear stress and exchange of particulate matter is parameterized as a function of sediment properties and bottom shear stress.

#### 3.5.1.1  Bottom Shear Stress

Bottom shear stress is a consequence of the nonslip condition at the bottom. Its value depends on the local turbulence intensity and velocity profile. Its parameterization depends on the type of model being used. In the case of a depth-integrated model, it has to be parameterized as a function of bottom properties (roughness), water column depth, and average velocity. In the case of a cross-section integrated model (one-dimensional), it has to be parameterized as a function of the discharge and the hydraulic radius. In both cases a Manning-type roughness coefficient can be used.

In the case of a three-dimensional model, bottom shear stress is usually parameterized based on the assumption of a logarithmic profile of velocity and bottom roughness.

#### 3.5.1.2  Other Bottom Processes

Bio-turbation is another important bottom boundary process. It can be accomplished by animals, bacteria, or even plants. Animals building galleries in sediment promote the mixing of the upper layer of sediment and are responsible for the destruction of gradients in the surface layer of sediment, modifying the exchanges mediated by bottom shear stress.

Bacteria act on the biogeochemical cycles, modifying concentrations and consequently fluxes of properties at the bottom–water interface (see Chapter 4 for details). Bacteria are also responsible for increasing sediment cohesiveness, contributing to reducing the erosion of cohesive sediment.

Filter feeders and plants play a major role in transferring matter from the water column into sediment. Aquatic plants absorb nutrients from the water column, generating organic matter, part of which is injected directly into sediment in the form of roots. The organic matter associated with the roots will be mineralized by bacteria and secondary producers, stimulating biological activity in sediment and modifying the mechanical properties of sediment and, consequently, its erodability. Plants also act on the bottom shear modifying the flow itself.

Filter feeders play two different roles. On one hand, they increase roughness of the flow, increasing bottom shear stress and reducing conditions for gravitational deposition of particulate matter. On the other hand, they extract particles from the water column and build pseudo-faeces that are deposited on the bottom, increasing the rate of accumulation of matter (including refractory organics) and also increasing cohesiveness of bottom sediment.

Detailed information on biological boundary processes is given in Chapter 5.

## 3.5.2 SOLID BOUNDARY PROCESSES

Lateral solid boundaries are usually artificial walls or created artificially in the model due to the incapacity of the grid to continuously describe the bottom until it reaches a negligible depth. Even if they are physical, there are no relevant biological processes. Along these boundaries momentum can only be transferred through shear diffusion. In a model these boundaries are represented imposing a null advective flux and most often no diffusive flux.

In fact, the consideration of horizontal shear requires the use of a grid step smaller than the boundary layer thickness; otherwise, the effect of the wall is propagated artificially into the model domain. The case of a coarse grid lateral diffusion is negligible when compared to the bottom shear diffusion due to the small value of the lateral surface of boundary cells when compared with values of their bottom surface. As a general rule, we can assume that horizontal diffusion in a model is relevant only when the grid size is of the same order of magnitude as the depth near solid boundaries.

## 3.5.3 FREE SURFACE PROCESSES

Through their free surfaces, aquatic systems can exchange mass, energy, and momentum, according to atmospheric conditions. These fluxes control the biological activity and the flow, except in tidal systems, where ocean-level oscillation can be the most important mechanism forcing the flow.

### 3.5.3.1 Mass Exchange

Gases, vapors, and solid matter can be exchanged across the free surface. The exchange of gases depends on relative partial pressures and can be in both senses. It is mostly a result of biological activity, which in coastal lagoons is very much related to land discharge of nutrients and organic matter. Globally, the aquatic system tends to import $CO_2$ and export $O_2$ and $N_2$. The rate of exchange depends on the surface concentration on both sides of the surface layer and that depends on turbulence intensities in the atmosphere and in the water column.

Mass can also be lost by the water in the form of vapor. This is mostly the case with water, but it can be the case with floating liquids such as hydrocarbons. The rate of evaporation also depends on partial pressure of the vapor in the atmosphere, the surface temperature, and the intensity of turbulence in the atmospheric boundary layer.

Particulate matter can also be exchanged through atmospheric deposition. This mechanism is the main source of particulate matter for the Mediterranean Sea, but it is negligible in coastal lagoons and estuaries when compared with the particulate matter transported by rivers and streams, due to the small dimension of the free surface of these water bodies.

Rain is a source of water and has implications for salinity. In the ocean, rain is the major source of salinity reduction in upper latitudes, and it can also be a source of heat. In the case of coastal systems, the amount of water entering the system through rainfall is negligible compared with the water discharged by the catchment areas.

### 3.5.3.2  Momentum Exchange

The exchange of momentum is a major mechanism generating the flow in coastal lagoons, except in case of tidal systems. This flux depends on the atmospheric intensity of turbulence (proportional to the square of the wind velocity) and on a friction coefficient.

In the case of bottom shear stress, the friction coefficient is a function of roughness. The difference is that along the bottom, roughness is a property of the bottom (and relative roughness, a function of that and of water column height), and at the free surface, roughness is also a function of wind speed, through wind waves, where amplitude increases with wind speed.

### 3.5.3.3  Energy Exchange

Energy is exchanged with the atmosphere in the form of latent and sensible heat, radiation, and mechanical energy. Part of the mechanical energy is the work of the shear stress transferring momentum, and part is exchanged in the form of turbulent kinetic energy.

Latent heat is lost through evaporation and is a direct function of the amount of moisture exchanged with the atmosphere. Sensible heat is a function of the temperature difference between the atmosphere and the water surface and of the turbulence intensity in the atmosphere.

Heat exchange by radiation has features of both latent and sensible heat exchange. The heat radiated by the water surface depends on water temperature, and the heat received from the atmosphere depends on the season, time of day, cloud cover, and atmospheric temperature.

Turbulent kinetic energy is the fraction of the mechanical energy lost by the atmosphere and not converted into work by the water velocity. This fraction of energy is accounted for by the turbulence models for computing vertical diffusivity and viscosity.

### 3.5.4  Cohesive and Noncohesive Sediment Processes

In coastal environments, solids can be found in dissolved, colloidal, or particulate forms. The limits defining these forms are somewhat arbitrary. A substance is considered as dissolved if it can pass through a 2-μm filter. Colloidal substances do not pass through the 2-μm filter, but individual particles cannot be identified. In favorable conditions colloidal substances can become solid. This can happen at the entrance of estuaries with very low salinity values. Colloids are very important for coastal water chemistry, but their contribution to the overall balance of solid material is small and usually negligible.

Particles are also generated by biological activity. Primary producers are "machines" converting dissolved matter (inorganic nutrients, $CO_2$, and water) into particulate material. Secondary producers feed on those particles and generate even larger particles (their own bodies and faecal pellets). Biological particles do have adhesive properties and act as gluing nodes favoring the formation of flocks formed by several small particles. These flocks, which are larger than individual particles, have a higher volume/surface ratio and settle more quickly than individual particles. Ionic attraction between inorganic particles is another flocculation mechanism and is enhanced by free ions into salt water. This is the dominant flocculation process at the entrance of the coastal lagoons at salinities higher than 2 psu. Details of these processes are presented in Chapter 4.

Particles are usually called sediment because they can settle and deposit on the bottom. While in suspension they are called suspended sediment or detritus. Sediment dynamics depends on transport by the currents, flocculation, sinking, and formation of new particles due to biological activity or originated by colloidal substances. The way particles are transported depends on their settling velocity and on the interaction between the water column and the bottom, described by the bottom shear stress.

Two classes of sediment, cohesive and noncohesive particles, and two regimes of transport can be considered for both types of particles, according to the concentration close to the bottom. In the case of cohesive sediment the flocks modify the settling velocity. In the case of very high concentrations close to the bottom, sediments modify the properties of the flow itself. Cohesive sediment dynamics differs from noncohesive sediment dynamics in terms of settling velocity and bottom exchange parameterization. Flocculation is specific to cohesive sediment, but the water-bottom dynamics is also different for noncohesive sediment. The freedom of movement of individual particles and their rapid sinking make the bed–load transport the most effective mechanism of noncohesive sediment dynamics.

The detailed description of cohesive and noncohesive sediment mechanisms and modeling of sediment dynamics is beyond the scope of this book.

## BIBLIOGRAPHY

Apel, J.R. *Principles of Ocean Physics*. Academic Press, London, 1987.
Batchelor, G.K. *An Introduction to Fluid Mechanics*. Cambridge University Press, Cambridge, 1967.
Defant, A. *Physical Oceanography*. Pergamon Press, New York, 1961.
Gill, A.E. *Atmosphere-Ocean Dynamics*. Academic Press, London, 1982.
Guyon, E., Hulin, J.-P., and Petit, L. *Hydrodynamique Physique*. CNRS Editions, Paris, 1991.
Holton, J.R. *An Introduction to Dynamic Meteorology*. Academic Press, London, 1992.

Lindzen, R. *Dynamics in Atmospheric Physics*. Cambridge University Press, Cambridge, 1990.
Pedlosky, J. *Geophysical Fluid Dynamics*. Springer-Verlag, Berlin, 1994.
Salby, M.L. *Fundamentals of Atmospheric Physics*. Academic Press, London, 1996.
UNESCO. The Practical Salinity Scale 1978 and the International Equation of State of Seawater 1980. UNESCO Technical Papers in Marine Science, No. 36, Paris, 1981.

# 4 Biogeochemical Cycles

## Melike Gürel, Aysegul Tanik, Rosemarie C. Russo, and I. Ethem Gönenç

## CONTENTS

1-56670-686-6/05/$0.00+$1.50
© 2005 by CRC Press

## 4.1   NUTRIENT CYCLES

Among the most productive ecosystems in the biosphere, coastal lagoons cover 13% of world's coastal zone[1] and constitute an interface between terrestrial and marine environments.[2,3] Nutrient loadings coming from both boundaries to lagoon ecosystems have increased considerably in recent years, and they have a major impact on water quality and ecology.[4,5] Control of nutrients is thus one of the major problems faced by those responsible for the management of these sensitive ecosystems. In order to develop appropriate modeling strategies for making scientifically sound approaches to reduce the risk of environmental degradation of these ecosystems, a better understanding of nutrient cycles is required.

In this section, nutrient cycles and their associated mechanisms and major reactions in coastal marine environments are described. Additional information on eutrophication caused by nutrient loading will be presented in Chapter 5.

### 4.1.1 NITROGEN CYCLE

Among nutrients, nitrogen is of particular importance because it is one of the major factors regulating primary production in coastal marine environments.[6-8] Nutrients are imported to coastal lagoons via atmosphere, agricultural lands, forests, rivers, urban and suburban run-off, domestic and industrial wastewater discharges, groundwater, and the sea. Nutrients are exported via tidal exchange, sediment accumulation, and denitrification. An additional source is nitrogen fixation. Internal sources of nitrogen include benthic and pelagic regeneration. In general, little is known about the supply of nutrients from the atmosphere and groundwater to coastal lagoons.[9]

The nitrogen forms that are important in aquatic environments are ammonia/ammonium ($NH_4^+/NH_3$), nitrate ($NO_3^-$), nitrite ($NO_2^-$), nitrogen gas ($N_2$), and organic nitrogen. These different forms of nitrogen, present in different oxidation states, undergo oxidation and reduction reactions. Ammonia and oxidized forms of nitrogen ($NO_2^-$, $NO_3^-$) constitute dissolved inorganic nitrogen (DIN), which can be utilized by phytoplankton for growth or by bacteria as an electron acceptor. Typical concentrations of $NH_4^+$ and $NO_3^-$ in coastal waters range from <1–10 μM and <2–25 μM, respectively.[10] The various nitrogen compounds and their oxidation states, together with their molecular formulas, are given in Table 4.1.

Ammonia exists in two forms: ammonium ion ($NH_4^+$) and unionized ammonia ($NH_3$). The latter form is toxic to aquatic organisms and is in equilibrium with the ammonium and hydrogen cations. The concentrations of these forms vary considerably as a function of pH and temperature in natural water bodies. The method of calculation of the percent of total ammonia that is unionized at different pH and temperature is given in Emerson et al.[11]

$$NH_4^+ \rightleftharpoons NH_3 + H^+ \qquad (4.1)$$

### TABLE 4.1
### Forms of Nitrogen and Their Oxidation States

| Forms of Nitrogen | Molecular Formula | Oxidation State of N |
|---|---|---|
| Ammonium | $NH_4^+$ | −3 |
| Unionized ammonia | $NH_3$ | −3 |
| Nitrogen gas | $N_2$ | 0 |
| Nitrite | $NO_2^-$ | +3 |
| Nitrate | $NO_3^-$ | +5 |

Nitrogen compounds can be classified into organic and inorganic nitrogen. Organic nitrogen in water bodies can be found in both dissolved and particulate forms. The particulate organic nitrogen (PON) is composed of organic detritus particles and phytoplankton and has two possible fates. Dead plant cells lyse and bacteria degrade the resulting dissolved organic nitrogen (DON) or protozoa/zooplankton to consume PON.[12] Most of the DON in seawater is still chemically uncharacterized, and its chemical and biological properties are becoming better known.[7] Except for amino acids and urea, which comprise only a small fraction of DON, most of the DON may be resistant to decomposers.[10] Excretion by animals also releases dissolved nitrogen. Zooplanktons excrete free amino acids, ammonia, and urea. Fish excrete ammonia, urea, and other organic compounds.[7]

In aquatic ecosystems, a very complex biogeochemical nitrogen cycle is observed (Figure 4.1). The following sections give information about the processes involved in the biogeochemical cycling of nitrogen in the aquatic environment.

### 4.1.1.1 Uptake of Nitrogen Forms

Primary production in coastal waters is largely regulated by the availability of $NH_4^+$ and $NO_3^-$ for growth. Ammonium is preferred by phytoplankton, as its oxidation

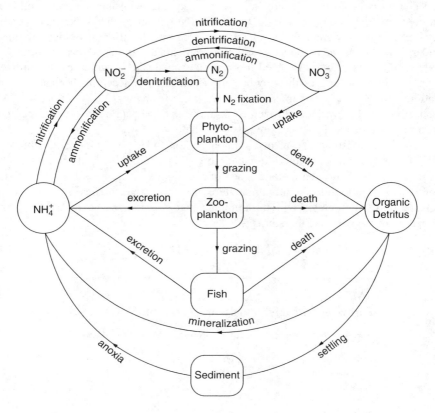

**FIGURE 4.1** Nitrogen cycle.

state is equivalent to that of cellular nitrogen (−3) and thus requires the least energy for assimilation.[12,13] Ammonia concentrations above 1–2 μM tend to inhibit assimilation of other nitrogen species.[10] On the other hand, if nitrate is to be assimilated for the synthesis of cellular materials, it should be reduced to ammonia with the aid of several enzymes including nitrate reductase (enzyme catalyzed reduction) within the cell. This reduction process is called "assimilatory nitrate reduction" and requires energy.[7,14]

Nitrogen uptake can be an important process. For example, in Basin d'Arcachon in southern France, due to the high nitrogen uptake rates of the seagrass *Zostera noltii,* nitrogen uptake is quantitatively more important than denitrification as a nitrogen sink.[15]

In shallow water systems, biological organisms larger than phytoplankton turn over slowly, and their metabolism is lower. Nevertheless, these organisms store large amounts of nitrogen, because a substantial amount of nitrogen is tied up in their biology. Thus, nitrogen concentrations in the shallow systems tend to be lower.[16]

Nutrient assimilation by macrophytes can be significantly different from that by phytoplankton because macrophytes have the ability to grow for long periods on stored nutrients. Rooted seagrasses can assimilate nutrients from sediment and possibly serve as nutrient pumps[10] (see Chapter 5 for details).

### 4.1.1.2  Nitrification

Nitrification is the microbiological oxidation of ammonium to nitrite and then to nitrate under aerobic conditions, to satisfy the energy requirements of autotrophic microorganisms. Much of the energy released by this oxidation is used to reduce the carbon present in $CO_2$ to the oxidation state of cellular carbon, during the formation of organic matter.

As indicated previously, the first step in nitrification is oxidation of ammonium to nitrite, which is accomplished by *Nitrosomonas* bacteria.

$$NH_4^+ + 1\tfrac{1}{2}O_2 \rightleftharpoons 2H^+ + NO_2^- + H_2O \qquad (4.2)$$

The second step is oxidation of nitrite to nitrate by *Nitrobacter*. This is a faster process.

$$NO_2^- + \tfrac{1}{2}O_2 \rightleftharpoons NO_3^- \qquad (4.3)$$

The overall nitrification reaction is therefore

$$NH_4^+ + 2O_2 \rightleftharpoons NO_3^- + H_2O + 2H^+ \qquad (4.4)$$

The nitrification process can influence marine primary production by competing with heterotrophs for the limited supply of dissolved oxygen and by decreasing the amount of $NH_4^+$ that is needed by phytoplankton for growth.[17] The coupling of the nitrification process with denitrification leads to loss of nitrogen from the atmosphere.

Nitrification can take place either in the water column or in the sediment. However, nitrification in the water column of shallow marine and estuarine systems appears to be relatively limited.[17] Nitrification rates in the water column are at least in order of magnitude smaller than the nitrification rates per unit volume in sediment. For example, in coastal waters, nitrification rates range from only ~0.001–0.1 $\mu mol\ l^{-1}\ h^{-1}$, whereas in coastal sediment nitrification rates are often 20 $\mu mol\ l^{-1}\ h^{-1}$.[18] Nitrification rates measured in coastal sediment are usually on the order of 30–100 $\mu mol\ m^{-2}\ h^{-1}$.[17,10]

Physico-chemical and biological factors regulating nitrification in coastal marine sediment include temperature, light, $NH_4^+$ concentration, dissolved oxygen concentration, pH, dissolved $CO_2$ concentration, salinity, the presence of any inhibitory compounds, macrofaunal activity, and the presence of macrophyte roots.[8,17]

Temperature influences the metabolic activities of nitrification bacteria. The optimum temperature is in the range of 25–35°C in pure cultures.[17] Due to both seasonal and diurnal changes in temperature in shallow coastal sediment, it is expected that nitrifying bacteria would exhibit optimal growth and/or activity during daytime and in the summer months when temperatures are maximum.[8] The effect of temperature on nitrification rates in pure cultures is usually expressed through Arrhenius type equations.[17] In addition, temperature also affects dissolved oxygen solubility and therefore the process rates. Light may influence the nitrification activity in shallow water sediment. Light availability and the penetration depth of light into sediment may affect benthic nitrification.[17]

Nitrification may be strongly impeded by hypoxia since it occurs only under aerobic conditions.[19] Nitrifying bacteria, therefore, have to compete with other heterotrophs for the limited supply of dissolved oxygen. The depth distribution of nitrifying bacteria in sediment is ultimately constrained by the downward dissolved oxygen diffusion, which is typically 1–6.5 mm. In Chesapeake Bay, U.S.A.,[18,20] Étang du Prévost in southern France,[21] and Danish coastal zones,[8] $O_2$ penetration into sediment declines due to increased temperature, organic inputs, and decreased macrofaunal activity in summer. Consequently, thinning of the surficial oxidized zone of sediment is responsible for the significant summer reduction in nitrification rates in these systems. The reported dissolved oxygen concentrations, which inhibit nitrification in sediment are in the range 1.1–6.2 $\mu M\ O_2$.[8]

Salinity is another factor influencing nitrification. Although nitrifying bacteria are able to acclimate to a wide range of salinities, such as those found in lagoon systems, short-term fluctuations may have strong regulating effects on nitrification. For example, a marine *Nitrosomonas* sp. isolated from the Ems-Dollard estuary at 15% salinity was able to adapt to the entire salinity range (0–35%) and grew at the same rate over the range after a lag phase of up to 12 days.[17] Rysgaard et al.[21] reported in their study conducted with the sediment from the Randers Fjord Estuary, Denmark that both nitrifying and denitrifying bacteria were physiologically influenced by the presence of sea salt, showing lower activities at higher salinities.

Salinity has another, nonphysiological effect on nitrification processes. As a consequence of higher salinity, the concentration of cations also increases. These compete with $NH_4^+$ for adsorption on the sediment. As a result, the residence time of $NH_4^+$ within the sediment decreases, and the $NH_4^+$ flux from the sediment increases. At higher salinities, $NH_4^+$ might diffuse out of the sediment before nitrification can take place.[21]

Rooted benthic macrophytes might also influence nitrification–denitrification processes in deeper sediment because they release $O_2$ via their roots. This release could stimulate nitrification and thus provide an additional $NO_3^-$ source for denitrification.[15]

Sulfide, the product of anaerobic sulfate reduction, is quantitatively the most important toxic sulfur compound in marine sediment.[17] Sulfide concentrations can significantly reduce the activity of nitrifying bacteria by lowering the redox potential,[20] and concentrations between 0.9 and 40 μM can inhibit nitrification completely.[18] $HS^-$ concentrations in estuarine sediment commonly range from 7–200 μM, which is much lower than those for organic-rich sediment (>1 mM). The range of $HS^-$ concentration in freshwater sediment pore water is much lower (0–30 μM).[22]

The presence of nitrifying bacteria in anaerobic sediment at depths well below the zone into which oxygen can penetrate is attributable to macrofaunal irrigation of sediment by physical resuspension and bioturbation. [15,20]

## 4.1.1.3  Denitrification

Denitrification is the microbiological reduction of nitrate to nitrogen gas, where facultative heterotrophic organisms use nitrate as the terminal electron acceptor under anoxic conditions:

$$NO_3^- \rightarrow NO_2^- \rightarrow NO \rightarrow N_2O \rightarrow N_2 \qquad (4.5)$$

Nitrogen gas is largely unavailable to support primary production; therefore, denitrification removes a substantial portion of the biologically available nitrogen and represents a mechanism for partial buffering against coastal eutrophication.[18,22]

The nitrification and denitrification processes taking place in the sediment and in the sediment–water interface are schematically shown in Figure 4.2. Several factors affect denitrification rate, including temperature, pH, redox potential, as well as concentrations of oxygen, nitrate, and organic matter.[7,8,13,14,18]

Denitrification rate is highly temperature dependent and generally increases with increasing temperature.[7] However, because of other factors such as nitrification rate and oxygen concentration, which also are temperature dependent it is difficult, especially in sediment, to separate the effect of temperature alone.[18]

The rate of denitrification decreases with acidity.[13,23] The pH range, where denitrifiers are most active, is given as 5.8–9.2.[7]

In marine systems, one of the most important environmental factors favoring denitrification is the availability of organic matter.[14] Simple organic compounds, such as formate, lactate, or glucose, usually serve as the electron donor in addition to their assimilation. Coastal marine environments act as centers of deposition for

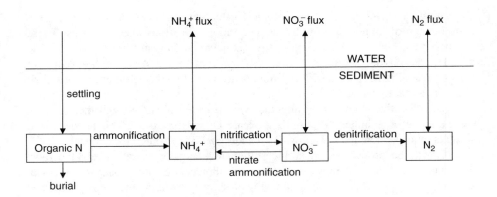

**FIGURE 4.2** Nitrification and denitrification in sediments.

continentally derived organic materials. Thus, most denitrification in marine sediment occurs in coastal regions rather than deep-sea environments.[22]

Oxygen concentrations can also affect denitrification rates. Denitrification is generally considered to occur only under low oxygen or anaerobic conditions.[7] To explain coupled nitrification–denitrification processes in sediment, it is often assumed that these processes are separated vertically within the sediment. However, denitrification can also occur within reduced microzones in the aerobic surface layer of sediment. In both freshwater and marine systems, an oxygen concentration of 0.2 mg $l^{-1}$ or less is required for denitrification to occur in the water or sediment.[18] Bonin and Raymond[24] studied the kinetics of denitrification under different oxygen concentrations using *Pseudomonas nautica* isolated from marine sediment. They reported that denitrification can take place in the presence of oxygen. However, enzymes associated with denitrification are affected by the presence of oxygen. Nitrate reductase enzyme was completely inhibited at oxygen concentrations greater than 4.05 mg $l^{-1}$, compared with 2.15 mg $l^{-1}$ and 0.25 mg $l^{-1}$ for nitrite and nitrous oxide reductase enzymes, respectively. Yet, these results must not be generalized to all denitrifying strains because some bacteria are inhibited by oxygen while other species are not.

In many coastal environments, seasonal trends in denitrification are determined largely by availability of $NO_3^-$ which is controlled by rates of nitrification.[20] The response of denitrification rates in sediment slurries to increasing nitrate concentrations can often be described by Michaelis-Menten type kinetics. The half-saturation constant for marine sediment generally ranges from 27–53 $\mu M$ $NO_3^-$.[18]

Supplies of $NO_3^-$ for denitrification in coastal marine sediment appear to be derived almost exclusively from sediment nitrification.[17] Diffusion of nitrate from the overlying water into the sediment is also a potential nitrate source for denitrification, and its rate in the sediment is 3–4 orders of magnitude greater than that of the overlying water. There is also evidence that the release rates of nitrate and ammonium from sediment are greater than their diffusion rates into the sediment. Nitrification is usually observed in the upper 5 cm of sediment, and the nitrate produced diffuses either up to the water or down to the anoxic zone, where denitrification takes place.[18,23]

Macrophytes, benthic algae, and certain macrofauna have been shown to influence denitrification rates in both freshwater and marine sediment by affecting the oxygen and/or the nitrate distribution in the sediment.[18]

A wide range of experimental methodologies has been developed to estimate denitrification rates in shallow marine environments. These techniques are based on different assumptions; therefore, care must be taken when comparing denitrification rates obtained using these different techniques. Seitzinger[18] has given the ranges of denitrification rates as 50–250 $\mu$mol N $m^{-2}$ $h^{-1}$ in estuarine and coastal marine sediment, 2–171 $\mu$mol N $m^{-2}$ $h^{-1}$ in lake sediment, and 0–345 $\mu$mol N $m^{-2}$ $h^{-1}$ in river and stream sediment. In low oxygen hypolimnetic lake waters, denitrification rates were generally 0.2–1.9 $\mu$mol N $l^{-1}d^{-1}$. The higher rates were from systems that receive substantial amounts of anthropogenic nutrient input. Groundwater is another nitrate source.[18]

### 4.1.1.4  Nitrate Ammonification

Denitrification is widely accepted as the dominant process of nitrate reduction in most shallow marine sediment. An alternate pathway to denitrification is nitrate ammonification, which is the reduction of $NO_3^-$ to $NH_4^+$ by heterotrophic bacteria. In contrast to denitrification, nitrogen is not lost from the system but converted to a readily available nitrogen form.[8] Nitrate ammonification can occur occasionally under anaerobic conditions.[2]

Nitrate ammonification is also called dissimilatory nitrate reduction, and it has been described as an important process in marine sediment.[24] In both Bassin d'Arcachon and Étang du Prévost, two coastal lagoons in southern France, rates of nitrate ammonification were quantitatively as important as denitrification.[15]

### 4.1.1.5  Mineralization of Organic Nitrogen (Ammonium Regeneration)

The process of transforming organic compounds back to inorganic compounds is generally referred to as mineralization.[25] Through the mineralization of organic nitrogen compounds, nitrogen recycling is accomplished. Recycled nitrogen is primarily in the form of ammonia and urea (a dissolved organic nitrogen compound). Urea is rapidly broken down to ammonia by bacteria or by the extracellular enzyme urease.[8,23,25] Ammonium is regenerated from organic compounds by animal excretion and by microbial decomposition of organic matter. It is presumed that excretion contributes to the largest part of $NH_4^+$ regeneration in the water column, while decomposition of organic matter is the most important in the sediment.[10]

It is widely accepted that shallow coastal sediments are important sites for the mineralization of organic matter. The difference of shallow coastal waters compared with open seas is that a much larger fraction of the organic matter is mineralized on the bottom rather than in the water column.[2,26] Because of the shallow depth of coastal areas (e.g., 2–20 m) and the relatively rapid settling rates, a significant portion of the primary production is transferred to the sediment. Thus, much of the mineralization of nutrients occurs in the upper layer of the sediment.[2,18,27,28]

Organic compounds are mineralized through both aerobic and anaerobic respiration processes. Aerobic respiration, which takes place in the surface sediment layers (typically 0–5 mm depth), results in a rapid depletion of oxygen. In the sediment, bacteria oxidize a significant fraction of the organic matter using terminal electron acceptors other than oxygen (e.g., nitrate, manganese and iron compounds, sulfate, and carbon dioxide).[8,29] The two dominant anaerobic processes are dissimilatory sulfate reduction and methanogenesis (methane production). Generally, sulfate reduction precedes methanogenesis because sulfate-reducing bacteria outcompete methanogens for substrates. Freshwater has lower sulfate concentrations (10–200 µM) than estuarine water (30 mM).[22] Decomposition through sulfate reduction occurs deeper in the sediment column (>10 cm) and provides an additional source of $NH_4^+$.[26] Sulfate reduction, and subsequent inhibition of nitrification and denitrification by $HS^-$, should lead to enhanced ammonium regeneration during summer, when sulfate reduction rates are high compared with those in winter.[22] In all cases, the mineralization of organic nitrogen compounds results in the production of $NH_3/NH_4^+$.

All living matter contains nitrogenous macromolecules, which become available to decomposer organisms upon the death of cells. Depending upon the structural complexity of the organic matter, mineralization can either be a simple deamination reaction or a complex series of metabolic steps involving a number of hydrolytic enzymes. Thus, mineralization rates depend on the degradability of the organic matter; i.e., whether it is labile or highly refractory. For example, seagrass detritus that has 25–30% lignin containing fibers, has a lower mineralization rate than phytoplankton cells, which contain more labile nitrogenous material.[8] Another parameter affecting the mineralization rate of organic matter is temperature.[7] Seasonal patterns of benthic nutrient regeneration generally exhibit strong summer maxima, which correlate well with water temperature. The effects of temperature can be represented by Arrhenius type expressions.[27]

Mineralization of organic nitrogen plays a central role in nitrogen recycling in coastal marine environments. Regeneration from the sediment regulates all productivity since inorganic nutrients are the limiting factors for primary production,[30] and much of the primary production of many coastal marine systems is supported by nutrient recycling rather than by nutrient inputs alone.[26] In shallow water ecosystems, benthic recycling may account for 20–80% of the nitrogen requirements of the phytoplankton.[8,27] Nixon[26] reported that nutrient inputs to Narragansett Bay, U.S.A. (without being recycled) could support, at the most, only 24–50% of the annual production, depending on the nutrient considered.

Ammonium produced during the deamination of organic nitrogen in sediment is not totally available to the primary producers; some of the ammonium remains dissolved in interstitial water, some is adsorbed and buried into deeper sediment,[7] some is consumed by benthic algae for cell synthesis,[13] and a fraction undergoes nitrification in the surficial oxic zone of the sediment.[8] Denitrification following nitrification produces gaseous forms of nitrogen ($N_2$, $N_2O$) essentially unavailable to most coastal phytoplankton.[2,31] Thus, the coupled processes of nitrification–denitrification represent a sink that shunts nitrogen away from recycling pathways.[20]

### 4.1.1.6 Ammonia Release from Sediment

Most of the nitrogen mineralized in the sediment is recycled by diffusion from the sediment to the overlying water as $NH_4^+$ or $NO_3^-$. Some nitrogen can also be released as urea and dissolved organic nitrogen, but the quantities of these fluxes are unknown.[2] There are a few important factors influencing the quantity of $NH_4^+$ and $NO_3^-$ release, which are explained below.

As stated previously, benthic regeneration is a function of temperature. $NH_4^+$ regeneration rates and pore-water concentrations tend to increase with temperature.[26] Factors such as sediment grain size and physical circulation also influence this process. It has been demonstrated that the activity of (meio- and macro-) fauna enhances the rate of $NH_4^+$ release from sediment.[10]

The amounts of $NH_4^+$ and $NO_3^-$ depend greatly on seasonal conditions. For example, the major part of nitrogen released from the sediment is $NH_4^+$ in summer when the mineralization rate is high and the aerobic zone depth is generally small. On the other hand, nitrification dominates during winter and spring when the aerobic zone is deeper, and, therefore, $NO_3^-$ is released from the sediment.

In the presence of anaerobic conditions, redox potential decreases significantly, resulting in the termination of nitrification in the sediment. The loss of the oxic microzone between the sediment and overlying water under anoxic conditions also causes a considerable decrease in adsorption capacity of the sediment, producing a significant increase in the release of $NH_4^+$ from the sediment.

Salinity is another factor influencing ammonium release from sediment. Ambient exchangeable ammonium concentrations in freshwater sediment are generally considerably greater than those reported for marine sediment.[32] Fluctuating salinity plays a major role in controlling the $NH_4^+$ adsorption capacity of the sediment.[21] Specifically, the total amount of cations (primarily $Na^+$, $Mg^{2+}$) increases with salinity, resulting in greater molecular competition with ammonium for the sediment cation exchange sites.[21,32,33]

The greater ammonium adsorption in freshwater sediment relative to marine sediment increases the amount of ammonium that can be nitrified. Seitzinger et al.[32] reported in their study that a larger percentage of net ammonium produced in aerobic freshwater sediment (Toms River, U.S.A.) was nitrified and denitrified (80–100%) compared with that in marine sediment (40–60%) (Barnegat Bay, U.S.A.).

Postma et al.[10] reported an $NH_4^+$ flux range of 50–800 $\mu$mol $m^{-2}$ $h^{-1}$ for estuarine and coastal sediment. According to Day et al.[27] the annual mean value of nutrient regeneration ranges from 20–300 $\mu$mol $m^{-2}$ $h^{-1}$ for $NH_4^+$ in estuarine sediment. The rate of release of ammonia by a wide variety of marine sediment during summer was given as 13–710 $\mu$mol $m^{-2}$ $h^{-1}$ by Nixon.[26] $NH_4^+$ fluxes in Chesapeake Bay, U.S.A., were reported to be 46 $\mu$mol $m^{-2}$ $h^{-1}$ in April, increasing to 753 $\mu$mol $m^{-2}$ $h^{-1}$ in August.[20] The calculated benthic $NH_4^+$ flux at Thau Lagoon, France, for a period of 10 days in August during anoxia was 600 $\mu$mol $m^{-2}$ $h^{-1}$.[28]

### 4.1.1.7 Nitrogen Fixation

The process that converts atmospheric nitrogen gas ($N_2$) into organic nitrogen compounds is known as nitrogen fixation. Most organisms, due to the significant amount of energy required to split NN triple bonds, cannot use the gaseous $N_2$ form. A few genera of bacteria, blue-green algae (cyanobacteria) that possess heterocytes (specialized cells present in most filamentous blue-green algae) and some unicellular forms of blue-green algae without heterocytes are capable of nitrogen fixation. Nitrogen fixation may be crucial to the acceleration of eutrophication in aquatic environments, since fixation can occur when other sources of nitrogen are not available or are insufficient for biological growth.[8,13,23] However, in almost all rivers, lakes, and coastal marine ecosystems, loss of nitrogen via denitrification exceeds the inputs of nitrogen via $N_2$ fixation.[18] For example, in Narragansett Bay, U.S.A., a small amount of nitrogen fixation occurs in the sediment, but it is insignificant compared with nitrogen losses through denitrification (less than 0.0007 mol N $m^{-2}$ $year^{-1}$ fixed in sediment compared with 0.52 mol N $m^{-2}$ $year^{-1}$ denitrified).[34]

Nitrogen fixation is less effective in making up nitrogen in marine systems than in freshwater systems as the rates of nitrogen fixation in marine waters are generally lower than those in freshwaters.[35] Many hypotheses have been proposed to explain the difference in the rates of $N_2$ fixation in fresh and marine waters.[7,10] The most likely explanation is the lower availability of two trace metals (iron and molybdenum) in seawater that are required for nitrogen fixation, compared with their availability in lakes.[34] Molybdenum is one of the active sites of the enzymes involved in nitrogen fixation. The abundant sulfate in seawater interferes with the uptake of molybdenum by fixers because of its steric similarity to molybdate; therefore, seawater sulfate could reduce the activity of $N_2$ fixers.[7] In addition, relatively large amounts of iron are required for the growth of cyanobacteria using $N_2$ rather than $NH_4^+$ or $NO_3^-$ as the nitrogen source. Thus, the rate of $N_2$ fixation might also be restricted by the low abundance of iron in seawater compared with freshwater.[7]

$N_2$ fixation by benthic cyanobacteria can be significant due to the direct supply of iron and molybdenum from the sediment. However, low light penetration can limit the growth of benthic $N_2$ fixers.[34] In unvegetated shallow coastal lagoons and intertidal sediment where light is not limiting, dense populations of benthic nitrogen-fixing cyanobacteria can locally develop and contribute to nitrogen fixation. However, even though the nitrogen fixed by cyanobacteria is locally important to the mat communities themselves, the contribution to the total nitrogen budget is minor in most shallow marine ecosystems due to the restricted distributions of mat communities.[8] Since the cyanobacteria population varies greatly in time and space, detailed measurements are required to estimate the total annual nitrogen fixation rate.[13,23]

Other physico-chemical parameters influencing nitrogen fixation activity in benthic sediment include carbon availability, temperature, pH, dissolved oxygen, inorganic nitrogen, and salinity.[8]

## 4.1.2 PHOSPHORUS CYCLE

Phosphorus is one of the limiting nutrients for the growth of microorganisms although the quantities of phosphorus needed are much smaller than those of C, Si, or N.[23]

There are many external sources of phosphorus for coastal ecosystems. Domestic wastewater discharges may often contain large quantities of phosphorus because many commercial cleaning products contain phosphorus. There are also industrial sources of phosphorus, such as wastewater discharge from boiler water treatment operations. Phosphates applied as fertilizers to agricultural or residential cultivated land are also transported into surface waters with surface run-off. Internal sources of phosphorus include benthic and pelagic regeneration.

Phosphorus in water can be classified into particulate and dissolved forms. Particulate phosphorus includes phosphorus in organisms in/sorbed to dead organic matter and in/sorbed to mineral phases of rock and soil. Dissolved phosphorus is composed of orthophosphate ($PO_4^{-3}$) polyphosphates (often originating from synthetic detergents), organic colloids, and phosphorus combined with adsorptive colloids and low-molecular-weight phosphate esters. Orthophosphate is the most significant form of phosphorus available for phytoplankton growth. Orthophosphate ions include phosphoric acid ($H_3PO_4$), its dissociation products ($H_2PO_4^-$, $HPO_4^{2-}$, $PO_4^{3-}$), and the ion pairs and complexes of these products with other constituents in seawater. The phosphorus atom has an oxidation state of +5 in orthophosphates.[7]

General phosphorus transformation mechanisms are illustrated in Figure 4.3. As seen from this figure, phosphorus can undergo various reactions, depending upon

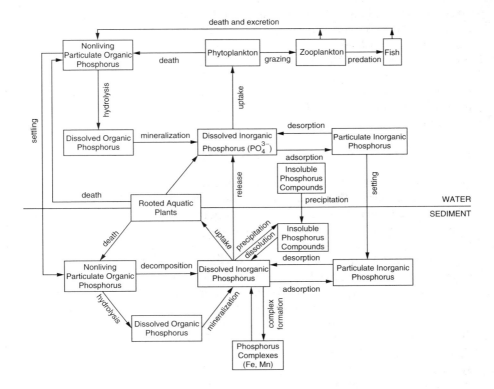

**FIGURE 4.3** Phosphorus cycle.

environmental conditions. Orthophosphate is taken up by phytoplankton and incorporated into cells during growth. Some fraction of the phosphorus taken up is released
in forms readily available to other algal cells. Other phosphorus compounds released
have to be mineralized and/or hydrolyzed into inorganic form before their use in
growth. It is likely that phosphorus is also excreted directly by invertebrates, similar
to zooplankton.[13] Algal growth produces an increase in particulate organic phosphorus
(POP), and the death of algae releases dissolved organic phosphorus (DOP). Some of
the POP is transformed to dissolved inorganic phosphorus (DIP) as particles decay in
the water column, but some POP settles onto the bottom sediment. In the sediment,
further degradation of settled organic phosphorus to DIP can take place, and some of
this formed DIP is subsequently precipitated or adsorbed.[7] Bacteria excrete some
phosphate, and DIP is also generated by microbial hydrolysis of the esters of the DOP.

### 4.1.2.1  Uptake of Phosphorus

DIP (orthophosphate) is the only form of phosphorus that can be assimilated by
bacteria, algae, and plants.[13,36] The growth process is usually represented by the
Monod equation, and the half-saturation concentration of nutrients ($Ks$) in the Monod
equation varies depending on the organisms involved.

   An organism with a lower $Ks$ value, has the advantage over other organisms
when the nutrient in question (here, phosphorus) is in short supply.[27] For example,
algae have lower $Ks$ concentrations (0.6–1.7 µM) while bacteria have higher $Ks$
values (6.7–11.3 µM).[7] The ability to store nutrients also makes algae a more
important reservoir than bacteria.

   Because aquatic macrophytes have the ability to use sedimentary nutrient
sources[37] and the nutrient concentrations in sediment are much higher than those in
the water column, high $Ks$ values for macrophytes do not result in severe nutrient
limitation.[27]

### 4.1.2.2  Phytoplankton Death and Mineralization

During respiration and death of phytoplankton, a fraction of the phosphorus released
is in inorganic form. The remaining fraction is in organic form, which must be
mineralized and/or hydrolyzed into inorganic form to be made available to other
organisms. This transformation usually occurs in the sediment in coastal ecosystems
because these systems are shallow and the settling process is more effective. After
the transformation, inorganic phosphorus is either released to the overlying water
and made available for growth or adsorbed or buried deeper into the sediment.

   In fish farms located in lagoon systems, the fish are fed with rich food distributed
in large quantities in the water column. This input could constitute an additional
source of organic matter to the sediment that might increase benthic nutrient fluxes.
On the other hand, the greater recycling rates of organic matter could explain the
significant increase in benthic nutrient fluxes observed in the lagoon fish culture
systems.[38]

   The release of available phosphorus from sediment to overlying water stimulates
primary production in the water column and is known as regeneration of phosphorus.

This can be a significant process in coastal waters. Across a variety of coastal sediment, the regeneration of phosphate provides an average of 28% of the phytoplankton requirements whereas, in Narragansett Bay, U.S.A., it provides 50%.[7]

Floating and submerged macrophytes also are important phosphorus sources since they release phosphorus rapidly (within days) from the decaying leaves and roots.[13]

Phosphorus cycles rapidly through the aquatic food chain, and is seldom limiting in the marine environment.[25] Howarth[34] states that during microbial decomposition, phosphorus is released faster than nitrogen, presumably because the ester bonds of phosphorus are more easily broken than are the covalent bonds of nitrogen.

### 4.1.2.3  Phosphorus Release from Sediment

Phosphorus retention and subsequent release from sediment to overlying waters may be important in preventing/delaying the improvement of water quality. Therefore, much study has been devoted to the phosphorus content of sediment and its movement into the overlying water.[39]

Exchanges across the sediment–water interface are regulated by mechanisms associated with mineral–water equilibria, sorption processes (notably ion exchange), oxygen-dependent redox reactions and microbial activities, as well as the environmental control of inorganic and organic compounds, i.e., enzymatic reactions. The release of adsorbed phosphorus from sediment is controlled by physical-chemical factors such as temperature, pH, and redox potential.[13] Lower redox potentials and high pH values in the surface sediment cause phosphorus release during summer, while low temperatures, high redox potentials, and neutral pH help to retain phosphorus in sediment in winter.[35] The rates of release of phosphate by a wide variety of marine sediment during summer are reported by Nixon,[26] as $-15$–$50$ $\mu$mol m$^{-2}$ h$^{-1}$. Release rates are also influenced by variable rates of turbulent diffusion and the burrowing activities of benthic invertebrates.[13]

The oxygen content of the sediment–water interface is one of the most important features of the interface. As long as a few millimeters of the sediment is aerobic, the phosphorus will be retained in the sediment efficiently. Phosphorus binding with ferric oxides is particularly strong. At a neutral pH and redox potential greater than 200 mV, $Fe(OH)_3$ is stable,[13] and sorption (chemisorption) of orthophosphate takes place. If the pore water becomes anaerobic due to respiratory activity in the sediment, ferric iron ($Fe^{+3}$) is reduced to ferrous iron ($Fe^{2+}$) and the binding is weakened.[36] This redox reaction causes the amorphous ferric oxyhydroxides to dissolve, making them unavailable to adsorb phosphates. The dissolved phosphate can leave the anaerobic sediment, but some of the phosphate may precipitate as $FePO_4$ at the oxic-anaerobic interface, and more is probably adsorbed onto the amorphous ferric oxyhydroxides, also at the oxic-anaerobic interphase. In any case, the release of phosphate from anaerobic sediment is faster than reoxidation and immobilization, resulting in a net phosphate regeneration from anaerobic sediment into overlying water.[7,13]

Caraco et al.[40] state that there is a correlation between sulfate abundance and phosphorus release from anaerobic sediment that is a result of the interaction between iron and sulfur. Enhanced sulfate reduction and the resulting formation of iron–sulfide mineral (e.g., FeS, $FeS_2$) were responsible for increased pore water phosphate accumulation. However, in sulfate-free sediment, much of the phosphate released during anaerobic microbial reduction of $Fe^{+3}$ was captured by solid phase reduced iron compounds (e.g., $Fe_3(PO_4)_2$). In freshwater systems, sulfate concentrations in sediment are generally low compared to coastal marine systems. Therefore, freshwater systems have a greater capacity of retaining phosphorus in the sediment. The phosphorus in freshwater sediment is bound more tightly and proportionally less is released back into the water column.[35]

### 4.1.2.4 Sorption of Phosphorus

Phosphates adsorb readily under aerobic conditions onto amorphous oxyhydroxides, calcium carbonate, and clay mineral particles. Thus, phosphate seldom travels far in sediment, except when transported by the movement of particles. Phosphate also precipitates with cations such as $Ca^{2+}$, $Al^{3+}$, and $Fe^{3+}$.[7] With a few exceptions, surface waters receive most of their phosphorus load from surface flows rather than from groundwater, since phosphates bind with most soils and sediment.[36]

There is an adsorption–desorption interaction between phosphates and suspended particulate matter in the water column. The subsequent settling of suspended solids together with the adsorbed inorganic phosphorus can be a significant phosphorus loss mechanism in the water column and is a major source of phosphorus to the sediment.

Although phosphorus exchange by adsorption–desorption within the sediment and between sediment particles and interstitial water can be as rapid as a few minutes, the rate of phosphorus exchange across the sediment–water interface depends on the state of the microzone.[13]

In Thau Lagoon, France, the release of phosphates adsorbed onto Fe(OOH) and also onto $CaCO_3$ is maximum in summer due to the low redox potential and low pH.[28]

### 4.1.2.5 Significance of N/P Ratio

Net primary production in many marine ecosystems is probably limited by nitrogen, but phosphorus also may limit production in some ecosystems.[6–8,34] There is a shift from phosphorus to nitrogen limitation in moving from freshwater to coastal waters. Some of the reasons for this are more efficient recycling of phosphorus,[36] the high losses of nitrogen to the atmosphere due to denitrification in coastal waters,[34,36] the role of sulfate in recycling phosphorus in coastal sediment,[36] and the low N:P ratio in nutrient inputs to many coastal waters with limited planktonic nitrogen fixation.[34]

Differences in nutrient limitation are the result of changes in the ratio of total nitrogen to total phosphorus in nutrient inputs and the dynamics of internal biogeochemical processes.[34] Analyses of algal cells show that the mean ratio of carbon to nitrogen to phosphorus is C:N:P = 106:16:1. It seems reasonable, therefore, to assume that the cells require these elements according to this ratio. Liebig's law of minimum states that if the ratio of elements in the water deviates widely from this ratio, elements present in excess cannot be utilized.[41] The observed N:P ratios that are less than 16:1 have been used to indicate that nitrogen is less abundant than phosphorus with respect to algal (usually phytoplankton) metabolic demand (Redfield Ratio). Boynton et al.[42] report that 22 of 27 estuaries surveyed were nitrogen limited with N:P ratios in the water column well below 16:1 during the time of peak algal growth. In the systems having low N:P ratios, blue-green algae may have a competitive advantage over other phytoplankton groups due to their ability to fix atmospheric nitrogen.[43]

There are, however, some problems with using the N:P ratio as an indicator of nutrient limitation. First, different types of algae have different N:P ratios ranging from 10:1 to 30:1.[41,42] Second, nutrient limitation is often assumed without testing, and other factors may be limiting. Third, the use of the water column N:P ratio is based on the assumption that nutrient loading is constant or at steady state; however, nutrients are often supplied in pulses and the N:P ratio is constantly altered, depending on both the pulse and uptake rates. Thus, nutrient ratios in the water column are insufficient for determining the limiting nutrient for algal growth, especially when more than one algal group is present.[44] Fourth, the limiting nutrient can change temporally in a system. For example, during winter, when algal crops are sparse and growth is slow, the amount of phosphorus present may be sufficient to be nonlimiting. As the growth of algae proceeds during spring, phosphorus is removed from the water by the algae and the supply becomes progressively depleted.[41] Fifth, in the case of phosphorus, the past history of cells must be considered. Many algae seem to be able to store phosphorus in excess of their present requirements. For this reason, sometimes algae may grow at dissolved phosphate concentrations that seem to be limiting. The problem is complicated further by the fact that phytoplanktons are free-moving plants, and thus, they may not have grown in and derived their nutrients from the water in which they were found (spatio-temporal organization). The time dimension is another problem, as discussed in Chapter 2. Nutrients are recycled rapidly through mineralization, and without additional input of phosphorus from external sources the recycled nutrients become available for growth.[41]

### 4.1.3 SILICON CYCLE

Although considered a minor nutrient, silicon is significant in the dynamics of phytoplankton because of its importance as a major structural element in the cells of diatoms, an important phytoplankton group in coastal waters.[7] Because silicon is used in large quantities for diatom cell walls, it can be a limiting element for phytoplankton growth where diatoms are the predominant algae.[23,45] Silicon limitation impacts the diatoms and silicoflagellates among the phytoplankton, and the radiolarians among the zooplankton.[25]

Silicon is present in coastal waters in three principal forms: detrital quartz, aluminosilicate clays, and dissolved silicon. Similar to phosphorus, silicon occurs primarily in one oxidation state (+4).[27] At the pH and ionic strength of seawater, the dominant dissolved species of silicon is silicic acid ($H_4SiO_4$).[12]

The dominant input of dissolved silicate to most aquatic systems occurs as riverine inputs, as a consequence of weathering reactions in the watershed. In order to become available for biological activity, silicate rocks must be broken down.[46-48]

Weathering is achieved by a combination of mechanical (physical processes of wind or ice) and chemical processes (reactions with acidic and oxidizing substances). The rate of chemical weathering varies with the physical conditions of temperature and rainfall amount, and the mineral composition of the rocks. Rainwater, springs, and the leachate from soils are all high in carbon dioxide (carbonic acid) that weathers rocks to release soluble silica.[23]

The ratios of the nutrients present and the availability of dissolved silicate (DSi) can regulate the species composition of phytoplankton assemblages. When concentrations of DSi become low, other types of algae that do not require DSi can dominate algal community composition and decrease the relative importance of diatoms in phytoplankton communities.[47] There is a considerable concern that altered nutrient ratios in coastal waters may favor blooms of nuisance flagellate species that could replace the normal spring and autumn bloom of siliceous diatoms.[49,50] Large-scale hydrologic alterations on land, such as river damming and river diversion, can cause reductions in silicate inputs to the sea.[49] This has already been observed in the Black and Baltic Seas. Changes in the nutrient composition of river discharges seem to be responsible for dramatic shifts in phytoplankton species composition in the Black Sea.[51,52] In the Baltic Sea, DSi concentrations and the DSi:N ratio have been decreasing since the end of the 1960s, and there are indications that the proportion of diatoms in the spring bloom has decreased while flagellates have increased. Changing phytoplankton species composition can have repercussions on the entire food web, and might have enormous economic impacts.[51]

Diatom phytoplankton populations are the usual food for zooplankton and filter-feeding fishes, and contribute directly to the large fishable populations in coastal zones. Diatoms grow very rapidly, have short lifetimes, are grazed heavily and are rarely a nuisance.[46] They tend to dominate ecosystems whenever silicate is abundant (concentrations greater than 2 $\mu M$).[53] Diatoms account for 60% of the world's primary production.[51] The diatom need for silicate is for the construction of their cell walls (known as frustules). This contrasts with other algae that construct their cell walls from organic material or from calcium carbonate.

Silicon does not have a gaseous phase, and its cycle is relatively simple because it involves only inorganic forms. In contrast to other nutrients, particulate and sedimentary silicon decay directly to dissolved inorganic silicon rather than passing through a dissolved organic phase (Figure 4.4).

### 4.1.3.1 Uptake of Silicon

In the silicon cycle, organisms utilize dissolved silicon ($H_4SiO_4$) to produce their skeletons, and this skeletal material dissolves following the death of the organisms.[25] The half-saturation constant for growth of several diatom species is about 0.5–5.0 $\mu M$.[46] Maximum *in situ* growth rates between 2–4 $d^{-1}$ have been repeatedly measured for diatoms, whereas the observed maximum growth rates for dinoflagellate, microflagellate, and eukaryotic nonmotile ultraplankton species or assemblages have generally been below 2.5 $d^{-1}$.[53]

Diatoms use the dissolved silicon together with nitrogen and phosphorus in an average Si:N:P ratio of 16:16:1.[27,46,54]

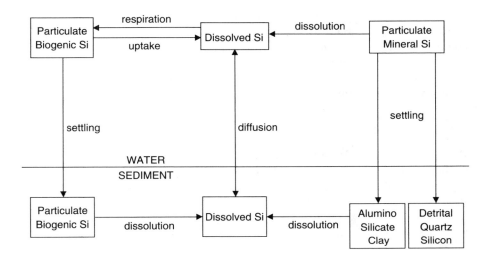

**FIGURE 4.4** Silicon cycle.

### 4.1.3.2 Settling of Diatoms

The mineral content of diatoms increases their specific gravity, which causes negative buoyancy. Diatoms, therefore, tend to sink.[7] Increased diatom production results in increased deposition of silica into sediment. Preservation of deposited material in sediment depends upon a variety of factors, including pH, salinity, temperature, type of sediment, and bulk sedimentation rate. In general, accumulation of biogenic silica in sediment mimics overlying water column productivity.[47]

### 4.1.3.3 Dissolution of Silica

Silicon is released by dissolution of diatom tests (skeletons) rather than by microbially mediated decomposition. It has been suggested by Officer and Ryther [46] and Conley et al.[47] that dissolution rates for particulate silicon are slow relative to regeneration of both nitrogen and phosphorus. Regeneration of biogenic silica is primarily a chemical phenomenon, whereas grazers and bacteria biologically mediate the regeneration of nitrogen and phosphorus. Both nitrogen and phosphorus will be recycled faster and reused on shorter time scales than Si. This could mean that midsummer shifts in algal species domination from diatoms to flagellates might result from silicon limitation due to relatively slow regeneration. However, Day et al.[27] report that measurements of benthic nutrient regeneration indicate that summer rates of silicon recycling are comparable to those for nitrogen. Nixon et al.[55] measured silica fluxes over 1 mmol $m^{-2}$ $h^{-1}$ from the sediment during summer in Narragansett Bay, U.S.A. This flux was higher than commonly supposed.

Salinity is a factor that directly affects the dissolution of siliceous minerals. The rate of dissolution of biogenic silica increases by a factor of 2 by changing the salinity from 1 to 5%. Thus, dissolution of siliceous minerals is more rapid in marine waters.[47]

An observed release of Si from sediment in Thau Lagoon, France, was attributed mainly to the dissolution of silica that was directly controlled by temperature. In contrast to $NH_4^+$ and soluble reactive phosphorus, anaerobic conditions are not supposed to enhance benthic Si fluxes but rather should decrease the release of Si by cessation of bioturbation. However, this behavior of Si, based on experimental results, was not confirmed by the observations in the Thau Lagoon.[28]

### 4.1.4 DISSOLVED OXYGEN

Dissolved oxygen (DO) is considered to be a very important and sensitive indicator of the health of aquatic systems. DO is necessary to support the life functions of higher organisms and to drive many redox reactions.[56] DO also provides key information about the system state,[57,58] e.g., insight into algal blooms, oxygen depletion rates, and zones of oxygen depletion. Changes in the shape of the DO depth curve, as well as the oxygen deficiency in bottom waters, are meaningful eutrophication indices.[59]

Low dissolved oxygen can cause the loss of aquatic animals. Most estuarine populations can tolerate short exposure to low dissolved oxygen concentrations without adverse effects. Extended exposures to dissolved oxygen concentrations less than 60% oxygen saturation can result in modified behavior, reduced abundance and productivity, adverse reproductive effects, and mortality.[60] For example, Thau Lagoon, France, has lost part of its shellfish production over the past decades due to mass mortality caused by the diffusion of hydrogen sulfide into the water column during anoxic periods.[28,38]

Many aquatic animals have adapted to a short period of hypoxia (dissolved oxygen concentrations below 3 mg l$^{-1}$) by taking up more oxygen and transporting it more effectively to their cells and mitochondria, that is, by ventilating their respiratory surfaces more intensely and increasing their heart rate. If these responses are insufficient to maintain the blood's pH, then the oxygen-carrying capacity will decrease.[61] An early behavioral response is the locomotory, i.e., the moving of organisms toward better-oxygenated waters even when other conditions there may be unfavorable.[61,62] Under hypoxic conditions, the animal may also reduce its swimming and feeding, which will reduce its need for energy and hence oxygen. Although reduced activity may make the animal more hypoxia tolerant for a short period, reduced swimming activity makes the animal more vulnerable to predators, and reduced feeding decreases its growth. If oxygen insufficiency persists, deaths will ultimately occur.[61]

Dissolved oxygen criteria developed by the U.S. EPA[61] for coastal waters apply to both continuous and cyclic DO conditions. If the DO conditions are always above the chronic criterion for growth (4.8 mg l$^{-1}$), most of the animals and plants can grow and reproduce unimpaired. DO conditions below the juvenile/adult survival criterion (2.3 mg l$^{-1}$) mean that there is not enough DO to protect aquatic life. When DO conditions are persistent between these two values, living organisms often become stressed and require further evaluation of duration and intensity of low DO events to determine whether the available levels of oxygen can support a healthy aquatic community.[61,63]

### 4.1.4.1 Processes Affecting the Dissolved Oxygen Balance in Water

The processes affecting the DO balance in aquatic ecosystems include atmospheric reaeration, photosynthesis and respiration, oxidation of organic matter, oxidation of inorganic matter, sediment oxygen demand, and nitrification.[56,64,65] These processes are shown in Figure 4.5.

#### *4.1.4.1.1 Reaeration*

Dissolved oxygen concentrations in unpolluted water are usually close to, but less than, 10 mg $l^{-1}$.[66] If oxygen concentrations decrease below the saturation level due to various physical, chemical, and biological processes, replenishment is provided via atmospheric reaeration.[67]

The dissolved oxygen saturation concentration is affected by several environmental factors including temperature, salinity, and partial pressure variations due to elevation. It decreases with increasing temperature and salinity and decreasing partial pressure of oxygen.[45] There are several empirical equations to express the effects of these factors on the oxygen saturation concentration. The following equation can be used to establish the dependence of oxygen saturation concentration on temperature.[68]

$$\ln(o_{sf}) = -139.34411 + \frac{1.575701 \times 10^5}{T_a} - \frac{6.642308 \times 10^7}{T_a^2}$$

$$+ \frac{1.243800 \times 10^{10}}{T_a^4} - \frac{8.621949 \times 10^{11}}{T_a^4}$$

(4.6)

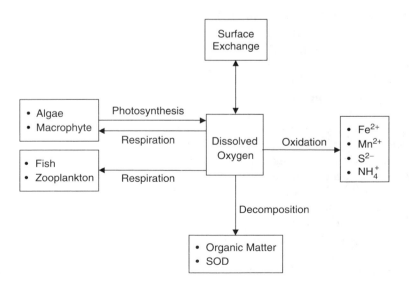

FIGURE 4.5 Processes effecting dissolved oxygen concentration.

where $o_{sf}$ is the saturation concentration of dissolved oxygen in fresh water at 1 atm [mg l$^{-1}$] and $T_a$ is absolute temperature [K].

The following equation can be used to establish the dependence of DO saturation concentration on salinity.[68]

$$\ln(o_{ss}) = \ln(o_{sf}) - S\left(1.7674 \times 10^{-2} - \frac{1.0754 \times 10^1}{T_a} + \frac{2.1407 \times 10^3}{T_a^2}\right) \quad (4.7)$$

where $o_{ss}$ is saturation concentration of dissolved oxygen in salt water at 1 atm [mg l$^{-1}$] and S is salinity [ppt].

Reaeration is represented by a rate coefficient and depends on average water velocity, water depth, and wind speed. No universal formula for estimation of the reaeration coefficient exists. Many investigators have developed formulae for predicting the reaeration coefficient in different aquatic systems. Water flow and/or wind speed can affect oxygen transfer rates. In shallow waters (free flowing streams) the reaeration coefficient depends largely on turbulence generated by bottom shear stress. In deeper systems (e.g., lakes and estuaries) effects of wind may dominate the reaeration process.[56,67] For estuaries, there are reaeration equations that combine the wind- and current-induced reaeration terms.[67,69] Chapra[45] describes the use of the commonly used formula for reaeration in estuaries.

$$k_a = 3.93 \frac{\sqrt{U_0}}{H^{3/2}} + \frac{0.728 U_w^{1/2} - 0.317 U_w + 0.0372 U_w^2}{H} \quad (4.8)$$

where

$k_a$ = reaeration rate coefficient [d$^{-1}$]
$U_o$ = mean current velocity [m s$^{-1}$]
$U_w$ = wind speed measured 10 m above the water surface [m s$^{-1}$]
$H$ = depth [m]

The reaeration rate coefficient can also be affected by temperature and salinity.[56] The temperature dependence is often expressed by an Arrhenius type equation.[69]

Stratification of the water column can prevent diffusion of atmospheric oxygen into bottom waters, and thus oxygen depletion may occur. [28,70-72] Hypoxic conditions, in bottom waters can cause mass mortality of benthic organisms and decline in fishery yields.[72] Similar conditions can also occur in semi-enclosed basins with limited water exchange, such as lagoons.[15]

### 4.1.4.1.2  Photosynthesis – Respiration

Photosynthesis is the conversion of simple inorganic nutrients into more complex organic molecules by autotrophic organisms. Via this reaction oxygen is liberated and $CO_2$ is consumed.[45,64]

$$106CO_2 + 16NH_4^+ + HPO_4^{2-} + 108H_2O \rightarrow C_{106}H_{263}O_{110}N_{16}P_1 + 107O_2 + 14H^+ \quad (4.9)$$

The rate of oxygen production is proportional to the growth rate of the phytoplankton because the stoichiometry is fixed. An additional source of oxygen is reduction of $NO_3^-$ instead of $NH_4^+$ for growth. When available ammonia nitrogen is exhausted and the nutrient source is nitrate, nitrate is initially reduced to ammonia, which produces oxygen.

Respiration is the reverse process of photosynthesis through which oxygen is diminished in the water column as a result of algal respiration. It is normally a small loss rate for the organism and is temperature dependent.[73]

Photosynthesis depends on solar radiation; therefore, the production of dissolved oxygen proceeds only during daylight hours. Concurrently with oxygen production via photosynthesis, aquatic plants require dissolved oxygen for respiration, which can be considered to proceed continuously. The photosynthesis and respiration processes can add and deplete significant quantities of dissolved oxygen, and the combined effect of these processes can cause diurnal and seasonal variations in dissolved oxygen concentrations in aquatic ecosystems.[45,64,66,67] From a seasonal perspective, photosynthesis will tend to dominate during the growing season, while respiration and decomposition will prevail during the nongrowing period of aquatic plants. Diurnal variations in dissolved oxygen can be induced by light, thus dissolved oxygen could be supersaturated during the afternoon and depleted severely just before dawn during the growing season.[45]

Heavy growth of rooted and attached macrophytes can cause oxygen supersaturation in water. Incidental to this supersaturation, a very high pH value can occur that is fatal to fish.[62] Rooted and attached macrophytes tend to have a greater impact on dissolved oxygen than phytoplanktons due to two factors.[45]

1. Since macrophytes are usually found in shallower water, for an equal growth or respiration rate as phytoplankton, the impact of rooted and attached macrophytes on a shallower system will be greater.
2. Because they are fixed in space, macrophytes tend to be more concentrated longitudinally.

### 4.1.4.1.3  Oxidation of Organic Matter

The primary loss mechanism associated with organic matter is oxidation (see Section 4.2 for details). Organic matter serves as an energy source for heterotrophic organisms in aerobic respiration and decomposition processes. These processes return organic matter to the simpler inorganic state. During breakdown, oxygen is consumed and $CO_2$ is liberated.[45] A principal source of organic matter, other than anthropogenic pollution and natural run-off, is detritus, produced as a result of death of organisms. The biodegradability of organic matter is a key factor affecting the oxidation rate. Thus, different types of organic matter; such as labile dissolved organic matter (labile DOM), labile particulate organic matter (labile POM), refractory dissolved organic matter (refractory DOM), and refractory particulate organic matter (refractory POM) are specified and used in models. Labile organic matter decomposes on a time scale of days to weeks, whereas refractory organic matter requires more time (e.g., 1 year).[45,56] Different oxidation rates are used for each of these forms of organic matter in water quality models.

Because the oxidation of organic matter is a bacterially mediated process, oxidation rate is a function of water temperature. Temperature correction of the rate can be made using an Arrhenius type of equation. Temperature correction coefficients ($\theta$) range from 1.02 to 1.09. A commonly used value for this coefficient is 1.047.[45]

### 4.1.4.1.4   Oxidation of Inorganic Matter

Under aerobic conditions, most common metals and some nonmetallic elements, like sulfur, are thermodynamically stable in their highest oxidation states. Ferric iron ($Fe^{3+}$) is more stable than ferrous iron ($Fe^{2+}$), and similarly, sulfate is more stable than sulfide. A series of chemical reductions occurs when the oxygen is consumed. All of these reactions tend to reverse if oxygen is reintroduced.[23]

#### 4.1.4.1.4.1   Sulfide Oxidation

Sulfide is produced in anaerobic sediment as a result of organic matter oxidation, in which sulfate is used as the electron acceptor. A portion of the sulfide precipitates as $FeS_{(S)}$, while the remaining dissolved sulfide diffuses into the aerobic zone, where it is oxidized back to sulfate. Dissolved oxygen is consumed during this last step.[27,74]

$$H_2S + 2O_2 \rightarrow H_2SO_4 \tag{4.10}$$

The particulate sulfide ($FeS_{(S)}$) can also be mixed into the aerobic zone, where it can be oxidized by oxygen to $Fe_2O_{3(S)}$. A portion of the $FeS_{(S)}$ is buried by sedimentation.[74]

$$FeS + \tfrac{9}{4}O_2 + H_2O \rightarrow \tfrac{1}{2}Fe_2O_3 + H_2SO_4 \tag{4.11}$$

#### 4.1.4.1.4.2   Iron Oxidation

$Fe^{2+}$ compounds are more soluble than $Fe^{3+}$ compounds, and thus exist in the low mg l$^{-1}$ range in sediment pore waters. As a consequence, $Fe^{2+}$ can diffuse to the oxic layer of the sediment, and via the loss of one electron, can be oxidized to $Fe^{3+}$ by the oxygen present there.[74]

$$Fe^{2+} \rightarrow Fe^{3+} + e^- \tag{4.12}$$

Half-reaction for oxygen is

$$O_2 + 4H^+ + 4e^- \rightarrow 2H_2O \tag{4.13}$$

followed by the precipitation of iron oxyhydroxide

$$Fe^{3+} + 2H_2O \rightarrow FeOOH_{(S)} + 3H^+ \tag{4.14}$$

to yield the overall redox reaction

$$Fe^{2+} + \tfrac{1}{4}O_2 + \tfrac{3}{2}H_2O \rightarrow FeOOH_{(S)} + 2H^+ \tag{4.15}$$

Iron oxyhydroxide can react with water.[74]

$$FeOOH_{(S)} + H_2O \rightarrow Fe(OH)_{3\ (S)} \tag{4.16}$$

### 4.1.4.1.4.3   Manganese Oxidation

$Mn^{2+}$ compounds are more soluble than $Mn^{4+}$ compounds, and therefore exist in the mg $l^{-1}$ range in sediment pore waters. As a consequence, $Mn^{2+}$ can diffuse to the oxic layer of the sediment, and via the loss of two electrons, can be oxidized to $Mn^{4+}$ by the oxygen present there.[74]

$$Mn^{2+} \rightarrow Mn^{4+} + 2e^- \tag{4.17}$$

Half-reaction for oxygen is

$$O_2 + 4H^+ + 4e^- \rightarrow 2H_2O \tag{4.18}$$

and the oxidation of manganese can be given as

$$Mn^{2+} + \tfrac{1}{2}O_2 + 2H^+ \rightarrow Mn^{4+} + H_2O \tag{4.19}$$

The $Mn^{4+}$ that is formed precipitates as manganese dioxide, and the overall redox reaction is

$$Mn^{2+} + \tfrac{1}{2}O_2 + H_2O \rightarrow MnO_{2(s)} + 2H^+ \tag{4.20}$$

This reaction occurs under aerobic conditions.[74]

### 4.1.4.1.5   Sediment Oxygen Demand

The decomposition of organic material and bioturbation activity of the macrofauna result in the exertion of an oxygen demand at the sediment–water interface.[38] This process can have profound effects on the dissolved oxygen concentration in the overlying waters. Oxygen diffuses from the overlying water into the sediment; therefore, organic matter decomposes aerobically in the upper sediment layers that are in direct contact with water.[64] The aerobic surface layer usually has a thickness of only a few millimeters. This is true even for sediment overlaid by oxygen-rich water because of the slow diffusion rates of oxygen.[19,27]

Oxygen uptake is regulated by the net deposition and degradability of organic matter.[19] In eutrophic systems, phytoplankton settling can have a significant effect on SOD levels.[64] Microbial benthic mineralization is a major recycling pathway of settled particles in shallow marine environments. Increased organic loading due to sedimentation increases oxygen consumption, which in turn leads to dissolved oxygen depletion within the sediment.[28] This depletion of oxygen causes SOD to cease.[64] In Thau Lagoon, France, the accumulation of organic matter in quantities that exceed the mineralization capacity of the sediment, high temperatures, and the absence of wind trigger anaerobic bottom conditions.[28,38]

SOD can be determined by *in situ* measurements or by model calibration if direct measurements from the field are not available. *In situ* measurements are usually conducted using a chamber. Oxygen uptake in the chamber is measured continuously over a prespecified period of time, providing the needed data to calculate the oxygen consumption rate as g $O_2$ m$^{-2}$ d$^{-1}$. In a modeling analysis, sediment oxygen demand is typically formulated as a zero-order process.[67] SOD is a function of temperature, and the Arrhenius equation can be used for temperature corrections in the 10–30°C range. Temperature correction coefficients ($\theta$) range from 1.040 to 1.130. A typical value of 1.065 is often used. Below 10°C, SOD probably decreases more rapidly and approaches zero for water in the temperature range of 0–5°C.[64] Actual temperature dependency of the sediment oxygen demand was observed in Sacca di Goro Lagoon, Italy. These followed a seasonal trend with pronounced peaks in the warmer months. Specifically, SOD was 80 mmol $O_2$ m$^2$ d$^{-1}$ in March, and increased to 365 mmol $O_2$ m$^2$ d$^{-1}$ in August at a sampling station close to a discharge.[75]

*4.1.4.1.6  Nitrification*

Nitrification may be a significant oxygen-demanding process as 4.57 mg $O_2$ per mg $NH_4^+$ is consumed during the oxidation of ammonia to nitrate.[76] Depressed oxygen levels in water can inhibit nitrification. Therefore, at least 1–2 mg l$^{-1}$ of dissolved oxygen is needed to promote nitrification. Usually in modeling studies, nitrification rates are multiplied by a factor that shuts down nitrification as the dissolved oxygen concentration approaches zero.

Changing concentrations of oxygen in the water just above the sediment has been shown to alter the penetration depth of oxygen within the sediment, and is believed to be a major controlling factor for nitrification and denitrification processes in sediment. Detailed information on nitrification is given in Section 4.1.1.2.

### 4.1.4.2   Redox Potential

Redox potential, $E_h$, refers to the relative degree of oxidation and reduction in an environment; high values indicate more oxidized conditions.[27] $E_h$ controls the change in the oxidation state of many metal ions and some nutrients.[23] In coastal marine sediment, redox potential is largely controlled by the sulfide concentration.[20]

Redox potential is one of the most important parameters characterizing surface sediment because it provides information about organic matter oxidation. The sequence of organic matter oxidation reactions is controlled by the energy gained from each particular reaction. Aerobic respiration is the most energetic redox reaction and proceeds first. When the oxygen is exhausted, denitrification, Mn-oxide reduction, Fe-oxide reduction, sulfate reduction, and methane production occur sequentially.[29] The progression to less energetic reactions also reflects a general decrease in sediment redox potential, with each reaction occurring within a certain redox potential range. Redox potential differs according to the sediment type and its location. In the Gulf of Gdansk, Poland, the sediment redox potentials were in the range of −365 – +246 mV within the system. As a result of better oxygenation, higher $E_h$ values were obtained in the sandy sediment as compared to the silty-clay sediment. These patterns in redox potentials were similar to those observed in other coastal areas like the Mediterranean and North Seas.[77]

The sediment depth at which oxygen is exhausted and the redox potential goes to zero has been defined as the redox potential discontinuity (RPD) layer. The RPD layer is visible due to its change in color and indicates the zone of habitability for benthic infauna. The closer this color change appears to the sediment surface, the lesser is the dissolved oxygen that exists in the sediment porewater.[27,61] Due to seasonal changes in the metabolic rates of benthic organisms, the boundary between the oxidized and the reduced sediment often occurs at different depths in the sediment throughout the year.[30]

## 4.1.5 MODELING OF NUTRIENT CYCLES

Nutrient cycles are presented in water quality models as a variety of mass balance equations that contain input, output, and reaction terms (kinetic process equations). In this section, mainly kinetic process equations in the EUTRO5 module of WASP5 (Water Quality Analysis Simulation Program) are presented. WASP is a dynamic compartment model that can be used to analyze a variety of water quality problems in such diverse water bodies as coastal waters, estuaries, ponds, streams, lakes, rivers, and reservoirs. This model was developed by the U.S. EPA[78] and is widely used for water quality modeling in the U.S. and in other countries.[67,79–89]

Due to the complexity of nutrient cycles in aquatic ecosystems, various models consider different processes and use different simplifications and assumptions. Therefore, the differences between WASP/EUTRO5 and some other water quality models that can be used for coastal lagoons are also evaluated and the kinetic parameters used in various models are presented in tables in this section.

EUTRO5 is a module of WASP applicable to modeling eutrophication. It simulates eight state variables (ammonium, nitrite/nitrate, orthophosphate, phytoplankton biomass, carbonaceous biochemical oxygen demand (CBOD), dissolved oxygen, nonliving organic nitrogen, and nonliving organic phosphorus) in the water column and sediment bed.[78]

In EUTRO5, phytoplankton kinetics and dissolved oxygen are considered as systems interacting with the nutrient cycles. The model characterizes the phytoplankton population as a whole by the total biomass of the phytoplankton present. Minimum formulation for nutrient limitation is used, and constant stoichiometry for algal biomass is assumed. The model does not simulate the kinetics of higher trophic level organisms such as zooplankton and fishes. Settling terms are not included in the kinetic equations in the EUTRO5 code, but are coded in the transport sections of WASP. Four types of model segments (surface water, subsurface water, upper benthic layer, and lower benthic layer) are defined in WASP. Six flow fields can be used for calculating the exchanges between segments, which are used for solving the mass balance equations.

### 4.1.5.1 Modeling Nitrogen Cycle

Four nitrogen variables are modeled in the EUTRO5 module of the WASP model: phytoplankton nitrogen (living), organic nitrogen (nonliving), ammonium, and nitrate. Both ammonium and nitrate are available for phytoplankton growth. However, ammonium is the preferred form because of physiological reasons. The ammonium preference factor used in the model is given in Equation (4.25). During phytoplankton respiration and death, a fraction of the phytoplankton biomass is recycled to the nonliving organic nitrogen pool, while the remaining fraction contributes to the ammonium pool. Although released ammonium is readily available for algal growth/nitrification, released organic nitrogen must undergo mineralization or bacterial decomposition into ammonium before utilization. EUTRO5 uses a saturating recycle rate, which is directly proportional to the phytoplankton biomass present (see Equation (4.22) and Equation (4.23)). Saturating recycle permits second-order dependency at low phytoplankton concentrations, and permits first-order recycle when the phytoplankton concentration greatly exceeds the half-saturation constant, $K_{mPC}$. Basically, this mechanism slows the recycle rate if the phytoplankton population is small but does not permit the rate to increase continuously as the phytoplankton concentration increases.

Equations describing the nitrification of ammonium in EUTRO5 module contain temperature and low dissolved oxygen correction terms. The latter is a Monod-type function, which represents the decline of nitrification as the dissolved oxygen concentration approaches zero. The denitrification equations also have a key term for dissolved oxygen. However, this term is designed to reduce the denitrification rate as a function of decreasing dissolved oxygen concentration above zero. However, in the benthic layer, where anaerobic conditions always exist, denitrification is always assumed to occur.

In AQUATOX,[73] the nitrogen cycle is modeled with two state variables: ammonium and nitrate. The atmospheric deposition of nitrate and biota excretion of ammonia are two nitrogen sources not modeled in WASP. Also in AQUATOX during the decomposition of detritus only inorganic nitrogen is released, while in EUTRO5 both inorganic and organic nitrogen are formed. Furthermore, three types of algae (blue-green, green, and diatoms) plus macrophytes are modeled in AQUATOX, and each of them has a different nitrogen uptake rate. For green algae and diatoms, the Redfield ratio may be used. Blue-green algae can supply nitrogen by fixation, and this is accounted for by smaller nitrogen uptake ratios in the model. Algae can be either phytoplankton or periphyton. Phytoplanktons are subject to sinking and washout, while periphytons are subject to substrate limitation and scour by currents because they are fixed in space. Macrophytes supply their nitrogen from sediment; therefore, nutrient limitation of macrophytes is not considered in the model. Macrophyte ammonium excretion to the water column is modeled.

Nitrification and denitrification processes described in AQUATOX have temperature and dissolved oxygen correction terms similar to those in EUTRO5; however, a pH correction term is also included in AQUATOX.

The model developed by Park and Kuo[69] uses three state variables—organic nitrogen, ammonium, and nitrite–nitrate–for modeling the nitrogen cycle. Similar to the EUTRO5, phytoplankton is considered as a single population biomass. The mineralization rate is modeled as a function of the organic nitrogen concentration. In EUTRO5 mineralization rate is modeled as a function of both the organic nitrogen and phytoplankton concentrations. In this model, the nitrification process description includes a Monod-type function to represent the effect of ammonium concentration on the nitrification rate. Similar to EUTRO5, the dissolved oxygen correction term is also a Monod-type function.

CE-QUAL-W2[58] has five different organic matter state variables. To calculate the amount of nitrogen incorporated into organic matter, two stoichiometric coefficients (nitrogen to organic matter and nitrogen to carbonaceous BOD) are used. Thus, organic nitrogen is not a state variable in this model. Two of the state variables used for modeling the nitrogen cycle are ammonium and nitrate. Organic matter decay produces ammonia. Decay of labile and refractory dissolved organic matter plus carbonaceous BOD form the elements of the dissolved organic nitrogen mineralization process, while decay of labile and refractory particulate organic matter corresponds to hydrolysis plus mineralization. Unlike EUTRO5, which simulates phytoplanktons as a single population biomass, CE-QUAL-W2 can simulate an unlimited number of algal and epiphyton groups, each group/member having different half-saturation constants and uptake rates for nitrogen. For the nitrification and denitrification, CE-QUAL-W2 and EUTRO5 both have a temperature correction term; however, their oxygen correction functions are different. In EUTRO5, the effect of dissolved oxygen concentration on nitrification and the denitrification rates are expressed as Monod-type functions. In CE-QUAL-W2, the nitrification and denitrification terms do not include such functions, but rather include discrete sign functions. These sign functions shut down nitrification if the dissolved oxygen concentration is less than a prespecified minimum value. On the contrary, denitrification shuts down when dissolved oxygen concentrations exceed the same minimum value. In CE-QUAL-W2, nitrogen fixation by blue-green algae can be modeled by setting the half-saturation concentration for nitrogen to zero for the corresponding algal group(s).

CE-QUAL-R1[57] also uses ammonium and nitrate as state variables. Similar to CE-QUAL-W2, the organic matter is represented by state variables, namely, labile dissolved organic matter, refractory dissolved organic matter, and detritus. Aerobic decomposition of the organic matter contributes to the ammonium pool. Respiration of three types of phytoplankton, one type of macrophyte, zooplankton, and fish are also included as ammonium sources. Nitrification is allowed only in layers with dissolved oxygen concentrations greater than a prespecified minimum value, which is also the case in CE-QUAL-W2.

The CE-QUAL-ICM[56] model has five state variables, namely ammonium, nitrate, dissolved organic nitrogen, labile particulate organic nitrogen, and refractory particulate organic nitrogen. Three algae groups are defined: green algae, diatoms, and blue-green algae. In CE-QUAL-ICM, the rate of dissolved organic nitrogen mineralization is related to algal biomass. When dissolved inorganic nitrogen (ammonium + nitrate) is scarce, algae stimulate production of an enzyme that mineralizes dissolved organic nitrogen to ammonium. Mineralization rate is highest when algae are strongly nitrogen limited; it is lowest when no limitation exists. Similar calculations are also made for hydrolysis of the labile and refractory particulate organic nitrogen. Similar to the case in the model developed by Park and Kuo,[69] the nitrification term in CE-QUAL-ICM includes a Monod-type function representing the effect of ammonium concentration on nitrification. Similar to EUTRO5, the dissolved oxygen correction term is also a Monod-type function.

The HEM-3D[90] model uses kinetic processes mostly from the CE-QUAL-ICM model. Macroalgae are added into HEM-3D as the fourth algal group.

The specific kinetic equations used in the EUTRO5 model for nitrogen (in almost the same or similar format as for other models) are presented below.

### 4.1.5.1.1  Phytoplankton Nitrogen
Phytoplankton nitrogen is the nitrogen contained in phytoplankton cells

$$\frac{\partial(C_4\,a_{nc})}{\partial t} = \underbrace{G_{P1}\,a_{nc}\,C_4}_{\text{Growth}} - \underbrace{D_{P1}\,a_{nc}\,C_4}_{\text{Death}} - \underbrace{\frac{v_{S4}}{D}\,a_{nc}\,C_4}_{\text{Settling}} \tag{4.21}$$

where
$C_4$ = phytoplankton carbon concentration [mg C l$^{-1}$]
$a_{nc}$ = nitrogen to carbon ratio [mg N/mg$^{-1}$ C]
$G_{P1}$ = specific growth rate constant of phytoplankton [day$^{-1}$]
$D_{P1}$ = death plus respiration rate constant of phytoplankton [day$^{-1}$]
$V_{s4}$ = net settling velocity of phytoplankton [m day$^{-1}$]
$D$ = depth of segment [m]

### 4.1.5.1.2  Organic Nitrogen

$$\frac{\partial C_7}{\partial t} = \underbrace{D_{P1}a_{nc}f_{on}C_4}_{\text{Death}} - \underbrace{k_{71}\theta_{71}^{(T-20)}\left(\frac{C_4}{K_{mPc}+C_4}\right)C_7}_{\text{Mineralization}} - \underbrace{\frac{v_{S3}(1-f_{D7})}{D}C_7}_{\text{Settling}} \tag{4.22}$$

where
$C_7$ = organic nitrogen concentration [mg N l$^{-1}$]
$D_{P1}$ = death plus respiration rate constant of phytoplankton [day$^{-1}$]
$a_{nc}$ = nitrogen to carbon ratio [mg N mg$^{-1}$ C]
$f_{on}$ = fraction of dead and respired phytoplankton recycled to the organic nitrogen pool [none]
$C_4$ = phytoplankton carbon concentration [mg C l$^{-1}$]
$k_{71}$ = organic nitrogen mineralization rate constant at 20°C [day$^{-1}$]
$\theta_{71}$ = organic nitrogen mineralization temperature coefficient [none]
$T$ = water temperature [°C]
$K_{mPC}$ = half-saturation constant for recycle [mg C l$^{-1}$]
$V_{S3}$ = organic matter settling velocity [m day$^{-1}$]
$f_{D7}$ = fraction of dissolved organic nitrogen [none]
$1 - f_{D7}$ = fraction of particulate organic nitrogen [none]
$D$ = depth of segment [m]

Literature values, obtained from various coastal areas, for the kinetic rate constants related to organic nitrogen transformations, their temperature correction coefficients, and settling velocities of organic nitrogen are given in Tables 4.2–4.5.

## TABLE 4.2
## PON to DON Rate Constant (day$^{-1}$)

| Min. | Max. | Avg. | Location | Reference | As Cited in | Explanation |
|------|------|------|----------|-----------|-------------|-------------|
| | | 0.075 | Chesapeake Bay | Cerco and Cole, 1994 | Cerco and Cole[91] | Labile PON to DON |
| | | 0.005 | Chesapeake Bay | Cerco and Cole, 1994 | Cerco and Cole[91] | Refractory PON to DON |
| 0.3 | 3 | | San Juan Bay Estuary | Bunch et al., 2000 | Bunch et al.[92] | Labile PON to DON |
| | | 0.005 | San Juan Bay Estuary | Bunch et al., 2000 | Bunch et al.[92] | Refractory PON to DON |

## TABLE 4.3
## Rate Constants for Decomposition of Organic Nitrogen to NH$_4^+$ (day$^{-1}$)

| Min. | Max. | Avg. | Location | Reference | As Cited in | Explanation |
|------|------|------|----------|-----------|-------------|-------------|
| | | 0.14 | Chesapeake Bay | Salas and Thomann, 1978 | Bowie et al.[93] | PON to NH$_4^+$ |
| | | 0.03 | Chesapeake Bay | Hydroqual Inc., 1987 | Lung and Paerl[79] | |
| 0.005 | 0.05 | | Texas bays and estuaries | Brandes, 1976 | Bowie et al.[93] | PON to NH$_4^+$ |
| | | 0.14 | Potomac River Estuary | O'Connor, 1981 | O'Connor[94] | |
| 0 | 0.3 | | James River Estuary | O'Connor, 1981 | O'Connor[94] | |
| 0.1 | 0.15 | | James River Estuary | Lung, 1986 | Lung and Paerl[79] | |
| | | 0.01 | Patuxent River Estuary | O'Connor, 1981 | O'Connor[94] | |
| | | 0.02 | Patuxent River Estuary | Lung, 1992 | Lung[80] | At 20°C, value used in modeling |
| | | 0.1 | Neuse River Estuary | Lung and Paerl, 1986 | Lung and Paerl[79] | |
| | | 0.08 | Manasquan Estuary | Najarian et al., 1984 | Najarian et al.[95] | Detritus to NH$_4^+$ |
| | | 0.015 | Chesapeake Bay | Cerco and Cole, 1994 | Cerco and Cole[91] | Given as minimum mineralization rate of DON |
| | | 0.075 | Maryland coastal waters | Lung and Hwang, 1994 | Lung and Hwang[81] | DON to NH$_4^+$, at 20°C, value used in modeling |
| 0.2 | 2 | | San Juan Bay Estuary | Bunch et al., 2000 | Bunch et al.[92] | DON to NH$_4^+$ |

**TABLE 4.4**
**Temperature Correction Factors for Decomposition of Organic Nitrogen to $NH_4^+$ ($\theta$)**

| Average | Location | Reference | As Cited in | Explanation |
|---|---|---|---|---|
| 1.03 | Texas bays and estuaries | Brandes, 1976 | Bowie et al.[93] | (1.02–1.04), PON to $NH_4^+$ |
| 1.045 | | O'Connor, 1981 | O'Connor[94] | |
| 1.051 | Manasquan Estuary | Najarian et al., 1984 | Najarian et al.[95] | Detritus to $NH_4^+$ |
| 1.08 | Patuxent River Estuary | Lung, 1992 | Lung[80] | |
| 1.08 | Maryland coastal waters | Lung and Hwang, 1994 | Lung and Hwang[81] | DON to $NH_4^+$, value used in modeling |

**TABLE 4.5**
**Settling Velocity of Organic Nitrogen (m day$^{-1}$)**

| Min. | Max. | Avg. | Location | Reference | As Cited in |
|---|---|---|---|---|---|
| | | 0.23 | James River | O'Connor, 1981 | O'Connor[94] |
| | | 0.4 | Patuxent River | O'Connor, 1981 | O'Connor[94] |
| | | 0.15 | Patuxent River Estuary | Lung, 1992 | Lung[80] |
| | | 0.6 | Sacramento San Joaquin Delta | O'Connor, 1981 | O'Connor[94] |

*4.1.5.1.3 Ammonium Nitrogen*

$$\frac{\partial C_1}{\partial t} = \underbrace{D_{P1}a_{nc}(1-f_{on})C_4}_{\text{Death}} - \underbrace{k_{71}\theta_{71}^{(T-20)}\left(\frac{C_4}{K_{mPc}+C_4}\right)C_7}_{\text{Mineralization}} - \underbrace{G_{P1}a_{nc}P_{NH3}C_4}_{\text{Nitrification}}$$

$$\underbrace{-k_{12}\theta_{12}^{(T-20)}\left(\frac{C_6}{K_{NIT}+C_6}\right)C_1}_{\text{Growth}}$$

(4.23)

where

$C_1$ = ammonium nitrogen concentration [mg N l$^{-1}$]

$D_{P1}$ = death plus respiration rate constant of phytoplankton [day$^{-1}$]

$a_{nc}$ = nitrogen to carbon ratio [mg N/mg$^{-1}$ C]

$1-f_{on}$ = fraction of dead and respired phytoplankton recycled to the ammonium nitrogen pool [none]

$C_4$ = phytoplankton carbon concentration [mg C l$^{-1}$]

$k_{71}$ = organic nitrogen mineralization rate constant at 20°C [day$^{-1}$]

$\theta_{71}$ = organic nitrogen mineralization temperature coefficient [none]

$T$ = water temperature [°C]

$K_{mPC}$ = half-saturation constant for recycle [mg C l$^{-1}$]

$C_7$ = organic nitrogen concentration [mg N l$^{-1}$]

$G_{P1}$ = specific growth rate constant of phytoplankton [day$^{-1}$]

$P_{NH3}$ = preference for ammonium uptake [none]

$k_{12}$ = nitrification rate constant at 20°C [day$^{-1}$]

$\theta_{12}$ = nitrification temperature coefficient [none]

$K_{NIT}$ = half-saturation constant for oxygen limitation of nitrification [mg O$_2$ l$^{-1}$]

$C_6$ = dissolved oxygen concentration [mg O$_2$ l$^{-1}$]

Table 4.6 and Table 4.7 contain nitrification rate constants and temperature correction factors for nitrification in coastal waters, respectively.

### 4.1.5.1.4 Nitrate Nitrogen

$$\frac{\partial C_2}{dt} = \underbrace{k_{12}\theta_{12}^{(T-20)}\left(\frac{C_6}{K_{NIT}+C_6}\right)C_1}_{\text{Nitrification}} - \underbrace{G_{P1}a_{nc}(1-P_{NH3})\,C_4}_{\text{Growth}} - \underbrace{k_{2D}\theta_{2D}^{(T-20)}\left(\frac{K_{NO3}}{K_{NO3}+C_6}\right)C_2}_{\text{Denitrification}}$$

$$(4.24)$$

where

$$P_{NH3} = C_1\left(\frac{C_2}{(K_{mN}+C_1)(K_{mN}+C_2)}\right) + C_1\left(\frac{K_{mN}}{(C_1+C_2)(K_{mN}+C_2)}\right) \qquad (4.25)$$

Ammonium preference factor

$C_2$ = nitrate–nitrogen concentration [mg N l$^{-1}$]

$k_{12}$ = nitrification rate constant at 20°C [day$^{-1}$]

$\theta_{12}$ = nitrification temperature coefficient [none]

$T$ = water temperature [°C]

$K_{NIT}$ = half-saturation constant for oxygen limitation of nitrification [mg O$_2$ l$^{-1}$]

$C_6$ = dissolved oxygen concentration [mg O$_2$ l$^{-1}$]

$C_1$ = ammonium nitrogen concentration [mg N l$^{-1}$]

$G_{P1}$ = specific growth rate constant of phytoplankton [day$^{-1}$]

$a_{nc}$ = nitrogen to carbon ratio [mg N mg$^{-1}$ C]

$C_4$ = phytoplankton carbon concentration [mg C l$^{-1}$]

$k_{2D}$ = denitrification rate constant at 20°C [day$^{-1}$]

$\theta_{2D}$ = denitrification temperature coefficient [none]

$K_{NO3}$ = half-saturation constant for oxygen limitation of denitrification [mg O$_2$ l$^{-1}$]

$K_{mN}$ = half-saturation constant for inorganic nitrogen [mg N l$^{-1}$]

**TABLE 4.6**
**Nitrification Rate Constant (day$^{-1}$)**

| Min. | Max. | Avg. | Location | Reference | As Cited in | Explanation |
|---|---|---|---|---|---|---|
| 0.1 | 0.14 | | Potomac River Estuary | Slayton and Trovato, 1978, 1979 | Bowie et al.[93] | |
| 0.09 | 0.13 | | Potomac River Estuary | Thomann and Fitzpatrick, 1982 | Bowie et al.[93] | |
| | | 0.2 | Potomac River Estuary | O'Connor, 1981 | O'Connor[94] | At 20°C |
| 0.09 | 0.54 | 0.3 | Delaware River Estuary | Bansal, 1976 | Bowie et al.[93] | |
| 0.004 | 0.11 | | Delaware River Estuary | Lipschultz et al., 1986 | Cerco and Cole[91] | |
| | | 0.2 | New York Bight | O'Connor, 1981 | O'Connor[94] | At 20°C |
| 0 | 0.15 | | James River Estuary | O'Connor, 1981 | O'Connor[94] | At 20°C |
| 0.05 | 0.15 | | James River Estuary | Lung, 1986 | Lung and Paerl[79] | At 20°C |
| | | 0.027 | James River Estuary | Kator, 1990 | Cerco and Cole[91] | |
| | | 0.05 | Neuse River Estuary | Lung and Paerl, 1986 | Lung and Paerl[79] | At 20°C |
| | | 0.05 | Chesapeake Bay | Hydroqual Inc., 1987 | Lung and Paerl[79] | At 20°C |
| | | 0.15 | Manasquan Estuary | Najarian et al., 1984 | Najarian et al.[95] | |
| | | 0.1 | Venice Lagoon | Melaku Canu et al., 2001 | Melaku Canu et al.[88] | At 20°C |
| | | 0.043 | Tamar Estuary | Owens, 1986 | Cerco and Cole[91] | |
| | | 0.03 | Patuxent River Estuary | Lung, 1992 | Lung[80] | At 20°C, value used in modeling |
| | | 0.08 | Maryland coastal waters | Lung and Hwang 1994 | Lung and Hwang[81] | At 20°C, value used in modeling |

## TABLE 4.7
## Nitrification Temperature Correction Factor ($\theta$)

| Value | Location | Reference | As Cited in |
|---|---|---|---|
| 1.045 | Potomac River Estuary | Thomann and Fitzpatrick, 1982 | Bowie et al.[93] |
| 1.05 | Manasquan Estuary | Najarian et al., 1984 | Najarian et al.[95] |
| 1.08 | Venice Lagoon | Melaku Canu et al., 2001 | Melaku Canu et al.[88] |
| 1.08 | Patuxent River Estuary | Lung, 1992 | Lung[80] |
| 1.08 | Maryland coastal waters | Lung and Hwang, 1994 | Lung and Hwang[81] |

Table 4.8 and Table 4.9 contain denitrification rate constants and temperature correction factors for denitrification in coastal waters, respectively. Nitrogen fixation rate constants used for coastal areas are given in Table 4.10.

The EUTRO5 model also includes benthic equations because the decomposition of organic material in benthic sediment can have profound effects on the concentrations of oxygen and nutrients in the overlying waters.

In CE-QUAL-W2,[58] two processes are defined to simulate ammonium release from sediment. One process is a first-order process that is coupled with the aerobic organic matter decay in sediment. The other is a zero-order process that occurs when the

## TABLE 4.8
## Denitrification Rate Constant (day$^{-1}$)

| Average | Location | Reference | As Cited in | Explanation |
|---|---|---|---|---|
| 0.09 | Potomac River Estuary | Thomann and Fitzpatrick, 1982 | Bowie et al.[93] | |
| 0.1 | New York Bight | O'Connor et al., 1981 | Bowie et al.[93] | |
| 0.1 | New York Bight | O'Connor, 1981 | O'Connor[94] | At 20°C |
| 0.09 | Patuxent River Estuary | Lung, 1992 | Lung[80] | At 20°C |
| 0.01 | Maryland coastal waters | Lung and Hwang, 1994 | Lung and Hwang[81] | At 20°C, model value |

## TABLE 4.9
## Denitrification Temperature Correction Factor ($\theta$)

| Average | Location | Reference | As Cited in |
|---|---|---|---|
| 1.045 | Potomac River Estuary | Thomann and Fitzpatrick, 1982 | Bowie et al.[93] |
| 1.045 | New York Bight | O'Connor et al., 1981 | Bowie et al.[93] |
| 1.045 | Patuxent River Estuary | Lung, 1992 | Lung[80] |
| 1.08 | Maryland coastal waters | Lung and Hwang, 1994 | Lung and Hwang[81] |

**TABLE 4.10**
**Nitrogen Fixation Rate Constant (g N m$^{-2}$ yr$^{-1}$)**

| Min. | Max. | Avg. | Location | Reference | As Cited in | Explanation |
|------|------|------|----------|-----------|-------------|-------------|
| | | 0.06 | S.W. Bothnian Sea (Baltic) | Lindahl and Wallstrom, 1985 | Howarth et al.[31] | |
| | | 0.07 | Aland Sea (Baltic) | Lindahl and Wallstrom, 1985 | Howarth et al.[31] | |
| | | 0.8 | Asko Area (Baltic) | Lindahl and Wallstrom, 1985 | Howarth et al.[31] | |
| 0.013 | 1.8 | | Stockholm Archipelago, Sweden | Brattberg, 1977 | Howarth et al.[31] | At 8 stations, 65 day yr$^{-1}$ bloom period |
| | | 1.2 | Harvey Estuary, Australia | Huber, 1986 | Howarth et al.[31] | Assumed to occur during Nodularia bloom |
| | | 0.002 | Vostok Bay, Japan | Odintsov, 1981 | Herbert[8] | |
| | | 0.01 | Upper Cook Inlet, Alaska | Haines et al., 1981 | Howarth et al.[31] | Calculated values for macrophyte-free sediments and well-developed cyanobacterial mats |
| | | 0.03 | Narragansett Bay, Rhode Island | Seitzinger and Garber, 1987 | Herbert[8] | Calculated values for macrophyte-free sediments and well-developed cyanobacterial mats |
| | | 0.03 | Kamishak Bay, Alaska | Haines et al., 1981 | Howarth et al.[31] | Calculated values for macrophyte-free sediments and well-developed cyanobacterial mats |
| | | 0.03 | Norton Sound, Alaska | Haines et al., 1981 | Howarth et al.[31] | Calculated values for macrophyte-free sediments and well-developed cyanobacterial mats |
| | | 0.03 | Beaufort Sea | Knowles and Wishart, 1977 | Howarth et al.[31] | Calculated values for macrophyte-free sediments and well-developed cyanobacterial mats |
| | | 0.07 | Elso Lagoon, Beaufort Sea | Haines et al., 1981 | Howarth et al.[31] | Calculated values for macrophyte-free sediments and well-developed cyanobacterial mats |
| | | 0.08 | Shelikoff Strait, Alaska | Haines et al., 1981 | Howarth et al.[31] | Calculated values for macrophyte-free sediments and well-developed cyanobacterial mats |

| | | | | |
|---|---|---|---|---|
| 0.13 | Rhode River Estuary, Maryland | Marsho et al., 1975 | Herbert[8] | |
| 0.14 | Lune Estuary, U.K. | Jones, 1982 | Howarth et al.[31] | Calculated values for macrophyte-free sediments and well-developed cyanobacterial mats |
| 0.37 | Waccasassa Estuary, Florida | Brooks et al., 1971 | Howarth et al.[31] | Calculated values for macrophyte-free sediments and well-developed cyanobacterial mats |
| 0.43 | Bank End, U.K. | Jones, 1974 | Herbert[8] | |
| 0.6 | Kaneohe Bay, Hawaii | Hanson and Gunderson, 1977 | Herbert[8] | |
| 1.56 | Barataria Basin, Louisiana | Casselman et al., 1981 | Howarth et al.[31] | Calculated values for macrophyte-free sediments and well-developed cyanobacterial mats |
| 0.24 | Chesapeake Bay | Marsho et al., 1975 | Howarth et al.[31] | In sediments of wetland and macrophyte beds—salt marshes |
| 1.7 | Lawrencetown Marsh, Nova Scotia | Patriquin and Keddy, 1978 | Howarth et al.[31] | In sediments of wetland and macrophyte beds—salt marshes |
| 2.3 | Horn Point, Maryland | Lipschultz, 1978 | Howarth et al.[31] | In sediments of wetland and macrophyte beds—salt marshes |
| 7.7 | Barataria Basin, Louisiana | Casselman et al., 1981 | Howarth et al.[31] | In sediments of wetland and macrophyte beds—salt marshes |

dissolved oxygen concentration is less than a prespecified minimum value at which anaerobic processes begin. The minimum dissolved oxygen default value is 0.1 mg l[-1].

CE-QUAL-R1[57] models macrophytes. Thus, ammonium and nitrate uptake by macrophytes from sediment is included. Anaerobic ammonium release, aerobic decomposition of organic matter, and settling of adsorbed ammonium are also considered in the sediment compartment model balance equations.

Both the CE-QUAL-ICM[56] and HEM-3D[90] models use the same predictive sediment submodel developed by Cerco and Cole[56] for the Chesapeake Bay eutrophication model study. This submodel consists of three basic processes: deposition of particulate organic matter to the sediment, organic matter diagenesis (decay), and release of nutrients (phosphate, ammonium, and nitrate) to the overlying water column or their burial into deeper sediment.

Benthic equations for organic, ammonium, and nitrate nitrogen are given in the following subsections.

### 4.1.5.1.5 Organic Nitrogen (Benthic):

$$\frac{\partial C_7}{\partial t} = \underbrace{k_{PZD} \; \theta_{PZD}^{(T-20)} a_{nc} \, f_{on} C_4}_{\text{Algal decomposition}} - \underbrace{k_{OND} \theta_{OND}^{(T-20)} C_7}_{\text{Mineralization}} \tag{4.26}$$

where

$C_7$ = organic nitrogen concentration [mg N l[-1]]
$k_{PZD}$ = anaerobic algal decomposition rate constant at 20°C [per day[-1]]
$\theta_{PZD}$ = anaerobic algal decomposition temperature coefficient [none]
$T$ = temperature [°C]
$a_{nc}$ = nitrogen to carbon ratio [mg N mg[-1] C]
$f_{on}$ = fraction of dead and respired phytoplankton recycled to the organic nitrogen pool [none]
$C_4$ = phytoplankton carbon concentration [mg C l[-1]]
$k_{OND}$ = organic nitrogen decomposition rate constant at 20°C [day[-1]]
$\theta_{OND}$ = organic nitrogen decomposition temperature coefficient [none]

### 4.1.5.1.6 Ammonia Nitrogen (Benthic)

$$\frac{\partial C_1}{\partial t} = \underbrace{k_{PZD} \; \theta_{PZD}^{(T-20)} a_{nc} (1 - f_{on}) C_4}_{\text{Algal decomposition}} + \underbrace{k_{OND} \theta_{OND}^{(T-20)} C_7}_{\text{Mineralization}} \tag{4.27}$$

where

$C_1$ = ammonium nitrogen concentration [mg N l[-1]]
$k_{PZD}$ = anaerobic algal decomposition rate constant at 20°C [day[-1]]
$\theta_{PZD}$ = anaerobic algal decomposition temperature coefficient [none]
$T$ = temperature [°C]
$a_{nc}$ = nitrogen to carbon ratio [mg N mg[-1] C]
$1 - f_{on}$ = fraction of dead and respired phytoplankton recycled to the ammonia nitrogen pool [none]

$C_4$ = phytoplankton carbon concentration [mg C l$^{-1}$]

$k_{OND}$ = organic nitrogen decomposition rate constant at 20°C [day$^{-1}$]

$\theta_{OND}$ = organic nitrogen decomposition temperature coefficient [none]

Ammonium nitrogen and organic nitrogen transformation rate constants in the sediment of coastal waters and related temperature correction factors are given in Tables 4.11–4.13. Although EUTRO5 does not include nitrification in sediment, the values given in Table 4.11 can be useful for other models.

**TABLE 4.11**
**Sediment Nitrification Rate (mg N m$^{-2}$ day$^{-1}$)**

| Min. | Max. | Avg. | Location | Reference | As Cited in |
|------|------|------|----------|-----------|-------------|
| 2 | 9 | | North Sea sediments | Goldhaber, 1977 | Herbert[8] |
| | | 20 | Kingoodie Bay, U.K. | Macfarlene and Herbert, 1984 | Herbert[8] |
| | | 38 | Lim Fjord sediments, Denmark | Jorgensen and Revsbech, 1989 | Herbert[8] |
| | | 112 | Norsminde Fjord, Denmark | Binnerup et al., 1992 | Herbert[8] |
| | | 84 | Ochlockonee Bay, Florida | Seitzinger, 1987 | Herbert[8] |
| | | 23 | Narragansett Bay, U.S.A. | Seitzinger et al., 1984 | Herbert[8] |
| 14 | 23 | | Chesapeake Bay, U.S.A. | Henriksen and Kemp, 1988 | Herbert[8] |
| | | 26 | Patuxent River, U.S.A. | Jenkins and Kemp, 1984 | Jenkins and Kemp[96] |
| 23 | 27 | | Kysing Fjord, Denmark | Hansen et al., 1981 | Herbert[8] |
| 38 | 42 | | Odawa Bay, Japan | Koike and Hattori, 1978 | Herbert[8] |

**TABLE 4.12**
**Rate Coefficient for Organic Nitrogen Decomposition in Sediments to NH$_4^+$ (day$^{-1}$)**

| Min. | Max. | Avg. | Location | Reference | As Cited in | Explanation |
|------|------|------|----------|-----------|-------------|-------------|
| 0.001 | 0.02 | | Texas bays and estuaries | Brandes, 1976 | Bowie et al.[93] | |
| | | 0.00013 | Sacca di Scardovri, Italy | Viel et al., 1991 | Viel et al.[97] | Anoxic conditions in the sediment |
| | | 0.02 | Patuxent River Estuary, U.S.A. | Lung, 1992 | Lung[80] | Given as org N decomposition rate, value used in modeling |

**TABLE 4.13**
**Temperature Correction Factor for Organic Nitrogen
Decomposition in Sediments to NH$_4^+$ ($\theta$)**

| Average | Location | Reference | As Cited in |
|---------|----------|-----------|-------------|
| 1.04 | Texas bays and estuaries | Brandes, 1976 | Bowie et al.[93] |
| 1.18 | Patuxent River Estuary | Lung, 1992 | Lung[80] |

Sediment NH$_4^+$ release rates for coastal marine environments are given in Table 4.14.

### 4.1.5.1.7  Nitrate Nitrogen (Benthic)

$$\frac{\partial C_2}{\partial t} = \underbrace{-k_{2D}\theta_{2D}^{(T-20)}C_2}_{\text{Denitrification}}$$

(4.28)

where

$C_2$ = nitrate nitrogen concentration [mg N l$^{-1}$]

$k_{2D}$ = denitrification rate constant at 20°C [day$^{-1}$]

$\theta_{2D}$ = denitrification temperature coefficient [none]

$T$ = water temperature [°C]

Sediment oxidized nitrogen release rates compiled from the literature are given in Table 4.15. Half-saturation concentrations for nitrate and sediment denitrification rates are presented in Tables 4.16–4.17.

## 4.1.5.2  Modeling of Phosphorus Cycle

The three phosphorus variables modeled in EUTRO5 are phytoplankton phosphorus (phosphorus incorporated in phytoplankton cells), organic phosphorus, and inorganic phosphorus (orthophosphate). Organic and inorganic phosphorus are divided into their particulate and dissolved fractions.

Dissolved inorganic phosphorus is incorporated into phytoplankton cells during growth. During phytoplankton respiration and death, organic and dissolved inorganic phosphorus are released. Upon release, dissolved inorganic phosphorus is readily available for algal growth and the released organic phosphorus must undergo mineralization or bacterial decomposition before it can be utilized by phytoplankton. Similar to organic nitrogen mineralization, the organic phosphorus mineralization term in EUTRO5 includes a saturating recycle rate that is directly proportional to the phytoplankton biomass present.

**TABLE 4.14**
**Sediment $NH_4^+$ Release Rate (mg N $m^{-2}$ $day^{-1}$)**

| Min. | Max. | Avg. | Location | Reference | As Cited in | Explanation |
|---|---|---|---|---|---|---|
| | | 0.24 | Manasquan Estuary | Najarian et al., 1984 | Najarian et al.[95] | |
| | | 8.736 | Ochlockonee Bay, Florida | Seitzinger, 1987b | Seitzinger[18] | Average measurements of all cores throughout the year |
| | | 30.576 | Narragansett Bay | Seitzinger et al., 1984; Nixon et al., 1976 | Seitzinger[18] | Annual average of mid-bay data |
| | | 24.528 | North Sea, Belgian coast | Billen, 1978 | Seitzinger[18] | Annual average |
| | | 156.912 | Patuxent Estuary | Henriksen and Kemp, 1988 | Seitzinger[18] | |
| | | 9.408 | Tejo Estuary, upper bay, Portugal | Seitzinger, unpublished data | Seitzinger[18] | Muddy sediments |
| | | 13.9 | St. Leonard's Creek, Patuxent River Estuary | Boynton et al., 1990 | Lung[80] | May 1985 |
| | | 9.27 | St. Leonard's Creek, Patuxent River Estuary | Boynton et al., 1990 | Lung[80] | June 1985 |
| | | 17.9 | St. Leonard's Creek, Patuxent River Estuary | Boynton et al., 1990 | Lung[80] | August 1985 |
| | | 8.77 | St. Leonard's Creek, Patuxent River Estuary | Boynton et al., 1990 | Lung[80] | October 1985 |
| | | 6.37 | Buena Vista, Patuxent River Estuary | Boynton et al., 1990 | Lung[80] | May 1985 |
| | | 31.5 | Buena Vista, Patuxent River Estuary | Boynton et al., 1990 | Lung[80] | June 1985 |
| | | 26.7 | Buena Vista, Patuxent River Estuary | Boynton et al., 1990 | Lung[80] | August 1985 |
| | | 21.5 | Buena Vista, Patuxent River Estuary | Boynton et al., 1990 | Lung[80] | October 1985 |
| | | 12.5 | Isle of Wight Bay, Maryland, U.S.A. | Lung and Hwang, 1994 | Lung and Hwang[81] | |
| | | 70 | Turnville Creek/St. Martin River, Maryland, U.S.A. | Lung and Hwang, 1994 | Lung and Hwang[81] | |
| | | 120 | Bishopsville Prong, Maryland, U.S.A. | Lung and Hwang, 1994 | Lung and Hwang[81] | |
| | | 90 | Assawoman Bay, Maryland, U.S.A. | Lung and Hwang, 1994 | Lung and Hwang[81] | |
| | | 120 | Upper Assawoman Bay, Maryland, U.S.A. | Lung and Hwang, 1994 | Lung and Hwang[81] | |
| 23.52 | 288.96 | 141.12 | Cape Lookout Bight, North Carolina, U.S.A. | Martens, 1993 | Martens[98] | |
| 0 | 15.79 | | Potomac River Estuary | Seitzinger, 1986 | Seitzinger[99] | At pH 8 |

## TABLE 4.15
## Sediment ($NO_2^-$ + $NO_3^-$) Release Rate (mg N m$^{-2}$ day$^{-1}$)

| Average | Location | Reference | As Cited in | Explanation |
|---|---|---|---|---|
| 6.048 | Ochlockonee Bay, Florida | Seitzinger, 1987b | Seitzinger[18] | Average measurements of all cores throughout the year |
| 3.36 | Narragansett Bay | Seitzinger et al., 1984; Nixon et al., 1976 | Seitzinger[18] | Annual average of mid-bay data |
| 32.928 | North Sea, Belgian coast | Billen, 1978 | Seitzinger[18] | Annual average |
| −14.112 | Patuxent Estuary | Henriksen and Kemp, 1988 | Seitzinger[18] | |
| 71.232 | Tejo Estuary, upper bay, Portugal | Seitzinger unpublished data | Seitzinger[18] | Muddy sediments |
| 0 | St. Leonard's Creek, Patuxent River Estuary | Boynton et al., 1990 | Lung[80] | May 1985 |
| 9.98 | St. Leonard's Creek, Patuxent River Estuary | Boynton et al., 1990 | Lung[80] | June 1985 |
| 4.93 | St. Leonard's Creek, Patuxent River Estuary | Boynton et al., 1990 | Lung[80] | August 1985 |
| 1.65 | St. Leonard's Creek, Patuxent River Estuary | Boynton et al., 1990 | Lung[80] | October 1985 |
| 3.13 | Buena Vista, Patuxent River Estuary | Boynton et al., 1990 | Lung[80] | May 1985 |
| 0 | Buena Vista, Patuxent River Estuary | Boynton et al., 1990 | Lung[80] | June 1985 |
| 2.37 | Buena Vista, Patuxent River Estuary | Boynton et al., 1990 | Lung[80] | October 1985 |

There is an adsorption–desorption interaction between dissolved inorganic phosphorus and suspended particulate matter in the water column. The subsequent settling of the suspended solids, together with the sorbed inorganic phosphorus, can act as a significant phosphorus loss mechanism from the water column and as a source of phosphorus to the sediment. Because the rates of adsorption–desorption are on the order of minutes vs. reaction rates on the order of days for the biological

**TABLE 4.16**
**Sediment Denitrification Half-Saturation Concentration**
**for Nitrate ($\mu M$ NO$_3^-$)**

| Min. | Max. | Avg. | Location | Method of Determination | Reference | As Cited in | Explanation |
|---|---|---|---|---|---|---|---|
| | | 24 | Tokyo Bay, Japan | Slurry technique | Koike et al., 1978 | Hattori[14] | |
| | | 50 | San Francisco Bay | Slurry technique | Oremland et al., 1984 | Seitzinger[18] | 20°C |
| | | 344 | Kysing Fjord, Denmark | Slurry technique | Oren and Blackburn, 1979 | Seitzinger[18] | 12°C |
| | | 53 | Izembek Lagoon, Alaska | Slurry technique | Iizumi et al., 1980 | Seitzinger[18] | 11°C–15°C |
| | | 50 | Belgian coast | Slurry technique | Billen, 1978 | Seitzinger[18] | |
| 27 | 42 | | Mangoku-Ura, Japan | | Koike et al., 1978 | Hattori[14] | |

processes, an equilibrium adsorption–desorption assumption can be made. This equilibrium assumption means that the dissolved and particulate phosphorus phases "instantaneously" react to any discharge sources of phosphorus, or run-off, or shoreline erosion of solids so as to redistribute the phosphorus to its "equilibrium" dissolved and solids phase concentrations. Sorption of inorganic phosphorus onto particulates and the subsequent particulate settling process are both included in the transport codes of WASP. The sorption term, therefore, is not included in the EUTRO5 kinetic equations.

In AQUATOX[73] the phosphorus cycle has only one essential state variable, orthophosphate. Atmospheric deposition and biota excretion are two phosphate sources in AQUATOX that are not modeled in WASP/EUTRO5. During decomposition of detritus, only inorganic phosphorus is released, while in EUTRO5 both inorganic and organic phosphorus are formed. Three types of algae (blue-green, green, and diatoms) and macrophytes are modeled, each with different phosphate uptake rates. Macrophytes supply their phosphorus from sediment and excrete phosphate to the water column.

The model developed by Park and Kuo[69] uses two state variables (organic and inorganic phosphorus) for modeling the phosphorus cycle. Similar to the EUTRO5 model, phytoplankton is considered as a single population biomass. Mineralization rate is a function of organic phosphorus concentration, while in EUTRO5 the mineralization rate is a function of both the organic phosphorus and phytoplankton concentrations present.

**TABLE 4.17**
**Sediment Denitrification Rate (mg N m$^{-2}$ day$^{-1}$)**

| Min. | Max. | Avg. | Location | Method of Determination | Reference | As Cited in | Explanation |
|---|---|---|---|---|---|---|---|
| | | 29.90 | St. Leonard's Creek, Patuxent River Estuary | | Jenkins and Kemp, 1984 | Jenkins and Kemp[96] | |
| | | 25.87 | Buena Vista, Patuxent River Estuary | | Jenkins and Kemp, 1984 | Jenkins and Kemp[96] | |
| 1.18 | 11.59 | | Izembek Lagoon, Alaska | | Iizumi et al., 1980 | Seitzinger[18] | |
| 2.02 | 4.03 | | Delaware Inlet, New Zealand | | Kaspar, 1983 | Seitzinger[18] | |
| 0.00 | 87.36 | 32.93 | Delaware Bay | | Seitzinger, unpublished data | Seitzinger[18] | |
| 2.02 | 11.76 | | Kenepuru Sound, New Zealand | | Kaspar et al., 1985b | Seitzinger[18] | |
| 48.72 | 199.58 | | Tamar Estuary | | Nishio et al., 1983 | Seitzinger[18] | |
| 43.68 | 265.44 | 164.64 | Tamar Estuary | Determined by continuous flow sediment–water system, $^{15}N$ technique | Nishio, 1982; Nishio et al., 1982 | Hattori[14] | May 1980–June 1981 |
| | | 204.96 | Tamar Estuary | Estimated from bottle incubation data | Nishio et al., 1981 | Hattori[14] | December 1978 |
| | | 12.43 | Odawa Bay, Japan | | Nishio et al., 1983 | Seitzinger[18] | |
| 0.07 | 13.10 | 4.37 | Odawa Bay, Japan | Determined by continuous flow sediment-water system, $^{15}N$ technique | Nishio, 1982; Nishio et al., 1982 | Hattori[14] | April 1980–June 1981 |

| | | | Location | Method | Reference | Source | Date |
|---|---|---|---|---|---|---|---|
| 5.38 | 11.09 | 8.40 | Tokyo Bay, Japan | Determined by continuous flow sediment–water system, [15]N technique | Nishio et al., 1982 | Hattori[14] | September 1980 |
| 24.86 | 40.32 | 32.59 | Tokyo Bay, Japan | Estimated from bottle incubation data | Nishio, 1982 | Hattori[14] | September 1980 |
| | 2.32 | | Kysing Fjord, Denmark | Estimated from bottle incubation data | Oren and Blackburn, 1979 | Hattori[14] | October 1977 |
| 0.34 | 110.88 | | Kysing Fjord, Denmark | Acetylene inhibition technique | Sorensen 1978b; Sorensen et al., 1979 | Seitzinger[18] | |
| 35.95 | 358.51 | 77.95 | Tejo Estuary, Portugal | | Seitzinger unpublished data | Seitzinger[18] | |
| 0.00 | 70.56 | 25.20 | Ochlockonee Bay, Florida | | Seitzinger, 1987b | Seitzinger[18] | |
| 13.10 | 36.62 | 19.82 | Narragansett Bay, Rhode Island | | Seitzinger et al., 1984 | Seitzinger et al.[100] | |
| 50.00 | 655.00 | | Narragansett Bay, Rhode Island | N$_2$ flux method | Nowicki, 1994 | Herbert[8] | |
| 1.01 | 5.38 | | Great Bay, Long Island | | Slater and Capone, 1987 | Seitzinger[18] | |
| | | 1.95 | Randers Fjord | Acetylene inhibition technique | Sorensen et al., 1979 | Hattori[14] | June 1978 |
| | | 13.78 | Randers Fjord | Acetylene inhibition technique | Sorensen, 1978b | Hattori[14] | January 1978 |
| 0.00 | 28.90 | 16.80 | Belgian coast, North Sea | Estimated from bottle incubation data | Billen, 1978 | Hattori[14] | March 1974–June 1976 |
| 4.03 | 71.57 | | Landrup Vig | | Andersen et al., 1984 | Seitzinger[18] | |
| 0.67 | 24.86 | | Four League Bay | | Smith et al., 1985 | Seitzinger[18] | |

(Continued)

**TABLE 4.17 (Continued)**
**Sediment Denitrification Rate (mg N m$^{-2}$ day$^{-1}$)**

| Min. | Max. | Avg. | Location | Method of Determination | Reference | As Cited in | Explanation |
|---|---|---|---|---|---|---|---|
| 102.48 | 298.37 | | MERL mesocosm | | Seitzinger and Nixon, 1985 | Seitzinger[18] | |
| 0.13 | 47.38 | | West coast, New Zealand | | Kaspar et al., 1985a | Seitzinger[18] | |
| 0.27 | 0.40 | | San Francisco Bay | | Oremland et al., 1984 | Seitzinger[18] | |
| | | 188.16 | Lim Fjord, Denmark | Estimated from bottle incubation data, and acetylene inhibition technique | Sorensen, 1978a | Hattori[14] | April 1977 |
| 0.00 | 120.96 | | New England Coast | Estimated from bottle incubation data, and settling a bell jar on sediments | Kaplan et al., 1979 | Hattori[14] | January 1975–March 1976 |
| 225 | 702 | | Chesapeake Bay | Acetylene inhibition technique | Caffrey and Kemp, 1990 | Herbert[8] | Zostera Marina vegetated |
| 20 | 739 | | Chesapeake Bay | Acetylene inhibition technique | Caffrey and Kemp, 1990 | Herbert[8] | Nonvegetated |
| 15 | 116 | | Guadalupe Estuary | $N_2$ flux method | Yoon and Benner, 1992 | Herbert[8] | |
| 248 | 1401 | | Norsminde Fjord | Acetylene inhibition technique | Jorgensen and Sorensen, 1988 | Herbert[8] | |
| 14 | 224 | | Norsminde Fjord | $^{15}N$ isotope pairing method | Nielsen et al., 1995 | Herbert[8] | |
| 0 | 59 | | Arcachon Bay | $^{15}N$ isotope pairing method | Rysgaard et al., 1996 | Herbert[8] | |

CE-QUAL-W2[58] has five different organic matter state variables. To calculate the amount of phosphorus incorporated into organic matter, two stoichiometric coefficients (phosphorus to organic matter and phosphorus to carbonaceous BOD) are used. Thus, organic phosphorus is not a state variable in this model. Organic matter decay produces orthophosphate, which is the main phosphorus state variable. Decay of labile and refractory dissolved organic matter and carbonaceous BOD corresponds to organic phosphorus mineralization process while decay of labile and refractory particulate organic matter corresponds to hydrolysis plus mineralization. Unlike EUTRO5, which simulates phytoplanktons as a single population biomass, CE-QUAL-W2 model can simulate an unlimited number of algal and epiphyton groups, each group/member having different half-saturation constants and uptake rates for phosphorus.

CE-QUAL-R1[57] has phosphate as the main state variable. Similar to CE-QUAL-W2, organic matter is represented by three state variables, namely, labile dissolved organic matter, refractory dissolved organic matter, and detritus. The aerobic decomposition of these three materials contributes to the phosphate pool. Respiration of three types of phytoplankton, one type of macrophyte, zooplankton, and fish are also included in phosphate sources.

The CE-QUAL-ICM[56] model has four state variables: total phosphate, dissolved organic phosphorus, labile particulate organic phosphorus, and refractory particulate organic phosphorus. Total phosphate comprises dissolved phosphate, particulate phosphate, and algal phosphorus. Three algae groups are defined: green algae, diatoms, and blue-greens. In CE-QUAL-ICM, the rate of dissolved organic phosphorus mineralization is related to algal biomass. When phosphate is scarce, algae stimulate production of an enzyme that mineralizes dissolved organic phosphorus to phosphate. Mineralization is highest when algae are strongly phosphorus limited and is the lowest when there is no limitation. Similar calculations are also made for hydrolysis of labile and refractory particulate organic phosphorus.

The HEM-3D[90] model uses kinetic process descriptions mostly from CE-QUAL-ICM model. Macroalgae are added to HEM-3D as the fourth algal group.

The mass balance model equations for various phosphorus forms are presented in the following subsections.

### 4.1.5.2.1  Inorganic Phosphorus

$$\frac{\partial(C_3)}{\partial t} = \underbrace{D_{P1}\, a_{\mathrm{PC}}(1 - f_{\mathrm{OP}})\, C_4}_{\text{Death}} - \underbrace{k_{83}\theta_{83}^{(T-20)}\left(\frac{C_4}{K_{m\mathrm{PC}} + C_4}\right)C_8}_{\text{Mineralization}} - \underbrace{G_{P1}\, a_{\mathrm{PC}}\, C_4}_{\text{Growth}} \qquad (4.29)$$

where

| | |
|---|---|
| $C_3$ | = inorganic phosphorus concentration [mg P l$^{-1}$] |
| $D_{P1}$ | = death plus respiration rate constant of phytoplankton [day$^{-1}$] |
| $a_{\mathrm{PC}}$ | = phosphorus to carbon ratio [mg P mg$^{-1}$ C] |
| $1 - f_{\mathrm{OP}}$ | = fraction of dead and respired phytoplankton recycled to the phosphate phosphorus pool [none] |
| $C_4$ | = phytoplankton carbon concentration [mg C l$^{-1}$] |

$k_{83}$ = dissolved organic phosphorus mineralization rate constant at 20°C [day$^{-1}$]
$\theta_{83}$ = dissolved organic phosphorus mineralization temperature coefficient [none]
$T$ = water temperature [°C]
$K_{mPC}$ = half-saturation constant for recycle [mg C l$^{-1}$]
$C_8$ = organic phosphorus concentration [mg P l$^{-1}$]
$G_{P1}$ = specific growth rate constant of phytoplankton [day$^{-1}$]

### 4.1.5.2.2  Phytoplankton Phosphorus

$$\frac{\partial (C_4\, a_{PC})}{\partial t} = \underbrace{G_{P1}\, a_{PC}\, C_4}_{\text{Growth}} - \underbrace{D_{P1}\, a_{PC}\, C_4}_{\text{Death}} - \underbrace{\frac{v_{S4}}{D}\, a_{PC}\, C_4}_{\text{Settling}} \qquad (4.30)$$

where

$C_4$ = phytoplankton carbon concentration [mg C l$^{-1}$]
$a_{PC}$ = phosphorus to carbon ratio [mg P mg$^{-1}$ C]
$G_{P1}$ = specific growth rate constant of phytoplankton [day$^{-1}$]
$D_{P1}$ = death plus respiration rate constant of phytoplankton [day$^{-1}$]
$V_{s4}$ = net settling velocity of phytoplankton [m day$^{-1}$]
$D$ = depth of segment [m]

### 4.1.5.2.3  Organic Phosphorus

$$\frac{\partial (C_8)}{\partial t} = \underbrace{D_{P1}\, a_{PC}\, f_{OP}\, C_4}_{\text{Death}} - \underbrace{k_{83}\, \theta_{83}^{(T-20)} \left( \frac{C_4}{K_{mPc} + C_4} \right) C_8}_{\text{Mineralization}} - \underbrace{\frac{v_{S3}(1-f_{D8})}{D}\, C_8}_{\text{Settling}} \qquad (4.31)$$

where

$C_8$ = organic phosphorus concentration [mg P l$^{-1}$]
$D_{P1}$ = death plus respiration rate constant of phytoplankton [day$^{-1}$]
$a_{PC}$ = phosphorus to carbon ratio [mg P mg$^{-1}$ C]
$f_{OP}$ = fraction of dead and respired phytoplankton recycled to the organic phosphorus pool [none]
$C_4$ = phytoplankton carbon concentration [mg C l$^{-1}$]
$k_{83}$ = dissolved organic phosphorus mineralization constant at 20°C [day$^{-1}$]
$\theta_{83}$ = dissolved organic phosphorus mineralization temperature coefficient [none]
$T$ = water temperature [°C]
$K_{mPc}$ = half-saturation constant for recycle [mg C l$^{-1}$]
$V_{S3}$ = organic matter settling velocity [m day$^{-1}$]
$f_{D8}$ = fraction dissolved organic phosphorus [none]
$1-f_{D8}$ = fraction particulate organic phosphorus [none]
$D$ = depth of segment [m]

**TABLE 4.18**
**POP to DOP Rate Constant (day$^{-1}$)**

| Avg. | Location | Reference | As Cited in | Explanation |
|------|----------|-----------|-------------|-------------|
| 0.22 | Potomac River Estuary | Thomann and Fitzpatrick, 1982 | Bowie et al.[93] | |
| 0.075 | Chesapeake Bay | Cerco and Cole, 1994 | Cerco and Cole[91] | Labile POP to DOP |
| 0.005 | Chesapeake Bay | Cerco and Cole, 1994 | Cerco and Cole[91] | Refractory POP to DOP |
| 0.075 | San Juan Bay Estuary | Bunch et al., 2000 | Bunch et al.[92] | Labile POP to DOP |
| 0.005 | San Juan Bay Estuary | Bunch et al., 2000 | Bunch et al.[92] | Refractory POP to DOP |

The literature values obtained from various coastal waters for kinetic rate constants related to organic phosphorus transformations, temperature correction coefficients, and settling velocities for organic phosphorus are given in Tables 4.18–4.21.

A complete analysis of the phosphorus fluxes from the sediment would require a rather complex and elaborate computation of solute-precipitate chemistry. Computation of solute-precipitate chemistry was outside the scope of WASP/EUTRO5. Instead, a simplified approach, which relies largely on empiricism, was taken. Anaerobic decomposition of detrital algal phosphorus yields both organic and inorganic phosphorus. Organic phosphorus then undergoes anaerobic decomposition and dissolved inorganic phosphorus is produced. DIP, which remains in the interstitial water, is not involved in the formation of precipitates and is not sorbed onto benthic solids. The effect of anaerobic conditions on sediment phosphorus flux was not included in WASP/EUTRO5 modeling.[78]

In CE-QUAL-W2,[58] two processes are defined for phosphate release from sediment. One is a first-order release process that is coupled with aerobic organic matter decay in the sediment. The second is a zero-order phosphate release process that is operative only when the oxygen concentration is less than a specified minimum value at which anaerobic processes are initiated. The default minimum dissolved oxygen concentration is 0.1 mg l$^{-1}$.

CE-QUAL-R1[57] models macrophytes. Thus, phosphate uptake of macrophytes from sediment is included. Anaerobic phosphate release, aerobic decomposition of organic matter, and settling of adsorbed phosphate are all simulated in the sediment compartment of the model.

Benthic mass balance equations for organic and inorganic phosphorus are given in the following subsections.

*4.1.5.2.4    Organic Phosphorus (Benthic)*

$$\frac{\partial(C_8)}{\partial t} = \underbrace{k_{PZD}\theta_{PZD}^{(T-20)}a_{PC}f_{OP}C_4}_{\text{Algal Decomposition}} - \underbrace{k_{OPD}\theta_{OPD}^{(T-20)}f_{D8}C_8}_{\text{Mineralization}} \qquad (4.32)$$

**TABLE 4.19**
**Organic Phosphorus to PO$_4{}^{3-}$ Rate Constant (day$^{-1}$)**

| Min. | Max. | Avg. | Location | Reference | As Cited in | Explanation |
|------|------|------|----------|-----------|-------------|-------------|
|      |      | 0.14 | Chesapeake Bay | Salas and Thomann, 1978 | Bowie et al.[93] | POP to PO$_4{}^{3-}$ |
|      |      | 0.03 | Chesapeake Bay | Hydroqual Inc., 1987 | Lung and Paerl[79] | |
| 0.001 | 0.02 |     | Texas bays and estuaries | Brandes, 1976 | Bowie et al.[93] | POP to PO$_4{}^{3-}$ |
|      |      | 0.22 | Potomac River Estuary | Thomann and Fitzpatrick, 1982 | Bowie et al.[93] | DOP to PO$_4{}^{3-}$ |
|      |      | 0.14 | Potomac River Estuary | O'Connor, 1981 | O'Connor[94] | |
| 0    | 0.15 |      | James River Estuary | O'Connor, 1981 | O'Connor[94] | |
| 0.05 | 0.1  |      | James River Estuary | Lung, 1986 | Lung and Paerl[79] | |
|      |      | 0.01 | Patuxent River Estuary | O'Connor, 1981 | O'Connor[94] | |
|      |      | 0.02 | Patuxent River Estuary | Lung, 1992 | Lung[80] | |
|      |      | 0.1  | Neuse River Estuary | Lung and Pearl, 1986 | Lung and Paerl[79] | |
|      |      | 0.1  | Chesapeake Bay | Cerco and Cole, 1994 | Cerco and Cole[91] | Given as minimum mineralization rate of DOP |
|      |      | 0.22 | Maryland coastal waters | Lung and Hwang, 1994 | Lung and Hwang[81] | DOP to PO$_4{}^{3-}$, at 20°C, value used in modeling |
|      |      | 0.05 | San Juan Bay Estuary | Bunch et al., 2000 | Bunch et al.[92] | Given as minimum mineralization rate of DOP |

where

$C_8$ = organic phosphorus concentration [mg P l$^{-1}$]

$k_{\text{PZD}}$ = anaerobic algal decomposition rate constant at 20°C [day$^{-1}$]

$\theta_{\text{PZD}}$ = anaerobic algal decomposition temperature coefficient [none]

$T$ = temperature [°C]

$a_{\text{PC}}$ = phosphorus to carbon ratio [mg P mg$^{-1}$ C]

$f_{\text{OP}}$ = fraction of dead and respired phytoplankton recycled to the organic phosphorus pool [none]

$C_4$ = phytoplankton carbon concentration [mg C l$^{-1}$]

## TABLE 4.20
## Organic Phosphorus to $PO_4^{3-}$ Temperature Correction Factor ($\theta$)

| Avg. | Location | Reference | As Cited in | Explanation |
|------|----------|-----------|-------------|-------------|
| 1.04 | Texas bays and estuaries | Brandes, 1976 | Bowie et al.[93] | POP to $PO_4^{3-}$, model documentation values |
| 1.08 | Potomac River Estuary | Thomann and Fitzpatrick, 1982 | Bowie et al.[93] | DOP to $PO_4^{3-}$ |
| 1.045 | | O'Connor, 1981 | O'Connor[94] | |
| 1.08 | Patuxent River Estuary | Lung, 1992 | Lung[80] | |
| 1.08 | Maryland coastal waters | Lung and Hwang, 1994 | Lung and Hwang[81] | DOP to $PO_4^{3-}$, value used in modeling |

## TABLE 4.21
## Settling Velocity of Organic Phosphorus (m day$^{-1}$)

| Avg. | Location | Reference | As Cited in | Explanation |
|------|----------|-----------|-------------|-------------|
| 0.23 | James River | O'Connor, 1981 | O'Connor[94] | |
| 0.4 | Patuxent River Estuary | O'Connor, 1981 | O'Connor[94] | |
| 0.30 | Patuxent River Estuary | Lung, 1992 | Lung[80] | Settling rate of POP |
| 0.6 | Sacramento San Joaquin Delta | O'Connor, 1981 | O'Connor[94] | |

## TABLE 4.22
## Rate Constants for Decomposition of Organic Phosphorus to $PO_4^{3-}$ in Sediments (day$^{-1}$)

| Min. | Max. | Avg. | Location | Reference | As Cited in | Explanation |
|------|------|------|----------|-----------|-------------|-------------|
| 0.00025 | 0.00038 | | Sacca di Scardovari, Italy | Viel et al., 1991 | Viel et al.[97] | Anoxic conditions in the sediment |
| | | 0.02 | Patuxent River Estuary | Lung, 1992 | Lung[80] | Given as Org P decomposition rate, value used in modeling |

$k_{OPD}$ = organic phosphorus decomposition rate constant at 20°C [day$^{-1}$]
$\theta_{OPD}$ = organic phosphorus decomposition temperature coefficient [none]
$f_{D8}$ = fraction dissolved organic phosphorus [none]

The rate constants for decomposition of organic phosphorus to $PO_4^{3-}$ in sediment are given in Table 4.22.

### 4.1.5.2.5  Inorganic Phosphorus (Benthic)

$$\frac{\partial(C_3)}{\partial t} = \underbrace{k_{PZD}\theta_{PZD}^{(T-20)}a_{PC}(1-f_{OP})C_4}_{\text{Algal decomposition}} + \underbrace{k_{OPD}\theta_{OPD}^{(T-20)}f_{D8}C_8}_{\text{Mineralization}} \qquad (4.33)$$

where

$C_3$ = inorganic phosphorus concentration [mg P l$^{-1}$]

$k_{PZD}$ = anaerobic algal decomposition rate constant at 20°C [day$^{-1}$]

$\theta_{PZD}$ = anaerobic algal decomposition rate temperature coefficient [none]

$T$ = temperature [°C]

$a_{PC}$ = phosphorus to carbon ratio [mg P mg$^{-1}$ C]

$1-f_{OP}$ = fraction of dead and respired phytoplankton recycled to the phosphate phosphorus pool [none]

$C_4$ = phytoplankton carbon concentration [mg C l$^{-1}$]

$k_{OPD}$ = organic phosphorus decomposition rate constant at 20°C [day$^{-1}$]

$\theta_{OPD}$ = organic phosphorus decomposition temperature coefficient [none]

$f_{D8}$ = fraction dissolved organic phosphorus [none]

$C_8$ = organic phosphorus concentration [mg P l$^{-1}$]

Sediment $PO_4^{3-}$ release rates for coastal marine environments are given in Table 4.23.

## 4.1.5.3  Modeling of Silicon Cycle

Because silicon is the limiting nutrient for diatoms, its concentration is included in eutrophication models as a state variable only if diatoms are simulated. The EUTRO5, Eutrophication Module of WASP5, simulates total phytoplankton. Therefore, silicon is not a state variable in EUTRO5. However, in the new version of the WASP eutrophication module, the advanced eutrophication kinetics section will allow modeling of three different algal species, one of which is diatoms. Therefore, silicon will be a new state variable in the newer versions of WASP.

In some other eutrophication models, such as CE-QUAL-ICM,[56] CE-QUAL-W2,[58] and CE-QUAL-R1,[57] silicon is a state variable.

CE-QUAL-W2[58] has two silicon state variables, namely, dissolved silicate and particulate biogenic silica. Dissolved silicon is taken up by algae (diatoms and epiphytes) for growth. In this model, no specific state variables are defined for diatoms, blue-green algae, green algae, or any other group. However, the user can assign algae groups by entering appropriate stoichiometric coefficients. For example, if a nonzero value is assigned as the stoichiometric ratio for silica for an algal group, that algal group will act as diatoms. In CE-QUAL-W2, the processes considered for dissolved silica are adsorption of dissolved silica onto suspended solids, particulate biogenic silica dissolution, respiration of algae/epiphyton, and release of dissolved silica from sediment. Adsorption of dissolved silica onto suspended solids is based on a partitioning coefficient. Dissolved silica is lost from

**TABLE 4.23**
**Sediment $PO_4^{3-}$ Release Rate (mg $PO_4^{3-}$ $m^{-2}$ $day^{-1}$)**

| Min. | Max. | Avg. | Location | Reference | As Cited in | Explanation |
|---|---|---|---|---|---|---|
| | | 1.7 | St. Leonard's Creek, Patuxent River Estuary, U.S.A. | Boynton et al., 1990 | Lung[80] | May 1985 |
| | | 6.6 | St. Leonard's Creek, Patuxent River Estuary, U.S.A. | Boynton et al., 1990 | Lung[80] | June 1985 |
| | | 0.92 | St. Leonard's Creek, Patuxent River Estuary, U.S.A. | Boynton et al., 1990 | Lung[80] | August 1985 |
| | | 4.15 | St. Leonard's Creek, Patuxent River Estuary, U.S.A. | Boynton et al., 1990 | Lung[80] | October 1985 |
| | | 0 | Buena Vista, Patuxent River Estuary, U.S.A. | Boynton et al., 1990 | Lung[80] | May 1985 |
| | | 12.3 | Buena Vista, Patuxent River Estuary, U.S.A. | Boynton et al., 1990 | Lung[80] | June 1985 |
| | | 1.78 | Buena Vista, Patuxent River Estuary, U.S.A. | Boynton et al., 1990 | Lung[80] | August 1985 |
| | | 14.3 | Buena Vista, Patuxent River Estuary, U.S.A. | Boynton et al., 1990 | Lung[80] | October 1985 |
| | | 2 | Isle of Wight Bay, Maryland, U.S.A. | Lung and Hwang, 1994 | Lung and Hwang[81] | |
| | | 4 | Turnville Creek/St. Martin River, Maryland, U.S.A. | Lung and Hwang, 1994 | Lung and Hwang[81] | |
| | | 8 | Bishopsville Prong, Maryland, U.S.A. | Lung and Hwang, 1994 | Lung and Hwang[81] | |
| | | 8 | Assawoman Bay, Maryland, U.S.A. | Lung and Hwang, 1994 | Lung and Hwang[81] | |
| | | 8 | Upper Assawoman Bay, Maryland, U.S.A. | Lung and Hwang, 1994 | Lung and Hwang[81] | |
| 0 | 54.312 | 14.88 | Cape Lookout Bight, North Carolina, U.S.A. | Martens, 1993 | Martens[98] | |

the system through adsorption and subsequent particulate settling. Two processes define dissolved silica release from sediment. The first process is a first-order silica release process that is coupled with aerobic organic matter decay in the sediment. The second is a zero-order release process that is triggered under anaerobic conditions. The processes simulated for particulate biogenic silica include dissolution, settling, and algal/epiphyton mortality. Biogenic silica is lost from the system through settling. Dissolution of biogenic silica and sediment release of dissolved silica processes include a temperature rate multiplier for temperature correction.

In CE-QUAL-R1,[57] there is only one state variable for the silicon cycle, dissolved silica, that is incorporated into the third algal group (diatoms) during photosynthesis. No silica recycling takes place in this model; thus, silica is lost from the system once used by diatoms.

CE-QUAL-ICM[56] incorporates two siliceous state variables, available silica and particulate biogenic silica. For practical purposes, available silica is equivalent to dissolved silica although sorption of available silica to inorganic solids occurs. Diatoms utilize dissolved silica and recycle both dissolved and particulate biogenic silica through metabolic activities and predation. Particulate silica dissolves in the water column or settles to the sediment. A portion of the settled particulate biogenic silica dissolves within the sediment and returns to the water column as dissolved silica. To present the adsorption of dissolved silica to the newly formed metal particles, dissolved silica is partitioned into dissolved and particulate fractions according to a linear adsorption isotherm.

The HEM-3D[90] model applies kinetic process descriptions from CE-QUAL-ICM. Both models use the same predictive sediment submodel, described in the nitrogen and phosphorus cycle modeling sections, to simulate the sediment release of dissolved silicon.

Dissolution rates of particulate biogenic silicon to dissolved silicon are presented in Table 4.24.

### 4.1.5.4  Modeling of Dissolved Oxygen

WASP5 simulates dissolved oxygen and associated variables using the EUTRO5 module. The user can operate EUTRO5 at various complexity levels for dissolved

**TABLE 4.24**
**Particulate Biogenic Si to Dissolved Si Dissolution Rate (day$^{-1}$)**

| Min. | Max. | Avg. | Location | Reference | As Cited in | Explanation |
|------|------|------|----------|-----------|-------------|-------------|
| 0.127 | 0.177 | | Sacca di Scardovari, Italy | Viel et al., 1991 | Viel et al.[97] | Anoxic conditions, in the sediment |
| | | 0.06 | Chesapeake Bay | Wollast, 1974 | Cerco and Cole[91] | |
| | | 0.04 | Chesapeake Bay | Vanderborght et al., 1977 | Cerco and Cole[91] | |
| | | 0.017 | Chesapeake Bay | Grill and Richards, 1964 | Cerco and Cole[91] | |
| | | 0.03 | Chesapeake Bay | Cerco and Cole, 1994 | Cerco and Cole[91] | Model value, 20°C |

oxygen simulation. Five state variables (dissolved oxygen, ultimate carbonaceous biochemical oxygen demand, ammonia, nitrate, and phytoplankton carbon) are used directly for dissolved oxygen balance calculations. Six processes related to dissolved oxygen are considered. These processes are atmospheric reaeration, oxidation of carbonaceous BOD, nitrification, sediment oxygen demand, phytoplankton growth, and phytoplankton respiration.

In EUTRO5, the user can specify a single reaeration rate for the entire model network, specify spatially variable reaeration rates, or let the model calculate the reaeration rates. The model calculates both a flow- and a wind-induced reaeration rate, and uses the larger value of the two. In the new version of WASP (WASP 6.x developed by Wool et al.[101]), another option for the reaeration calculation is provided that sums up the flow- and wind-induced reaeration rates.

The kinetic expression for carbonaceous BOD oxidation in EUTRO5 contains a first-order rate constant, an Arrhenius-type temperature correction factor, and a Monod-type low dissolved oxygen correction function. The low dissolved oxygen correction simulates the decline of aerobic oxidation rate as the dissolved oxygen concentration approaches zero. The user specifies the half-saturation concentration for dissolved oxygen, $K_{BOD}$. Otherwise, the default value for this concentration is zero in the model, allowing the reaction to proceed fully even under anaerobic conditions.

AQUATOX[73] can simulate the effects of algae (either phytoplankton or periphyton), macrophytes, and animals on dissolved oxygen concentration. Algae and macrophytes produce oxygen during photosynthesis and consume oxygen via respiration. Oxygen utilization during respiration of animals (zooplankton, benthic invertebrates, benthic insects, and fish) is expressed as the sum of endogenous respiration and dynamic action terms. Detritus is divided into eight compartments, each of them contributing to dissolved oxygen consumption. The sum of organism respiration plus detritus decomposition is defined as BOD in this model.

The processes included in the model developed by Park and Kuo[69] are the same as those in EUTRO5. Although there is no low dissolved oxygen correction function for carbonaceous BOD oxidation, such a correction function is included in the SOD calculations in this model.

In CE-QUAL-W2,[58] epiphyte oxygen production via photosynthesis is also considered in addition to algal dissolved oxygen production. Five state variables used to simulate organic matter include labile dissolved organic matter, refractory dissolved organic matter, labile particulate organic matter, refractory particulate organic matter, and carbonaceous BOD. Each of these five has different decay rates. Decay of organic matter is allowed only when the dissolved oxygen concentration is greater than zero.

In CE-QUAL-R1,[57] oxidation of reduced anaerobic by-products ($Fe^{2+}$, $S^{2-}$, $Mn^{2+}$), macrophyte photosynthesis and respiration, and zooplankton and fish respiration are modeled along with atmospheric reaeration, decomposition of labile and refractory dissolved organic matter, algal photosynthesis and respiration, nitrification, sediment decomposition, and detrital decomposition. Simultaneous oxidation and reduction processes are initiated or suspended by the model when a given dissolved oxygen concentration is reached. If the dissolved oxygen concentration falls below this given concentration, reduction processes and sediment release of ammonium, phosphate,

$Mn^{2+}$, $Fe^{2+}$, $S^{2-}$ start. When the oxygen concentration exceeds this specified value, reduced anaerobic byproducts are oxidized.

Dissolved oxygen kinetics in CE-QUAL-ICM[56] and HEM-3D[90] are similar to EUTRO5 except that the two former models also consider the oxidation of inorganic substances such as sulfide. In the predictive sediment submodel, sulfide and methane can stay in the sediment layer, be exported to the water column, or be buried into deeper sediment. If not buried into deeper inactive sediment, the sulfide and methane will exert an oxygen demand either in the sediment or in the water column.

### 4.1.5.4.1 Dissolved Oxygen

The mass balance equation for dissolved oxygen given in EUTRO5 is given

$$
\frac{\partial C_6}{\partial t} = \underbrace{k_2(C_S - C_6)}_{\text{Reaeration}} - \underbrace{k_d\theta_d^{(T-20)}\left(\frac{C_6}{K_{BOD}+C_6}\right)C_5}_{\text{Oxidation}} - \underbrace{\frac{64}{14}k_{12}\theta_{12}^{(T-20)}\left(\frac{C_6}{K_{NIT}+C_6}\right)C_1}_{\text{Nitrification}}
$$

$$
\underbrace{-\frac{SOD}{D}\theta_S^{(T-20)}}_{\substack{\text{Sediment Oxygen}\\\text{Demand}}} + \underbrace{G_{P1}\left(\frac{32}{12}+\frac{48}{14}a_{nc}(1-P_{NH3})\right)C_4}_{\text{Phytoplankton Growth}} - \underbrace{\frac{32}{12}k_{1R}\theta_{1R}^{(T-20)}C_4}_{\text{Respiration}} \quad (4.34)
$$

where

$C_6$ = dissolved oxygen concentration [mg $O_2$ $l^{-1}$]

$k_2$ = reaeration rate constant at 20°C [$day^{-1}$]

$C_S$ = dissolved oxygen saturation concentration [mg $O_2$ $l^{-1}$]

$k_d$ = deoxygenation rate constant at 20°C [$day^{-1}$]

$\theta_d$ = temperature coefficient [none]

$T$ = water temperature [°C]

$K_{BOD}$ = half-saturation constant for oxygen limitation [mg $O_2$ $l^{-1}$]

$C_5$ = carbonaceous biochemical oxygen demand (CBOD) concentration [mg $O_2$ $l^{-1}$]

$k_{12}$ = nitrification rate constant at 20°C [$day^{-1}$]

$\theta_{12}$ = nitrification temperature coefficient [none]

$K_{NIT}$ = half-saturation constant for oxygen limitation of nitrification [mg $O_2$ $l^{-1}$]

$C_1$ = ammonium nitrogen concentration [mg N $l^{-1}$]

SOD = sediment oxygen demand [g $m^{-2}$ $day^{-1}$]

$D$ = depth of segment [m]

$\theta_S$ = SOD temperature coefficient [none]

$G_{P1}$ = specific growth rate constant of phytoplankton [$day^{-1}$]

$P_{NH3}$ = preference for ammonium uptake term [none]

$C_4$ = phytoplankton carbon concentration [mg C $l^{-1}$]

$k_{1R}$ = phytoplankton respiration rate constant at 20°C [$day^{-1}$]

$\theta_{1R}$ = phytoplankton respiration temperature coefficient [none]

$a_{nc}$ = nitrogen to carbon ratio [mg N $mg^{-1}$ C]

## TABLE 4.25
## CBOD Decay Rate Constants (day⁻¹)

| Avg. | Location | Reference | As Cited in | Explanation |
|---|---|---|---|---|
| 0.2 | San Francisco Bay Estuary | Chen, 1970 | Bowie et al.[93] | |
| 0.31 | Delaware River Estuary | Hydroscience, 1972 | Bowie et al.[93] | |
| 0.31 | Wappinger Creek Estuary | Hydroscience, 1972 | Bowie et al.[93] | |
| 0.1 | New York Bight | O'Connor, 1981 | O'Connor[94] | Given as organic carbon mineralization rate |
| 0.2 | James River | O'Connor, 1981 | O'Connor[94] | Given as organic carbon mineralization rate |
| 0.2 | Patuxent River Estuary | O'Connor, 1981 | O'Connor[94] | Given as organic carbon mineralization rate |
| 0.2 | Patuxent River Estuary | Lung, 1992 | Lung[80] | At 20°C |
| 0.2 | Venice Lagoon | Melaku Canu et al., 2001 | Melaku Canu et al.[88] | |
| 0.01 | Chesapeake Bay | Cerco and Cole, 1994 | Cerco and Cole[91] | Given as DOC mineralization rate |
| 0.007 | Maryland coastal waters | Lung and Hwang, 1994 | Lung and Hwang[81] | At 20°C |

Rate constants for carbonaceous BOD decay, related temperature correction factors, half-saturation concentrations for dissolved oxygen, and transformation rate constant of POC to DOC obtained from various coastal waters are given in Tables 4.25–4.28, respectively.

## TABLE 4.26
## CBOD Decay Temperature Correction Factor ($\theta$)

| Value | Location | Reference | As Cited in | Explanation |
|---|---|---|---|---|
| 1.047 | | Chen, 1970 | Bowie et al.[93] | |
| 1.047 | | Genet et al., 1974 | Bowie et al.[93] | |
| 1.047 | | Thomann and Fitzpatrick, 1982 | Bowie et al.[93] | |
| 1.045 | | O'Connor, 1981 | O'Connor[94] | Given as organic carbon mineralization rate temperature correction |
| 1.047 | Venice Lagoon | Melaku Canu, et al., 2001 | Melaku Canu et al.[88] | |
| 1.05 | Maryland coastal waters | Lung and Hwang, 1994 | Lung and Hwang[81] | |

**TABLE 4.27**
**Half-Saturation Concentrations of Dissolved Oxygen for Organic Matter Degradation (mg $O_2$ $l^{-1}$)**

| Average | Location | Reference | As Cited in |
|---|---|---|---|
| 0.1 | Venice Lagoon | Melaku Canu et al., 2001 | Melaku Canu et al.[88] |
| 0.5 | Chesapeake Bay | Cerco and Cole, 1994 | Cerco and Cole[91] |

**TABLE 4.28**
**POC to DOC Rate Constant (day$^{-1}$)**

| Min. | Max. | Avg. | Location | Reference | As Cited in | Explanation |
|---|---|---|---|---|---|---|
| | | 0.075 | Chesapeake Bay | Cerco and Cole, 1994 | Cerco and Cole[91] | Labile POC to DOC |
| | | 0.005 | Chesapeake Bay | Cerco and Cole, 1994 | Cerco and Cole[91] | Refractory POC to DOC |
| 0.15 | 1.5 | | San Juan Bay Estuary | Bunch et al., 2000 | Bunch et al.[92] | Labile POC to DOC |
| | | 0.005 | San Juan Bay Estuary | Bunch et al., 2000 | Bunch et al.[92] | Refractory POC to DOC |

Due to their effects on the dissolved oxygen and nutrient cycles, related literature values about algal kinetics and stoichiometric parameters are also provided in Tables 4.29–4.36.

### 4.1.5.4.2 Dissolved Oxygen (Benthic)

The benthic layer oxygen balance equations are given as

$$\frac{\partial C_{6j}}{\partial t} = \underbrace{-k_{DS}\,\theta_{DS}^{(T-20)}\,C_5}_{\text{Oxidation}} + \underbrace{\frac{E_{DIF}}{D_j^2}(C_{6i} - C_{6j})}_{\text{Diffusion}} \qquad (4.35)$$

where for the benthic segment $j$ and the water segment $i$

$C_6$ = dissolved oxygen concentration [mg $O_2$ $l^{-1}$]

$k_{DS}$ = organic carbon as (CBOD) decomposition rate constant at 20°C [day$^{-1}$]

$\theta_{DS}$ = CBOD decomposition temperature coefficient [none]

$T$ = water temperature [°C]

$E_{DIF}$ = diffusive exchange coefficient [m$^2$ day$^{-1}$]

$D_J$ = benthic layer depth [m]

**TABLE 4.29**
**Algal Maximum Growth Rates (day⁻¹)**

| Min. | Max. | Avg. | Location | Reference | As Cited in | Explanation |
|---|---|---|---|---|---|---|
| 1.3 | 2.5 | | Potomac River Estuary | Thomann and Fitzpatrick, 1982 | Bowie et al.[93] | 20°C total phytoplankton |
| | | 2 | Potomac River Estuary | O'Connor, 1981 | O'Connor[94] | 20°C total phytoplankton |
| 1.3 | 2.5 | | Chesapeake Bay | Salas and Thomann, 1978 | Bowie et al.[93] | 20°C total phytoplankton |
| 2 | 2.5 | | Chesapeake Bay | Hydroqual Inc., 1987 | Lung and Paerl[79] | 20°C total phytoplankton |
| 1.3 | 2.5 | | Sacramento San Joaquin Delta | Di Toro, 1971 | Bowie et al.[93] | 20°C total phytoplankton |
| | | 2.5 | Sacramento San Joaquin Delta | O'Connor, 1981 | O'Connor[94] | 20°C total phytoplankton |
| 1.3 | 2.5 | | New York Bight | O'Connor et al., 1981 | Bowie et al.[93] | 20°C total phytoplankton |
| | | 2.5 | New York Bight | O'Connor, 1981 | O'Connor[94] | 20°C total phytoplankton |
| 1 | 2 | | Grays Harbor and Chehalis River | Battelle, 1974 | Bowie et al.[93] | 20°C total phytoplankton |
| 2 | 3 | | James River | O'Connor, 1981 | O'Connor[94] | 20°C total phytoplankton |
| 2 | 2.5 | | James River | Lung, 1986 | Lung and Paerl[79] | 20°C total phytoplankton |
| | | 2 | Patuxent River Estuary | O'Connor, 1981 | O'Connor[94] | 20°C total phytoplankton |
| | | 2.03 | Maryland coastal waters | Lung and Hwang, 1994 | Lung and Hwang[81] | 20°C total phytoplankton, value used in modeling |
| 1.5 | 2 | | Texas bays and estuaries | Brandes, 1976 | Bowie et al.[93] | 20°C total phytoplankton, model documentation value |
| | | 1.75 | Sacramento San Joaquin Delta | Di Toro, 1971 | Bowie et al.[93] | 27°C diatom, literature values |
| | | 2 | Neuse Estuary | Lung and Paerl, 1986 | Lung and Paerl[79] | 20°C diatom |
| | | 2 | Patuxent River Estuary | Lung, 1992 | Lung[80] | 20°C diatom, value used in modeling |
| | | 2.25 | Chesapeake Bay | Cerco and Cole, 1994 | Cerco and Cole[91] | 20°C diatom |
| | | 1.3 | Neuse Estuary | Lung and Paerl, 1986 | Lung and Paerl[79] | 20°C blue-green algae |

(Continued)

**TABLE 4.29 (Continued)**
**Algal Maximum Growth Rates (day$^{-1}$)**

| Min. | Max. | Avg. | Location | Reference | As Cited in | Explanation |
|------|------|------|----------|-----------|-------------|-------------|
| | | 2.5 | Chesapeake Bay | Cerco and Cole, 1994 | Cerco and Cole[91] | 27.5°C cyanobacteria, rate is valid only for upper Potomac River |
| | | 1.8 | Neuse Estuary | Lung and Paerl, 1986 | Lung and Paerl[79] | 20°C green algae |
| | | 1.8 | Patuxent River Estuary | Lung, 1992 | Lung[80] | 20°C green algae, value used in modeling |
| | | 2.5 | Chesapeake Bay | Cerco and Cole, 1994 | Cerco and Cole[91] | 25°C green algae |
| 1.5 | 3.9 | 2.16 | Sacramento San Joaquin Delta | Di Toro, 1971 | Bowie et al.[93] | 25°C green algae, literature values |
| | | | Sacramento San Joaquin Delta | Di Toro, 1971 | Bowie et al.[93] | 20°C dinoflagellates, literature values |
| 0.2 | 0.28 | 2.25 | New York Bight | O'Connor et al., 1981 | Bowie et al.[93] | 20°C dinoflagellates |
| | | | Indian River-Rehoboth Bay, Delaware, U.S.A. | Cerco and Seitzinger, 1997 | Cerco and Seitzinger[102] | 20°C benthic algae |
| | | 2.4 | Venice Lagoon | Bergamasco, 1998 | Bergamasco et al.[103] | |

## TABLE 4.30
### Half-Saturation Constants for Algal Growth (mg l⁻¹)

| Nitrogen | | | Phosphorus | | | Location | Reference | As Cited in | Explanation |
|---|---|---|---|---|---|---|---|---|---|
| Min. | Max. | Avg. | Min. | Max. | Avg. | | | | |
| | | 0.025 | 0.0005 | 0.03 | | Potomac River Estuary | Thomann and Fitzpatrick, 1982 | Bowie et al.[93] | 20°C total phytoplankton |
| | | 0.025 | | | 0.005 | Potomac River Estuary | O'Connor, 1981 | O'Connor[94] | 20°C total phytoplankton |
| | | 0.025 | 0.0005 | 0.03 | | Chesapeake Bay | Salas and Thomann, 1978 | Bowie et al.[93] | 20°C total phytoplankton |
| | | 0.015 | | | 0.0015 | Chesapeake Bay | Hydroqual Inc., 1987 | Lung and Paerl[79] | Total phytoplankton |
| | | 0.01 | | | 0.001 | Chesapeake Bay | Cerco and Cole, 1994 | Cerco and Cole[91] | Total phytoplankton, model value |
| 0.001 | 0.008 | | | | | Chesapeake Bay | Wheeler et al., 1982 | Cerco and Cole[91] | Total phytoplankton, NH₄ uptake |
| | | | | | | Chesapeake Bay | Taft et al., 1975 | Cerco and Cole[91] | Total phytoplankton |
| | | 0.025 | 0.0005 | 0.03 | | Sacramento San Joaquin Delta | Di Toro et al., 1977 | Bowie et al.[93] | 20°C total phytoplankton |
| | | 0.025 | | | 0.005 | Sacramento San Joaquin Delta | O'Connor, 1981 | O'Connor[94] | 20°C total phytoplankton |
| | | 0.025 | 0.0005 | 0.03 | | New York Bight | O'Connor et al., 1981 | Bowie et al.[93] | 20°C total phytoplankton |
| | | 0.025 | | | | New York Bight | O'Connor, 1981 | O'Connor[94] | 20°C total phytoplankton |
| | | 0.005 | | | 0.001 | James River Estuary | O'Connor, 1981 | O'Connor[94] | 20°C total phytoplankton |
| | | 0.005 | | | 0.001 | James River Estuary | Lung, 1986 | Lung and Paerl[79] | Total phytoplankton |
| | | 0.005 | | | 0.001 | Patuxent River Estuary | O'Connor, 1981 | O'Connor[94] | 20°C total phytoplankton |
| | | 0.025 | | | 0.001 | Patuxent River Estuary | Lung, 1992 | Lung[80] | 20°C total phytoplankton |

(Continued)

**TABLE 4.30 (Continued)**
**Half-Saturation Constants for Algal Growth (mg l⁻¹)**

| Nitrogen | | | Phosphorus | | | Location | Reference | As Cited in | Explanation |
|---|---|---|---|---|---|---|---|---|---|
| Min. | Max. | Avg. | Min. | Max. | Avg. | | | | |
| | | 0.025 | | | 0.001 | Maryland coastal waters, U.S.A. | Lung and Hwang, 1994 | Lung and Hwang[81] | Total phytoplankton |
| | | 0.01 | | | 0.001 | San Juan Bay Estuary | Bunch et al., 2000 | Bunch et al.[92] | Total phytoplankton |
| | | 0.025 | 0.006 | 0.025 | | Grays Harbor and Chehalis River | Battelle, 1974 | Bowie et al.[93] | 20°C total phytoplankton |
| | | | | | 0.005 | Neuse River Estuary | Lung and Paerl, 1986 | Lung and Paerl[79] | Total phytoplankton |
| 0.1 | 0.4 | | 0.03 | 0.05 | | Texas bays and estuaries | Brandes, 1976 | Bowie et al.[93] | 20°C total phytoplankton |
| 0.0014 | 0.018 | | | | 0.006 | Sacramento San Joaquin Delta | Di Toro et al, 1971 | Bowie et al.[93] | Total phytoplankton, literature |
| 0.0063 | 0.12 | | | 0.025 | | Sacramento San Joaquin Delta | Di Toro et al., 1971 | Bowie et al.[93] | 27°C diatom, literature values |
| | | 0.025 | | | | Neuse River Estuary | Lung and Paerl, 1986 | Lung and Paerl[79] | Diatom |
| | | 0.15 | | | 0.01 | Sacramento San Joaquin Delta | Di Toro et al., 1971 | Bowie et al.[93] | 25°C green algae |
| | | 0.025 | | | | Neuse River Estuary | Lung and Paerl, 1986 | Lung and Paerl[79] | Green algae |
| 0.007 | 0.13 | | | | | Sacramento San Joaquin Delta | Di Toro et al., 1971 | Bowie et al.[93] | 20°C dinoflagellates, literature |
| | | 0.005 | | | | New York Bight | O'Connor et al., 1981 | Bowie et al.[93] | 20°C dinoflagellates |
| | | | | | 0.01 | Sacramento San Joaquin Delta | Di Toro et al., 1971 | Bowie et al.[93] | Blue-green algae |
| | | 0.015 | | | | Neuse River Estuary | Lung and Paerl, 1986 | Lung and Paerl[79] | Blue-green algae (non-N₂ fixing) |
| | | 0 | | | | Neuse River Estuary | Lung and Paerl, 1986 | Lung and Paerl[79] | Blue-green algae (N₂ fixing) |
| | | 0.01 | | | 0.001 | Indian River-Rehoboth Bay, Delaware | Cerco and Seitzinger, 1997 | Cerco and Seitzinger[102] | Benthic algae, units in g m⁻² |

## TABLE 4.31
## Respiration Rate Constants (day$^{-1}$)

| Min. | Max. | Avg. | Location | Reference | As Cited in | Explanation |
|------|------|------|----------|-----------|-------------|-------------|
| 0.05 | 0.15 |      | Potomac River Estuary | Thomann and Fitzpatrick, 1982 | Bowie et al.[93] | 20°C total phytoplankton |
|      |      | 0.1  | Potomac River Estuary | O'Connor, 1981 | O'Connor[94] | 20°C total phytoplankton |
| 0.05 | 0.15 |      | Sacramento San Joaquin Delta | Di Toro, 1977 | Bowie et al.[93] | 20°C total phytoplankton |
|      |      | 0.1  | Sacramento San Joaquin Delta | O'Connor, 1981 | O'Connor[94] | 20°C total phytoplankton |
| 0.05 | 0.15 |      | New York Bight | O'Connor et al., 1981 | Bowie et al.[93] | 20°C total phytoplankton |
|      |      | 0.1  | New York Bight | O'Connor, 1981 | O'Connor[94] | 20°C total phytoplankton |
|      |      | 0.1  | James River | O'Connor, 1981 | O'Connor[94] | 20°C total phytoplankton |
|      |      | 0.1  | James River | Lung, 1986 | Lung and Paerl[79] | 20°C total phytoplankton |
|      |      | 0.2  | Patuxent River Estuary | O'Connor, 1981 | O'Connor[94] | 20°C total phytoplankton |
|      |      | 0.125 | Patuxent River Estuary | Lung, 1992 | Lung[80] | 20°C total phytoplankton, value used during modeling |
|      |      | 0.125 | Maryland coastal waters | Lung and Hwang, 1994 | Lung and Hwang[81] | 20°C total phytoplankton, value used during modeling |
| 0.1  | 0.125 |      | Chesapeake Bay | Hydroqual Inc., 1987 | Lung and Paerl[79] | 20°C total phytoplankton |
|      |      | 0.05 | Neuse Estuary | Lung and Paerl, 1986 | Lung and Paerl[79] | 20°C total phytoplankton |
|      |      | 0.01 | San Juan Bay Estuary | Bunch et al., 2000 | Bunch et al.[92] | 30°C, total given as basal metabolic rate which also includes excretion of dissolved carbon |
|      |      | 0.051 | Texas bays and estuaries | Brandes, 1976 | Bowie et al.[93] | 20°C total phytoplankton, model documentation value |
| 0.04 | 0.08 |      | Sacramento San Joaquin Delta | Di Toro, 1971 | Bowie et al.[93] | 27°C diatom, literature values |
|      |      | 0.01 | Chesapeake Bay | Cerco and Cole, 1994 | Cerco and Cole[91] | 20°C diatom, given as basal metabolic rate which also includes excretion of dissolved carbon |
|      |      | 0.04 | Chesapeake Bay | Cerco and Cole, 1994 | Cerco and Cole[91] | 20°C Cyanobacteria, given as basal metabolic rate which also includes excretion of dissolved carbon |
|      |      | 0.01 | Chesapeake Bay | Cerco and Cole, 1994 | Cerco and Cole[91] | 20°C green algae, given as basal metabolic rate, which also includes excretion of dissolved carbon |
|      |      | 0.047 | New York Bight | O'Connor et al., 1981 | Bowie et al.[93] | 20°C dinoflagellates |
|      |      | 0.01 | Indian River-Rehoboth Bay, Delaware | Cerco and Seitzinger, 1997 | Cerco and Seitzinger[102] | 20°C benthic algae |
|      |      | 0.004 | Venice Lagoon | Bergamasco, 1998 | Bergamasco et al.[103] | |

## TABLE 4.32
## Algal Respiration Rates Temperature Correction Factor (θ)

| Min. | Max. | Avg. | Location | Reference | As Cited in | Explanation |
|------|------|------|----------|-----------|-------------|-------------|
| 1.04 | 1.08 |      |          | O'Connor, 1981 | O'Connor[94] | |
|      |      | 1.08 | Patuxent River Estuary | Lung, 1992 | Lung[80] | Value used in modeling |
|      |      | 1.045 | Maryland coastal waters | Lung and Hwang, 1994 | Lung and Hwang[81] | Value used in modeling |

## TABLE 4.33
## Algal Nonpredatory Mortality Rates (day$^{-1}$)

| Min. | Max. | Avg. | Location | Reference | As Cited in | Explanation |
|------|------|------|----------|-----------|-------------|-------------|
|       |      | 0.02 | Potomac River Estuary | Thomann and Fitzpatrick, 1982 | Bowie et al.[93] | Total phytoplankton |
| 0.005 | 0.1 |      | Chesapeake Bay | Salas and Thomann, 1978 | Bowie et al.[93] | Total phytoplankton |
| 0.1   | 0.2 |      | Chesapeake Bay | Hydroqual Inc., 1987 | Lung and Paerl[79] | Total phytoplankton given as death rate |
|       |      | 0.1  | James River Estuary | Lung, 1986 | Lung and Paerl[79] | Total phytoplankton given as death rate |
|       |      | 0.05 | Neuse River Estuary | Lung and Paerl, 1986 | Lung and Paerl[79] | Total phytoplankton given as death rate |
|       |      | 0.007 | Venice Lagoon | Bergamasco, 1998 | Bergamasco et al.[103] | Value used in the modeling |
|       |      | 0.125 | Patuxent River Estuary | Lung, 1992 | Lung[80] | Value used in the modeling, given as death rate |
|       |      | 0.02 | Maryland coastal waters | Lung and Hwang, 1994 | Lung and Hwang[81] | Value used in the modeling |

**TABLE 4.34**
**Settling Velocity of Algae (m day⁻¹)**

| Min. | Max. | Avg. | Location | Reference | As Cited in | Explanation |
|---|---|---|---|---|---|---|
| 0.05 | 0.2 | | Potomac River Estuary | Thomann and Fitzpatrick, 1982 | Bowie et al.[93] | Total phytoplankton |
| 0.05 | 0.2 | | New York Bight | O'Connor et al., 1981 | Bowie et al.[93] | Total phytoplankton |
| | | 0.1 | New York Bight | O'Connor, 1981 | O'Connor[94] | Total phytoplankton |
| | | 0.4 | Patuxent River | O'Connor, 1981 | O'Connor[94] | Total phytoplankton |
| | | 0.23 | James River Estuary | O'Connor, 1981 | O'Connor[94] | Total phytoplankton |
| | | 0.23 | James River Estuary | Lung, 1986 | Lung and Paerl[79] | Total phytoplankton |
| | | 0.1 | Chesapeake Bay | Hydroqual Inc., 1987 | Lung and Paerl[79] | Total phytoplankton |
| | | 0.3 | Maryland coastal waters | Lung and Hwang, 1994 | Lung and Hwang[81] | Total phytoplankton, value used in modeling |
| 0 | 0.2 | | Texas bays and estuaries | Brandes, 1976 | Bowie et al.[93] | Total phytoplankton, model documentation value |
| | | 8 | New York Bight | O'Connor et al., 1981 | Bowie et al.[93] | Dinoflagellates |
| | | 0.46 | Neuse River Estuary | Lung and Paerl, 1986 | Lung and Paerl[79] | Diatom |
| | | 0.30 | Patuxent River Estuary | Lung, 1992 | Lung[80] | Diatom, value used in modeling |
| | | 0.1 | Chesapeake Bay | Cerco and Cole, 1994 | Cerco and Cole[91] | Diatom, base diatom settling velocity |
| | | 0.25 | Chesapeake Bay | Cerco and Cole, 1994 | Cerco and Cole[91] | Diatom, enhanced settling of large diatoms |
| | | 0.015 | Neuse River Estuary | Lung and Paerl, 1986 | Lung and Paerl[79] | Blue-green algae (non-nitrogen fixing) |
| | | 0.058 | Neuse River Estuary | Lung and Paerl, 1986 | Lung and Paerl[79] | Blue-green algae (nitrogen fixing) |
| | | 0 | Chesapeake Bay | Cerco and Cole, 1994 | Cerco and Cole[91] | Cyanobacteria |
| | | 0.46 | Neuse River Estuary | Lung and Paerl, 1986 | Lung and Paerl[79] | Green-algae |
| | | 0.1 | Chesapeake Bay | Cerco and Cole, 1994 | Cerco and Cole[91] | Green-algae |
| | | 0.09 | Patuxent River Estuary | Lung, 1992 | Lung[80] | Green-algae, value used in modeling |

**TABLE 4.35**
**Nutrient Composition of Algal Cells—Ratio to Chl a**

| C:Chl a | | | N:Chl a | | | P:Chl a | | | Si:Chl a | | | O:Chl a | | | Location | References | As Cited in | Explanation |
|---|---|---|---|---|---|---|---|---|---|---|---|---|---|---|---|---|---|---|
| Min. | Max. | Avg. | Min. | Max. | Avg. | Min. | Max. | Avg. | Min. | Max. | Avg. | Min. | Max. | Avg. | | | | |
| 50 | 100 | | 7 | 15 | | 0.5 | 1 | | | | | | | | Chesapeake Bay | Salas and Thomann, 1978 | Bowie et al.[93] | Total phytoplankton |
| 30 | 50 | | 8 | 17.5 | | 0.55 | 1.23 | | | | | 80 | 133 | | Chesapeake Bay | Hydroqual Inc., 1987 | Lung and Paerl[79] | Total phytoplankton |
| | | 60 | | | | | | | | | | | | | Chesapeake Bay | Cerco and Cole, 1994 | Cerco and Cole[91] | Total phytoplankton, model value |
| | | 90 | | | | | | | | | | | | | Chesapeake Bay | Harding et al., 1986 | Cerco and Cole[91] | Total phytoplankton |
| 30 | 143 | | | | | | | | | | | | | | Chesapeake Bay | Malone et al., 1988 | Cerco and Cole[91] | Total phytoplankton, mesohaline Chesapeake Bay |
| 10 | 100 | | 2.7 | 9.1 | | | | | | | | | | | New York Bight | O'Connor et al., 1981 | Bowie et al.[93] | Total phytoplankton, literature values |
| | | 50 | | | 7 | | | | | | | | | 133 | New York Bight | O'Connor, 1981 | O'Connor[94] | Total phytoplankton |
| | | 50 | | | 10 | | | 1 | | | | | | | Potomac River Estuary | O'Connor, 1981 | O'Connor[94] | Total phytoplankton |

| | | | | | | | | | | | | Location | Reference | Reference | Note |
|---|---|---|---|---|---|---|---|---|---|---|---|---|---|---|---|
| | | 50 | | | | 7 | | | 0.5 | 70 | 133 | Sacramento San Joaquin Delta | O'Connor, 1981 | O'Connor[94] | Total phytoplankton |
| | | 33 | | | | 7 | | | 1 | | 88 | Patuxent River Estuary | O'Connor, 1981 | O'Connor[94] | Total phytoplankton |
| | | 50 | | | | 7 | | | 1 | | 133.5 | Patuxent River Estuary | Lung, 1992 | Lung[80] | Total phytoplankton, values used in modeling |
| | | 25 | | | | 7 | | | 1 | | 67 | James River Estuary | O'Connor, 1981 | O'Connor[94] | Total phytoplankton |
| | | 25 | | | | 7 | | | 1 | | 66.75 | James River Estuary | Lung, 1986 | Lung and Paerl[79] | Total phytoplankton |
| | | 50 | | | | 7 | | | 1 | | 133 | Neuse River Estuary | Lung and Paerl, 1986 | Lung and Paerl[79] | Total phytoplankton |
| | | 60 | | | | | | | | | | San Juan Bay Estuary | Bunch et al., 2000 | Bunch et al.[92] | Total phytoplankton |
| | | 50 | | | | | | | | | | Maryland coastal waters | Lung and Hwang, 1994 | Lung and Hwang[81] | Total phytoplankton, values used in modeling |
| 18 | 500 | | 2.2 | 74.6 | 19.2 | | 0.27 | 2.4 | | | | Sacramento San Joaquin Delta | DiToro et al., 1971 | Bowie et al.[93] | Diatom, literature values |
| | 275 | | | | 19.3 | | | 50.7 | | | | New York Bight | O'Connor et al., 1981 | Bowie et al.[93] | Dinoflagellates |

**TABLE 4.36**
**Nutrient Composition of Algal Cells—Ratio to Carbon**

| N:C (mg N/ mg C) | P:C (mg P/ mg C) | Si:C (mg Si/ mg C) | O:C (mg O₂/ mg C) | Location | References | As Cited in | Explanation |
|---|---|---|---|---|---|---|---|
| 0.167 | 0.0167 | | 2.67 | Indian River– Rehoboth Bay, Delaware, U.S.A. | Cerco and Seitzinger, 1997 | Cerco and Seitzinger[102] | Benthic algae |
| 0.150 | | | 2.66 | Venice Lagoon | Bergamasco et al., 1998 | Bergamasco et al.[103] | |
| | | 0.4 | | Chesapeake Bay | Cerco and Cole, 1994 | Cerco and Cole[91] | Regression of 38 observations at a station |
| | | 0.3 | | Chesapeake Bay | D'Elia et al., 1983 | Cerco and Cole[91] | |
| 0.167 | | 0.5 | | Chesapeake Bay | Cerco and Cole, 1994 | Cerco and Cole[91] | Model value |
| 0.250 | 0.0250 | | 2.67 | Maryland coastal waters | Lung and Hwang, 1994 | Lung and Hwang[81] | |
| 0.167 | | | 2.67 | San Juan Bay Estuary | Bunch et al., 2000 | Bunch et al.[92] | |

*4.1.5.4.3  Sediment Oxygen Demand*

$$\text{SOD} = \frac{E_{\text{DIF}}}{D_j}(C_{6i} - C_{6j}) \tag{4.36}$$

where for the benthic segment *j* and the water segment *i*
  SOD = sediment oxygen demand [g m⁻² day⁻¹]
  $E_{\text{DIF}}$ = diffusive exchange coefficient [m² day⁻¹]
  $D_j$ = benthic layer depth [m]
  $C_6$ = dissolved oxygen concentration [mg O₂ l⁻¹]

Sediment oxygen demand and temperature correction factors for SOD obtained from coastal waters are given in Table 4.37 and Table 4.38, respectively.

**TABLE 4.37**
**SOD Rate Constant (g $O_2$ m$^{-2}$ day$^{-1}$)**

| Min. | Max. | Avg. | Location | Method of Determination | Reference | As Cited in | Explanation |
|---|---|---|---|---|---|---|---|
| 1 | 2 | 1.5 | | | Thomann, 1972 | Bowie et al.[93] | Estuarine mud |
| 0.07 | 0.13 | 0.1 | | | NCASI, 1981 | Bowie et al.[93] | 12°C A North Carolina Estuary |
| 0.15 | 0.25 | 0.2 | | 45-day incubation of 0.6-l sediment in 3.85 BOD dilution water, light | NCASI, 1981 | Bowie et al.[93] | 20°C A North Carolina Estuary |
| 0.13 | 0.31 | 0.22 | | 45-day incubation of 0.6-l sediment in 3.85 BOD dilution water, light | NCASI, 1981 | Bowie et al.[93] | 28°C A North Carolina Estuary |
| 0.22 | 0.52 | 0.37 | | 45-day incubation of 0.6-l sediment in 3.85 BOD dilution water, light | NCASI, 1981 | Bowie et al.[93] | 36°C A North Carolina Estuary |
| 2.16 | 2.48 | 2.32 | Buzzard Bay | In situ dark respirometers stirred 1–3 days, temperature unknown | Smith et al., 1973 | Bowie et al.[93] | Near raw sewage outfall |
| 1.86 | 1.89 | 1.88 | Buzzard Bay | In situ dark respirometers stirred 1–3 days, temperature unknown | Smith et al., 1973 | Bowie et al.[93] | |
| 0.14 | 0.68 | | Puget Sound | Laboratory incubations | Pamatmat et al., 1973 | Bowie et al.[93] | Sediment cores 5°C |
| 0.2 | 0.76 | | Puget Sound | Laboratory incubations | Pamatmat et al., 1973 | Bowie et al.[93] | Sediment cores 10°C |
| 0.3 | 1.52 | | Puget Sound | Laboratory incubations | Pamatmat et al., 1973 | Bowie et al.[93] | Sediment cores 15°C |
| 1.25 | 3.9 | | Yaquina River Estuary, Oregon | Dark laboratory incubators stirred at 20°C | Martin and Bella, 1971 | Bowie et al.[93] | |
| 0 | 10.7 | | Delaware Estuary | In situ dark respirometers 13–14°C | Albert, 1983 | Bowie et al.[93] | 22 stations |
| 0.3 | 3 | | | In situ respirometry 0–18°C, laboratory cores 5–13°C | Edberg and Hofsten, 1973 | Bowie et al.[93] | Fresh and brackish water, Sweden |
| | | 0.2 | Venice Lagoon | | Melaku Canu et al., 2001 | Melaku Canu et al.[88] | Value used in the model |

*(Continued)*

**TABLE 4.37 (Continued)**
**SOD Rate Constant (g O$_2$ m$^{-2}$ day$^{-1}$)**

| Min. | Max. | Avg. | Location | Method of Determination | Reference | As Cited in | Explanation |
|------|------|------|----------|-------------------------|-----------|-------------|-------------|
| | | 2.71 | St. Leonard's Creek, Patuxent River Estuary | | Boynton et al., 1990 | Lung[80] | May 1985 |
| | | 3.54 | St. Leonard's Creek, Patuxent River Estuary | | Boynton et al., 1990 | Lung[80] | June 1985 |
| | | 1.02 | St. Leonard's Creek, Patuxent River Estuary | | Boynton et al., 1990 | Lung[80] | August 1985 |
| | | 0.79 | St. Leonard's Creek, Patuxent River Estuary | | Boynton et al., 1990 | Lung[80] | October 1985 |
| | | 1.5 | Buena Vista, Patuxent River Estuary | | Boynton et al., 1990 | Lung[80] | May 1985 |
| | | 1.69 | Buena Vista, Patuxent River Estuary | | Boynton et al., 1990 | Lung[80] | June 1985 |
| | | 0.7 | Buena Vista, Patuxent River Estuary | | Boynton et al., 1990 | Lung[80] | August 1985 |
| | | 1.79 | Buena Vista, Patuxent River Estuary | | Boynton et al., 1990 | Lung[80] | October 1985 |

**TABLE 4.38**
**SOD Temperature Correction Factor ($\theta$)**

| Average | Location | Reference | As Cited in | Explanation |
|---------|----------|-----------|-------------|-------------|
| 1.08 | Potomac River Estuary | Thomann and Fitzpatrick, 1982 | Bowie et al.[93] | |
| 1.1 | New York Bight | O'Connor et al., 1981 | Bowie et al.[93] | |
| 1.04 | | Genet et al., 1974 | Bowie et al.[93] | For an estuary |

## 4.2  ORGANIC CHEMICALS

On the earth, there are almost 4 million organic chemicals with names given and catalogued by the International Union of Pure and Applied Chemistry (IUPAC). Each year approximately 1000 new organic chemicals are synthesized and used commercially. The vast majority of these compounds break down in the environment, and only a fraction prove to be toxic and/or carcinogenic; some are known to be persistent and bioaccumulating. In toxicology, any toxic effect is related to two factors: the exposure dose and the susceptibility of the organism to the specific chemical being considered. The dose received by an organism is influenced by the available concentration of the contaminant in the environment and by the length of exposure. Therefore, for some compounds, a long-term exposure to a low available concentration may cause a toxic effect.[104] Uptake of pollutants by biota within the water column can result in bioaccumulation and transfer of the pollutants into the food chain.

To determine the exposure concentrations and doses of chemical compounds to living organisms, it is often helpful to use mathematical models to estimate the fate of the chemicals and their transport in the aquatic environment. Models can also be used to perform waste load allocations to meet water quality standards. According to the Organization for Economic Cooperation and Development (OECD), there are about 70,000 synthetic chemicals (mostly organic) in daily use. Numerous compounds are continuously introduced into the environment in large quantities, e.g., solvents, components of detergents, dyes and varnishes, additives in plastics and textiles, chemicals used for construction, antifouling agents, and pesticides.[105]

In order to assess the risk and rate of pollution of coastal waters, e.g., lagoons, attention must be given to the input of pollutants, their transport and transformations, the transfer of pollutants between the water and sediment, and the effects of the different pollutants on the aquatic organisms.

### 4.2.1  SOURCES OF ORGANIC CHEMICALS

There is widespread environmental contamination by organic chemicals from both natural sources and human activities. The main sources of organic chemicals in water are the degradation of naturally found organic substances, reactions occurring during drinking water and wastewater treatment and transmission, and land-based sources (point and nonpoint) of pollutants of domestic, industrial, and agricultural origin through atmospheric deposition, silviculture, leaching, etc. Coastal lagoons can receive some or all of these sources.

The most abundant organic chemicals are humic substances, which are biodegraded products of natural organic matter, humus. Humus is rich in nutrients and remains an important food supply for microorganisms for long periods.

Organic compounds formed during water treatment consist of disinfection by-products, such as trihalomethane and chloramines. Other chemicals, e.g., acrylamide, are coagulants used in water treatment. In addition, compounds, such as polyhalogenated aromatic hydrocarbons from transmission pipes and pipe connections are considered as secondary sources. Wastewaters of point sources include many compounds such as phenols that also are found in nonpoint sources. Domestic and industrial chemicals diffuse into aquatic systems via contaminated soil, urban surface run-off, surface run-off from agricultural areas, and from construction sites. Organic chemicals having adverse effects on the aquatic environment mostly belong to this latter group. Nonpoint sources include pesticides, such as chlordane and carbofuran; solvents, such as trichlorobenzene and tetrachloroethylene; metal redactors, such as trichloroethylene and trichloroethene; monomers and plasticizers, such as polychlorinated biphenyls; and polycyclic aromatic hydrocarbons from coal combustion units.[106]

### 4.2.2  CLASSIFICATION OF ORGANIC CHEMICALS THAT MIGHT APPEAR IN AQUATIC ENVIRONMENTS

*Combustion products of liquid petroleum products and coal.* The global annual production of liquid petroleum products (e.g., gasoline, kerosene, heating oils) is about 3 billion metric tons. Producing, transporting, processing, handling, storing, using, and disposing of these hydrocarbons pose numerous problems. Properties of various forms and compounds that interact with the environment include some easily vaporize; some bind to solids; some oil hydrocarbons are extremely nonreactive; some interact with light; some are quite nontoxic; and some are carcinogenic. The various properties of these chemicals are factors in the environmental threat.

*Halogenated methanes, ethanes, and ethenes.* Some of these compounds are inert and nonflammable, depending on the type and number of halogen substituents. Some exhibit physical properties that make them useful as aerobic propellants, refrigerants, blowing agents for plastic foams, and solvents for such purposes as dry cleaning and metal degreasing. Some of these chemicals are flammable and toxic, and a few are used as pesticides.

Over 85% of fluorocarbons (freons) manufactured are released into the atmosphere. These compounds are thought to be responsible for depleting the stratospheric ozone layer. The chlorinated solvents, dichloromethane and tri- and tetrachloroethane, are among the top ten organic groundwater contaminants. These compounds are quite persistent and mobile in groundwater, and their contamination of groundwater can, in turn, lead to contamination of coastal waters.

*Polyhalogenated aromatic hydrocarbons.* Because of their tendency to bioaccumulate, these chemicals, for example polychlorinated biphenyls (PCBs), have attracted considerable attention. More than 1 million metric tons of PCBs have been produced and used as capacitor dielectrics, transformer coolants, hydraulic fluids, heat transfer fluids, and plasticizers. PCBs are commonly applied as complex mixtures of congeners (i.e., isomers and compounds exhibiting different numbers of chlorine atoms but originating from the same source) and enter the environment mostly from production, storage, and disposal sites. PCB use has been restricted in numerous countries; however, PCBs are still ubiquitous in the environment.

*Phthalates.* These compounds are diesters of phthalic acid. The annual world production of phthalates exceeds more than 1 million metric tons. The phthalates are mainly used as plasticizers, particularly to make polyvinylchloride (PVC) flexible, and they are ubiquitous in the environment. In aqueous solution, like all carboxylic acid esters, phthalates can hydrolyze to form the corresponding acid (i.e., phthalic acid) and alcohol(s) (i.e., ROH, $R^IOH$).

*Surfactants:* Commercially available surfactants are not uniform substances but mixtures of compounds of different carbon chain lengths. They have an amphiphilic (partly hydrophilic and partly hydrophobic) characteristic and special properties that render them unique among environmental chemicals. In aqueous solutions, they distribute in such a manner that their concentration at the interfaces of water with gas or solids is higher than it is in the inner regions of the solution. This results in a change of system properties, e.g., a lowering of the interfacial tension between water and an adjacent nonaqueous phase, and in a change of wetting properties. Surfactants are also widely used as wetting agents, dispersing agents, and emulsifiers in all kinds of consumer products and industrial applications.[107] Because of direct use in water, a large portion of over 3 million metric tons of surfactants is annually discharged into domestic and industrial wastewater. Since these compounds often constitute a significant part of the organic carbon loading of wastewater, their biological degradation is of particular interest. Giger et al.[108] found that during biological treatment, toxic products were formed from nonionic surfactants of the alkylphenol polyethyleneglycol ether type. Some of the toxic products were detected in very high concentrations in sewage sludge. Examples of commercially important surfactants include anionic surfactants (soaps, linear alkylbenzene sulfonates, secondary alkyl sulfonates, fatty alcohol sulfates), cationic surfactants (quaternary ammonium chloride), and nonionic surfactants (alkylphenol polyethyleneglycol ethers, fatty alcohol polyethyleneglycol ethers).

*Pesticides (herbicides, insecticides, fungicides).* In general, pesticides are synthesized so as to have a specific detrimental biological effect on one or several target organisms (e.g., plants, insects, and fungi), but (ideally) they should have little impact on the rest of the living environment in which they are applied. In order to achieve this goal, compounds with very special structures, often exhibiting several functional groups, have to be employed. World pesticide consumption was estimated to be around 5 million tons in 2000.[109] Because several functional groups are frequently present in a pesticide molecule, the prediction of the distribution, mobility, and reactivity of such compounds in the environment is often more difficult than that for less complex chemicals. Pesticides, because they are toxic, are of utmost importance when they reach water resources by means of percolation or surface run-off. Although certain characteristics of pesticides are well known, their exhibited characteristics upon reaching a waterbody are extremely difficult to estimate.[110] WHO and the U.S. EPA classify pesticides according to their toxicological effects. The classifications are based on increasing acute $LD_{50}$ values in male mice (lethal dose in 50% of the test animals).[111]

The time lapse between pesticide application and the first rainfall event and the rainfall intensity are the most important factors affecting the loss of pesticides from agricultural land. A considerable part of total pesticide loss occurs during the first rainfall after application. In addition to the chemical properties of the pesticides and the weather conditions, soil properties have a significant effect on pesticide leaching and surface run-off.[112]

The most important processes to be taken into account when evaluating the effects of pesticides on aquatic ecosystems are leaching, degradation, accumulation, and toxicity. Pesticides with low solubility in water and which are tightly adsorbed to soil particles can be washed out only by surface run-off. However, many pesticides with high water solubility are currently widely applied. These pesticides can also be leached from the soil by subsurface drainage systems and can infiltrate to ground-water.[112]

Pesticides are a chemically diverse group of compounds. Some are produced naturally by certain species of plants, but chemists have synthesized the great majority of organic pesticides. Prominent organic pesticides include natural organic compounds (alkaloid nicotine, pyrethrum), synthetic organometallic chemicals (organomercurials), and phenols (tri-, tetra-, and penta-cholorophenol). Chlorinated hydrocarbons are a diverse subset of synthetic pesticides; prominent examples are DDT and lindane. Other pesticides classes include cyclodienes, organic phosphorus compounds, carbamates, triazine herbicides, and synthetic pyrethroids.

Detailed information on pesticides is given in Montgomery,[113] Edit et al.,[114] Clark et al.,[110] and Line et al.[115] BCPC,[116] DPR,[117] EXTOXNET,[118] Watts and Moore,[119] Jorgensen and Gromiec,[120] Miyamoto et al.,[121] Bennett,[122] Larson et al.,[123] Rao et al.,[124] and Hemond and Fechner-Levy[125] discuss fate and transport mechanisms of pesticides in the aquatic environment.

### 4.2.3 FATE OF ORGANIC CHEMICALS IN AQUATIC ENVIRONMENTS

After a chemical substance enters the water, a number of processes will act to decrease its concentration. Part of the substance can be transferred to the atmosphere via volatilization, and the rest will distribute itself in water, suspended solids, and living biota where chemical and microbiological transformations can occur. Particles, with their associated chemicals, can sink through the water column and be incorporated into the sediment. Macrobiota can metabolize a chemical or accumulate it within their tissues. The latter, however, results in a very minor loss. The same processes will also govern the fate of the products of chemical and biological transformation. In some cases, the products of such reactions may be of greater concern than the precursor compound, but in others, the ultimate result of these transformations may be that the compound is completely mineralized. These transformations are caused by indirect and direct photolysis, oxidation, hydrolysis, and microbial biotransformation.[119] Sorption can occur in both suspended particles and bottom sediment; the former maintains the pollutant in the water column, while the latter effectively immobilizes it. There can also be some association with dissolved organic material present in the water body, effectively increasing the solubility of the pollutant in water.

In coastal waters including lagoons, salinity-related changes in chemical reaction rates are important, and are generated both by mixing-controlled changes in the relative concentrations of reactants and by the influence of ambient ionic strength on the activities of the reacting species. It is, therefore, expected that solubility, speciation, and partition behavior of organic contaminants might vary as a result of salinity-induced changes in pH. Occasionally, extreme pH changes might occur as a result of input of acidic or alkaline waste into coastal waters. When salinity increases, organic compounds become less soluble in water and, hence, become more sorbable, leading to increased sorption on sediment particles.[126] The pH strongly influences adsorption, since hydrogen and hydroxide ions are adsorbed and the charges of other ions are influenced by the pH of the water. For typical organic pollutants, adsorption increases with increasing pH. This is important for coastal lagoons where the pH is usually higher than 8. Normally, the adsorption reactions are exothermic, which means that adsorption will increase with decreasing temperature, although small variations in temperature do not tend to alter the adsorption process to a significant extent.

A schematic fate model for organic chemicals in water and sediment is given in Figure 4.6.

Ney,[127] Kolset et al.,[128] Saleh,[129] Rajar,[130] and Schnoor[131] report evaluation of some chemical fate and transport models. Bowmer,[132] Day,[133] Miles,[134] Saleh,[129] Inaba et al.,[135] and Jaskulke et al.[136] emphasize the evaluation of pesticide residues in water.

### 4.2.3.1  Volatilization

Volatilization is the transfer of chemicals from the liquid phase to the atmosphere. It does not result in the breakdown of a substance, only its movement from the liquid to gas phase, or vice versa. Volatilization is one of the key processes affecting the transport and distribution of many organic compounds in the environment.[106] Here are few examples of the importance of this process, especially in coastal lagoons, are:

- When the oxygen concentration in water is higher than the temperature-dependent equilibrium concentration between water and the atmosphere, oxygen is transferred from the water to the air. This condition can exist in eutrophic lagoons during summer when photosynthesis is dominant.
- When the carbon dioxide concentration in water is higher than the equilibrium concentration between water and atmosphere, surplus carbon dioxide will be transferred from the hydrosphere to the atmosphere. However, the transfer of carbon dioxide from atmosphere to water is, in general, more normal and important than the opposite process. Carbon dioxide is produced by the combustion of fossil fuel, and about 60% of the carbon dioxide produced is dissolved in the sea. However, the transfer of carbon dioxide from water to air is environmentally significant for water bodies with low pH.[106] Acidification of the water environment means that bicarbonate is transformed into carbon dioxide, which then escapes from the water. This implies that carbon can become a limiting nutrient in those circumstances and the water environment can turn oligotrophic.

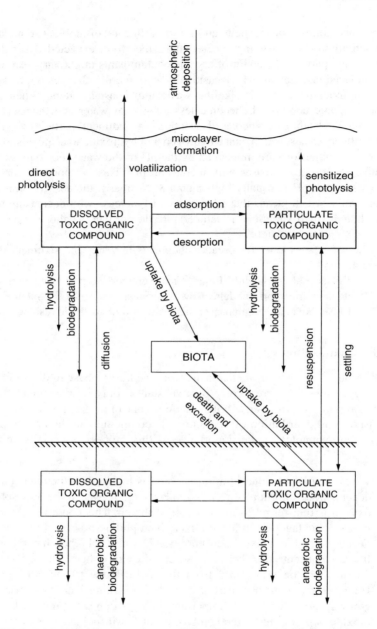

**FIGURE 4.6** Schematic fate model for organic chemicals in water and sediment.

- Ammonia is very soluble in water, but because the concentration of ammonia in the atmosphere is small, volatilization of ammonia from water to air will always take place. At neutral pH, most of the total ammonia is in the form of ammonium ion; however, during summer in eutrophic coastal environments, such as lagoons, a pH of 8.5 or even higher is observed, producing a greater ratio of ammonia to ammonium ion. This ratio can be calculated from the mass balance equation, and the influence of salinity on this ratio can be accounted for by use of the concept of ionic strength, I.[131] For example, at pH 9.3, about half of the total ammonia is ammonia. Therefore, the volatilization of ammonia can be significant in very eutrophic environments during the summer months.
- Volatilization is an important process for many toxic organic compounds in the environment. When discharged into the water environment even in small concentrations, they often have sufficient vapor pressure to give rise to significant volatilization.

In many cases, volatilization is the most important removal process for toxic substances from aquatic ecosystems, since other processes, including biodegradation, can be very slow.

Some of the factors that affect volatilization in the environment, apart from vapor pressure, are climate, sorption, hydrolysis, and phototransformation. A chemical with a low vapor pressure (VP), high adsorptive capacity, or high water solubility is less likely to volatilize into the air. A chemical with a high VP, low sorptive capacity, or very low water solubility is more likely to volatilize into the air. Chemicals that are gases at ambient temperatures will get into the air.[131,137]

During volatilization, the dissolved concentration of the unionized molecule attempts to equilibrate with the gas phase concentration. Equilibrium occurs when the partial pressure exerted by the chemical in solution equals the partial pressure of the chemical in the overlying atmosphere. The rate of exchange is proportional to the gradient between the dissolved concentration and the concentration in the overlying atmosphere, and to the conductivity across the interface between the liquid and gas phase. The conductivity (mass transfer coefficient) is influenced both by chemical properties, such as molecular weight and Henry's law constant, and by environmental conditions at the air–water interface, e.g., turbulence-controlled by wind speed, current velocity, and water depth.[78]

The governing volatilization reaction equation is presented and discussed later in Section 4.2.4. Liquid–gas transfer models are often based on the two-film theory, as illustrated in Figure 4.7. The mass transfer rate is governed by molecular diffusion through a stagnant liquid and gas film at the interface. Mass moves from areas of high concentration to areas of low concentration. The transfer rate can be limited at either the gas film or liquid film side of the interface. Oxygen, for example, is controlled by the liquid film resistance. Nitrogen gas, although approximately four times more abundant in the atmosphere than oxygen, still has a greater liquid film resistance than oxygen.[131] Concentration differences serve as the driving force for the water layer diffusion. Pressure differences drive the diffusion for the air layer. From mass balance considerations, it is obvious that the same mass must pass

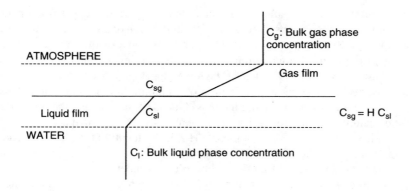

**FIGURE 4.7** Two-film theory of gas–liquid transfer.

through both films; therefore, the two resistances combine in series, and the total conductivity (transfer rate constant) is the reciprocal of the total resistances.[78] In addition to the liquid and gas film resistance, there is actually another resistance involved, the transport resistance between the two interfaces, which is assumed to be negligible.

The value of the conductivity (transfer rate constant) depends on the intensity of turbulence in the water body and in the overlying atmosphere. As the compound's Henry's law constant increases, the conductivity tends to be increasingly influenced by the intensity of turbulence in water. Conversely, as the Henry's law constant decreases, the value of the conductivity tends to be increasingly influenced by the intensity of atmospheric turbulence. Volatilization, as described by the two-film theory, is a function of the compound's Henry's law constant, the gas film resistance, and the liquid film resistance. As described previously, film resistances depend on diffusion and mixing. The Henry's law constant is the ratio of a chemical's vapor pressure to its solubility. It is also thermodynamically the ratio of the fugacity of the chemical (escaping tendency from air and water) in the air to that in water.

$$H = p_g/C_{sl} \qquad (4.37)$$

where

$H$ = Henry's law constant for the air–water partitioning of the chemical [atm m$^3$ mole$^{-1}$]

$p_g$ = the partial vapor pressure of the chemical of interest in the gas phase (air)

$C_{sl}$ = the chemical's saturation solubility in water

Henry's law constant (H) serves as a measure of a chemical's volatility caused by water; the larger a chemical's H value, the more volatile it is, and the more easily it will transfer from the aqueous phase to the gas phase.[138] The H value of a compound can be used to develop simplifying assumptions for modeling volatilization. If the resistance of either the liquid film or the gas film control is significantly greater than the other, the lesser resistance can be neglected. The threshold of H for gas or liquid film control is approximately 0.1 for the dimensionless H or about $2.2 \times 10^{-3}$ atm-m$^3$·mol$^{-1}$ for the water–air system. Above this threshold value, chemical volatilization is liquid-film controlled, and below it, it is often gas film controlled.[131] However, in actual cases, an intermediate (mixed) stage usually occurs.

Because H generally increases with increasing vapor pressure and generally decreases with increasing solubility of a compound, highly volatile low solubility compounds are more likely to exhibit mass transfer limitations on the water side, whereas relatively nonvolatile, high solubility compounds are more likely to exhibit mass transfer limitations on the air side.[78]

### 4.2.3.2 Ionization

Ionization is the dissociation of a chemical into multiple charged species. In an aquatic environment, some chemicals occur only in their neutral form, while others react with water molecules to form positively (cationic) or negatively (anionic) charged ions. These reactions are rapid and are generally assumed to be at (local) equilibrium. At equilibrium, the pH and temperature of water along with the ionization constant control the distribution of a chemical's form between its neutral and ionized species.

Ionization is important because of the different toxicological and chemical properties of the neutral and ionized species. For example, in some cases, only the neutral form of the chemical can transform or be transported through biotic membranes, resulting in toxicity. As a result, it is often necessary to compute the distribution of a chemical's form among its ionic and neutral forms as well as to allow each species to react, transform, or sorb at different rates.[78]

Several studies have shown that there is a relationship among salinity, pH, temperature, and the concentration of organic compound on the toxicity of heavy metals. The ionic forms of heavy metals are more toxic, because these forms can easily permeate the cell membranes of aquatic organisms. Observed changes in the biological availability of metals with pH, temperature, and alkalinity are the result of several factors. In order to predict heavy metal effects, it is clear that pH, temperature, and alkalinity of the water must be known.[139,140] It is also noted that the presence of the organic complexes invariably decreases the concentration of the ionic forms of heavy metals.[141] For example, there is a negative effect of increased salinity on the uptake of cadmium by aquatic organisms. The increased availability of cadmium at low salinity is actually due to increased free cadmium ion level.[142]

### 4.2.3.3 Sorption

Soluble organic materials in natural waters can sorb onto particulate suspended material or bed sediment.[131] Sorption is a process whereby a dissolved substance is transferred to and becomes associated with solid material. It includes both the accumulation of dissolved substances on the surface of solids (adsorption) and the interpenetrating or intermingling of substances with solids (absorption). The substance that is sorbed is called the sorbate, and the solid is called the sorbent. Desorption is the process whereby a sorbed substance is released from a particle.[45]

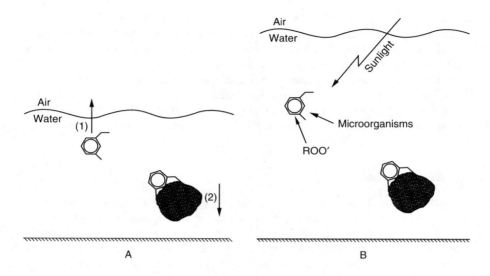

**FIGURE 4.8** Some processes in which sorbed species behave differently than dissolved molecules of the same substance. (A) Dissolved species can volatilize (1), whereas sorbed species can settle to the sediment (2). (B) Dissolved and sorbed species react at different rates.

Sorption is extremely important, because it may dramatically affect the fate and impact of a chemical in the aquatic environment. Structurally identical molecules behave very differently if they are surrounded by water molecules and ions as opposed to adsorbing onto the exterior of solids or being absorbed within a solid matrix.[106] As illustrated in Figure 4.8, sorption can cause a compound to accumulate in bed sediment or bioconcentrate in fish. It can retard processes such as volatilization and base hydrolysis, or enhance others such as photolysis and acid-catalyzed hydrolysis.[78] The sorption of toxicants to suspended particulates and bed sediment is a significant transfer mechanism. Partitioning of a chemical between particulate matter and the dissolved phase is not a transformation pathway; it only relates its concentration in the dissolved phase to that in the solid phase.[45] Figure 4.8A is an example of dissolved species that can volatilize while sorbed species may settle to sediment, whereas Figure 4.8B presents dissolved and sorbed species that react at different rates.

A significant portion of the impurities in water is found in the suspended matter, where the concentration can be much higher than in the water. Transport of pollutants in aquatic systems often takes place on suspended matter, either clay particles or organic matter (cohesive particles). This implies that many pollutants, otherwise adsorbed or fixed by ion exchange on the sediment and, therefore, not transported by clear water, are transported to a greater distance by water with high turbidity (see Section 4.2.4).

For neutral organics, several mechanisms are involved in the sorption process. These include hydrophobic effects that cause the sorbate to associate with organic matter in the particulate phase because of an unfavorable free-energy cost of staying in solution; weak surface interactions via van der Waals, dipole-dipole, induced dipole, and other weak intermolecular forces; and surface reactions where the sorbate actually binds with the solid.[106]

For charged pollutants, the additional mechanism of ion exchange can occur to promote sorption. It should also be mentioned that, aside from organically rich materials, solids with little or no organic content also sorb neutral organic chemicals. For such cases, the sorbent consists of inorganic matter, such as clay. Such inorganic solid sorption is usually significant only when the organic carbon content of the solids is quite low.[45]

Adsorption–desorption in natural waters is in some cases a reversible reaction. The chemical is initially dissolved (e.g., discharged into) in water in the presence of various concentrations of suspended solids. After an initial kinetic reaction, a dynamic equilibrium is established in which the rate of the forward reaction (sorption) is exactly equal to the rate of the reverse reaction (desorption). After equilibrium is attained, samples taken from the filtered water and/or solids and analyzed for chemical concentration can be used to obtain the partition coefficient. Sorption reactions usually reach equilibrium quickly, and the kinetic relationships can often be assumed to be at steady state. This is sometimes referred to as the "local equilibrium" assumption, when the kinetics of adsorption and desorption are rapid relative to other kinetic and transport processes in the system.[45] The rate expressions of these sorption–desorption processes are given in Section 4.2.4.

Adsorption and ion exchange are significant processes in the environmental context. Whenever water is in contact with suspended matter (organic matter or clay particles), sediment, or biota, a significant transfer of chemicals by adsorption and ion exchange can take place. Pure adsorption and ion exchange are rarely observed in nature. A mixture of the two processes is most often observed.[143] Adsorption is the transfer of components from the liquid phase onto the surface of a solid phase. It is often explained as an electrical attraction to the solid surface of chemicals possessing a (minor) electrical charge. Adsorption results in the formation of a molecular layer of adsorbate on the solid surface. Often, an equilibrium adsorbate concentration is rapidly formed on the solid surface that is sometimes followed by a slow diffusion of the adsorbate into the particles of the adsorbent. In contrast, the ion exchange process is an actual exchange of ions between a liquid in contact with the solid phase. The ion exchange process can be explained in the same way as any other chemical process. That is, the chemical energy at equilibrium after the process has occurred is lower than it was before the process was initiated. If pure ion exchange occurs, the number of ions released is equivalent to the number of ions taken up by the process.

Adsorption and ion exchange are fast processes, and both have great importance in water quality modeling whenever the concentration of suspended matter in the water is sufficiently high so that significant amounts of the modeled chemical compounds are adsorbed or present on the exchange complex. The normal model simulation time interval is weeks, days, or hours. This implies that these two processes can be simulated using equilibrium equations. Modeling of surfactants, pesticides, and heavy metals often emphasizes adsorption and ion exchange processes as these materials represent easily sorbed chemicals.[120]

The concentrations of pollutants in other environmental compartments can be estimated by applying appropriate partition coefficients to the water concentrations. However, reliable adsorption isotherms and equilibrium constants for ion exchange are available only for a very limited number of chemicals and sorbents of importance for environmental modeling. Therefore, scientifically defensible estimation methods are needed. To a certain extent, it is possible to predict the adsorption behavior of a particular sorbate–sorbent combination. The solubility of the dissolved chemical is by far the most significant factor in determining the intensity of the driving forces. The greater the affinity of the chemical compound for the solvent the less likely it is to move toward an interface to be adsorbed. For an aqueous solution this means that the more hydrophilic the chemical is, the less likely it is to be adsorbed. Conversely, hydrophobic substances will easily be adsorbed from aqueous solutions. Many organic compounds, e.g., sulfonated alicyclic benzenes, have a molecular structure consisting of both hydrophilic and hydrophobic groups. The hydrophobic parts will be adsorbed at the surface and the hydrophilic parts will tend to stay in the water phase. Hydrophobic organic chemicals, such as DDE and PCB, strongly sorb to solids and tend to concentrate in the bottom sediment.[144]

The value of the partition coefficient is dependent on numerous factors in addition to the fraction of organic carbon of the sorbing particles. Of these, perhaps the most potentially significant and controversial is the effect of particle (suspended solids) concentration. Based on empirical evidence, the partition coefficient appears to be inversely related to the (suspended) solids concentration. The partition coefficient can be calculated by

$$K_p = f_{oc} \cdot K_{oc} \qquad (4.38)$$

where

$K_p$ = sediment/water partition coefficient suitable for natural waters

$f_{oc}$ = the decimal fraction of organic carbon present in the particulate matter [mass/mass]

$K_{oc}$ = organic carbon-normalized partition coefficient [(mg chemical/kg organic carbon)/(mg chemical/l of $H_2O$)]

Once an estimate of $K_{oc}$ is obtained, the calculation of $K_p$ is straightforward.

For a wide variety of organic chemicals, the octanol/water partition coefficient $K_{ow}$ is a good estimator of the organic carbon-normalized partition coefficient $K_{oc}$. Octanol was chosen as a reference because it is a model solvent with properties that make it similar to organic matter and lipids in nature.

The octanol/water partition coefficient $K_{ow}$ is related to the solubility of a chemical in water.[137] The less soluble a chemical is in water, the higher is its octanol/water partition coefficient (log $K_{ow}$), and the more likely it is to sorb to the surfaces of sediment or suspended particles.[119] $K_{ow}$ values for a few organic compounds are reported by Schnoor.[131] If the octanol/water partition coefficient cannot be reliably measured, or is not available in databases, it can be estimated from solubility and molecular weight and structure information.[131] More detailed information about estimating these coefficients is available in SPARC.[145]

Values for partition coefficients can be obtained from laboratory experiments. As stated earlier, laboratory studies have shown that the partition coefficient is related to the hydrophobicity of the chemical and organic matter content of the sediment for organic chemicals. Many organic pollutants of current interest are nonpolar, hydrophobic compounds whose partition coefficients correlate quite well with the organic fraction of the sediment. Karickoff et al.[146] and Schwarzenbach and Westall[147] have developed empirical expressions relating equilibrium partition coefficients to laboratory measurements, leading to fairly reliable means of estimating appropriate values for $K_{oc}$ as given below:

Karickhoff et al.[146]                    $\log K_{oc}: 1.00 \log K_{ow} - 0.21$    (4.39)

Schwarzenbach and Westall[147]            $\log K_{oc}: 0.72 \log K_{ow} + 0.49$    (4.40)

These relationships have proven very useful for predicting equilibrium partition coefficients of a great number of neutral hydrophobic organic compounds between water and natural sorbents of very different origins. Other chemical property estimation methods are reported by Lyman et al.[148] Table 4.39 provides a list of measured $\log K_{ow}$ values for a number of organic chemicals of environmental interest. Determination of $K_{ow}$ values for organic chemicals needs experimental analysis. Recent data on such values may be found in BCPC,[116] EXTOXNET,[118] and the Merck Index.[149]

**TABLE 4.39**
**Octanol/Water Partition Coefficients of Some Selected Organics**

| Chemical | $\log K_{ow}$ | As Cited in |
| --- | --- | --- |
| Bromoform | 2.3 | Schnoor et al.[150] |
| Carbofuran | 1.60 | Schnoor[131] |
| Carbofuran | 17–26 | DPR[117] |
| Chloroform | 1.90–1.97 | Schnoor[131] |
| Chloroform | 1.97 | Schnoor et al.[150] |
| Chloroform | 6.22 | Merck Index[149] |
| 2-Choloroethyl vinyl ether | 1.28 | Schnoor[131] |
| 2-Choloroethyl vinyl ether | 4.08–4.28 | Kidd and James[151] |
| DDT | 6.4 | Schnoor[131] |
| DDT | 3.98–6.36 | Pontolillo and Eganhouse[152] |
| Ethylbenzene | 3.34 | Schnoor et al.[150] |
| Hexachlorobenzene | 6.41 | Schnoor et al.[150] |
| Pentachlorophenol | 5.01 | Schnoor[131] |
| 2-Nitrophenol | 1.75 | Schnoor et al.[150] |
| Tetrachloroethylene | 2.88 | Schnoor et al.[150] |

Adsorption isotherms are used to define the equilibrium relationship of sorption between organic chemicals and metals and particulate matter at constant temperature. Several models have been proposed to mathematically represent these isotherms.[45] Among the most popular are the Langmuir and Freundlich isotherms, which are mentioned in Section 4.2.4.

Examples of models of especially toxic organic compounds in water include the Mirex behavior model developed by Halfon;[153] the model of hydrophobic organic pollutants by Schwarzenbach and Imboden;[154] and the model of toxins in the Tamar Estuary, U.K., developed by Harris et al.[155] The effects of toxic substances on aquatic biota are often determined by the concentration of the dissolved fractions. It is, therefore, of some significance to correctly formulate the equilibrium between the sorbed and the dissolved components.[120] This was clearly demonstrated in the model developed by Schwarzenbach and Imboden.[154]

The extent of adsorption is also proportional to the sorbent surface area. To be able to compare different adsorbents, a specific surface area, defined as that portion of the total surface area available for adsorption per unit of adsorbent, is used. This means that the adsorption capacity of a nonporous adsorbent should vary inversely with the particle diameter, whereas for highly porous adsorbents the capacity should be almost independent of the particle diameter. Because flocculation and consequently increasing floc size and pore characteristics cannot occur at salinities below 2%, adsorption will indirectly be affected by salinity. This is an important phenomenon for coastal lagoons in which salinities vary over in a wide range. The nature of the adsorbate also influences adsorption. In general, an inverse relationship can be anticipated between the adsorption of a solute and its solubility in the solvent (water) from which adsorption occurs. This is the so-called Lundilius' rule, which may be used for the semi-quantitative prediction of the effect of the chemical character of a solute on its uptake from solution (water). Ordinarily, the solubility of any organic compound in water decreases with increasing chain length because the compound becomes more hydrophobic as the number of carbon atoms increases. This is Traube's rule. Together, these rules also suggest that increasing ionization means decreasing adsorption (when water is the solvent).

### 4.2.3.4  Hydrolysis

Hydrolysis of organic compounds in the aquatic environment is of major environmental interest since it is one of the most important mechanisms for the breakdown of pollutants. Organic pollutants can undergo reactions with water, resulting in the introduction of a hydroxyl group into the chemical structure:

$$RX + H_2O \rightleftharpoons ROX + HX \tag{4.41}$$

$$RCOX + H_2O \rightleftharpoons RCOOH + HX \tag{4.42}$$

Oxonium and/or hydroxyl ions catalyze these reactions. Hydrolysis refers to the reaction of an organic compound (RX) or (RCOX) with water, resulting in a net exchange of group X for the OH group from the water at the reaction center.

Dominant reaction pathways can be determined quantitatively from data on hydrolysis rate constants and half-lives. Hydrolysis data will generally be important in assessing risks from organic chemicals that have hydrolyzable functional groups (e.g., esters, amides, alkyl halides, epoxides, and phosphoric esters).

As stated previously, a hydrolysis reaction can be catalyzed by acidic or basic species, including $H_3O^+$ ($H^+$) and $OH^-$. The promotion of hydrolysis by $H^+$ or $OH^-$ is known as specific acid or specific base catalysis. Hydrolysis rates are the same in natural freshwaters and in buffered distilled water at the same temperature and pH. Thus, only specific acid or base catalysis, together with the neutral reaction, needs to be considered. Although other chemical species can catalyze hydrolysis reactions, the available concentrations of these species in the environment are usually too low to have any effect and are not expected to contribute significantly to the rate of hydrolysis.[45,156] Hydrolysis can be simulated using rates that are first order for the neutral chemical and second order for its ionic forms. The second-order rates are pH and temperature dependent.[78] Hydrolysis and chemical oxidation of natural organic matter are usually insignificant compared to microbiological decomposition. The influence of hydrolysis on the oxygen budget is, therefore, mostly unimportant. However, these processes can be of great importance for specific, toxic organic compounds. Therefore, many fate transport and exposure models of toxic substances in aquatic ecosystems include these processes.[120,157] The hydrolysis rate equations commonly used are presented in Section 4.2.4.

In the environment, hydrolysis of specific organic chemicals (pollutants) occurs in dilute solution. Under these conditions, water is present in excess and the concentration of chemicals is essentially constant during hydrolysis. Hence, the kinetics of hydrolysis is rather pseudo first order at a fixed pH. Many environmental factors influence the rate of hydrolysis, such as temperature, pH, solubility, sunlight, adsorption or absorption, and volatility.[137]

Mabey and Mill,[158] Wolfe et al.,[159] and Tinsley[160] have studied the mechanisms of hydrolysis and provide predictive methods to estimate the kinetic rates of hydrolysis of various compounds. Table 4.40 gives examples of the hydrolysis rates of selected halogenated compounds.

Some chemicals, including alkyl halides, were found to have hydrolysis rates that were independent of pH over the usual environmental pH range of 4–9, while other chemicals, such as carboxylic acid esters, had hydrolysis rates that were highly dependent on pH (see Figure 4.9).[131]

Recent studies have shown that the hydrolysis of organic compounds actually proceeds in the sediment-sorbed state. It has been reported that on sediment surfaces alkaline hydrolysis is retarded, neutral hydrolysis is unaltered, and acid hydrolysis is unaltered or accelerated as compared to the water phase process. It is difficult to predict the magnitude of rate constants for hydrolysis reactions in the sediment-sorbed state.[121]

### 4.2.3.5  Oxidation

All organic matter will undergo oxidation if present in a natural aerobic aquatic environment for a sufficiently long period of time. If reduced material is sufficiently abundant, however, the oxygen dissolved in interstitial water and/or in the lower

**TABLE 4.40**
**Hydrolysis Rates at 25°C and pH 7 of Selected Halogenated Compounds**

| | Rate Constants (1/s) | | | |
|---|---|---|---|---|
| Compound | $K_N$, Neutral-Catalyzed Hydrolysis Reaction Rate Constant | $K_B$ [OH$^-$], Base-Catalyzed Hydrolysis Reaction Rate Constant | KH, Pseudo-First-Order Reaction Rate Constant | $T_{1/2}$ Half-Lives |
| CH$_3$F [a] | $7.44 \times 10^{-10}$ | $5.82 \times 10^{-14}$ | $7.44 \times 10^{-10}$ | 30 years |
| CH$_3$Cl [a] | $2.37 \times 10^{-8}$ | $6.18 \times 10^{-13}$ | $2.37 \times 10^{-8}$ | 339 days |
| CH$_3$Br [a] | $4.09 \times 10^{-7}$ | $1.41 \times 10^{-11}$ | $4.09 \times 10^{-7}$ | 20 days |
| CH$_3$I [a] | $7.28 \times 10^{-8}$ | $6.47 \times 10^{-12}$ | $7.28 \times 10^{-8}$ | 110 days |
| CH$_3$CHClCH$_3$ [a] | $2.12 \times 10^{-7}$ | — | $2.12 \times 10^{-7}$ | 38 days |
| CH$_3$CH$_2$CH$_2$Br [a] | $3.86 \times 10^{-6}$ | — | $3.86 \times 10^{-6}$ | 26 days |
| CH$_2$Cl$_2$ [a] | $3.2 \times 10^{-11}$ | $2.13 \times 10^{-15}$ | $3.2 \times 10^{-11}$ | 704 years |
| CHCl$_3$ [b] | — | $6.9 \times 10^{-12}$ | $6.9 \times 10^{-12}$ | 3500 years |
| CHBr$_3$ [b] | — | $3.2 \times 10^{-11}$ | $3.2 \times 10^{-11}$ | 686 years |
| CCl$_4$ [b,c] | — | $4.8 \times 10^{-7}$ | $4.8 \times 10^{-7}$ | 7000 years (1 ppm) |
| C$_6$H$_5$CH$_2$Cl | $1.28 \times 10^{-5}$ | — | $1.28 \times 10^{-5}$ | 15 h |

[a]$K_H = K_N$; $K_B \ll K_N$

[b]$K_H = K_B$; $K_N \ll K_B$

[c]Rate second order with respect to [CCl$_4$], $K_H$ (1 mol$^{-1}$ s$^{-1}$)

*Source:* Jorgensen, S.E. and Gromiec, M.J., *Mathematical Submodels in Water Quality Systems*, Elsevier, Amsterdam, 1989. With permission.

layers of the aquatic ecosystem will be exhausted. Oxidation of the organic matter, however, will continue after the dissolved oxygen is exhausted by denitrification and sulfur reduction. All these processes can, in principle, occur by pure chemical oxidation, but in general microbiological oxidation plays a far more important role.

The source of free radicals in natural waters is the photolysis of naturally occurring organic molecules. If a water body is turbid, or very deep, these free radicals are likely to be generated only near the air–water interface; consequently, chemical oxidation will be relatively less important.

Oxidation processes are not always effective for all compounds, particularly so for organochlorines.[161] Although chemical oxidation of natural organic matter is usually insignificant compared to microbial decomposition, chemical oxidation can be of great importance for specific, toxic organic compounds. A chain sequence of processes, some chemical and some biochemical, usually decompose specific organic compounds. For modeling purposes, it is important to know at least the slowest reaction and the type of kinetics valid for each because they will determine the overall decomposition rate.[143]

Aquatic systems contain several oxidants, such as radicals, singlet oxygen, and triplet diradicals, with half-lives of only a few milliseconds, in addition to more

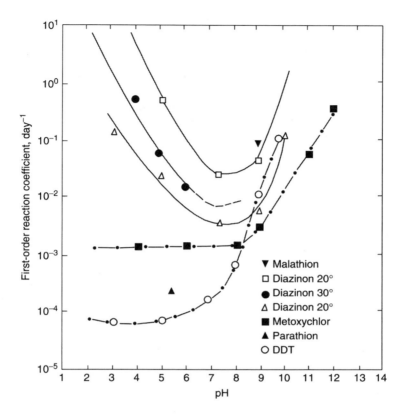

**FIGURE 4.9** Effect of pH on hydrolysis rate constants. (From Schnoor, C.J.L., *Environmental Modeling: Fate and Transport of Pollutants in Water, Air, and Soil*, Wiley-Interscience, New York, 1996. With permission.)

stable oxidants, such as peroxides, peracids, and ozone. Table 4.41 summarizes some representative organic chemical oxidation rates.[120]

Chemical oxidation reactions occur in natural waters whenever a sufficient level of oxidant is present. Common oxidants other than dissolved oxygen are chlorine and ozone. The general form of the chemical oxidation equation is given in Section 4.2.4. According to Ambrose et al.[78] chemical oxidation can be modeled as a general second-order process for the various species and phases of each chemical as referred to in Section 4.2.4.

### 4.2.3.6  Photolysis

Photolysis refers to the breakdown of chemicals due to the radiant energy of light. Light transforms compounds by two general modes. The first, called direct photolysis, occurs through absorption of light by the compound itself. The second, called sensitized or indirect photolysis, represents a group of processes that are initiated through light

**TABLE 4.41**
**Rate Constants of Oxidation by Singlet Oxygen in Water at 25°C**

| Compound | $\log k_{Ox}$ $M^{-1}$ $s^{-1}$ |
|---|---|
| Unsubstituted aliphatic carbons | 3.5 |
| Cyclic olefins | 5.3 |
| Substituted olefins | 6.0 |
| Dialkyde sulfide | 6.8 |
| Diene | 7.0 |
| Imidazole | 7.6 |
| Furan | 8.2 |
| Trialyleneamine | 8.9 |

The concentration of singlet oxygen was estimated as 10–12 M.

*Source:* Jorgensen, S.E. and Gromiec, M.J., *Mathematical SubModels in Water Quality Systems*, Elsevier, Amsterdam, 1989, p. 183. With permission.

absorption by intermediary compounds.[162] Direct decomposition predominates in systems with little extraneous dissolved organic matter or particulates.

Indirect photolysis was discovered when researchers noticed that some compounds degraded faster in natural water than in distilled water. The scattering of light by reflection from particulate matter, and its absorption by non-target molecules causes light disappearance. Absorbed energy can be converted to heat or can cause photolysis. Light disappearance is a function of wavelength and water quality (e.g., color, suspended solids, and dissolved organic carbon). Thus, in more turbid or highly colored systems, sensitized photolysis, as compared to direct photolysis, could be a very significant decomposition mechanism for certain contaminants.[45] Photolysis rate is a function of the quantity and wavelength distribution of the incident light, the light absorption characteristics of the compound to be photolyzed, and the efficiency at which the absorbed light produces a chemical reaction.[78]

Only light that is absorbed by a molecule can produce chemical changes. Photochemical energy reaches the Earth from the sun in the form of photons with wavelengths covering the spectra from infrared to the far ultraviolet, including the visible. The principal energy sources of photons for environmental degradation reactions are those in the ultraviolet (UV) region.[156] This first law of photochemistry, the Grotthus–Draper law, has been recognized since the early 19th century.[143] The photolysis rate expression is given in Section 4.2.4.

At low latitudes, under a cloudless sky and the summer sun, the irradiance at sea level is of the order of 1000 W·m$^{-2}$. An overwhelming fraction of this energy input is converted to heat. However, even a small fraction (about 0.04% on an average) entering the water is captured by photosynthetic processes is enough to be the driving force shaping the entire aquatic biological domain. Considering such a massive input of energy into the complex aquatic environment, the possibility of occurrence of biological light-induced chemical reactions and their importance in determining the chemical composition of natural waters can hardly be overlooked.

The direct photolysis rate depends on a number of factors related to the characteristics of the compound and the environment:[45]

- Solar radiation (intensity and wavelength)—Depending on the time of the year, the weather, and the geographical position of the water body, different levels of incoming solar radiation will be delivered to the surface of water.
- Light attenuation in the water—Suspended matter, surface films, color, and other factors influence the penetration and attenuation of light. Photolysis in quiescent, turbid water bodies might be limited to a thin surface layer, whereas it could extend to great depths in relatively clear water.
- Absorption spectrum of the chemical—Due to its chemical structure, each compound will absorb light energy to different degrees from various wavelengths.
- Quantum yield—This refers to the fraction of absorbed photons that result in a chemical reaction.

When light strikes a pollutant molecule, the energy content of the molecule is increased and the molecule is elevated into an excited electron state. This excited state is unstable, and the molecule returns to its normal (lower) energy level by one of the two paths:[131]

- It loses its "extra" energy through energy emission; that is, fluorescence or phosphorescence.

or

- It is converted to a different molecule through the new electron distribution produced in the excited state. Usually the organic chemical is oxidized.

Photolysis rate constants can be measured in the field with sunlight or under laboratory conditions. Some representative photolysis rate constants, quantum yields, and wavelengths at which they were measured are given in Table 4.42.[131]

Photolysis will not be an important fate process unless sunlight is absorbed in the visible or near-ultraviolet wavelength ranges (above 290 nm) by either the organic chemical or its sensitized agent.

The intensity of incident light varies over the depth of the water column and may be modeled by

$$I_z = I_0 \, e^{-Ke.z} \tag{4.43}$$

where

$I_z$ = intensity at depth z
$I_0$ = intensity at the water surface
$Ke$ = extinction coefficient for light disappearance

Indirect, or sensitized, photolysis occurs when a nontarget molecule is transformed directly by light, which, in turn, transmits its energy to the pollutant molecule. Transformation then occurs in the pollutant molecule as a result of the increased energy content.

**TABLE 4.42**
**Selected Photolysis Reaction Parameters**

| | Photolysis Reaction Parameters | | |
|---|---|---|---|
| Compound | Rate Constant (Near Surface) $k_p$, $h^{-1}$ | Quantum Yield | Wavelength, nm |
| Pesticides | | | |
| Carbaryl | 0.002–0.004 | 0.01 | Sunlight |
| DDT | $<5 \times 10^{-7}$ | 0.16 | 254 |
| Parathion | 0.0024–0.003 | <0.001 | >280 |
| Atrazine | 0.0002 | — | Sunlight |
| Trifluarin | 0.03 | 0.002 | 313 |
| Monocyclic aromatics | | | |
| Pentachlorophenol | 0.23–1.2 | — | 290–330 |
| 2,4-Dinitrotoluene | 0.016 | 0.00075 | 313 |
| Polycyclic aromatic hydrocarbons | | | |
| Anthracene | 0.15 | 0.003 | 360 |
| Benzo[a]pyrene | 0.58 | 0.0009 | 313 |

*Source:* Schnoor, C.J.L., *Environmental Modeling: Fate and Transport of Pollutants in Water, Air, and Soil*, Wiley-Interscience, New York, 1996. With permission.

Rate constants and half-lives for photolysis in sunlight can be calculated as functions of season, latitude, time of day, depth of water body, and the ozone layer condition. Environmental influences, such as depth of the chemical in water, sensitizers, quenchers, and pH, can also affect the rate of phototransformation.[137]

Water is not as good a matrix as the atmosphere for photochemical reactions because of the attenuation of the incident light in the water column. Nevertheless, photochemical transformations in water have been shown to be important for some compounds.[156] Some chemical species in natural waters can absorb light energy directly and undergo direct photochemical reactions. Examples include the direct photolysis of nitrate, nitrite, and methyl iodide in seawater. These, in turn, can react with compounds that are not photoreactive resulting in indirect, or sensitized photochemical reactions. If these reaction products are still reactive, as are singlet oxygen and hydrogen peroxide, further reactions can occur.

The kinetics of photochemistry are discussed in detail in Leifer,[163] Matsumura and Katayama,[164] and Nubbe et al.[162]

### 4.2.3.7  Biodegradation

Biodegradation is a commonly accepted term that describes microbially mediated alterations of a compound that are sufficient to change its identity. All biodegradable compounds are ultimately converted to one of a very few key intermediates, such as acetyl-coenzyme A or to one of the tricarboxylic acid cycle intermediates prior to total oxidation. As such these compounds are used for the resynthesis of larger molecules, only a portion of biodegradable compounds being totally oxidized to inorganic constituents. Biodegradation of organic compounds appears to be the most environmentally desirable reaction because it generally results in completely mineralized end products (inorganic compounds). In contrast, photochemical and other nonmetabolic processes usually result in only slight modifications in the parent compound.[156]

Biodegradation is the process by which aerobic or anaerobic microorganisms break down organic chemicals to either a higher- or a lower-molecular-weight chemical(s) called a biodegrade(s). Soil and aquatic microorganisms can detoxify chemicals, but can also sometimes form more toxic chemical compounds. Typical degrading microorganisms are heterotrophic bacteria, actinomycetes, autotrophic bacteria, fungi, and protozoa.[45] Organisms responsible for biodegradation display a spectrum of interactions, ranging from predation to synergism. The presence/absence of predators, parasites, or hosts causes microbial population to fluctuate over a wide range.

Guidelines used to predict whether biodegradation can occur include chemicals that are highly water soluble can biodegrade, but those with low water solubility will not; chemicals that adsorb in soil will usually not biodegrade, but those that do not adsorb can; chemicals with a high $K_{ow}$ will usually not biodegrade, but those with a low $K_{ow}$ can; and chemicals that leach in soil will usually biodegrade, but those that do not leach usually will not.[137]

Two general types of biodegradation are recognized: growth metabolism and cometabolism. Growth metabolism occurs when the organic compound serves as a food source for bacteria. Adaptation times from 2 to 20 days were suggested by Mills et al.[165] Adaptation may not be required for some chemicals or in chronically exposed environments. Adaptation times can be lengthy in environments with a low initial density of degraders.[165] For cases where biodegradation is limited by the degrader population size, adaptation is faster in high density microbial populations and slower in the low initial density populations. Following adaptation, biodegradation proceeds at the first-order rate. Co-metabolism occurs when the organic compound is not a food source for the bacteria. Here, adaptation is seldom necessary, and the transformation rates are slow compared to growth metabolism.

Biodegradation also encompasses the broad and complex processes of enzymatic attack by organisms on organic chemicals. Dehalogenation, dealkylation, hydrolysis, oxidation, reduction, ring cleavage, and condensation reactions are all known to occur either metabolically or via organisms that are not capable of utilizing the chemical as a substrate for growth.[78] It also encompasses a number of distinctly different processes denoted by[45]

- Mineralization—the conversion of an organic compound to inorganic products;
- Detoxication—the conversion of a toxicant to innocuous byproducts;
- Co-metabolism—the metabolism of a compound that organisms cannot use as a nutrient that does not result in mineralization (organic metabolites remain);
- Activation—the conversion of a nontoxic substance to a toxic one or an increase in a substance's toxicity, due to microbial action;
- Defusion—the conversion of a potential toxicant into a harmless metabolite before its potential is realized.

Within aquatic ecosystems, microbial decomposition of organic matter plays a prominent role in the energy and mass transformation processes. Organic wastewater discharge in a water body results in dissolved oxygen uptake, either directly by chemical oxidation of the reducing pollutants or by their metabolism by microorganisms, i.e., their biodegradation. The increase in water temperature accelerates the kinetics of chemical and microbiological oxygen use and pollution load reduction.

Growth kinetics of the bacterial population degrading toxic chemicals is not well understood.[78] The presence of competing substrates and other bacteria, the toxicity of the chemical to the degrading bacteria, and the possibilities of adaptation to the chemical or co-metabolism make quantification of population changes difficult. As a result, toxic chemical fate models assume a constant biological activity rather than modeling the bacteria directly. Often, measured first-order biodegradation rate constants from other aquatic systems are used directly. Table 4.43 is a

## TABLE 4.43
## Selected Biotransformation Rate Constants

| Chemical | Pseudo-First-Order Rate Constant Range, day$^{-1}$ | Estimated Second-Order Rate Constants, ml cell$^{-1}$ h$^{-1}$ |
|---|---|---|
| Pesticides | | |
| Carbofuran | 0.03 | $1 \times 10^{-8}$ |
| DDT | 0.0–0.10 | $3 \times 10^{-12}$ |
| Parathion | 0.0–0.12 | — |
| Dioxin TCDD | <0.01 | $1 \times 10^{-10}$ |
| Atrazine | 0.02–0.03 | $1 \times 10^{-8}$ |
| Alachlor | 0.05–0.06 | $3 \times 10^{-8}$ |
| PCBs | | |
| Aroclor 1248 | 0.0–0.007 | $10^{-9}-10^{-12}$ |
| Halogenated aliphatic hydrocarbons | | |
| Chloroform | 0.09–0.10 | $1 \times 10^{-10}$ |
| Halogenated ethers | | |
| 2-Choloroethyl vinyl ether | 0.0–0.20 | $1 \times 10^{-10}$ |
| Monocyclic aromatics | | |
| 2,4-Dimethylphenol | 0.24–0.66 | $1 \times 10^{-7}$ |
| Pentachlorophenol | 0.00–33.6 | $3 \times 10^{-8}$ |
| Benzene 14 | 0.006–0.01 | $1 \times 10^{-7}$ |
| Toluene 14 | 0.01–0.019 | $1 \times 10^{-7}$ |
| Phenol | — | $3 \times 10^{-6}$ |
| Phthalate esters | | |
| bis(2-ethylhexyl)phtalate | 0.00–0.14 | $1 \times 10^{-7}$ |
| Polycyclic aromatic hydrocarbons | | |
| Anthracene | 0.007–15.0 | $3 \times 10^{-9}$ |
| Benzo[a]pyrene | 0.0–0.075 | $3 \times 10^{-12}$ |

*Source:* Schnoor, C.J.L., *Environmental Modeling: Fate and Transport of Pollutants in Water, Air, and Soil*, John Wiley & Sons, New York, 1996. With permission.

summary of pseudo-first-order and second-order rate constants for the disappearance of representative toxic organic compounds from natural waters via biotransformation.[131]

For modeling, first-order biodegradation rate constants or half-lives for the water column and the benthos may be specified. If these rate constants have been measured under similar conditions, this first-order approach is likely to be as accurate as more complicated approaches. If first-order rate constants are unavailable, or if they must be extrapolated to different bacterial conditions, then the second-order approach may be preferred. It is assumed that bacterial populations are unaffected by the presence of the compound at low concentrations.

Certain organic pollutants such as PCBs, dioxin, and some pesticides (e.g., DDT) are not biodegradable, and thus persist in the environment despite their use being banned. Therefore, current environmental policy in most countries dictates that only readily biodegradable compounds be used in commerce. Additionally, a large part of wastewater disposal regulations is based on the ability of various organic wastes to be biodegraded in the ecosystem.

Various procedures have been developed to measure the biodegradability of chemical compounds. One procedure is the inoculated die-away test where water, salts, and biomass are incubated for 5 to 40 days with the test compound. Another is the activated sludge test method where large amounts of biomass and several test organic substrates are continuously fed with air or oxygen to provide for better adaptation to degrade the multiple substrates.

Many factors can affect the rate of biodegradation: pH, sorption, populations and types of microorganisms, moisture, the presence of other chemicals, and the concentration of chemicals present.[166] Biodegradation is also affected by numerous factors that influence biological growth.[131]

- Temperature effect on the biodegradation of toxic compounds is similar to that on biochemical oxygen demand (BOD), and is quantified using an Arrhenius-type relationship,
- Nutrients are necessary for growth and often limit growth rate. Other organic compounds can serve as the primary substrate so that the chemical of interest is degraded via co-metabolism or as secondary substrate (the utilization of organic chemicals at concentrations lower than are required for growth in the presence of one or more primary substrates that are used as the carbon and energy sources),
- Acclimation is the process of adaptation necessary to express repressed enzymes or foster those organisms that can degrade the toxicant through gradual exposure to the toxicant over time. A shock load of toxicant can kill a culture that would otherwise adapt if gradually exposed. On the other hand, the chronic presence of some toxic organic chemical sometimes spurs on the induction or expression of an enzyme that helps to degrade other organic chemicals.
- Population density or biomass concentration dictates that organisms must be present in numbers large enough to significantly degrade the toxicant (a time lag often occurs if the organisms are too few).

In some environments, such as lagoons or low-energy beaches and sediment, rates of petroleum biodegradation are limited by oxygen availability, and biodegradation in such cases can be enhanced by providing oxygen.[167]

Because organic chemical pollution control has been a major focus of water quality studies and regulations, parameters relating to biodegradation are relatively well defined. Dissolved oxygen (DO) is easily determined, either by absolute chemical methods, such as the Winkler volumetric titration using manganese and iodide salts in basic media, or by electrochemical methods using a Clark-type electrode, i.e., catalytic reduction of oxygen diffusing through a hydrophobic membrane. Although DO can be easily and rapidly monitored with a precision better than 1–2%, organic matter (OM) represents a large variety of compounds and cannot be determined without ambiguity. The biochemical oxygen demand (BOD) test assesses the pollution potential of wastewater that contains available organic carbon source(s) for aerobic microorganisms by measuring the amount of oxygen utilized during the growth of microbial organisms seeded into a sample of wastewater. Nevertheless, since nitrogenous compounds are often specifically measured and their microbial decomposition kinetics are taken into account during this test, the reactive carbonaceous organic matter can be estimated by the 5 days BOD, denoted as $BOD_5$. The amount of organic matter can alternatively be determined by either chemical oxidation at standard conditions, e.g., 2-hr ebullition with excess dichromate in sulfuric acid (chemical oxygen demand, COD) and persulfate in the presence of heavy UV radiation (total oxygen demand, TOD), or by total oxidation combustion at 900°C. The carbon dioxide released during such total oxidation procedures is monitored in various ways. The total organic carbon (TOC) value includes carbonaceous compounds present in organic pollutants. Because the organic matter content of surface water and of waters generally is of great economic and ecological importance, BOD, COD, TOD, and TOC have been thoroughly studied.

## 4.2.4  Governing Equations of Reactions To Be Used in Modeling

In this subsection, the reaction and mass transfer process equations referred to in Section 4.2.3 are presented to help those interested in predicting the concentrations of organic chemicals in the aquatic environment.

### 4.2.4.1  Volatilization

The exchange of organic chemicals between the liquid and vapor phase through volatilization is illustrated in Figure 4.10. The relationship quantifying volatilization from water is given as:

$$\frac{\partial c_w}{\partial_t} = -\frac{K_v}{D}\left(c_w - \frac{c_a}{H/RT}\right)$$

(4.44)

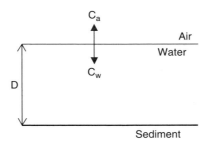

**FIGURE 4.10** Volatilization.

where

$c_w$ = dissolved organic chemical concentration in water [$\mu$g l$^{-1}$]

$c_a$ = concentration of organic chemical in air [$\mu$g l$^{-1}$].

$H$ = Henry's law constant [atm m$^3$ mole$^{-1}$]

$R$ = universal gas constant (8.206 × 10$^{-5}$) [atm m$^3$ mole$^{-1}$ K$^{-1}$]

$T$ = water temperature [K]

$D$ = water depth [m]

$K_v$ = volatilization transfer rate constant [m day$^{-1}$]

The rate constant $K_v$ is the reciprocal of the total interface film resistances, and depends on the intensity of turbulence in the water body and the overlying atmosphere.[78]

$$K_V = (R_L + R_G)^{-1} = [K_L^{-1} + (K_G \cdot (H/RT_K)^{-1}]^{-1} \qquad (4.45)$$

where

$K_V$ = volatilization transfer rate constant [m day$^{-1}$]

$R_L$ = liquid phase resistance [day m$^{-1}$]

$R_G$ = gas phase resistance [day m$^{-1}$]

$K_L$ = liquid phase transfer coefficient [m day$^{-1}$]

$K_G$ = gas phase transfer coefficient [m day$^{-1}$]

$H$ = Henry's law coefficient for the air–water partitioning of the chemical [atm m$^3$ mole$^{-1}$]

$R$ = universal gas constant (8.206 × 10$^{-5}$) [atm m$^3$ mole$^{-1}$ K$^{-1}$]

$T_K$ = water temperature [K$^{-1}$]

### 4.2.4.2 Sorption

Several models have been proposed to mathematically represent sorption isotherms.[45] Among the most popular are:

- Langmuir isotherm:

$$v = (v_m \, b \, c_d)/(1 + b \, c_d) \qquad (4.46)$$

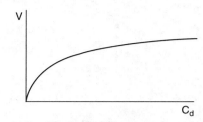

**FIGURE 4.11** Sorption data and isotherm.

- Freundlich isotherm:

$$v = K_f \, c_d^{\,1/n} \qquad (4.47)$$

- BET isotherm:

$$v = (v_m B \, c_d)/[(c_s - c_d) \, (1 + (B - 1) \, (c_d/c_s)] \qquad (4.48)$$

where
  $v$ = concentration of pollutant on the solids, [mg g$^{-1}$]
  $c_d$ = dissolved concentration [mg m$^{-3}$]
  $v_m$ = maximum concentration attained [mg m$^{-3}$]
  $b$, $B$, $c_s$, $K_f$, and $n$ are coefficients used to calibrate the curves to the measured
    data; the curve given in Figure 4.11 commonly occurs

The simplest equilibrium expression uses a linear adsorption isotherm

$$C_s' = K_{ps} \cdot C_w' \qquad (4.49)$$

where
  $C_s'$ = concentration of sorbed chemicals on sediment in segment
  $C_w'$ = concentration of dissolved chemical in water in segment
  $K_{ps}$ = partition coefficient of chemical on sediment in segment

At equilibrium, the distribution between the phases is controlled by the partition coefficient $K_{ps}$, and the amount of solid phase present controls the total mass of chemical in each phase. An overview of sorption equations used in WASP/TOXI module follows.[78]

Dissolved chemical in the water column and benthic segments interacts with sediment particles and dissolved organic carbon to form five phases: dissolved, DOC-sorbed, and sediment-sorbed (three sediment types "s"). The reactions can be written with respect to a unit volume of water as:

$$M_s' + C_w' \rightleftharpoons C_s/n \qquad (4.50)$$

$$B' + C_w' \rightleftharpoons C_B/n \qquad (4.51)$$

where

$M'_s$ = concentration of sediment in water in segment [kg$_s$ l$_w$$^{-1}$]

$B'$ = concentration of dissolved organic carbon in water in segment [kg$_B$ l$_w$$^{-1}$]

$C'_w$ = concentration of dissolved chemical in water in segment [mg l$_w$$^{-1}$]

$C_s$ = concentration of sorbed chemical on sediment in segment [mg l$^{-1}$]

$C_B$ = concentration of DOC-sorbed chemical in segment [mg l$^{-1}$]

$n$ = porosity or volume water per volume segment [l$_w$ l$^{-1}$]

The forward reaction is sorption and the backward reaction is desorption. These reactions are usually fast in comparison with the model time step, and can usually be considered in local equilibrium. The phase concentrations $C_w$, $C_s$, and $C_B$ are governed by the equilibrium partition coefficients $K_{ps0}$ and $K_{pB}$ (l kg$^{-1}$):

$$K_{ps0} = \frac{C_s/n}{M'_s \cdot C'_w} = \frac{C'_s}{C'_w} \qquad (4.52)$$

$$K_{pB} = \frac{C_B/n}{B' \cdot C'_w} = \frac{C'_B}{C'_w} \qquad (4.53)$$

$C_B$ = concentration of DOC-sorbed chemical in segment [mg l$^{-1}$]

$C'_B$ = concentration of DOC-sorbed chemical in segment ($C_B/B$) [mg kg$^{-1}$]

$B$ = concentration of DOC in segment [kg l$^{-1}$]

$B'$ = concentration of DOC in water in segment ($B/n$) [kg l$_w$$^{-1}$]

$n$ = porosity or volume water per volume segment [l$_w$$^{l}$ l$^{-1}$]

These two equations describe the linear form of the Freundlich isotherm, applicable when sorption sites on sediment and DOC are plentiful:

$$C'_s = K_{ps} \cdot C'_w \qquad (4.54)$$

$$C'_B = K_{pB} \cdot C'_w \qquad (4.55)$$

where $K_{ps}$ is the partition coefficient of chemical on sediment in segment [l$_w$ kg$_s$$^{-1}$] and $K_{pB}$ is the partition coefficient of chemical on DOC in segment [l$_w$ kg$_B$$^{-1}$]

The total chemical concentration is the sum of the five phase concentrations:

$$C = C'_w \cdot n + \sum_s C'_s \cdot M_s + C'_B \cdot B \qquad (4.56)$$

Substituting in equations, factoring, and rearranging terms gives the dissolved fraction $f_D$:

$$f_D = \frac{C'_w \cdot n}{C} = \frac{n}{n + K_{pB} \cdot B + \sum_s K_{ps} \cdot M_s} \qquad (4.57)$$

where

$C$ = concentration of total organic chemical [mg l$^{-1}$]

$B$ = concentration of DOC [kg$_B$ l$^{-1}$]

$M_s$ = concentration of sediment in segment [kg$_s$ l$^{-1}$]

Similarly, the sediment-sorbed ($f_S$) and DOC-sorbed fractions ($f_B$) are:

$$f_s = \frac{C_w' \cdot n}{C} = \frac{n}{n + K_{pB} \cdot B + \sum_s K_{ps} \cdot M_s} \tag{4.58}$$

$$f_B = \frac{C_B' \cdot B}{C} = \frac{K_{pB} \cdot B}{n + K_{pB} \cdot B + \sum_s K_{ps} \cdot M_s} \tag{4.59}$$

These fractions are determined in time and space throughout a simulation from the partition coefficients, internally calculated porosities, simulated sediment concentrations, and specified DOC concentrations. Given the total concentration and the five-phase fractions, the dissolved, sorbed, and biosorbed concentrations are uniquely determined.

$$C_w = C \cdot f_D \tag{4.60}$$

$$C_s = C \cdot f_s \tag{4.61}$$

$$C_B = C \cdot f_B \tag{4.62}$$

These five concentrations have units of mg l$^{-1}$ and can be expressed as concentrations within each phase:

$$C_w' = C_w/n \tag{4.63}$$

$$C_s' = C_s/M_s \tag{4.64}$$

$$C_B' = C_B/B \tag{4.65}$$

These concentrations have units of mg/l$_w$, mg/kg$_s$, and mg/kg$_B$, respectively.

### 4.2.4.3  Computation of Partition Coefficients

If the octanol/water coefficient cannot be reliably measured or is not available in databases, it can be estimated from solubility and molecular weight and structure information[131] (see Equation 4.66). More detailed information about estimating these coefficients is available within SPARC.[145]

$$\log S = -1.08 \log K_{ow} + 3.70 + \log MW \tag{4.66}$$

where MW is the molecular weight of the pollutant [g·mol⁻¹] and $S$ is the solubility, in units of ppm, for organic compounds that are liquid in their pure state at 25°C.

For organic chemicals that are solids in their pure state at 25°C:

$$\log S = -1.08 \log K_{ow} + 3.70 + \log MW - (\Delta S_f/1360)(MP - 25) \quad (4.67)$$

where MP is the melting point of the pollutant [°C] and $\Delta S_f$ the entropy of fusion of the pollutant [cal.mol⁻¹·deg⁻¹].

Karickhoff et al.[146] have developed empirical expressions relating environmental equilibrium partition coefficients to standard laboratory measurements, resulting in fairly reliable estimations. The correlation is

$$K_{ps()} = f_{ocs} \cdot K_{oc} \quad (4.68)$$

$$K_{pB} = 1.0 \, K_{oc} \quad (4.69)$$

where

$K_{oc}$ = organic carbon partition coefficient [l_w/kg_oc]
$f_{ocs}$ = organic carbon fraction of the sediment
1.0 = organic carbon fraction of DOC

Correlation of $K_{oc}$ with the octanol/water partition coefficient of the chemical (and through that to solubility) has yielded successful predictive tools for incorporating the hydrophobicity of a chemical in an estimate of its partitioning. If no log $K_{oc}$ values are available, they can be generated internally using the following correlation with the octanol/water partition coefficient $K_{ow}$ (l_w/l_oct):

$$\log K_{oc} = a_0 + a_1 \log K_{ow} \quad (4.70)$$

where $a_0$ and $a_1$ are typically considered to be log 0.6 and 1.0, respectively. Once the value of $K_{oc}$ is determined, the computation of partition coefficient proceeds as per Equation 4.68 and Equation 4.69.

The value of the partition coefficient is also dependent on other factors in addition to the fraction of organic carbon of the sorbing particles. Of these, perhaps the most potentially significant and the most controversial is the effect of particle concentration. Based on empirical evidence, the partition coefficient is inversely related to the solids concentration. The equation defining partition coefficient is

$$K_{ps} = \frac{K_{ps0}}{1 + M_s K_{ps0}/U_x} \quad (4.71)$$

where

$K_{ps0}$ = limiting partition coefficient with no particle interaction ($f_{ocs} K_{oc}$ for neutral organic chemicals)
$M_s$ = solids concentration [kg l⁻¹]
$U_x$ = ratio of adsorption to particle-induced desorption rate

Empirical mathematical relationships among $K_{ps}$, $f_{oc}$, and $K_{ow}$ have been derived for various combinations of compounds and natural sorbents:

$$K_{ps} = f_{oc} \cdot K_{oc} = f_{oc.} \ b \cdot (K_{ow})^b \tag{4.72}$$

Values reported for a and b include a = 1.00, b = 0.48 × 10⁻⁶ for polycyclic aromatic hydrocarbons; a = 0.52, b = 4.4 × 10⁻⁶ for a variety of pesticides; and a = 0.72, b = 3.2 × 10⁻⁶ for alkylated and chlorinated benzenes.[120,146,168] From currently available data, it can be concluded that the values of a and b are primarily determined by the type of compounds (i.e., the compound class(es)) on which the relationship was established and only to a much smaller extent by the type of natural sorbents used. Thus, the reported relationships are deemed reliable for predicting the equilibrium partition coefficients of a great number of neutral hydrophobic organic compounds between water and natural sorbents of very different origins.

### 4.2.4.4  Hydrolysis

Reactions catalyzed by oxonium and/or hydroxyl ions are called hydrolysis. The rate of reaction $R$ is given by the equation

$$R = \sim dc/dt \cdot K_H - [A] = K_A \cdot [H^+] \cdot [A] + K_B \cdot [OH^-] \cdot [A] + K_N \cdot [H_2O] \cdot [A] \tag{4.73}$$

where

$K_H$ = pseudo-first-order rate constant at a given pH
$K_A$ and $K_B$ = second-order rate constants
$K_N$ = second-order rate constant for neutral reaction of a chemical compound
      with water, which may be expressed as a pseudo-first-order rate constant

This equation indicates that the rate of hydrolysis is strongly dependent on pH, unless $K_A$ and $K_B$ are equal to zero.

The kinetic expression for hydrolysis is

$$dC/dt = -k_a [H^+] \ C - k_n \ C - k_b [OH^-] \ C \tag{4.74}$$

where

$C$      = concentration of the toxicant being hydrolyzed [mole l⁻¹]
$k_a$      = acid-catalyzed hydrolysis reaction rates constant [l mole⁻¹ day⁻¹]
$k_b$      = base-catalyzed hydrolysis reaction rates constant [l mole⁻¹ day⁻¹]
$k_n$      = neutral-catalyzed hydrolysis reaction rates constant [day⁻¹]
[H⁺]    = hydrogen ion concentration [mole l⁻¹]
[OH⁻] = hydroxyl ion concentration [mole l⁻¹]

In order to evaluate $k_a$ and $k_b$, several nonneutral pH hydrolysis experiments must be conducted as depicted in Figure 4.8. In general, hydrolysis reaction rates are highly dependent on pH.

Often, the hydrolysis reaction rate expression in Equation 4.74 is simplified to a pseudo-first-order reaction rate expression at a given pH and temperature:

$$dC/dt = -k_h\, C \tag{4.75}$$

where

$$k_h = k_b\, [OH^-] + k_a\, [H^+] + k_n \tag{4.76}$$

### 4.2.4.5 Oxidation

The oxidation rate of an organic pollutant $R_{Ox}$ can be written as follows:

$$R_{Ox} = dC/dt = k_{Ox} \cdot [C] \cdot [Ox] \tag{4.77}$$

where
$k_{OX}$ = specific second-order rate constant for oxidation at a specific temperature
$[C]$ = molar concentrations of the chemical compound
$[Ox]$ = molar concentrations of the oxidant

The total rate of oxidation, if more than one oxidant is working simultaneously, is the sum of the rates of reaction for each oxidant:

$$R_{Ox} = (k_{Ox1}\, [Ox_1] + k_{Ox2}\, [Ox_2] + \dots k_{Oxn}\, [Ox_n]) \cdot [C] = (k_{Oxn}\, [Ox_n]) \cdot [C] \tag{4.78}$$

Integration of Equation 4.77 between the time limits 0 and $t$ gives

$$\ln[C_0]/[C_t] = \sum_{n=1}^{n=n} k_{Oxn}[OX_n] \cdot t \tag{4.79}$$

assuming $[Ox_n]$ for all values of $n$ remain constant during the reaction time. If the time to oxidize half of the chemical, or the reactions half-life, $t_{1/2}$, is desired:

$$t_{1/2} = \ln 2 \sum_{n=1}^{n=n} k_{Oxn}[Ox_n] \tag{4.80}$$

If the oxidant(s) concentration remains constant, $k_{Ox}$ in Equation 4.77 becomes a first-order rate.

According to Ambrose et al.[78] oxidation can be modeled as a general second-order process for the various species and phases of each chemical:

$$K_O = [RO_2]\, \Sigma_i\, \Sigma_j\, k_{Oij}\, f_{ij} \tag{4.81}$$

where

$K_O$     = total oxidation rate constant [day$^{-1}$]

$[RO_2]$ = concentration of oxidant [moles l$^{-1}$]

$k_{Oij}$     = second-order oxidation rate constant for the polluting chemical as species
        $i$ in phase $j$ [l mole$^{-1}$ day$^{-1}$]

The individual rate constant can be specified as independent of temperature, with activation energy constants set to 0. If it is necessary to determine rates as a function of temperature based on the Arrhenius function, then nonzero activation energies, specified as constants, will invoke the following calculation for each rate constant $k$:

$$k\ (T_K) = k\ (T_R)\ \exp\ [1000\ E_{ao}\ (T_K - T_R)/(RT_K T_R)]$$     (4.82)

where $E_{ao}$ is the Arrhenius activation energy for the oxidation reaction [kcal mole$^{-1}$ K$^{-1}$]. Activation energies can be specified for each ionic species simulated.

### 4.2.4.6 Photolysis

The basic equation for direct photolysis is of the form

$$dC/dt = -k_p\ C$$     (4.83)

where $C$ is the concentration of organic chemical being photolyzed and $k_p$ is the rate constant for direct photolysis.

The first-order rate constant can be estimated directly as

$$k_p = 2.303\ J^{-1}\ \Phi\ \Sigma\ I_\Lambda\ e_\Lambda$$     (4.84)

where

$k_p$ = rate constant for direct photolysis [s$^{-1}$]

J  = conversion constant [$6.02 \times 10^{20}$]

$\Phi$ = quantum yield [number of moles of chemical reacted/number of einsteins
     absorbed]

$I_\Lambda$ = sunlight intensity at wavelength $\Lambda$ [photons cm$^{-2}$ s$^{-1}$]

$e_\Lambda$ = molar absorbtivity or molar extinction coefficient at wavelength $\Lambda$ [molarity$^{-1}$
     cm$^{-1}$]

An einstein is the unit of light on a molar basis (a quantum or photon is the unit of light on a molecular basis). The quantum yield may be thought of as the efficiency of photoreaction. Incoming radiation is measured in units of energy per unit area per time (e.g., cal cm$^{-2}$ s$^{-1}$). The incident light in units of einsteins cm$^{-1}$ s$^{-1}$ nm$^{-1}$ can be converted to watts cm$^{-2}$ nm$^{-1}$ by multiplying with the wavelength (nm) and $3.03 \times 10^{39}$.

The kinetic equation for indirect photolysis is

$$dC/dt = -k_2 CX = -k_p C \qquad (4.85)$$

where
   $k_2$ = indirect photolysis rate constant
   $X$ = concentration of the nontarget intermediary
   $k_p$ = overall pseudo-first-order rate constant for sensitized photolysis

### 4.2.4.7 Biodegradation

If the microbial community has adapted to the contaminant, the Michaelis-Menten equation can be used to represent the rate of biotransformation as

$$k_b = (\mu_{max} X)/(Y (k_s + c)) \qquad (4.86)$$

where
   $c$    = concentration of chemical being biodegraded [$\mu$g m$^{-3}$]
   $\mu_{max}$ = maximum growth rate of the culture [yr$^{-1}$]
   $X$    = biomass concentration of the microorganisms [cells.m$^{-3}$]
   $Y$    = yield coefficient [cells produced per mass toxicant removed]
   $k_s$    = half-saturation constant [$\mu$g.m$^{-3}$]
   $k_b$    = biological transformation rate constant [m$^3$ cells$^{-1}$ year$^{-1}$]

When $c \ll k_s$, then

$$k_b = (\mu_{max} X)/(Y \cdot k_s) = k_{b2} \cdot X \qquad (4.87)$$

where $k_{b2}$ is the second-order biotransformation rate constant [m$^3$ cells$^{-1}$ year$^{-1}$].
    For systems where the microbial population is relatively constant, the transformation rate equation reduces to a pseudo-first-order rate expression.[45]
    If pseudo-first-order rates are unavailable, or if they must be extrapolated to different bacterial conditions, then the second-order approach can be used. It is generally assumed that bacterial populations are unaffected by the presence of the pollutant compound at low concentrations. Second-order kinetics for dissolved, DOC-sorbed, and sediment-sorbed chemical can also be considered:

$$K_{BW} = P_{ac}(t) \Sigma_i \Sigma_j k_{bij} f_{ij} \qquad j = 1, 2 \qquad (4.88)$$

$$K_{BS} = P_{ac}(t) \Sigma_i \Sigma_j k_{bij} f_{ij} \qquad j = 3 \qquad (4.89)$$

where
   $K_{BW}$  = total biodegradation rate constant in the water column [day$^{-1}$]
   $K_{BS}$   = total biodegradation rate constant on sediment [day$^{-1}$]
   $k_{bij}$   = second-order biodegradation rate constant for species $i$, phase $j$ [ml cell$^{-1}$ day$^{-1}$]
   $P_{ac}(t)$ = active bacterial population density in segment [cell ml$^{-1}$]
   $f_{ij}$    = fraction of chemical as species $i$ in phase $j$

The individual species phase biodegradation rate constant can be adjusted for temperature as:

$$k_{bij\ (T)} = k_{bij\ (20)}\ Q_{Tij}^{(T-20)/10} \qquad (4.90)$$

where $Q_{Tij}$ is "Q–10" temperature correction factor for biodegradation of species i, in phase $j$ and $T$ is the ambient temperature in segment [°C].

The temperature correction factors represent the increase in the biodegradation rate constants resulting from a 10°C temperature increase. Values for $QT_{ij}$ in the range of 1.5 to 2 are common.

## ACKNOWLEDGMENTS

The authors express their sincere thanks to Natalia Kazantseva, Ph.D. (Turkmenistan), for her valuable comments, Mr. Robert Swank (U.S. EPA) for his editorial comments, and Research Assistant Ali Ertürk, M.Sc. (Turkey), for his help and contribution during the preparation of this chapter.

## REFERENCES

1. Picot, B., Péna, G., Casellas, C., Bondon, D., and Bontoux, J., Interpretation of the seasonal variations of nutrients in a Mediterranean lagoon: Étang de Thau, *Hydrobiologia*, 207, 105, 1990.
2. Wollast, R., Interaction of carbon and nitrogen cycles in the coastal zone, in *Interactions of C, N, P, and S Biogeochemical Cycles and Global Change*, Wollast, R., Mackenzie, F.T., and Chou, L., Eds., NATO ASI Series, Series I: Global Environmental Change, Vol. 4, Springer-Verlag, Berlin, 1993, p. 195.
3. Castel, J., Caumette, P., and Herbert, R., Eutrophication gradients in coastal lagoons as exemplified by the Bassin d'Arcachon and the Étang du Prévost, in *Coastal Lagoon Eutrophication and Anaerobic Processes (C.L.E.A.N)*, Caumette, P., Caster, J., and Herbert, R., Eds., *Hydrobiologia*, 329, ix, 1996.
4. Nixon, S.W., Coastal marine eutrophication: A definition, social causes, and future concerns, *Ophelia*, 41, 199, 1995.
5. Howarth, R.W., Anderson, D.M., Church, T.M., Greening, H., Hopkinson, C.S., Huber, W.C., Marcus, N., Nainman, R.J., Segerson, K., Sharpley, A.N., and Wiseman, W.J., Understanding nutrient over-enrichment: an introduction, in *Clean Coastal Waters: Understanding and Reducing the Effects of Nutrient Pollution*, Committee on the Causes and Management of Eutrophication, Ocean Studies Board, Water Science and Technology Board, National Research Council, National Academy Press, Washington, D.C., 2000, p. 13.
6. Smith, S.V., Phosphorus versus nitrogen limitation in the marine environment, *Limnol. Oceanogr.*, 29(6), 1149, 1984.
7. Valiela, I., Nutrient cycles and ecosystem stoichiometry, in *Marine Ecological Processes*, 2nd ed., Springer-Verlag, New York, 1995, chap. 14.
8. Herbert, R.A., Nitrogen cycling in coastal marine ecosystems, *FEMS Microbiol. Rev.*, 23, 563, 1999.

9. Knoppers, B., Aquatic primary production in coastal lagoons, in *Coastal Lagoon Processes*, Kjerfve, B., Ed., Elsevier Oceanography Series 60, 1994, p. 243.

10. Postma, H., Kemp, W.M., Colebrook, J.M., Horwood, J., Joint, I.R., Lampitt, R., Nixon, S.W., Pilson, M.E.Q, and Wulff, F., Nutrient cycling in estuarine and coastal marine ecosystems, in *Flows of Energy and Materials in Marine Ecosystems Theory and Practice*, Fasham, M.J.R., Ed., Plenum Press, New York, 1984, p. 651.

11. Emerson, K., Russo, R.C., Lund, R.E., and Thurston, R.V., Aqueous ammonia equilibrium calculations—effect of pH and temperature, *J. Fish. Res. B Canada*, 32(12), 2379, 1975.

12. Libes, S.M., *An Introduction to Marine Biogeochemistry*, John Wiley & Sons, New York, 1992, chap. 16, 24.

13. Wetzel, R.G., *Limnology*, 2nd ed., Saunders College Publishing, Philadelphia, 1983, chap. 12, 13.

14. Hattori, A., Denitrification and dissimilatory nitrate reduction, in *Nitrogen in the Marine Environment*, Carpenter, E.J. and Capone, D.G., Eds., Academic Press, New York, 1983, p. 191.

15. Rysgaard, S., Risgaard-Petersen, N., and Sloth, N.P., Nitrification, denitrification and nitrate ammonification in sediments of two coastal lagoons in southern France, in *Coastal Lagoon Eutrophication and Anaerobic Processes (C.L.E.A.N)*, Caumette, P., Castel, J., and Herberts, R., Eds., *Hydrobiologia*, 329, Kluwer, Brussels, 1996, p. 133.

16. Nixon, S.W., Nutrient enrichment of shallow marine ecosystems, in *Proc. of Delmarva Coastal Bay Conference III: Tri-State Approaches to Preserving Aquatic Resources*, Kutz, F.W., Koenings, P., and Adelhardt, L., Eds., USEPA, National Health and Environmental Effects Research Laboratory, Atlantic Ecology Division, Narragansett, RI, EPA/620/R-00/001, 1999.

17. Henriksen, K. and Kemp, W.M., Nitrification in estuarine and coastal marine sediments, in *Nitrogen Cycling in Coastal Marine Environments*, Blackburn, T.H. and Sorensen, J., Eds., SCOPE, John Wiley & Sons, New York, 1988, p. 207.

18. Seitzinger, S.P., Denitrification in freshwater and coastal marine ecosystems: ecological and geochemical significance, *Limnol. Oceanogr.*, 33(4), Part 2, 702, 1988.

19. Jorgensen, B.B., Material flux in sediments, in *Eutrophication in Coastal Marine Ecosystems, Coastal and Estuarine Studies*, Vol. 52, American Geophysical Union, Washington, D.C., 1996, p. 115.

20. Kemp, W.M., Sampou, P., Caffrey, J., and Mayer, M., Ammonium recycling versus denitrification in Chesapeake Bay sediment, *Limnol. Oceanogr.*, 35(7), 1545, 1990.

21. Rysgaard, S., Thastum P., Dalsgaard, T., Christensen, P.B., and Sloth, N.P., Effects of salinity on $NH_4^+$ adsorption capacity, nitrification, and denitrification in Danish estuarine sediment, *Estuaries*, 22(1), 21, 1999.

22. Joye, S.B. and Hollibaugh, J.T., Influence of sulfide inhibition of nitrification on nitrogen regeneration in sediment, *Science*, 270, 623, 1995.

23. Horne, J.A. and Goldman, C.R., *Limnology*, McGraw-Hill, New York, 1994, chap. 8, p. 9.

24. Bonin, P. and Raymond, N., Effects of oxygen on denitrification in marine sediments, *Hydrobiologia*, 207, 115, 1990.

25. Lalli, C.M. and Parsons, T.R., Energy, flow and mineral cycling, in *Biological Oceanography: An Introduction*, Pergamon Press, New York, 1993, chap. 5.

26. Nixon, S.W., Remineralization and nutrient cycling in coastal marine ecosystems, in *Estuaries and Nutrients*, Neilson, B.J. and Cronin, L.E., Eds., Humana Press, Totowa, NJ, 1981, p. 111.

27. Day, J.W., Jr., Hall, C.A.S., Kemp, W.M., and Yanez-Arancibia, A., Estuarine chemistry, in *Estuarine Ecology*, John Wiley & Sons, New York, 1989, chap. 3.

28. Souchu, P., Gasc, A., Collos, Y., Vaquer, A., Tournier, H., Bibent, B., and Deslous-Paoli, J.M., Biogeochemical aspects of bottom anoxia in a Mediterranean lagoon (Thau, France), *Mar. Ecol-Prog. Ser.*, 164, 135, 1998.

29. Canfield, D.E., Organic matter oxidation in marine sediment, in *Interactions of C, N, P, and S Biogeochemical Cycles and Global Change*, Wollast, R., Mackenzie, F.T., and Chou, L., Eds., NATO ASI Series, Series I: Global Environmental Change, Vol. 4, Springer-Verlag, Berlin, 1993, p. 333.

30. Jorgensen, B.B., Mineralization in marine sediment, in *Coastal Pollution Control*, Vol. 1, DANIDA&WHO, Polyteknisk Forlag/Special-Trykkeriet Viborg, 1976.

31. Howarth, R.W., Marino, R., and Lane, J., Nitrogen fixation in freshwater, estuarine, and marine ecosystems, 1. Rates and importance, *Limnol. Oceanogr.*, 33(4, part 2), 669, 1988.

32. Seitzinger, S.P., Gardner, W.S., and Spratt, A.K., The effect of salinity on ammonium sorption in aquatic sediment—Implications for benthic nutrient recycling, *Estuaries*, 14(2), 167, 1991.

33. Morlock, S., Taylor, D., Giblin, A., and Hopkinson, C., Effect of salinity on the fate of inorganic nitrogen in sediments of the Parker River Estuary, Massachusetts, *Biol. Bull. U.S.*, 193, 2, 290, 1997.

34. Howarth, R.W., Nutrient limitation of net primary production in marine ecosystems, *Ann. Rev. Ecol.*, 19, 89, 1988.

35. USEPA, *Nutrient Criteria Technical Guidance Manual—Estuarine and Coastal Marine Waters*, U.S. Environmental Protection Agency, Office of Water, EPA822-B-01-003, 2001.

36. Correll, D.L., The role of phosphorus in the eutrophication of receiving waters: a review, *J. Environ. Qual.*, 27, 261, 1998.

37. Rattray, M.R., Howard-Williams, C., and Brown, J.M.A., Sediment and water as sources of nitrogen and phosphorus for submerged rooted aquatic macrophytes, *Aquat. Bot.*, 40, 225, 1991.

38. Mazouni, N., Gaertner, J.C., Deslous-Paoli, J.M., Landrein, S., and Geringer d'Oedenberg, M., Nutrient and oxygen exchanges at the water-sediment interface in a shellfish farming lagoon (Thau, France), *J. Exp. Mar. Biol. Ecol.*, 205, 91, 1996.

39. Boers, P.C.M., Van Raaphorst, W., and Van der Molen, D.T., Phosphorus retention in sediment, *Water Sci. Technol.*, 37(3), 31, 1998.

40. Caraco, N., Cole, J.J., and Likens, G.E., Evidence for sulfate-controlled phosphorus release from sediment of aquatic systems, *Nature*, 341, 316, 1989.

41. Gibson, C.E., Nutrient limitations, *J. Water Poll. Cont. Fed.*, 43(12), 2346, 1971.

42. Boynton, W.R., Kemp, W.M., and Keefe, C., A comparative analysis of nutrients and other factors influencing estuarine phytoplankton production, in *Estuarine Comparisons*, Kennedy, V., Ed., Academic Press, New York, 1982, p. 69.

43. Plinski, M. and Jóźwiak, T., Temperature and N:P ratio as factors causing blooms of blue green algae in the Gulf of Gdańsk, *Oceanologia*, 41(1), 73, 1999.

44. Fong, P., Zedler, J.B., and Donohoe, R.M., Nitrogen vs. phosphorus limitation of algal biomass in shallow coastal lagoons, *Limnol. Oceanogr.*, 38(5), 906, 1993.

45. Chapra, S.C., *Surface Water Quality Modelling*, WBC/McGraw-Hill, New York, 1997.

46. Officer, C.B. and Ryther, J.H., The possible importance of silicon in marine eutrophication, *Mar. Ecol-Prog. Ser.*, 3, 83, 1980.

47. Conley, D.J., Schelske, C.L., and Stoermer, E.F., Modification of the biogeochemical cycle of silica with eutrophication, *Mar. Ecol-Prog. Ser.*, 101, 179, 1993.

48. Dudgale, R.C. and Wilkerson, F.P., Sources and fates of silicon in the ocean: the role of diatoms in the climate and glacial cycles, *Scientia Marina*, 65(Suppl. 2), 141, 2001.

49. Ittekkot, V., Humborg, C., and Schäfer, P., Hydrological alterations on land and marine biogeochemistry: a silicate issue? *BioScience*, 50(9), 776, 2000.

50. Livingston, R.J., *Eutrophication Processes in Coastal Systems: Origin and Succession of Plankton Blooms and Effects on Secondary Production in Gulf Coast Estuaries*, CRC Press, Boca Raton, FL, 2001.

51. Humborg, C., Conley, D.J., Rahm, L., Wulff, F., Cociasu, A., and Ittekkot, V., Silicon retention in river basins: far-reaching effects on biogeochemistry and aquatic food webs in coastal marine environments, *Ambio*, 29(1), 45, 2000.

52. Konovalov, S.K. and Murray, J.W., Variations in the chemistry of the Black Sea on a time scale of decades (1960–1995), *J. Marine Syst.*, 31, 217, 2001.

53. Egge, J.K. and Aksnes, D.L., Silicate as regulating nutrient in phytoplankton competition, *Mar. Ecol.-Prog. Ser.*, 83, 281, 1992.

54. Justic, D., Rabalais, N.N., Turner, R.E., and Dortch, Q., Changes in nutrient structure of river-dominated coastal waters: stoichiometric nutrient balance and its consequences, *Estuar. Coast. Shelf S.*, 40, 339, 1995.

55. Nixon, S.W., Pilson, M.E.Q., Oviatt, C.A., Donaghay, P., Sullivan, B., Seitzinger, S., Rudnick, D., and Frithsen, J., Eutrophication of a coastal marine ecosystem – an experimental study using the MERL microcosms, in *Flows of Energy and Materials in Marine Ecosystems Theory and Practice*, Fasham, J.R.M., Ed., Plenum Press, New York, 1984, p. 105.

56. Cerco, C.F. and Cole, T., User's Guide to the CE-QUAL-ICM Three Dimensional Eutrophication Model, Release Version 1.0, Technical Report EL-95-15, U.S. Army Corps of Engineers Waterways Experiment Station, Vicksburg, MS, 1995.

57. Environmental Laboratory, CE-QUAL-R1: A Numerical One Dimensional Model of Reservoir Water Quality; User's Manual, Instruction Report E-82-1, Rev. Ed., U.S. Army Engineer Waterways Experiment Station, Vicksburg, MS, 1995.

58. Cole, T.M. and Wells, S.A., CE-QUAL-W2: A Two Dimensional, Laterally Averaged, Hydrodynamic and Water Quality Model, Version 3.1, Instruction Report EL-2002-1, U.S. Army Engineering and Research Development Center, Vicksburg, MS, 2002.

59. Hooper, F.F., Eutrophication indices and their relation to other indices of ecosystem change, in *Eutrophication: Causes, Consequences, Correctives*, National Academy of Sciences, Washington, D.C., 1969, p. 225.

60. U.S. EPA, Estuarine and Coastal Marine Waters: Bioassessment and Biocriteria Technical Guidance, U.S. EPA, Office of Water, Washington, D.C., December, EPA-822-B-00-024, 2000.

61. U.S. EPA, Ambient Aquatic Life Water Quality Criteria for Dissolved Oxygen (Saltwater): Cape Cod to Cape Hatteras, U.S. EPA, Office of Water, Washington, D.C., November, EPA-822-R-00-012, 2000.

62. Fry, F.E.J., Some possible physiological stresses induced by eutrophication, in *Eutrophication: Causes, Consequences, Correctives*, National Academy of Sciences, Washington, D.C., 1969, p. 531.

63. U.S. EPA at http://www.epa.gov/nep/monitor/chptr09.html, EPA National Estuary Program, *Volunteer Estuary Monitoring: A Methods Manual*, Ohrel, R.L., Jr., and Register, K.M., Eds., 9.

64. Thomann, R.V. and Mueller, J.A., *Principles of Surface Water Quality Modeling and Control*, HarperCollins, New York, 1987.

65. Chapelle, A., Ménesguen, A., Deslous-Paoli, J.M., Souchu, P., Mazouni, N., Vaquer, A., and Millet, B., Modeling nitrogen, primary production and oxygen in a Mediterranean lagoon: impact of oyster farming and input from the watershed, *Ecol. Model.*, 127, 161, 2000.
66. UNESCO/WHO/UNEP, *Water Quality Assessment*, Chapman, D.E. and Spon, F.N, London, 1996.
67. Lung, W., *Water Quality Modeling, Volume III, Application to Estuaries*, CRC Press, Boca Raton, FL, 1993.
68. APHA, *Standard Methods for the Examination of Water and Wastewater*, 20th ed., American Public Health Association, Washington, D.C., 1998.
69. Park, K. and Kuo, A.Y., A Vertical Two Dimensional Model of Estuarine Hydrodynamics and Water Quality, Special Report in Applied Marine Science and Ocean Engineering, No. 321, Virginia Institute of Marine Science, Gloucester Point, VA, 1993.
70. Martin, S.C., Effler, S.W., DePinto, J.V., Trama, F.B., Rodgers, P.W., Dobi, J.S., and Wodka, M.J., Dissolved oxygen model for a dynamic reservoir, *J. Environ. Eng.-ASCE*, 111(5), 647, 1985.
71. Rysgaard, S., Risgaard, P.N., Sloth, N.P., Jensen, K., and Nielsen, L.P., Oxygen regulation of nitrification and denitrification in sediment, *Limnol. Oceanogr.*, 39(7), 1643, 1994.
72. Alongi, D.M., The coastal ocean, I. The coastal zone, in *Coastal Ecosystem Processes*, CRC Press, Boca Raton, FL, 1998, chap. 6.
73. U.S. EPA, AQUATOX for WINDOWS, A Modular Fate and Effects Model for Aquatic Ecosystems, Release 1, Volume 2: Technical Documentation, U.S. EPA, Office of Water, Washington, D.C., September, EPA-823-R-00-007, 2000.
74. DiToro, D.M., *Sediment Flux Modeling*, John Wiley & Sons, New York, 2001, chap. 8, 18, 20.
75. Bartoli, M., Cattadori, M., Giordani, G., and Viaroli, P., Benthic oxygen respiration, ammonium and phosphorus regeneration in surficial sediment of the Sacca di Goro (Northern Italy) and two French coastal lagoons: a comparative study, in *Coastal Lagoon Eutrophication and Anaerobic Processes (C.L.E.A.N)*, Caumette, P., Castel, J., and Herberts, R., Eds., *Hydrobiologia*, 329, Kluwer, Brussels, 1996, p. 143.
76. Shultz, D.J., Nitrogen Dynamics in the Tidal Freshwater Potomac River, Maryland and Virginia, Water Years 1979–81, United States Geological Survey Water Supply Paper 2234-J, 1989.
77. Maksymowska-Brossard, D. and Piekarek-Jankowska, H., Seasonal variability of benthic ammonium release in the surface sediments of the Gulf of Gdansk (southern Baltic Sea), *Oceanologia*, 43(1), 113, 2001.
78. Ambrose, R.B., Wool, T.A., and Martin, J.L., The Water Quality Analysis Simulation Program, WASP5, Part A: Model Documentation, U.S. Environmental Protection Agency, Athens, GA, 1993.
79. Lung, W. and Paerl, H.W., Modeling blue-green algal blooms in the lower Neuse River, *Wat. Res.*, 22(7), 895, 1988.
80. Lung, W., A Water Quality Model for the Patuxent Estuary, Final report submitted to Maryland Department of the Environment, Environmental and Water Resources Engineering Research Report, No. 8, School of Engineering and Applied Science, University of Virginia, Charlottesville, 1992.
81. Lung, W. and Hwang, B.G., Water Quality Modeling of the St. Martin River, Assawoman and Isle of Wight Bays, Final report submitted to Maryland Department of the Environment, Environmental and Water Resources Engineering Research Report, No. 15, School of Engineering and Applied Science, University of Virginia, Charlottesville, 1994.

82. Lung, W. and Larson, C.E., Water quality modeling of upper Mississippi River and Lake Pepin, *J. Environ. Eng.-ASCE*, 121(10), 691, 1995.
83. Benaman, J., Armstrong, N.E., and Maidment, D.R., Modeling of Dissolved Oxygen in the Houston Ship Channel Using WASP5 and Geographic Information Systems, Center for Research in Water Resources (CRWR), Bureau of Engineering Research, The University of Texas at Austin, 1996.
84. Picket, P. J., Pollutant loading capacity for the Black River, Chehalis River System, Washington, *J. Am. Water Resour. Assoc.*, 33(2), 465, 1997.
85. Warwick, J.J., Cockrum, D., and Horvath, M., Estimating non-point source loads and associated water quality impacts, *J. Water Res. Pl.-ASCE*, 123(5), 302, 1997.
86. Delaware Department of Natural Resources and Environmental Control, Total Maximum Daily Load (TMDL) Analysis for Nanticoke River and Broad Creek, Delaware, Final Report, Watershed Assessment Section, Division of Water Resources Delaware Department of Natural Resources and Environmental Control, Dover, 1998.
87. Tufford, D.L. and McKellar, H.N., Spatial and temporal hydrodynamic and water quality modeling analysis of a large reservoir on the South Carolina (USA) coastal plain, *Ecol. Model.*, 114, 137, 1999.
88. Melaku Canu, D., Umigesser, G., and Solidoro, C., Short-term simulations under winter conditions in the Lagoon of Venice: a contribution to the environmental impact assessment of temporary closure of the inlets, *Ecol. Model.*, 138, 215, 2001.
89. Johnston, R.K., Wang, P.F., and Ritchter, K.E., A Partnership for Modeling the Marine Environment of Puget Sound, Washington—Puget Sound Naval Shipyard Report, Award Number N000140210502, 2002.
90. Park, K., Kuo, A.Y., Shen, J., and Hamrick, J.M., A Three-Dimensional Hydrodynamic-Eutrophication Model (HEM-3D): Description of Water Quality and Sediment Process Submodels, Special Report in Applied Marine Science and Ocean Engineering No. 327, Virginia Institute of Marine Science, Gloucester Point, VA, 1995.
91. Cerco, C.F. and Cole, T., Three Dimensional Eutrophication Model of Chesapeake Bay; Volume 1, Main Report, Technical Report EL 94-4, U.S. Army Corps of Engineers Waterways Experiment Station, Vicksburg, MS, 1994.
92. Bunch, B.W., Cerco, C.F., Dortch, M.S., Johnson, B.H., and Kim, K.W., Hydrodynamic and Water Quality Model Study of San Juan Bay Estuary, Final Report, U.S. Army Corps of Engineers, Engineer Research and Development Center, Jacksonville District, ERDC TR-00-1, 2000.
93. Bowie, G.L., Mills, W.B., Porcella, C.L., Campbell, C.L., Pagenkopf, J.R., Rupp, G.L., Johnson, K.M., Chan, P.W.H., Gherini, S.A., and Chamberlin, C.E., Rates, Constants and Kinetics Formulations in Surface Water Quality Modeling, 2nd ed., U.S. Environmental Protection Agency, Athens, GA, EPA-600/3-85-040, 1985.
94. O'Connor, D.J., Modeling of eutrophication in estuaries, in *Estuaries and Nutrients*, Neilson, B.J. and Cronin, L.E., Eds., Humana Press, Totowa, NJ, 1981, p. 183.
95. Najarian, T.O., Kaneta, P.J., Taft, J.L., and Thatcher, M.L., Application of nitrogen cycle model to Manasquan Estuary, *J. Environ. Eng.-ASCE*, 110(1), 190, 1984.
96. Jenkins, M.C., and Kemp, W.M., The coupling of nitrification and denitrification in two estuarine sediment, *Limnol. Oceanogr.*, 29(3), 609, 1984.
97. Viel, M., Barbanti, A., Langone, L., Buffoni, G., Paltrinieri, D., and Rosso, G., Nutrient profiles in the pore water of a deltaic lagoon: methodological considerations and evaluation of benthic fluxes, *Estuar. Coast. Shelf Sci.*, 33, 361, 1991.

98.  Martens, C.S., Recycling efficiencies of organic carbon, nitrogen, phosphorus and reduced sulfur in rapidly depositing coastal sediment, in *Interactions of C, N, P, and S Biogeochemical Cycles and Global Change*, Wollast, R., Mackenzie, F.T., and Chou, L., Eds., NATO ASI Series, Series I: Global Environmental Change, Vol. 4, Springer-Verlag, Berlin, 1993, p. 379.

99.  Seitzinger, S.P., The Effect of pH on the Release of Phosphorus from Potomac River Sediment, Final Project Report No. 86-8F, Division of Environmental Research, Academy of Natural Sciences of Philadelphia, 1986.

100. Seitzinger, S.P., Nixon, S.W., and Pilson, M.E.Q., Denitrification and nitrous oxide production in a coastal marine ecosystem, *Limnol. Oceanogr.*, 33, 702, 1984.

101. Wool, T.A., Ambrose, R.B., Martin, J.L., and Comer, E.A., Water Quality Analysis Simulation Program WASP, Version 6.0.0.12, U.S. Environmental Protection Agency, Athens, GA, 2001.

102. Cerco, C.F. and Seitzinger, S.P., Measured and modeled effects of benthic algae on eutrophication in Indian River-Rehoboth Bay, Delaware, *Estuaries*, 20, 1, 231, 1997.

103. Bergamasco, A., Carniel, S., Pastres, R., and Pecenic, G., A unified approach to the modeling of the Venice Lagoon-Adriatic Sea Ecosystem, *Estuar. Coast. Shelf Sci.*, 46, 483, 1998.

104. Freedman, B., *Environmental Ecology: The Ecological Effects of Pollution, Disturbance, and Other Stresses*, 2nd ed., Academic Press, San Diego, CA, 1995.

105. Hansen, P.E. and Jorgensen, S.E., *Introduction to Environmental Management*, Elsevier, Amsterdam, 1991.

106. Schwarzenbach, R.P., Gschwend, P.M., and Imboden, D.M., *Environmental Organic Chemistry*, Wiley-Interscience, Canada, 1993.

107. Piorr, R., Structure and application of surfactants, in *Surfactants in Consumer Products*, Springer-Verlag, Berlin, 1987, p. 5.

108. Giger, W., Brunner, P.H., and Schaffaner, C., 4-Nonylphenol in sewage sludge accumulation of toxic metabolites from non-ionic surfactants, *Science*, 225, 623, 1984.

109. Serge-Chiron, A., Fernandez, A.A., and Rodriques, E.G., Review paper, Pesticide chemical oxidation: state-of-the-art, *Wat. Res.*, 34(2), 366, 2000.

110. Clark, J.R., Lewis, M.A., and Pait, A.S., Pesticide inputs and risks in coastal wetlands, *Environ. Toxicol. Chem.*, 12, 2225, 1993.

111. Ware, J.B., *The Pesticide Book*, 4th ed., Thomson Publications, Fresno, CA, 1994.

112. Rekolainen, S., Occurrence and leaching of pesticides in waters draining from agricultural land, in *Proc. 5th European Symp. Organic Micropollutants in the Aquatic Environment*, Rome, Italy, October 1987, p. 195.

113. Montgomery, J.H., *Agrochemical Desk Reference: Environmental Data*, Lewis Publishers, Chelsea, MI, 1983.

114. Edit, D.C., Hollebone, V.E., Lockhart, W.L., Kingsbury, P.D., Gadsby, M.C., and Ernst, W.R., Pesticides in forestry and agriculture: effects on aquatic habitats, in *Aquatic Toxicology and Water Quality Management*, Wiley-Interscience, Canada, 1989, p. 245.

115. Line, D.E., Jennings, G.D., McLaughlin, R.A., Osmond, D.L., Harman, W.A., Lombardo, C.A., Tweedy, K.L., and Spooner, J., Nonpoint sources, *Water Environ. Res.*, 71, 1054, 1999.

116. BCPC (British Crop Protection Council), *The Electronic Pesticide Manual*-Version 2.2, Latest Edition, 2002. (http://www.bcpcbookshop.co.uk/catalog/BCPC)

117. DPR (Department of Pesticide Regulation), *Pesticide Chemistry Database*, Environmental Monitoring Branch, CA, 2002. (http://www.cdpr.ca.gov/docs/pur/purmain.htm)

118. EXTOXNET, Extension Toxicology Network, *Pesticide Information Profiles (PIP's)*, *Movement of Pesticides in the Environment*, 2003 (http://www.ace.orst.edu/info.extoxnet/pip/ghindex.htm).

119. Watts, C.D. and Moore, K., Fate and transport of organic compounds in rivers, in *Proc. 5th European Symp., Organic Micropollutants in the Aquatic Environment*, Rome, Italy, October 1987, p. 154.

120. Jorgensen, S.E. and Gromiec, M.J., *Mathematical Submodels in Water Quality Systems*, Elsevier, Amsterdam, 1989.

121. Miyamoto, J., Mikami, N., and Takimoto, Y., The fate of pesticides in aquatic ecosystems, in *Environmental Fate of Pesticides*, Hutson, D.H. and Roberts, T.R., Eds., Wiley-Interscience, Chichester, U.K., 1990, p. 123.

122. Bennett, D., Evaluation of the fate of pesticides in water and sediment, in *Environmental Fate of Pesticides*, Hutson, D.H. and Roberts, T.R., Eds., Wiley-Interscience, Chichester, U.K., 1990, p. 149.

123. Larson, S.J., Capel, P.E., and Majewski, M.S., Pesticides in surface waters: Distribution, trends and governing factors, in *Pesticides in the Hydrologic System*, Gilliom, R.J., Ed., Ann Arbor Science, Ann Arbor, MI, 1997.

124. Rao, P.S.C., Mansell, R.S., Baldwin, L.B., and Laurent, M.F., Pesticides and Their Behavior in Soil and Water, Florida Cooperative Extension Service, Institute of Food and Agricultural Sciences, University of Florida, 1999. (http://www.pmep.cce.cornell.edu/ facts-slides-self/facts/gen-pubre-soil-water.html)

125. Hemond, H.F. and Fechner-Levy, E.J., *Chemical Fate and Transport in the Environment*, 2nd ed., Academic Press, San Diego, 2000.

126. Zhou, J.L. and Rowland, S.J., Evaluation of the interactions between hydrophobic organic pollutants and suspended particles in estuarine waters, *Wat. Res.*, 31, 1708, 1997.

127. Ney, R.E., Fate, transport and prediction model application to environmental pollutants, *Proc. Spring Res. Symp.*, James Madison University, Harrisonburg, VA, 1981.

128. Kolset, K., Aschjem, B.F., Christophersen, N., Heiberg, A., and Vigerust, B., Evaluation of some chemical fate and transport models: a case study on the pollution of the Norrsundet Bay (Sweden), *Proc. 5th European Symp. Organic Micropollutants in the Aquatic Environment*, Rome, Italy, October 1987, p. 372.

129. Saleh, M.A., Computer assisted molecular prediction of metabolism and environmental fate of agrochemical, in *Pesticide Transformation Products: Fate and Significance in the Environment*, Somasundaram, L. and Coats, J.R., Eds., American Chemical Society, Washington, D.C., 1991, p. 148.

130. Rajar, R., Application of the three-dimensional model to Slovenian Coastal Sea, in *Computer Modeling of Seas and Coastal Regions*, Hilbradt, P. and Holz, K.P., Eds., Computational Mechanics Publications, Southampton, U.K., 1992, p. 413.

131. Schnoor, J.L., *Environmental Modeling: Fate and Transport of Pollutants in Water, Air, and Soil*, Wiley-Interscience, New York, 1996.

132. Bowmer, K.H., Herbicides in surface water, in *Herbicides*, Hutson, D.H., and Roberts, T.R., Eds., Wiley-Interscience, Chichester, U.K., 1987, p. 272.

133. Day, K.E., Pesticide transformation products in surface waters, in *Pesticide Transformation Products: Fate and Significance in the Environment*, Somasundaram, L., and Coats, J.R., Eds., American Chemical Society, Washington, D.C., 1991, p. 217.

134. Miles, C.J., Degradation products of sulfur-containing pesticides in soil and water, in *Pesticide Transformation Products: Fate and Significance in the Environment*, Somasundaram, L. and Coats, J.R., Eds., American Chemical Society, Washington, D.C., 1991, p. 61.

135. Inaba, K., Shiraishi, H., and Soma, Y., Effects of salinity, pH, and temperature on aqueous solubility of four organotin compounds, *Wat. Res.*, 29, 1415, 1995.

136. Jaskulke, E., Patty, L., and Bruchet, A., Evaluation of pesticide residues in water, in *Pesticide Chemistry and Bioscience*, Brooks, G.T. and Roberts, T., Eds., Royal Society of Chemistry, Cambridge, U.K., 1999, p. 368.

137. Ney, R.E., *Fate and Transport of Organic Chemicals in the Environment: A Practical Guide*, 2nd ed., Government Institutes, Rockville, MD, 1995.

138. Brennan, R.A., Nirmalakhandan, N., and Speece, R.E., Comparison of predictive methods for Henry's law coefficients of organic chemicals, *Wat. Res.*, 32, 1901, 1998.

139. Blust, R., Kinden, A.V., Verhegen, E., and Decleir, W., Effect of pH on the biological availability of copper to the Brine Shrimp *Artemia franciscana*, *Mar. Biol.*, 98, 31, 1988.

140. Blust, R., Ginneken, L.V., and Decleir, W., Effect of temperature on the uptake of copper by the Brine Shrimp Artemia Franciscana, *Aquatic Toxicology*, 30, 343, 1994.

141. Blust, R., Verhegen, E., Doumen, C., and Decleir, W., Effect of complexation by organic ligands on the bioavailability of copper to the Brine Shrimp, *Artemia* sp., *Aquat. Toxicol.*, 8, 211, 1986.

142. Blust, R., Kockelbergh, E., and Boillieul, M., Effect of salinity on the uptake of cadmium by the Brine Shrimp Artemia Franciscana, *Mar. Ecol. Prog. Ser.*, 84, 245, 1992.

143. Jorgensen, S.E., *Modeling in Environmental Chemistry*, Elsevier, Amsterdam, 1991.

144. Dortch, M., Ruiz, C., Gerald, T., and Hall, R., Three-dimensional contaminant transport/fate model, in *Proc. Estuarine and Coastal Modeling Symp.*, VA, October 1997, p. 75.

145. SPARC web site: http://ibmlc2.chem.uga.edu/sparc/

146. Karickhoff, S.W., Brown, D.S., and Scott, T.A., Sorption of hydrophobic pollutants on natural sediment, *Wat. Res.*, 13, 241, 1979.

147. Schwarzenbach, R.P. and Westal, J., Transport of nonpolar organic compounds from surface water to groundwater—laboratory sorption studies, *Environ. Sci. Technol.*, 15, 1360, 1981.

148. Lyman, W.J., Reehl, W.F., and Rosenblatt, D.H., *Handbook of Chemical Property Estimation Methods*, McGraw-Hill, New York, 1982.

149. Schnoor, J.L. et al., Processes, Coefficients and Models for Simulation Toxic Organics and Heavy Metals in Surface Waters, EPA/600/3-87-015, U.S. EPA, Environmental Research Laboratory, Athens, GA, 1987.

150. Merck Index, *An Encyclopedia of Chemicals, Drugs and Biologicals*, 13th ed., O'Neil, M.J. et al., Eds., Merck & Co. Inc., White House Station, NJ, 2001.

151. Kidd, H. and James, D.R., *The Agrochemicals Handbook*, 3rd ed., Royal Society of Chemistry Information Services, Cambridge, U.K., 1991, p. 2.

152. Pontolillo, J. and Eganhouse, R.P., The Search for Reliable Aqueous Solubility (Sw) and Octanol/Water Partition Coefficients Data for Hydrophobic Organic Compounds: DDT and DDE as a Case Study, USGS, Water Resources Investigations, Report 01-4202, U.S. Department of the Interior, Reston, VA, 2001.

153. Halfon, E., Error-analysis and simulation of Mirex Behaviour in Lake Ontario, *Ecol. Model.*, 22, 213, 1984.

154. Schwarzenbach, R.P. and Imboden, D.M., Modeling concepts for hydrophobic pollutants in lakes, *Ecol. Model.*, 22, 171, 1984.

155. Harris, J.R.W., Bale, A.J., Bayne, B.L., and Mantoura, R.C.F., A preliminary model of the dispersal and biological effect of toxins in the Tamar Estuary, *Ecol. Model.*, 22, 253, 1984.

156. Verschueren, K., *Handbook of Environmental Data on Organic Chemicals*, Van Nostrand Reinhold, New York, 1983.

157 Dowling, K.C. and Lemney, A.T., Evaluation of organophosphorus insecticide hydrolysis by conventional means and reactive ion exchange, in *Pesticide Waste Management: Technology and Regulation*, Bourke, J., Jensen, J.K., Felsot, A.S., Gilding, T.J., and Seiber, J.N., Eds., American Chemical Society, Washington, D.C., 1992, p. 177.

158. Mabey, W.M. and Mill, T., Critical review of hydrolysis of organic compounds in water environmental conditions, *J. Phys. Chem. Ref. Data*, 7, 383, 1978.

159. Wolfe, N.L., Zepp, R.G., and Paris, D.F., Use of structure reactivity relationships to estimate hydrolytic persistence of carbamate pesticides, *Wat. Res.*, 12, 516, 1978.

160. Tinsley, I.J., *Chemical Concepts in Pollution Behaviour*, Wiley-Interscience, New York, 1979.

161. Hapeman-Somich, C.J., Chemical degradation of pesticide wastes, in *Pesticide Waste Management: Technology and Regulation*, Bourke, J., Jensen, J.K., Felsot, A.S., Gilding, T.J., and Seiber, J.N., Eds., American Chemical Society, Washington, D.C., 1992, p. 157.

162. Nubbe, M.E., Adams, V.D., and Moore, W.M., The direct and sensitised photooxidation of hexacholorocyclopentadiene, *Wat. Res.*, 29, 1287, 1995.

163. Leifer, A., *The Kinetics of Environmental Aquatic Photochemistry*, American Chemical Society, Washington, D.C., 1988.

164. Matsumura, F. and Katayama, A., Photochemical and microbial degradation technologies to remove toxic chemicals, in *Pesticide Waste Management: Technology and Regulation*, Bourke, J., Jensen, J.K., Felsot, A.S., Gilding, T.J., and Seiber, J.N., Eds., American Chemical Society, Washington, D.C., 1992, p. 201.

165. Mills, W.B., Porcella, D.B., Ungs, M.J., Gherini, S.A., Summers, K.V., Mok, L., Rupp, G.L., Bowie, G.L., and Haith, D.A., *Water Quality Assessment: A Screening Procedure for Toxic and Conventional Pollutants*, Parts 1 and 2, U.S. Environmental Protection Agency, Athens, GA, 1985.

166. Matsumura, F. and Murti, C.R.K., *Biodegradation of Pesticides*, Plenum Press, New York, 1982.

167. Lee, K. and Levy, E.M., Biodegradation of petroleum in the marine environment and its enhancement, in *Aquatic Toxicology and Water Quality Management*, Wiley-Interscience, Canada, 1989, p. 217.

168. Briggs, G.G., Theoretical and experimental relationships between soil adsorption, octanol/water partition coefficients, water solubilities, bioconcentration factors, and the parachlor, *J. Agric. Food Chem.*, 29, 1050, 1981.

# 5 Effects of Changing Environmental Conditions on Lagoon Ecology

*Sofia Gamito, Javier Gilabert, Concepción Marcos Diego, and Angel Pérez-Ruzafa*

## CONTENTS

1-56670-686-6/05/$0.00+$1.50
© 2005 by CRC Press

## 5.1 INTRODUCTION

Coastal lagoon ecosystems are dynamic and open systems, dominated and subsidized by physical energies, and characterized by particular features (such as shallowness, presence of physical and ecological boundaries, and isolation) that distinguish them from other marine ecosystems.[1] Shallowness usually provides a lighted bottom, and the wind affects the entire water column, promoting resuspension of materials, nutrients, and small organisms from the sediment to the surface layer. The large number of boundaries (between water and sediment, pelagic and benthic communities, and among lagoon, marine, freshwater, and terrestrial systems and with the atmosphere) involve the existence of intense gradients and, consequently, a high potential to do work.[1] (Figure 5.1). Because of that, coastal lagoons are usually among the marine habitats with the highest biological productivity.[2] Nutrient input from both run-off and irrigated land waters and from currents through tidal channels contribute to increase the primary productivity affecting the structure of the communities. On the one hand, due to their relatively high degree of isolation, outlets usually have a total surface of less than 20% of the barrier closing the lagoon,[3] and the water exchange between lagoons and the open sea is limited, resulting in a series of physical, chemical, and hydrodynamic boundaries.[4] On the other hand, the generated environmental stress regulates the structure of biological assemblages and leads to complex interactions among physical (light, temperature, mixing, flow), chemical (organic and inorganic carbon, oxygen, nutrients), and biological parameters and processes (nutrients uptake, predation, competition).

As a consequence of high levels of biological productivity, lagoons play an important ecological role among the coastal zone ecosystems, providing a collection of habitat types for many species[5] and maintaining high levels of biological diversity. Most lagoons are subjected to human exploitation through fishing, aquaculture, tourism, and urban, industrial and agricultural developments, inducing changes that affect their ecology.

Under the designation of lagoons a high diversity of environments can be found. Size can vary from a few hundred square meters to extensive areas of shallow coastal sea. The salinity range can go from nearly fresh to hyperhaline waters, with concentrations of salt reaching three times the salinity of the adjacent sea.[6] Salt balance relies on several factors such as the exchange of water with the open sea, the inputs of continental waters from rivers, watercourses and groundwater, and on the rainfall-evaporation balance. The variability of salinity can also be observed inside the lagoon both spatially and temporally. From a hydrographical point of view, most of this variability between lagoons can be summarized by a set of quantitative parameters or indexes that describe both lagoon orientation and structure, as well as spatial variability and the potential sea influence (see Chapter 6 for details).

In biological terms, heterogeneity can be applied to both the structure (species composition and abundance) and functioning (productivity, trophic webs, and fluxes) of the lagoon ecosystem at a wide range of spatial and temporal scales, from biogeographic (thousands of kilometers) to regional (hundred to thousands

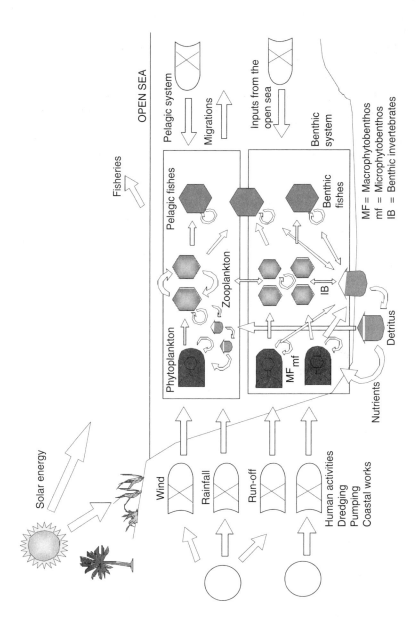

**FIGURE 5.1** Diagram showing the main components and ecological relationships in a coastal lagoon. External inputs of matter and energy and intense gradients at different interfaces result in ecosystems with high levels of biological productivity.

of kilometers, including distinct lagoons in the same area) and local (the inside of the lagoon) (Table 5.1).

From both structural and functional points of view, it is possible to categorize two extreme types of lagoons, one with a stable and predictable environment and the other with frequent physical and chemical disturbances and fluctuations[1] or, according to Sanders,[7] biologically adapted lagoons and physically controlled ones, respectively. Species strategies respond to these situations according to a continuum of life-history strategies, $r$ vs. $K$ ($r$ refers to the rate of increase in the exponential population growth curve and $K$ refers to the carrying capacity of the population in the logistic growth model).[8] The $r$-strategy involves increased reproductive effort through early reproduction, small and numerous offspring with large dispersive capability, short life span, and small body size of adults. This provides a selective advantage in unpredictable or short-lived environments. At the other extreme, $K$-strategy species spend more energy on maintenance structures and adaptations, in a predictable environment, than on reproduction. Species with this kind of strategy usually are larger, long living, less abundant, and show higher biomass/reproductive ratios.

The models used to simulate lagoons dynamics can work at different spatial and temporal scales depending on the process considered, the grid size used, and the quality of input data. Physical and hydrodynamic numerical models can provide quantitative descriptions in a continuum of spatial and temporal scales because of the linearity of many of the involved processes (see Chapters 3 and 6 for details). However, biological processes are complex and show nonlinear

## TABLE 5.1
## Main Sources of Variability to Explain Differences at Hierarchical Spatial and Temporal Scales, in Coastal Lagoons

| Spatial Scales | Main Source of Spatial Variability | Temporal Scales | Main Source of Temporal Variability |
|---|---|---|---|
| $10^3$ km | Biogeographical climatic differences | $>10^4$ years | Global climatic change (ecosystem level) |
| $10^2 - 10^3$ km | Hydrographic features and geomorphology of lagoons (mainly isolation degree); trophic status | $10^1 - 10^4$ years | Changes in hydrographic and geomorphological features and trophic status (sucessional level) |
| $10^{-3} - 10^2$ km | Substrate type; confinement gradient; hydrodynamics; trophic status | $10^0 - 10^1$ years | Interannual fluctuations in populations; changes in recruitment; colonization of species and migrations; predation and competition processes (community level) |
| $10^{-5} - 10^{-3}$ km | Vertical zonation; patchiness of species distribution and population density; microhabitat heterogeneity | $< 10^0$ years | Seasonal fluctuations of populations; predation and competition processes (population and community level) |

relationships, and the scale of the scenarios to be modeled should be previously defined (see Chapter 2 for details).

In coastal lagoons, the relevant spatial and temporal scales for management purposes mainly affect the overall lagoon ecosystem and its successional stages. At these scales, heterogeneity can be explained mainly by variations in two principal factors: water renewal rate related to isolation degree, and trophic status related to availability of nutrients.

The focus of this chapter is to outline changes in the main biological features and processes in lagoons under different eutrophication states and water renewal rates that must be considered when implementing ecological modeling as a decision support tool for sustainable use and development.

## 5.2  EUTROPHICATION PROCESS

As explained in Chapter 2, human activity is responsible for extensive modifications of many of the global element cycles, to the extent that more elements/nutrients are fixed annually by human-driven activities than by natural processes.[9] Coastal lagoons may receive nutrients from a wide range of sources such as domestic sewage, agricultural activities, industrial wastewater, and atmospheric fall-out. The process in which there is an increase in the rate of addition of nitrogen and phosphorus, considered as the two main limiting factors for primary production to a natural system, usually aquatic, is called eutrophication.

Eutrophication is a process,[10] not a trophic state, meaning "an increase in the rate of supply of organic matter to an ecosystem."[11] It is mainly identified with an increase in the input of inorganic nutrients in the ecosystem. It must be taken into account, however, that if the level of primary production, even though it is high, remains constant over time, it does not imply that eutrophication will occur because there will not be any change in the carbon supply rate.

It is well known that small amounts of nutrients usually stimulate primary production. This does not automatically imply a linear increase of the whole production of the ecosystem, but it frequently produces changes in the biological structure and functioning of the whole ecosystem. This leads to the progressive replacement of seagrasses and slow-growing macroalgae by fast-growing macroalgae and phytoplankton, with a final dominance of phytoplankton at high nutrient loads.[12,13] Competition of primary producers for nutrients is one of the responsible processes, but not the only one. Alteration in water turbidity, changes in the hydraulic conditions resulting in modifications of water residence time and a decline of grazing pressure, are also factors that promote shifts in the dominant plant communities. A comprehensive sequence of changes in major plant groups following nutrient enrichment in a wide range of ecosystems has been given by Harlin.[14] These changes in submerged vegetation during eutrophication appear to occur as a step process, with sudden shifts in submerged vegetation, not directly coupled to increased nutrient loading alone, but occurring due to many indirect and feedback mechanisms.[12]

Changes in the primary producers' structure affect secondary producers, as they are the basis of the trophic food web. The trophic status of a coastal lagoon, however, does not depend exclusively on the nutrient load but on the hydrodynamics, which, in turn, determine the residence time of nutrients in the lagoon. For example, discharging the same amount of nutrients into a leaky lagoon with strong tidal currents will not have the same local effects as will a similar discharge into a choked lagoon with a low water exchange (see Chapter 6 for classifications of lagoons).

Four successional stages can be identified in the eutrophication process: oligotrophic, mesotrophic, eutrophic, and hypertrophic. Nixon[11] provides some ranges of carbon supply (primary production) in the ecosystem for each stage (Table 5.2).

Abundant seagrasses (such as Eelgrass *Zostera, Cymodocea, Posidonia,* or *Thalassia*) and transparent water at relatively low nutrient concentrations generally characterize the oligotrophic state of coastal lagoons. The mesotrophic state, characterized by moderate nutrient concentrations, is associated with the presence of benthic macroalgae at the bottom level and some higher phytoplankton concentration in the water column. At this stage, complex interactions among these primary producers (macroalgae and phytoplankton) and with primary consumers (grazers) lead, in some systems, to cycles of alternate dominance by either submerged vegetation or phytoplankton. These cycles can be relatively stable. However, a large disturbance, with the ability to affect different parts of the ecosystem, can override the self-stabilizing capacities, causing a shift from a benthic to a planktonic dominated system.[15,16]

A lagoon is considered eutrophic when high nutrient concentrations can be found in the water column. The biomass and production of phytoplankton communities that are greatly stimulated with nutrients produce highly turbid waters until the point that the phytoplankton biomass becomes dense enough to limit light access to the bottom,[17] thus preventing growth of benthic vegetation seagrasses. Benthic vegetation is then restricted to shallower areas, mostly disappearing in the deepest zones. Oxygen consumption from degradation of produced organic material increases, especially in the sediment, thus causing anoxic periods. The lack of oxygen and production of toxic gasses, such as hydrogen sulphide, due to the anaerobic condition in the sediment (see Chapter 4), has detrimental effects on the bottom-living fauna and in the recruitment of species (mainly fishes and crustaceans) that enter into the lagoon as larvae and juvenile stages. Hypereutrophy is generally considered an extreme case of eutrophy in which the above-mentioned characteristics are heavily enhanced. An idealized sequence of the main features of the eutrophication processes is summarized in Figure 5.2 and will be described in the following sections of this chapter.

**TABLE 5.2**
**Successional Stages in Eutrophication Processes Related to Organic Carbon Supply**

| Successional Stages | Organic Carbon Supply (g C m$^{-2}$ y$^{-1}$) |
|---|---|
| Oligotrophic | <100 |
| Mesotrophic | 100–300 |
| Eutrophic | 301–500 |
| Hypertrophic | >500 |

*Source*: Nixon, S.W., *Ophelia*, 41, 199, 1995. With permission.

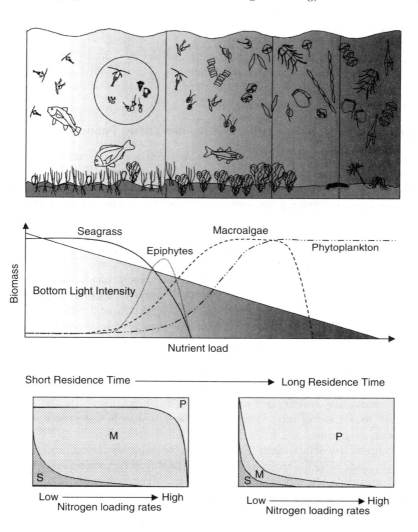

**FIGURE 5.2** Representation of changes in the lagoon ecosystem with increasing nutrient loads. Top: In the oligotrophic state submerged aquatic vegetation is dominated by seagrasses and the planktonic food web is based on the microbial loop. At moderate nutrient loads— mesotrophic state—macroalgae outcompete seagrasses and small phytoplankton grow. At high nutrient loads, large-celled phytoplankton dominates in the water column. Light is strongly trapped becoming the limiting factor for macroalgae and benthic fauna turns to deposit feeders. Middle: Evolution of the abundance of submerged aquatic vegetation, epiphytes and phytoplankton, with nutrient load and light reaching the bottom. (Adapted from Nienhuis, P.H., *Vie Milieu*, 42, 59, 1992. With permission.) Bottom: Relative changes from benthic to pelagic dominated vegetation with nutrient load and residence time of the water in the lagoon (S = seagrass; M = macroalgae; P = phytoplankton). (Adapted from Valiela, I. et al., *Limnol. Oceanogr.*, 42, 1105, 1997. Copyright 1997 by the American Society of Limnology and Oceanography, Inc. With permission.)

## 5.2.1 OLIGOTROPHIC STATE

Oligotrophic lagoons have low levels of nutrient concentrations in the water column. The first consequence of this lack of nutrients is to restrict phytoplankton growth, keeping water at high transparency levels. Light can easily reach the bottom, and is not a limiting factor for benthic vegetation.

### 5.2.1.1 Submerged Vegetation and Related Energy Pathways

Under oligotrophic conditions, with very low concentrations of nutrients in the water column, nutrients are mainly available at the sediment level. Therefore, the stimulation of growth of aquatic plants that take up nutrients from roots vs. algae that take up nutrients directly from water is enhanced.[18] Seagrass can develop within coastal lagoons for a long time (decades to centuries) based on slowly accumulating nutrient pools which are efficiently recycled.[19] This long-term development is also supported by self-stabilizing mechanisms. Seagrass influences the water transparency, decreasing sediment resuspension by retention in the water-sediment interface. Benthic microalgae also contribute to keep sediment oxygenated through photosynthesis. Sediment mineralization (see Chapter 4) usually supplies enough nutrients to benthic micro algae to make them relatively independent of nutrient concentration in the water.[20,21] Sediment maintained at high oxygenation levels provide a suitable environment for both benthic filter feeders and detritivorous organisms. Low levels of nutrients in water, moreover, prevent the presence of epiphytes on seagrasses that can cause a detrimental effect on their growth by reducing light at the leaf level.[16,22]

Benthic rooted vegetation seagrass is the main primary producer in oligotrophic lagoons, providing food to many organisms such as benthic invertebrates and fishes. However, the energy of most seagrasses becomes available to secondary producers after being fragmented and processed through the detrital pathway.[23] The process of decomposition of leaf litter usually starts with autolysis leaching out soluble materials, such as dissolved organic matter (DOM) with bacteria colonizing fragmented material. Macrobenthic organisms, mainly debris-eating amphipods and isopods, can also tear off pieces of plant material with its attached community of microorganisms. Other macrobenthic organisms, such as herbivorous gastropods, can enhance seagrass growth and production by grazing on epiphytes.[24,25] Populations of predators such as ciliates, nematodes, and some polychaetes can develop and their feces may be re-colonized by microorganisms and reingested again, thus reducing progressively the size of the debris.[26-28]

Part of the dissolved organic matter is released to the water column and some is aggregated into amorphous organic particles (approximately from a few μm to 500 μm in diameter).[29] Many biotic and abiotic mechanisms are involved in the aggregate formation, providing a microenvironment that facilitates growth of bacteria and very small phytoplankton in nutrient-deficient waters.[30] Both environments, plant debris and amorphous organic particles, can be colonized by bacteria, making them available to larger consumers that are not effective in capturing free bacteria. Initial colonization of plant debris by bacteria is subsequently completed by a community of protozoa and ciliates feeding on them even though bacteria also can be released to the water column.

Bacteria in the water column are the base of the so-called microbial loop.[31] They can be effectively grazed by heterotrophic flagellates and ciliates and then by other zooplankton that in turn provide available food for larger pelagic organisms such as fish larvae and juveniles. Although the energetic transfer efficiency of the microbial loop is relatively low, because many trophic steps are involved, it remains as one of the most characteristic food web structure in the pelagic environment of oligotrophic lagoons[32] based on recycled nutrients.

### 5.2.1.2    Phytoplankton

Photosynthesis, the process allowing phytoplankton cells to grow, is regulated by the adaptation of cells to varying environmental conditions at a certain range of space and time scales (see last row in Table 5.1). The main environmental variables affecting the physiological state of the algae are light, temperature, and nutrient concentrations. Others, such as salinity, can be determinant for the presence of certain species. The adaptative response of phytoplankton cells varies widely, depending on their ecophysiology and on the environmental conditions of the area where they have been growing.

For a specific lagoon habitat, some phytoplankton species would find better conditions to grow on the basis of their ability to compete for resources at characteristic ranges. As mentioned above, light is not usually a problem for phytoplankton in oligotrophic waters, but lack of nutrients may constitute a serious limitation. Phytoplankton takes up nutrients from the water following the carrier-mediated transport of Michaelis-Menten kinetics, in which nutrient uptake ($V$) is a hyperbolic function of substrate concentration ($S$), with the half-saturation constant $K_s$ equivalent to the concentration necessary to achieve half of the maximum rate of uptake ($V_{max}$):[33]

$$V = V_{max} \; S/(K_s + S) \qquad (5.1)$$

$K_s$ can vary, depending on temperature, light, and $V_{max}$, making this parameter characteristic for species from different areas, either oceanic or coastal (see Chapter 4, Section 4.1.5 for details). If algal cells are under steady-state conditions of nutrient limitation, then the Michaelis-Menten expression can be assumed to reflect the growth kinetics in the form[33]

$$\mu = \mu_{max} \; S/(K_s + S) \qquad (5.2)$$

where $\mu$ and $\mu_{max}$ are the growth rate and maximum growth rate, respectively, and $K_s$ is now the half-saturation constant for growth which is very similar to the half-saturation constant for nutrient uptake (Figure 5.3).

The steady-state condition of nutrient limitation assumed by this kinetics is not generally fulfilled in oligotrophic waters as phytoplankton cells can store nutrients in internal pools to be used when they are scarce. This phenomenon, known as *luxury uptake*, was described by Droop[34–36] in his model for intracellular content of the limiting nutrient:

$$\mu = \mu_{max} \; [1-(Q_o/Q)] \qquad (5.3)$$

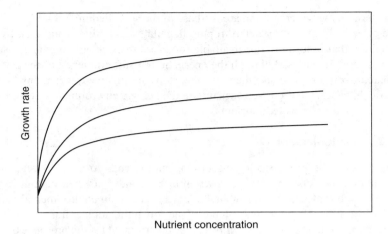

Nutrient concentration

**FIGURE 5.3** Relation between phytoplankton growth rate and nutrient concentration. Phytoplankton takes up nutrients from the water following the Michaelis-Menten kinetics ($V = V_{max} S/(K_s + S)$). Different lines show kinetics with different $K_s$ and $V_{max}$.

where $Q$ is the nutrient content per cell ($Q$ = uptake rate/cell division rate) and $Q_o$ is the minimal nutrient content allowing cell division. Growth rate is thus dependent on cell nutrient content which, at the same time, depends on the adaptation of cells to nutrient concentration in water (Figure 5.4).

Implications for the luxury uptake of nutrients by phytoplankton are relevant in nutrient-limited water, as nutrients are not homogeneously distributed in the water column. Zooplankton, or excretion of other organisms, creates small patches of recycled nutrients that phytoplankton can go through, taking up and storing them,[37] thus providing the chance to grow even at very low nutrient concentration in the water.

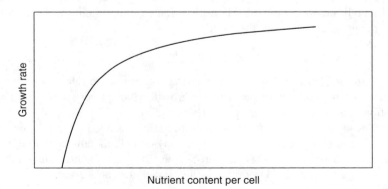

Nutrient content per cell

**FIGURE 5.4** Luxury uptake: Growth rate is dependent on cell nutrient content which, at the same time, depends on the adaptation of cells to nutrient concentration in water.

The Michaelis-Menten kinetic parameters can be interpreted, with caution, as ecological indicators of the physiological adaptation states of the phytoplankton. $K_s$ values for $NO_3^-$ and $NH_4^+$ uptake vary with different eutrophication levels, from higher values in eutrophic water to lower values in oligotrophic water.[38] Some relationships between $K_s$ and cell size also have been traced.[39,40] Comparatively, larger cells exhibit lower surface to volume ratios per unit of biomass than smaller ones. Higher surface to volume ratios indicate a relatively higher number of uptake sites in smaller cells, thus providing some advantage in nutrient uptake in oligotrophic water.[38] The dominance of small-celled phytoplankton in oligotrophic waters also might result from the fact that the acquisition of nutrients by large cells can be limited by molecular diffusion at very low nutrient concentrations. It has been estimated that for a non-swimming osmotrophic cell, with density higher than seawater, the minimum concentration of limiting nutrient at which it can maintain a stable population is a fourth power of cell radius.[41]

Based on these principles, the oligotrophic lagoons are expected to show low cell concentration, dominated by small sized and motile phytoplankton cells (such as cryptophytes and prasinophytes), whereas larger algae (diatoms and dinoflagellates) are expected to dominate in nutrient enriched water.[42]

For modeling purposes, however, the high diversity and species richness of phytoplankton make it unrealistic to know the growth kinetics for individual algae species according to their adaptive stage. A rather reasonable and common alternative to reduce this problem is to group the phytoplankton cells into functional groups related to size (small flagellates, diatoms, dinoflagellates, etc.) rather than use strictly taxonomic criterion classes.

Phytoplankton is usually classified by size into three groups: pico- (0.2–2 μm diameter), nano- (2–20 μm), and micro-phytoplankton (>20 μm).[43]

As stated above, the standing stock of phytoplankton is influenced not only by competition for resources (bottom-up control mechanisms) but also by grazing processes (top-down control mechanism). Zooplankton (mainly crustaceans, such as copepods and cladocerans, in marine and brackish water lagoons, respectively) and fishes are generally the most important phytoplankton grazers. Competition for resources, the bottom-up mechanism control of the food web, provides the phytoplankton with some characteristic species. These species are, at the same time, available food for some particular type of zooplankton or higher consumers. Grazing pressure is highly dependent on temperature, concentration, and size of the food, as well as the consumer size. As a result, the biomass of phytoplankton assemblages is a trade-off between nutrient competition (bottom-up control mechanisms) and herbivory.[44]

### 5.2.1.3    Zooplankton

Zooplankton assemblages in coastal lagoons also can widely vary, depending on the water features. Usually, as in the open sea, the most abundant taxonomic group is copepods, grazing both on phytoplankton and microzooplankton. Although a distinction among herbivores, carnivores, and omnivores can be traced between copepods, they can change their strategy depending on the availability of food.[45] In oligotrophic waters high densities of small omnivorous copepods would be expected as they can effectively feed on small phytoplankton cells, small flagellates from the microbial loop, and some detritus. The rate of filtering generally increases with body size, both within and between species. It decreases for some species when food concentration is above or below certain limits.

The food consumed ($R$) by zooplankton can be expressed in terms of amount of food available by the Ivlev model:

$$R = R_{max} (1 - e^{-kp})$$ (5.4)

where $R_{max}$ is the maximum ration (maximum amount of food that an organism can consume during a certain period of time), $p$ is the prey concentration, and $k$ is a proportionality constant. Although the main factors determining the grazing rate are the size of the organisms and their oral pieces,[46–48] they have a certain ability to choose the food based on its quality, with each species grazing preferentially over a certain particle size range.

From an ecological point of view, other approaches to population dynamics are frequently used to explain the evolution of both phytoplankton and zooplankton populations based on predator-prey type of models. Basic equations describing predator–prey models can be found in any ecology textbook. A minimal model applicable to the eutrophication process should consider logistic growth for preys, in this case phytoplankton, but with carrying capacity depending on nutrient concentration in water. When considering models with three variables (nutrients, phytoplankton, and zooplankton) the outputs show limit cycles and unstable equilibrium points with complex dynamics introducing some degree of uncertainty in predictions.[16]

### 5.2.1.4    Benthic Fauna

As mentioned above, seagrass meadows subsidize not only the microbial webs and zooplankton community but also the benthic filter and detritus feeder organisms. These organisms also benefit from seagrass meadows by efficiently retrieving their energetic requirements from its fragmented debris. The equilibrium between stability of the substrate, refuge provided, and moderate supply of organic matter, all together favored by aerobic conditions, gives the benthic fauna community a balance between their filter feeders and detritivorous organisms. Benthic fauna may vary depending on the lagoon characteristics as they adapt to substrate, with major groups being deposit feeders and other detritivorous, filter feeders and predatory annelids, molluscs, and crustaceans. High abundance of filter feeders including mainly sponges, bivalves, and ascidians are frequently found in oligotrophic lagoons, providing the water with high filtering rates that retrieve many of its particles, including phytoplankton. It has been reported that in shallow waters the entire water column may be turned over in a few days by filter feeders.[8,49]

Benthic macrofauna play an important role, by bioturbation, in the microbial communities in sediment, directly due to burrowing and ventilation[50] and indirectly by feeding on detritus and microorganisms.[51,52] Increased transport due to ventilation of burrow water usually enhances reaction rates and solute fluxes whereas reworking during burrowing is responsible for displacement of organic particles.[8]

## 5.2.1.5  Fish Assemblages

Oligotrophic lagoons also have extensive areas of sandy and muddy bottoms without vegetation coverage that provide extensive feeding resources to several fishes (such as grey mullet and sparid). Open sandy areas are frequently inhabited by small fishes such as *Pomatoschistus* spp., *Gobius* spp., *Callyonimus* spp. and *Solea* spp. Seagrasses not only provide small fishes (mainly gobiids, singnatids, and some blennies) with food as well as shelter from large predators, but also cover the needs of migratory species that require protected habitats for breeding and nursery.[53] Additionally, more intense waterfowl grazing may occur at sheltered areas than at exposed sites.

The growth rate of fish is influenced by several factors, such as temperature, food availability, population density, and competition.[54] In oligotrophic waters primary and secondary production is low, so the food will be scarce, increasing the competition among the different species present and also among individuals of the same species.

Several growth models can be used to describe fish growth, the most commonly used being the von Bertalanffy model.[55] However, in lagoon environments, the majority of fishes only stay inside the lagoon during their first years. Other models, such as the parabolic or the Gompertz models, are more appropriate to describe their growth.[56]

At the ecosystem level, the relationship among the several trophic groups is generally given by

$$d \text{ trophic group}_i/dt = U_i \pm J_i - U_{i+1} - L_i - E_i \qquad (5.5)$$

where $U_i$ represents the food uptake by trophic group $i$, $J_i$ denotes migration rates of group $i$, $U_{i+1}$ represents group losses due to group $i$ consumption by higher trophic groups, $L_i$ denotes loss rates by defecation and natural mortality and $E_i$ corresponds to excretion rates of group $i$ (adapted from Gurney and Nisbet).[57] If $i$ corresponded to zooplankton, then we would have phytoplankton uptake by zooplankton, and zooplankton consumption by the carnivorous group, which could be composed, for example, of fishes and benthic invertebrates. Migration rates for some trophic groups might be equal to zero.

This equation is relatively similar to the net growth equation of the "standard organism" proposed by Baretta-Bekker et al.:[58]

$$STc/dt = (\text{uptake} - (\text{respiration} + \text{mortality} + \text{excretion} + \text{grazing}))STc \quad (5.6)$$

where STc is the carbon biomass of the standard organism. Despite the simplicity of their biological representation, these models need an enormous number of biological parameters. Each term of these equations is described by a relatively complex equation, which might be related to some environmental or biotic factors such as temperature, dissolved oxygen content, or food availability.

## 5.2.2 MESOTROPHIC STATE

Mesotrophic systems are characterized by a medium level nutrient concentration in the water high enough to allow growth of macroalgae, together with phytoplankton, as the major primary producers. Therefore, it is understood that nutrients at this stage can still be assimilated by organisms, hence introducing major changes in the community structure, but keeping the water at relatively high levels of transparency. Main human-induced sources for mesotrophy are agricultural run-off and urban or industrial sewage. However, rich nutrient river and groundwater inputs, atmospheric deposition or the exchange of nutrient-enriched seawater from upwelling areas outside the lagoon also can provide significant loads of nutrients. The nutrient increase in the water column affects both the planktonic system and the benthic system, by increasing competition between seagrass and macroalgae. In this subsection, the general structure and function of mesotrophic lagoons is summarized.

### 5.2.2.1 Competition between Rooted Vegetation and Macroalgae

The increase of competition between seagrass and macroalgae on the lagoon bottom is one of the primary effects derived from water nutrient enrichment. While seagrass take up nutrients by their roots, macroalgae do it efficiently from water by their fronds. Taking up nutrients from water provides macroalgae with a competitive advantage over seagrass, allowing them to spread extensively on the lagoon bottom.[59] Seagrasses and slow-growing macroalgae have nutrient contents much lower than those of phytoplankton.[60,61] It has been estimated that nitrogen and phosphorus requirements of phytoplankton and macroalgae are about 50- and 100-fold higher, and 8- and 1.5-fold higher, respectively, than those of seagrasses.[12]

Occasionally water nutrient enrichment can stimulate blooms of opportunistic species of green algae such as *Enteromorpha, Cladophora, Caulerpa, Chaetomorpha* or *Ulva*,[62] but light also plays an important role in regulating vegetation. Seagrass, as well as thick macroalgae, have low chlorophyll concentrations per unit plant weight with light absorption per unit plant weight much lower than that of phytoplankton and fast-growing macroalgae.[12] In fact, successional sequence of submerged vegetation during eutrophication is largely dependent on an associated shift from nutrient to light limitation, with phytoplankton and free-floating macroalgae being superior competitors under light limitation.

Moderate densities of macroalgae still keep some of the advantages of seagrass meadows as refuge and shelter for fish larvae and juveniles and other planktonic feeders, supply of food for many benthic organisms and sediment stabilization. However, dense algae mats can induce catastrophic effects on the underlying invertebrate fish and bird assemblages through deoxygenation at the sediment level.[63–66] Some long-term studies on eutrophication[14,67] report an increase of macroalgal biomass. In South Quay, in the Ythan Estuary, Scotland, it increased from a few hundred g m$^{-2}$ wet weight in the 1960s to about 1 kg m$^{-2}$ in the 1970s and reached more than 2 kg m$^{-2}$ in the 1980s.[67] On the other hand, the availability of nutrients in water also favors the growth of epiphytes on seagrasses and macro-algae, thus regulating their abundance and distribution. Nevertheless, some mac-roalgae (as *Caulerpa prolifera*) can actively avoid epiphytes by releasing toxic substances, preventing its development. In addition, distribution of submerged vegetation in shelter areas of the lagoons can also be efficiently regulated by grazing, mainly due to fishes and waterfowl.[68]

### 5.2.2.2   Microbial vs. Herbivorous Food Web

The planktonic food web configuration of coastal lagoons depends on the environ-mental conditions. A continuum between the classical "herbivorous" food web and the "microbial loop" web can be observed. The herbivorous food web occurs when medium- to large-size microplankton can grow, thus transferring its production to herbivorous zooplankton and to large invertebrates and fishes. In contrast, the pro-duction of small (pico- and nano-) phytoplankton leads to "microbial food webs," which comprise phototrophic cells (eukaryotic algae and cyanobacteria) as well as heterotrophic bacteria and protozoa. Both "herbivorous" and "microbial loop" food webs would be at each extreme of the continuum.

According to Rassoulzadegan[69] the term *microbial loop*, coined by Azam et al.,[70] designates the almost closed system of heterotrophic bacteria and zooflagellate herbivores, in which the latter release dissolved organic matter (DOM) used as substrate by the former.

On the other hand, the phytoplankton exude dissolved organic carbon that stimulates the bacterial growth, thus fueling the microbial loop. At the same time, bacteria remineralize nutrients that can be taken up by small-celled phytoplankton, thus fueling the microbial food web.[40,71] Figure 5.5 represents the changes in the planktonic food web with an increasing nutrient load in lagoons.

As described above, the planktonic food web structure in oligotrophic lagoons is mainly based on both the detrital pathway and the subsequent microbial loop. Phy-toplankton in these lagoons can be scarce, even in a very small amount, because of the process explained above. Its growth is based on regenerated nutrients, mainly ammo-nium, from both submerged vegetation and fauna. In contrast, in meso to eutrophic lagoons, phytoplankton increases, and microbial and herbivorous food webs become more important, with the major source of nitrogen being nitrate entering mainly from agricultural run-off and/or urban and industrial sewage.[72] Phytoplankton assimilate ammonium first (because of the easily obtainable energy compared to nitrate) and nitrate is utilized only after ammonium has been consumed. There are several biochemical reactions for this preferential assimilation of nitrogen forms involving both repression of the enzyme responsible for nitrate uptake in the presence of ammonium and activation of its synthesis process by exposure to nitrate and absence of ammonia[73] (see Chapter 4 for details).

Availability of nutrients in the water also relieves the competitive disadvantage of large phytoplankton cells against the small-celled ones, promoting the shift to larger cells through the eutrophic gradient. Thus, whereas in oligotrophic water small flagellates with low $K_s$ and $\mu_{max}$ would tend to dominate, in more eutrophic water larger cell diatoms with higher $K_s$ and $\mu_{max}$ would generally dominate. From these

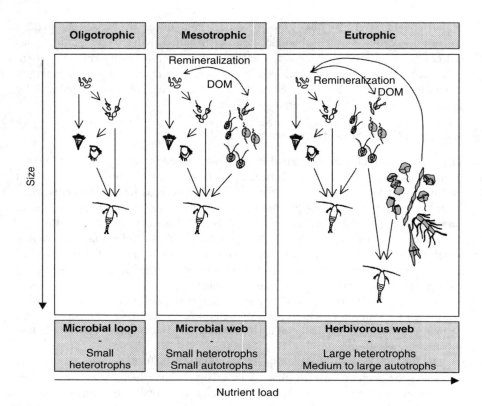

**FIGURE 5.5** Representation of the changes in the planktonic food web with increasing nutrient load in coastal lagoons. Microbial loop is characteristic of oligotrophic waters. In mesotrophic waters, small phytoplankton (autotrophs) are directly eaten by small crustaceans that release dissolved organic matter (DOM) used as substrate by bacteria. Bacteria, on the other hand, remineralize nutrients that can be effectively uptaken by autotrophs. At the eutrophic state, large-celled phytoplankton is the main food source for herbivores. DOM released by large cells can maintain both systems, the microbial loop and the microbial web.

competitive principles (bottom-up control mechanism) a change from small flagellates to large diatoms can be expected with increasing eutrophication. However, the complex dynamics induced by predation processes may greatly modify this trend and it cannot be automatically deduced that the eutrophication process leads to a final stage dominated by large diatoms because of the many indirect effects involved.

The uptake efficiency also depends on other environmental factors such as light and temperature. Temperature imposes both a higher and a lower limit for algae growth. The relationship between growth rate and temperature describes a typical asymmetric parabola, where optimum temperature for each species is displaced to the maximum temperature supported.[74] Light dependence of nutrient uptake, on the other hand, exhibits a truncated hyperbolic function.[75]

Photosynthesis rate is also strongly affected by light intensity. The photosynthesis:light (irradiance) curve (P:I curve) shows that photosynthesis increases with increasing light intensity up to some asymptotic value $P_{max}$, where the system becomes light saturated (Figure 5.6). Light is collected by units composed of accessory pigments and a reaction center containing chlorophyll. At low light regime, photosynthesis increases linearly with light. The initial slope of this part of the P:I curve is the photosynthesis efficiency factor ($\alpha$), which means the number of moles of carbon incorporated per unit of light intensity (quantum yield).

The maximum photosynthesis ($P_{max}$, or assimilation number) is reached at a light intensity in which the enzymes involved in photosynthesis cannot act fast enough to proceed with the excess of light. Several mathematical formulations for the P:I relationships have been described.[76] One of the most commonly used equations is the negative exponential with photoinhibition:

$$P = P_m(1 - e^{\frac{-I\alpha}{P_m}})e^{\frac{-\beta I\alpha}{P_m}} \tag{5.7}$$

where the normalized photosynthetic rate $P$ is a function of incident irradiance $I$, normalized photosynthetic capacity $P_m$, normalized photosynthetic efficiency $\alpha$, and a dimensionless photo-inhibition parameter ($\beta$). Respiration is often subtracted from the right-hand side of the equation.

Both the photosynthesis efficiency factor and the assimilation number depend on the algae taxon (e.g., dinoflagellates are generally expected to show higher $P_{max}$ than diatoms or green algae) but also vary within one species depending on environmental factors such as temperature, nutritional environment, and light regime

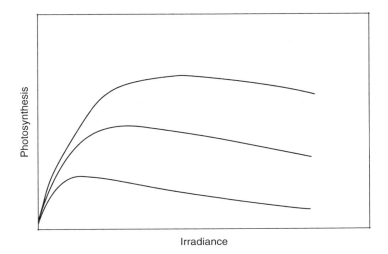

FIGURE 5.6 Relationship between photosynthesis and light intensity (irradiance) showing the photo-inhibition effect of high irradiance. Different lines show P:I curves with different constants.

recent history. Adaptation to light intensity mainly consists of varying the amount of pigments involved in capturing photons, thus varying their carbon to chlorophyll ratios. More chlorophyll is developed as an adaptation to low light conditions to improve their ability to collect light.[77]

A primary consequence of the increased phytoplankton assemblages due to nutrient enrichment is the increase of the light attenuation coefficient in water. Light is efficiently trapped by phytoplankton thus decreasing the amount reaching the bottom for macrophytes growth. In mesotrophic lagoons, phytoplankton growth may be enhanced seasonally by some environmental variables such as temperature, with a detrimental effect on macro-algae due to light limitation.

Increasing the trophic state of lagoons not only causes changes in the phytoplankton assemblages structure but also in the whole planktonic food web, eventually moving from a microbial loop-based web to a herbivorous food web. However, predicting changes at the whole food web is an intricate task as a result of the complex interactions among organisms. As supported by theory and frequently reported, the increased nutrient concentration facilitates growth of larger cells. Furthermore, heavy grazing on small cells, mainly due to microzooplankton (ciliates and crustaceans larvae stages) and small and medium-sized crustaceans, enhances large cells to take up nutrients. As a result, increased nutrients and increased large zooplankton tend to decrease the relative abundance of small phytoplankton and to increase the average phytoplankton size.[78,79] Increased average size of plankton assemblages can, in some cases, lead to an increase of gelatinous zooplankton populations (e.g., ctenophores, jellyfishes),[80,81] and fish larvae because large phytoplankton cells and crustaceans are an important component of their diet.

An alternation of domination by either submerged macrophytes or phytoplankton has also been reported for some lagoons.[15,82] Each of the states remains stable until a disturbance large enough to override the self-stabilizing capacities, even affecting only different parts of the ecosystem, causes a shift to the other state.[16] Population dynamics models have successfully explained such alternation in shallow lakes but its application to coastal lagoons still needs further development.

It can be said, in short, that the resulting food web structure is a trade-off between growth rate variation of organisms caused by resource availability ("bottom-up" controls) and loss rate variations caused by predation ("top-down" controls).[44] However, prediction on such trade-offs seems difficult as small changes in the structure of the primary producers may result in quite unpredictable and large changes at the whole lagoon level.

### 5.2.2.3  Benthic Fauna

Mesotrophic lagoons combine phytoplankton, especially at the plankton-benthos interface, microphytobenthos, and bacteria-enriched detritus, originating from submerged aquatic vegetation, as major food sources for benthic fauna.[83]

According to the Pearson-Rosenberg model, inputs of organic enrichment usually involve changes in abundance, biomass, and species richness of macrobenthic assemblages.[8,84] (Figure 5.7). There is an initial increase in the three parameters and a

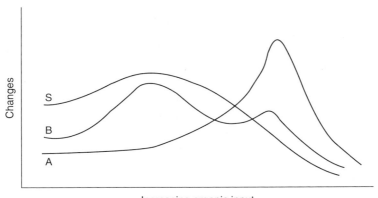

**FIGURE 5.7**  Changes in abundance (A), biomass (B), and species richness (S) of macrobenthic assemblages with increasing organic input, according to the Pearson-Rosenberg model. (Adapted from Pearson, T.H. and Rosenberg, R., *Oceanogr. Mar. Biol. Annu. Rev.*, 16, 229, 1978. With permission.)

progressive decline in species richness and biomass when eutrophication increases, while abundance (mainly of opportunistic species) continues to rise.

Effects of nutrient enrichment can usually be detected earlier and more obviously in meiofaunal populations, with a lifecycle that is usually shorter than in the macrofauna, thus responding more quickly to small changes in the organic content of sediment. On the other hand, compared to macrobenthic fauna, meiofauna is also less affected by physical disturbance of sediment (such as humans digging up sediment to capture bivalves).[85,86] The joint study of macro- and meiofauna can be useful to discriminate between two disturbance factors: pollution or organic content increase and physical disturbances.

A moderate organic enrichment usually increases the secondary production of the affected communities, supporting larger numbers of fishes and waterfowl. However, when the origin of the load is sewage it can produce significant public health problems. Filter feeders, such as mussels and clams, may accumulate and increase concentrations of sewage-derived pathogens. Many beds of shellfish cannot be exploited or can only be eaten after depuration. Pollution, whether chronic or acute, usually tends to favor short-lived opportunistic species.[66]

### 5.2.2.4  Fish Assemblages (Pelagic/Benthic)

As a result of increasing benthic invertebrate biomass and planktonic biomass, fish production can also increase. This phenomenon was clearly observed in the Baltic Sea. Elmgren[87] estimated that the pelagic energy flows in the Baltic increased by 30 to 70% in the 20th century and the fish catches increased more than tenfold. In some areas of the Mediterranean, in the Adriatic and the Aegean Seas, eutrophication extended offshore, increasing primary production and associated production of pelagic fish, such as anchovies.[88]

However, changes in the benthic meadows, with the substitution of seagrasses by algae, can induce changes in species distribution and abundance. In the Mar Menor Lagoon (Spain), the replacement of scant seagrass meadows of *Cymodocea nodosa* by dense meadows of the macroalgae *Caulerpa prolifera* caused a displacement of the benthic fish *Gobius niger* from the sandy bottoms to the *Caulerpa* meadow, which provides a better refuge.[89] At the same time, changes induced at the sediment–water interface by the dense covering of the *Caulerpa* meadows and by the high annual supply of organic matter caused oxygen depletion, $H_2S$ release, and changes in organoleptic characteristics of the organic matter. These changes have been considered as one of the causes of the strong reduction in mugilid fisheries.[63]

## 5.2.3 EUTROPHIC STATE

Coastal lagoons are considered eutrophic when they have high levels of nutrients in water and sustain large assemblages of phytoplankton as dominant primary producers. Hypereutrophy is considered the state in which phytoplankton assemblages increase up to the self-shadow level, preventing light from reaching the bottom and not allowing macroalgae to grow. This subsection outlines characteristics of the eutrophic state.

### 5.2.3.1 Phytoplankton

The previously discussed ecological bases of phytoplankton ecology provide an explanation of the expected changes in nutrient-enriched water. Several consequences for the planktonic food webs, and hence for the whole ecology of the lagoon, have already been outlined. Many lagoons in the eutrophic state meet the environmental conditions required for phytoplankton to bloom: high nutrient loads and a relatively stable hydrological regime. Phytoplankton blooms can reach very high numbers of cells up to a point that they become light limited by self-shading.[90] In very nutrient-rich waters the phytoplankton blooms sometimes are unable to completely absorb the amount of dissolved nutrients. Large diatoms such as *Chaetoceros* or *Coscinodiscus* and dinoflagellates such as *Ceratium* or *Dinophysis*, some of them toxic, frequently bloom in eutrophic waters, and colonial cianobacteria are usually found in nutrient-overenriched water.

### 5.2.3.2 Benthic Vegetation

As a result of nutrient loading, fast-growing macroalgae develop in large quantities up to 1000–2000 g DW m$^{-2}$.[21] With such densities, the algae cover the bottom in thick mats. The photosynthetic production in the lower part of the mats is limited by light, where the respiratory oxygen demand easily exceeds the photosynthetic oxygen production. When this situation lasts for several days, oxygen depletion and $H_2S$ poisoning kill the algae. Growth-limiting nutrients, which previously were bound in the macrophytes, become available for further growth of phytoplankton and opportunistic macroalgae.[91] Eutrophication does not stimulate the total primary production per unit area but shifts the main productivity from the benthic to the planktonic community.[92]

As discussed previously, the result of the competition between macroalgae and rooted submerged vegetation seagrasses is a widespread and usually thick seaweed layer, implying a large amount of organic matter retained at the sediment level. Under hypereutrophic conditions, seaweed starts to decompose, as light limitation by phytoplankton imposes a serious constraint on growth, liberating organic matter both into the water column and into sediment, producing severe anoxia.

### 5.2.3.3  Benthic Fauna

As the degree of eutrophication increases, the effects of anoxia cause strong changes in the benthic community by limiting the growth of benthic filter feeders and thus causing providing a shift in species composition from bivalves toward polychaetes and oligochaetes on sand flats and mudflats.[84,93] The benthic communities can be perturbed by sediment organic enrichment to the point of disappearance. As mentioned above, it is generally assumed that the benthic communities subjected to an increasing load of organic matter show a decrease in species richness, an increase in the total number of individuals (due to high densities of a reduced number of opportunistic species), a general reduction of the biomass, a decrease in the average size of the species and individuals, and a decrease in the sediment thickness occupied by the fauna. All of these alterations cause a change in the trophic web.[84,94]

The depth to which benthic animal life occurs is limited by food availability. In eutrophic sediment oxygen availability and sulfide concentrations also are limiting factors. Animals feeding at depth must have access to oxygen, either in the surface layer or from the overlying water. When oxygen availability is limited, or when the sediment oxygen demand at the feeding depth is great, animals are presumably no longer able to withstand this environmental stress.[8] In shelter areas where large-celled phytoplankton assemblages are controlled by the top-down mechanisms of benthic filter feeders, anoxia implies the loss of one of the important control mechanisms of eutrophication at the entire lagoon level.[83] On the other hand, in addition to moving around within sediment, the fauna build tubes, construct burrows and feeding pits, and transport sediment. When the fauna disappear, due to anoxia events, the physical and chemical characteristics of the top sediment layers are greatly changed.[8] Increased organic loadings will lead to upward movement of reducing conditions and anoxia in sediment and ultimately in the water column. This will shift sediment from aerobic to anaerobic pathways, which may ultimately lead to the disappearance of the fauna.[95]

### 5.2.3.4  Fish and Bird Assemblages

There is no direct evidence that eutrophication has damaged the fisheries in some eutrophic waters. This is largely due to the belief that the decline in commercially important fish populations may be due to over-exploitation.[96] However, as the increase in primary production due to eutrophication has not been followed by an increase in decomposition, the net result has been the production of anoxic conditions in deeper waters, either impoverishing or completely eliminating benthic communities and fish populations.[97] Strongly eutrophic areas can be subjected to algal blooms, including toxic dinoflagellates, resulting in anoxic conditions near the bottom, with associated fish kills and possible viral infections from consumption of contaminated shellfish.[88]

Fish yield can increase to a certain extent due to an increase of nutrient load, but if the conditions are too extreme, massive fish mortalities can result from oxygen depletion and $H_2S$ release. These effects will also be reflected in the total catches in the surrounding coastal sea, as most of the species are migratory.[63] A reduction in total catches is then expected, which can also be a consequence of over-fishing. For instance, Parsons[98] has questioned if phytoplankton blooms may be driven not only by bottom-up processes of nutrient enrichment, but also as a result of overfishing of commercial fish stocks and removal of top predator control. These predators may be replaced by another predator, such as the jellyfish, which feeds on the same food organisms (see Chapter 9, Section 9.3, Mar Menor Lagoon case study).

The birds feed essentially on the intertidal areas or in very shallow waters. In moderately eutrophic lagoons, the amphipods and crabs proliferate in the algal mats, as long as important algal-free areas exist that ensure a healthy population. Many species of birds, which feed on these preys, can be enhanced.[99] However, if macro-algae mats continued to proliferate and spread over larger areas, there will be a much-reduced overall area for bird feeding[100] and, as a consequence, there will also be a general decrease in bird populations.

## 5.3  WATER RENEWAL RATES

Two components underlie the concept of water renewal, and both are important for understanding aquatic ecosystem function: changes in the vertical mixing and exchange with adjacent systems. As stated above, shallowness is one of the main features characterizing coastal lagoons, ensuring that effects of wind usually act over the entire water column and vertical mixing of waters occurs. Only in shelter areas and in areas covered by dense meadows (e.g., of *Caulerpa prolifera* or *Ruppia cirrhosa*) does water remain stagnant at the bottom level, leading to anoxic environments and consequently to fauna impoverishment (see Chapter 9, Section 9.5, Koycegiz–Dalyan Lagoon case study).

The main aspects of the renewal concept applied to lagoons therefore relate to their exchange with the adjacent aquatic systems and its magnitude reflects the time-scale and the range of hydrological variability.[101–103] The renewal rate affects, in the first place, the physico-chemical characteristics of water mass, mainly salinity. This parameter has been frequently used to classify the type of lagoons (for example, the Venice system;[104] see also Petit[105] and Por[106]), explaining composition of their communities on the basis of autoecological criteria and indicator species.[107] At this point it is important to differentiate between salinity and ionic composition. Salinity is the total amount of dissolved salts in the water and depends mainly on the hydrological balance in the lagoon (rain and other inputs of fresh water and evaporation processes). Ionic composition refers to the qualitative composition of the salts. Thalasogenic waters are waters of marine origin that maintain constant proportion of constituents. Limnogenic waters have variable composition, depending on the nature of the geologic materials in the watercourses and drainage area.

Another classification system, also related to water renewal rate and water constituents, but independent of salinity, was suggested by Guelorget and Perthuisot.[108] They define the term "confinement" as a measure of the degree of isolation related to the renewal rates and the impoverishment of substances ("vitamins") of marine origin (Figure 5.8). Biological gradients such as biomass, abundance, and diversity that can be found in coastal lagoons are explained as a function of confinement (see also Guelorget et al.,[109] Bianchi,[107,110] and Carrada and Fresi[111]).

However, the distribution of organisms in coastal lagoons responds to more than one parameter.[89,107,110,112,113] For example, the distribution of some communities and

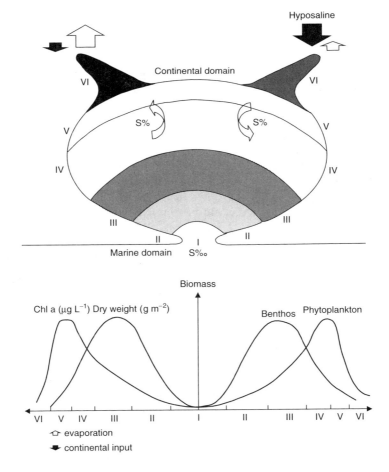

**FIGURE 5.8** Diagram showing the biological zonation (I to VI ) in coastal lagoons according to the confinement model and the associated variation in benthic and phytoplankton biomass. (Adapted from Guelorget, O. and Perthuisot, J.P., *Trav. Lab. Geol.*, 16: 1, 1983. With permission.)

macrophyte meadows display patch distributions related to the nature of the bottom, the physical and chemical composition of sediment, temperature ranges, wave-energy, or hydrodynamism and depth rather than only horizontal gradients.

The horizontal zonation and the structure of lagoon communities are, therefore, the result of interactions between biotic and abiotic factors. Species survival depends on physiological adaptations to these extreme and fluctuating environments, and such adaptations involve energy costs that limit their growth and reproductive rates. It also must be taken into account that for many species the ionic composition of water can be a factor.[114,115]

Another aspect that must be emphasized is the role played by the immigration of species from coastal open ecosystems into the structure of the lagoon communities, and the advantages and difficulties of different ecological and reproductive strategies (e.g., $r$ vs. $K$) in an environment that can be considered stable from some points of view but stressed from others. In this sense, Pérez-Ruzafa[89] and Pérez-Ruzafa and Marcos[112,113] redefined the term confinement as the capability of open-sea organisms to colonize lagoon environments on the basis of their arrival rates depending on water renewal rates and the circulation model (i.e., as planktonic or juvenile stages and other passive dispersal forms) and swimming capabilities, and survival, reproductive, and competition capabilities, depending on environmental conditions inside the lagoon. In this context, in the equations for modeling the population dynamics of two competitors based on Lotka-Volterra type models, the disadvantage in the competition of allocthonous species can be compensated if their immigration rates are high enough (see Levinton[116]). On the other hand, models of species equilibrium (MacArthur and Wilson[117]) also could be applied to explain seasonal changes in specific richness.

Given that water exchange with the coastal ocean is the main factor affecting both hydrological characteristics and biological structure of a lagoon's assemblages, Kjerfve's classification[118] into three types (choked, restricted, and leaky; see Chapter 6 for additional details) is useful here to explain lagoon variability.

## 5.3.1 CHOKED LAGOONS

Choked lagoons are those connected to the open sea by a single long narrow entrance channel with reduced tidal oscillations when compared to the adjacent coastal sea. They are characterized by long flushing times, dominant wind forcing, and intermittent stratification events that result from changes in the hydrological balance (mainly evaporation and run-off),[119] which also determines the salinity range. This type of lagoon is characteristic of physically controlled ecosystems with strong fluctuations in environmental parameters depending on weather conditions. It corresponds to zones V and VI in Figure 5.8, with faunal composition characterized by a few, but very abundant, small-sized opportunistic species. Under such extreme environmental conditions food chains are shortened and competitive weakness increases[106] (Figure 5.9).

In extreme isolation faunal and flora assemblages are highly impoverished. However, with moderate renewal rates, it is possible to find maximum concentrations

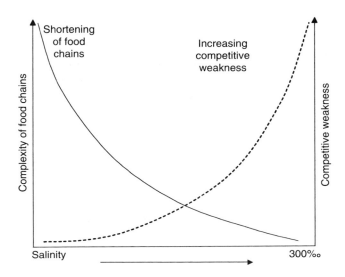

**FIGURE 5.9** Effects of increasing salinity on food chains and the competitive capability of species. Under extreme environmental conditions food chains are shortened and competitive weakness increases. (Adapted from Por, F.D., *Mar. Ecol.*, 1, 121, 1980. With permission.)

of chlorophyll *a* and high phytoplanktonic biomass (see Figure 5.8). In this kind of lagoon, primary production mainly remains in phytoplankton assemblages, generally characterized by Cyanophytes, and the benthic biomass is low.

Sedimentation is mainly organic, and the interstitial environment in sediment is reductor. Faunal assemblages are mainly composed by detritivorous and some herbivorous crustaceans (belonging to genera such as *Sphaeroma, Corophium, Idotea, Gammarus, Microdeutopus*), grazer gastropods (e.g., *Hydrobia*), carnivorous polychaetes (*Nereis*), and detritivorous insect larvae (Chironomids). In hypersaline lagoons, *Artemia*, Tricoptera, Oligochaeta, Odonates can easily be found on a vegetation stand of *Potamogeton* and Characeae (mainly in limnogenic waters). In general, the communities of these lagoons are dominated by *r*-selected species, characterized by a short generation time and a high reproductive effort, producing many small offspring. The diversity is low and many ecological niches may remain unoccupied because of lack of immigration and colonization. The top-down control by predator–prey relationships is low or nonexistent.

In choked lagoons, which are only temporarily open to the sea (during storms, for instance), the migration of marine organisms is intermittent and depends on the temporary openings to the sea.[64,65] These marine organisms rely on the maintenance of marine conditions inside the lagoon for their subsistence. The presence of marine organisms inside the lagoon can be temporary, and it may lead to a situation where only few species survive throughout the year.

As pointed out previously, coastal lagoons are more productive than other ecosystems in terms of fisheries yield.[120] They are important in the life cycles of many coastal fishes, allowing a high standing stock. Three main groups of fishes may occur in coastal lagoons:

1. Sedentary species—Those that spend their entire life cycle within coastal lagoons. This group is very limited, especially when species with planktonic stages are considered. Most lagoon fishes spawn outside of the lagoon.
2. Seasonal migrants—Those that enter the lagoon during a more or less well-defined season. This group dominates the fauna of coastal lagoons.
3. Occasional visitors—Those that enter and leave a lagoon without a clear pattern within the same years and from year to year.

The relative level of recruitment into coastal lagoons is determined by the facility with which fish can penetrate through the channels and by the total number of potential recruits along the coast. This implies that recruitment to coastal lagoons can be artificially increased by keeping the inlets of lagoons open during periods when juveniles of preferred species occur along the coast, or by deepening the sill of lagoons with very shallow mouths.[120]

Some fisheries management practices are based, in general, on taking advantage of fish migration by opening or digging outlets to increase the water renewal rate, increasing at the same time the chance of fishes entering into the lagoon. These practices, however, can drastically change the composition and structure of natural assemblages, especially in choked lagoons, inducing high mortality rates of naturally occurring organisms and replacing them by others from the open sea. There is evidence showing that in lagoons the increase in renewal rates with the open sea has involved an increase in the number of species and diversity, accompanied by a dramatic decrease in the harvest of traditional fish species (mainly Mugilidae and Sparidae).[65] Further, when the artificial temporary openings are later closed, there are usually high mortality rates of marine organisms that had entered,[121] especially if the environment inside the lagoon becomes brackish or hypersaline.

The low migration rates in choked lagoons lead to the populations inside them potentially remaining isolated for long periods of time with loss of some genotypes. Under such isolated conditions, populations usually show low genetic diversity and high homocigosys, although in coastal lagoons the extreme environmental variability can induce high genetic diversity and heterocigosys.[122]

## 5.3.2 Restricted Lagoons

Restricted lagoons have two or more entrance channels or inlets, and flushing times are considerably shorter than for choked coastal lagoons. As a result, they have a well-defined tidal circulation and exhibit salinity gradients from open sea to brackish or hypersaline water. Restricted lagoons are also influenced by winds and are vertically well mixed. Because of their productivity and shelter facilities, restricted lagoons are used as nursery areas by nektonic crustaceans and fishes.

Restricted lagoons exhibit a well-defined pattern of horizontal zonation of biological assemblages according to the model proposed by Guelorget et al.[109] Biological diversity shows a gradient from maximum to minimum, from the communication channels to the confined areas. Meanwhile, benthic biomass and abundance of individuals exhibit the inverse pattern, with maximum biomass and abundance of individuals in the intermediate areas (Figure 5.8)

These gradients also correspond to changes in benthic taxonomic composition. Molluscs usually are the dominant groups close to the sea, where some echinoderms can also be found (such as *Asterina*, *Holothuria* or *Paracentrotus*). Polychaetes, crustaceans, and chironomids successively increase their relative abundance with confinement.

In zones where the environmental conditions are more stable (near the channels due to the buffered influence of the open sea environment) the settlement of species exhibiting $K$-selection strategies is possible. As stated above, these species are larger, reproduce later in life, and develop more slowly. They are, however, superior competitors to $r$-species in such stable environments. The $K$-strategists can control, through predator–prey relationships, the $r$-strategists, which would proliferate exponentially in their absence (top–down control), allowing the establishment of several species spread by the different trophic levels. The diversity is usually high with the use of all available resources, and this leads to a highly diverse community.

### 5.3.3 LEAKY LAGOONS

Leaky lagoons have many wide ocean entrance channels and tidal currents are sufficiently strong to overcome the trend of being closed by wave action and littoral drift. They are characterized by unimpaired water exchange with the ocean, strong tidal currents, and salinity similar to that of the coastal ocean.[119] In these lagoons biologic assemblages are similar to marine environments in sheltered open coastal areas, and the horizontal zonation is less evident or completely absent. The relative level of recruitment into coastal lagoons is determined by the total number of potential recruits along the coast.[120] Vegetation on sandy or muddy bottoms is characterized by meadows of *Caulerpa prolifera*, *Zostera* spp., *Cymodocea nodosa*, *Thalassia* spp., or *Posidonia oceanica* (in the Mediterranean lagoons).

These lagoons offer a protected shallow marine habitat, which can be highly diversified and productive, comparable to other marine shallow water ecosystems. Well-structured communities, controlled by $K$-strategists, can develop and settle down in leaky lagoons. In the absence of exogenous disturbance, biomass can accumulate in large organisms, while the choked lagoons, characterized by variable or persistent physical stress, appear to be dominated by communities of small-sized $r$-strategists organisms. The restricted lagoons have characteristics of both leaky and choked lagoons.[101]

### 5.3.4 WATER RENEWAL RATE AND EUTROPHICATION

In eutrophic choked lagoons the instability of the environment increases by several orders of magnitude. Frequent algal blooms can occur, followed by "crashes" or dystrophic crises, resulting in high ammonia and low oxygen concentrations[123,124] and subsequent organism mortality. These crises are followed by a rapid development of the $r$-strategist species. In eutrophic lagoons with high hydrodynamics the accumulation of nutrients and subsequent problems of development of phytoplankton and macroalgae are minimized as the excessive production may be exported out of the lagoon and eventually sink in deeper waters. Furthermore, critical situations are avoided because oxygenated conditions are re-established by water exchange and mixing.[91]

## 5.4 CHANGES IN LAGOON PROCESSES AND MANAGEMENT OF LIVING RESOURCES

Because of their high biological productivity, lagoons are usually used for fisheries and aquaculture exploitation.

Coastal lagoons constitute an integral part of marine fisheries and provide important spawning and nursery grounds for many fish species. These lagoons and their wetlands provide valuable products and services, which include supporting the fisheries and protecting biodiversity.[125] The fishery yield is dependent upon a lagoon's geographical location and morphometry. The significant environmental and anthropogenic factors determining the yield are the exchanges of water between lagoons and the ocean, the physico-chemical properties of the water, the extent of aquatic vegetation, and the fishing pressure. The single influence of freshwater input may be considered negligible while the influence of oceanic tide through the inlets is significant. The single most influential factor of all is the fishing pressure.[126]

Small, intermittently open lagoons present a complex management problem. These water bodies are easily degraded in terms of water and sediment quality, and are often opened mechanically because of flooding or health concerns. However, fish assemblages change after opening, with several species being recruited from the ocean. The importance of these areas as fish habitats and the recorded changes in fish assemblages after opening suggest that careful management is required.[127] Improvement of fisheries implies enhancement of fish production by manipulation of the existing fish population, and/or optimization of the aquatic environment for fish production. Regular stocking with fish may be necessary to compensate low natural recruitment, or disruption of recruitment in the case of land-locked lagoons.

High pollution levels in coastal lagoon waters caused significant changes in fish species composition. For instance, in the Curonian Lagoon, for many commercial migratory and predatory fish species such as salmon (*Salmo salar*), twaite shad (*Alosa fallax*), zanthe (*Vimba vimba*), eel, pikeperch, pike, and turbot harvests are much lower than 40–50 years ago.[128]

The aquaculture systems more suitable to lagoon environments include cages for fish and shrimp and systems for rearing bivalve molluscs or seagrasses on artificial substrates.[103] Other important systems are fish and shrimp culture in earth ponds, a growing economic activity in Mediterranean countries, and bivalve culture in tidal flats in leaky lagoons such as the Ria Formosa (Portugal). In this lagoon it is estimated that 5 to 8% of the natural wet area is occupied by bivalve culture (between 500 to 800 ha), with a production of clams of more than 2000 tons per year. This is an important economic activity, representing more than 90% of income of aquaculture products.[129] However, extreme and fluctuating environmental conditions of choked and restricted lagoons make aquaculture activity infeasible because of the low growth rates and fecundity of the individuals.[130] Furthermore, aquaculture may be both a polluter of the littoral and a petitioner for clean water.[130,131] In addition, the intensive farming and the monocultures may diminish genetic and species diversity, increasing their vulnerability to infections and invasion.[132]

Mussel rearing represents a more or less self-regulated extensive aquaculture system that is, to a large extent, integrated with the natural marine ecosystem. Cage-farming of fishes adds to eutrophication due to nutrient input through the supplied food, while mussel culturing counteracts eutrophication since no feed is added and nutrients are removed when mussels are harvested. Since mussels belong to a low trophic level, and therefore demand less primary production, it is possible to harvest a much larger biomass of mussels per unit of water area when compared with high trophic level fish cultures.[133]

In systems with abundant bivalve populations, top-down control (grazing) will dominate, which means that increasing nutrient loading may result in only a limited response of phytoplankton.[83] From an aquaculture point of view, this also means that the effect of fertilization of lagoons will be limited. From a water management point of view, this implies that bivalves could be used for eutrophication control.[134,135] However, cultured molluscs are highly susceptible to the effects of eutrophication and pollution in lagoons, as they will rapidly accumulate pathogens, toxins, heavy metals, and other contaminants, thereby posing a potential risk to human health.[103]

## 5.5   REMARKS

For modeling biological assemblages and communities of lagoons it is necessary to avoid single-factor approaches and work with multifactorial models (Figure 5.10) where ionic composition of the waters, trophic status, and water renewal rates are

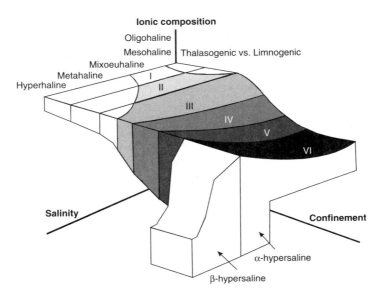

**FIGURE 5.10** Diagram showing the possible classification of coastal lagoons and lagoon water bodies in function of the three main parameters (ionic composition of salts, salinity, and confinement) that affect biological communities. (Redrawn from Pérez-Ruzafa, A. and Marcos, C., *Rapp. Comm. Int. Mer Mediter.*, 33, 100, 1992. With permission.)

some of the main factors to be considered, together with other environmental parameters (wave exposure, sediment types, oxygen level, etc.) that also play an important role in the distribution of the species.[89,110,112,113,136] On the other hand, the old concepts of salinity gradients[106] and confinement[108] related only to water characteristics should be changed to a model related to the colonization rates of species (Figure 5.11).[112,113,137,138] In restricted and leaky lagoons, high water renewal rates and low residence time facilitate the introduction of species by means of trophic or reproductive migratory activities in swimming stages and current drift transport in pelagic and planktonic stages such as eggs, larvae, and juveniles. This leads to a more or less intense horizontal gradient and a zonation of the species assemblages, from open sea to typical lagoon communities. The penetration of much colonization inside the lagoon depends on the mobility of the species and on the intensity of the currents and renewal rates. After this penetration in the lagoon environment, colonizing species must survive in the new environmental condition. Some species will be

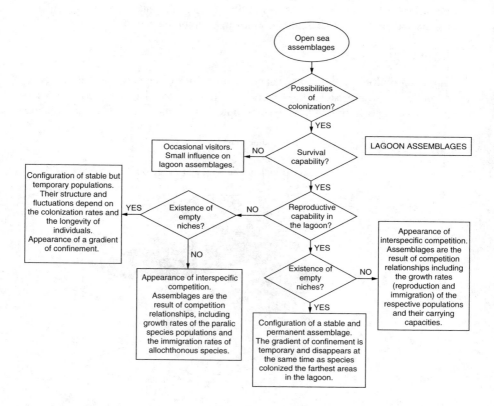

**FIGURE 5.11** Conceptual model to explain the structure of lagoon benthic assemblages and the zonation patterns and gradients observed. According to this model, the confinement would be related with colonization rates of marine species and indirectly with water renewal rates. (Modified from Pérez-Ruzafa, A. and Marcos, C., *Rapp. Comm. Int. Mer Mediter.*, 33, 100, 1992. With permission.)

able to survive and develop into adults but not reproduce inside the lagoon, and individuals of colonizing species must compete with lagoon species for resources.

In summary, lagoon species assemblages are the result of the continuous or seasonal interaction between native lagoon inhabitants (physiologically adapted to stressed and changing environments, and with reproductive capability in such conditions), and sporadic, accidental, or periodic colonization. Some species also adapt to the lagoon environment but are incapable of reproduction, while others survive, reproduce, and become established but only under favorable conditions. In both cases such species are at a disadvantage in competition with native lagoon species because they spend most of their energy budget in physiological adaptations. However, from the point of view of the species population stability, the resulting high mortality rates and loss of reproduction could be compensated for by high recruitment rates from larval stages or from adults from the open coastal sea.

## REFERENCES

1. UNESCO, Coastal lagoons research, present and future. *UNESCO Technical Papers in Marine Science*, 33, 1981, pp. 51–79.
2. Alongi, D.M., *Coastal Ecosystem Processes,* CRC Press, Boca Raton, FL, 1998, p. 188.
3. Bird, E.C.F., Changes on barriers and spits enclosing coastal lagoons, *Oceanol. Acta,* N.S.P. 45, 1982.
4. UNESCO, Coastal lagoons survey. *UNESCO Technical Papers in Marine Science,* 31, 1980, p. 7.
5. Clark, J.R., *Coastal Seas: The Conservation Challenge,* Blackwell Science, Oxford, U.K. 1998, p. 10.
6. Barnes, R.S.K., *Coastal Lagoons,* Cambridge University Press, Cambridge, 1980.
7. Sanders, H.L., Marine benthic diversity: a comparative study, *Am. Nat.,* 102, 243, 1968.
8. Heip, C., Eutrophication and zoobenthos dynamics, *Ophelia,* 41, 113, 1995.
9. Vitousek, P.M., Beyond global warming: ecology and global change, *Ecology,* 75, 1861, 1994.
10. Likens, G.E., Eutrophication and aquatic ecosystems, *Limnol. Oceanogr.,* 1 (Special Symposia), 3, 1972.
11. Nixon, S.W., Coastal marine eutrophication: a definition, social causes, and future concerns, *Ophelia,* 41, 199, 1995.
12. Duarte, C.M., Submerged aquatic vegetation in relation to different nutrient regimes, *Ophelia,* 41, 87, 1995.
13. Cloern, J.E., Our evolving conceptual model of the coastal eutrophication problem, *Mar. Ecol. Prog. Ser.,* 210, 223, 2001.
14. Harlin, M.M., Changes in major plant groups following nutrient enrichment, in *Eutrophic Shallow Estuaries and Lagoons,* McComb, A.J., Ed., CRC Press, Boca Raton, FL, 1995, chap. 11.
15. Nienhuis, P.H., Ecology of coastal lagoons in The Netherlands (Veerse Meer and Grevelingen), *Vie Milieu,* 42, 59, 1992.
16. Scheffer, M., *Ecology of Shallow Lakes,* Chapman & Hall, London, 1998.
17. Nixon, S.W. and Pilson, M.E.Q., Nitrogen in estuarine and coastal marine ecosystems, in *Nitrogen in the Marine Environment,* Carpenter, E.J. and Capone, D.G., Eds., Academic Press, New York, 1983, chap. 16.

18. Raven, J.A., Nutritional strategies of submerged benthic plants: the acquisition of C, N and P by rhizophytes and haptophytes, *New Phytol.,* 88, 1, 1981.
19. Borum, J. and Sand-Jensen, K., Is total primary production in shallow coastal marine waters stimulated by nitrogen loading? *Oikos,* 76, 406, 1996.
20. Valiela, I. et al., Coupling of watersheds and coastal waters: sources and consequences of nutrient enrichment in Waquoit Bay, Massachusetts, *Estuaries,* 15, 443, 1992.
21. Sfriso, A. et al., Macroalgae, nutrient cycles, and pollutants on the Lagoon of Venice, *Estuaries,* 15, 517, 1992.
22. Knoppers, B., Aquatic primary production in coastal lagoons, in *Coastal Lagoon Processes,* Kjerfve, B., Ed., Elsevier, Amsterdam, 1994, chap. 9.
23. Newell, R.C., The energetics of detritus utilisation in coastal lagoons and nearshore waters, *Oceanol. Acta,* 4, 347, 1982.
24. Cattaneo, A., Grazing on epiphytes, *Limnol. Oceanogr.,* 28, 124, 1983.
25. Orth, R.J. and van Montfrans, J., Epiphyte-seagrass relationships with an emphasis on the role of micrograzing: a review, *Aquat. Bot.,* 18, 43, 1984.
26. Mann, K.H., Macrophite production and detritus food chain in coastal waters, *Mem. Ist. Ital. Idrobiol.,* 29, 353, 1972.
27. Newell, R.C., The biological role of detritus in the marine environment, in *Flows of Energy and Materials in Marine Ecosystems: Theory and Practice,* Fasham, M.J.R., Ed., NATO Advanced Research Institute, Plenum Press, New York, 1984, p. 317.
28. Raffaelli, D. and Hawkins, S., *Intertidal Ecology,* Chapman & Hall, London, 1996, chap. 6.
29. Alber, M. and Valiela, I., Production of microbial organic aggregates from macrophyte-derived dissolved organic matter, *Limnol. Oceanogr.,* 39, 37, 1994.
30. Goldman, J.C., Oceanic nutrient cycles, in *Flows of Energy and Materials in Marine Ecosystems: Theory and Practice,* Fasham, M.J.R., Ed., NATO Advanced Research Institute, Plenum Press, New York, 1984, p. 137.
31. Azam, F. et al., The ecological role of water column microbes in the sea, *Mar. Ecol. Prog. Ser.,* 10, 257, 1983.
32. Wiegert, R.G., The past, present, and future of ecological energetics, in *Concepts of Ecosystem Ecology,* Pomeroy, L.R. and Alberts, J.J., Eds., Ecological Studies 67, Springer-Verlag, New York, 1988, p. 29.
33. Valiela, I., *Marine Ecological Processes,* 2nd ed., Springer-Verlag, New York, 1995, chap. 2.
34. Droop, M.R., Vitamin B12 and marine ecology. IV. The kinetics of uptake, growth and inhibition in Monochrysis lutheri, *J. Mar. Biol. Assoc. U.K.,* 48, 689, 1968.
35. Droop, M.R., Some thoughts on nutrient limitation in algae, *J. Phycol.,* 9, 264, 1973.
36. Droop, M.R., The nutrient status of algal cells in continuous culture, *J. Mar. Biol. Assoc. U.K.,* 54, 825, 1974.
37. Lehman, J.T. and Scavia, D., Microscale patchiness of nutrient in plankton communities, *Science,* 216, 729, 1982.
38. Malone, T.C., Algal size, in *The Physiological Ecology of Phytoplankton,* Morris, I., Ed., Blackwell, Oxford, U.K., 1980, p. 433.
39. Hein, M., Pedersen, M.F., and Sand-Jensen, K., Size-dependent nitrogen uptake in micro- and macroalgae, *Mar. Ecol. Prog. Ser.,* 118, 247, 1995.
40. Legendre, L. and Rassoulzadegan, F., Plankton and nutrient dynamics in marine waters, *Ophelia,* 41, 153, 1995.
41. Thingstad, T.F., A theoretical approach to structuring mechanisms in the pelagic food chain, *Arch. Hydrobiol.,* 363, 59, 1998.

42. Schriver, P. et al., Impact of submerged macrophytes on fish-zooplankton-phytoplankton interaction: large-scale enclosure experiments in a shallow eutrophic lake, *Freshwat. Biol.*, 33, 255, 1995.

43. Sieburth, J..McN., *Sea Microbes*, Oxford University Press, New York, 1979, chap.1.

44. Lehman, J.T., Interacting growth and loss rates: the balance of top-down and bottom-up controls in plankton communities, *Limnol. Oceanogr.*, 36, 1546, 1991.

45. Landry, M.R., Switching between herbivory and carnivory by the planktonic marine copepod *Calanus pacificus*, *Mar. Biol.*, 65, 77, 1981.

46. Frost, B.W., Effects of size and concentration of food particles on the feeding behavior of the marine planktonic copepod *Calanus pacificus*, *Limnol. Oceanogr.*, 17, 805, 1972.

47. Frost, B.W., A threshold feeding behavior in *Calanus pacificus*, *Limnol. Oceanogr.*, 20, 263, 1975.

48. Bartram, W.C., Experimental development of a model for the feeding of neritic copepods on plankton, *J. Plankton Res.*, 3, 25, 1980.

49. Herman, P.M.J. and Scholten, H., Can suspension-feeders stabilise estuarine ecosystems? in *Proc. 24th EMBS Trophic Relationships in the Marine Environment*, Barnes, M. and Gibson, R.N., Eds., Aberdeen University Press, Aberdeen, 1990, p. 104.

50. Kristensen, E., Benthic fauna and biogeochemical processes in marine sediments: microbial activities and fluxes, in *Nitrogen Cycling in Coastal Marine Environments*, Blackburn, T.H. and Sorensen, J., Eds., John Wiley & Sons, New York, 1988, p. 275.

51. Cammen, L.M., The significance of microbial carbon in the nutrition of the deposit feeding polychaete *Nereis succinea*, *Mar. Biol.*, 61, 9, 1980.

52. Hanson, R.B. and Tenore, K.R., Microbial metaboplism and incorporation by the polychaete *Capitella capitata* of aerobically and anaerobically decomposed detritus, *Mar. Ecol. Prog. Ser.*, 6, 299, 1981.

53. Werner, E.E. and Gilliam, J.F., The ontogenetic niche and species interactions in size-structured populations, *Annu. Rev. Ecol. Syst.*, 15, 393, 1984.

54. Weatherley, A.H. and Gill, H.S., *The Biology of Fish Growth*, Academic Press, London, 1987, chap.1.

55. von Bertalanffy, L., Principles and theory of growth, in *Fundamental Aspects of Normal and Malignant Growth*, Nowinski, W.W., Ed., Elsevier, Amsterdam, 1960, chap. 2.

56. Gamito, S., Growth models and their use in ecological modelling: an application to a fish population, *Ecol. Model.*, 113, 83, 1998.

57. Gurney, W.S.C. and Nisbet, R.M., *Ecological Dynamics*, Oxford University Press, Oxford, U.K., 1998, p. 204.

58. Baretta-Bekker, J.G., Baretta, J.W. and Rasmussen, E.K., The microbial food web in the European regional seas ecosystem model, *Neth. J. Sea Res.*, 33, 363, 1995.

59. Blindow, I., Long- and short-term dynamics of submerged macrophytes on two shallow eutrophic lakes, *Freshwat. Biol.*, 28, 15, 1992.

60. Duarte, C.M., Seagrass nutrient content, *Mar. Ecol. Prog. Ser.*, 67, 201, 1990.

61. Duarte, C.M., Nutrient concentration of aquatic plants: patterns across species, *Limnol. Oceanogr.*, 37, 882, 1992.

62. Knox, G.A., *Estuarine Ecosystems: A Systems Approach*, Vol. II, CRC Press, Boca Raton, FL, 1986, chap. 4.

63. Pérez-Ruzafa, A. and Marcos, C., Los sustratos arenosos y fangosos del Mar Menor (Murcia), su cubiertavegetal y su posible relación con la disminución del mújol en la laguna, *Cuad. Marisq. Publ. Tec.*, 11, 111, 1987.

64. Pérez-Ruzafa, A. et al., Evolución de las características ambientales y de los poblamientos del Mar Menor (Murcia, SE de España), *An. Biol.,* 12, 53, 1987.
65. Pérez-Ruzafa, A., Marcos, C., and Ros, J., Environmental and biological changes related to recent human activities in the Mar Menor, *Mar. Pollut. Bull.,* 23, 747, 1991.
66. Raffaelli, D. and Hawkins, S., *Intertidal Ecology,* Chapman & Hall, London, 1996, p. 224.
67. Raffaelli, D.J., Hull, S., and Milne, H., Long-term changes in nutrients, weed mats and shorebirds in an estuarine system, *Cah. Biol. Mar.,* 30, 259, 1989.
68. Weisner, S.E.B., Strand, J.A., and Sandsten, H., Mechanisms regulating abundance of submerged vegetation in shallow eutrophic lakes, *Oecologia,* 109, 592, 1997.
69. Rassoulzadegan, F., Protozoan patterns in the Azam-Ammerman's bacteria-phytoplankton mutualism, in *Trends in Microbial Ecology,* Guerrero, R. and Pedrós-Alió, C., Eds., Spanish Society of Microbiology, Barcelona, 1993, p. 435.
70. Azam, F., et al., The ecological role of water column microbes in the sea, *Mar. Ecol. Prog. Ser.,* 10, 257, 1983.
71. Dortch, Q., The interaction between ammonium and nitrate uptake in phytoplankton, *Mar. Ecol. Prog. Ser.,* 61, 183, 1990.
72. Gilabert, J., Short-term variability of the planktonic size structure in a Mediterranean Coastal Lagoon, *J. Plankton Res.,* 23, 219, 2001.
73. McCarthy, J.J., The kinetics of nutrient utilization, in *Physiological Bases of Phytoplankton Ecology,* Platt., T., Ed., *Can. Bull. Fish. Aquat, Sci.,* 210, 211, 1981.
74. Eppley, R.W., Temperature and phytoplankton growth in the sea, *Fish. Bull.,* 70, 1063, 1972.
75. Dugdale, R.C. et al., Adaptation of nutrient asimilation, *Can. Bull. Fish. Aquat. Sci.,* 210, 234, 1981.
76. Jassby, A.D. and Platt, T., Mathematical formulation of the relationship between photosynthesis and light for phytoplankton, *Limnol. Oceanogr.,* 21, 540, 1976.
77. Perry, M.J., Talbot, M.C., and Alberte, R.S., Photoadaptation in marine phytoplankton: response of the photosynthesis unit, *Mar. Biol.,* 62, 91, 1981.
78. Gliwicz, Z.M., Effects of zooplankton grazing on photosynthetic activity and composition of zooplankton, *Verh. Internat. Verein Limnol.,* 19, 1490, 1975.
79. Harris, G.P., The measurement of photosynthesis in natural populations of phytoplankton, in *The Physiological Ecology of Phytoplankton,* Morris, I., Ed., Blackwell, Oxford, U.K., 1980, p. 129.
80. Kingsford, M.J., Pitt, K.A, and Gillanders, B.M., Management of jellyfish fisheries, with special reference to the O. Rhizostomeae, *Oceanogr. Mar. Biol. Ann. Rev.,* 38, 85, 2000.
81. Pérez-Ruzafa, A. et al., Evidence of a planktonic food web response to changes in nutrient input dynamics in the Mar Menor coastal lagoon, Spain, *Hydrobiologia,* 475, 359, 2002.
82. Nienhuis, P.H., Eutrophication of estuaries and brackish lagoons in the South-West Netherlands, in *Hydroecological Relations in the Delta Waters of the South-West, Netherlands,* Hooghart, J.C. and Posthumus, C.W.S., Eds., TNO Com., *Hydrol. Res. Proceed. Inform.,* 41, 49, 1989.
83. Prins, T.C., Smaal, A.C., and Dame, R.F., A review of the feedbacks between bivalve grazing and ecosystem processes, *Aquat. Ecol.,* 31, 349, 1998.
84. Pearson, T.H. and Rosenberg, R., Macrobenthic succession in relation to organic enrichment and pollution of the marine environment, *Oceanogr. Mar. Biol. Ann. Rev.,* 16, 229, 1978.

85. Warwick, R.M. et al., Analysis of macrobenthic and meiobenthic community structure in relation to pollution and disturbance in Hamilton Harbour, Bermuda, *J. Exp. Mar. Biol. Ecol.*, 138, 119, 1990.

86. Raffaelli, D., Conservation of Scottish estuaries, *Proc. R. Soc. Edinburgh*, 100B, 55, 1992.

87. Elmgren, R., Man's impact on the ecosystem of the Baltic Sea: energy flows today and at the turn of the century, *Ambio*, 18, 326, 1989.

88. Laevastu, T., *Marine Climate, Weather and Fisheries*, Halsted Press, New York, 1993, p. 149.

89. Pérez-Ruzafa, A., Estudio ecológico y bionómico de los poblamientos bentónicos del Mar Menor (Murcia, SE de España), Ph.D. thesis, University of Murcia, Murcia, Spain, 1989.

90. Agustí, S., Duarte, C.M., and Kalf, J., Algal cell size and the maximum density and biomass of phytoplankton, *Limnol. Oceanogr.*, 32, 983, 1987.

91. Flindt, M.R. et al., Nutrient cycling and plant dynamics in estuaries: a brief review, *Acta Oecol.*, 20, 237, 1999.

92. Sand-Jensen, K. and Borum, J., Interactions among phytoplankton, periphyton and macrophytes in temperate freshwaters and estuaries, *Aquatic Bot.*, 41, 137, 1991.

93. Warwick, R.M. and Clarke, K.R., Relearning the ABC: taxonomic changes and abundance/biomass relationships in disturbed benthic communities, *Mar. Biol.*, 118, 739, 1994.

94. Weston, D.P., Qualitative examination of macrobenthic community changes along an organic enrichment gradient, *Mar. Ecol. Prog. Ser.*, 61, 233, 1990.

95. Gray, J.S., Wu, R.S., and Or, Y.Y., Effects of hypoxia and organic enrichment on the coastal marine environment, *Mar. Ecol. Prog. Ser.*, 238, 249, 2002.

96. Clark, R.B., *Marine Pollution*, 4th ed., Clarendon Press, Oxford, U.K., 1997, p. 135.

97. McNeely, J.A. et al., Human influences on biodiversity, in *Global Biodiversity Assessment*, Heywood, V.H. and Watson, R.T., Eds., Cambridge University Press, Cambridge, U.K., 1995, chap. 11.

98. Parsons, T.R., The impact of industrial fisheries on the trophic structure of marine ecosystems, in *Food Webs. Integration of Patterns and Dynamics*, Polis, G.A. and Winemiller, K.O., Eds, Chapman & Hall, New York, 1996, chap. 33.

99. Cabral, J.A. et al., The impact of macroalgal blooms on the use of the intertidal area and feeding behaviour of waders (Charadrii) in the Mondego estuary (west Portugal), *Acta Oecol.*, 20, 417, 1999.

100. Raffaelli, D., Nutrient enrichment and trophic organisation in an estuarine food web, *Acta Oecol.*, 20, 449, 1999.

101. Gamito, S., The benthic ecology of some Ria Formosa lagoons, with reference to the potential for production of the gilthead seabream (*Sparus aurata* L.), Ph.D. thesis, University of Algarve, Faro, Portugal, 1994.

102. Gamito, S., Application of canonical correspondence analysis to environmental and benthic macrofauna data of four sites in the Ria Formosa (Portugal), *Publ. Espec. Inst. Esp. Oceanogr.*, 23, 41, 1997.

103. MacIntosh, D.J., Aquaculture in coastal lagoons, in *Coastal Lagoon Processes*, Kjerfve, B., Ed., Elsevier Oceanography series, 1994, chap. 14.

104. Anonymous, Final resolution. The Venice System for the Classification of Marine Waters According to Salinity, *Arch. Oceanogr. Limnol.*, 11, Suppl. 243, 1959.

105. Petit, G., Introduction à l'étude écologique des étangs méditerranéens, *Vie Milieu*, 4, 569, 1953.

106. Por, F.D., A classification of hypersaline waters, based on trophic criteria, *Mar. Ecol.*, 1, 121, 1980.

107. Bianchi, C.N., Caratterizzazione bionomica delle lagune costiere italiane, *Acqua & Aria*, 4, 15, 1988.
108. Guelorget, O. and Perthuisot, J.P., Le domaine paralique. Expressions géologiques, biologiques et économiques du confinement, *Trav. Lab. Geol.*, 16, 1, 1983.
109. Guelorget, O., Frisoni, G.F., and Perthuisot, J.P., Zonation biologique des milieux lagunaires: definition d'une echelle de confinement dans le domaine paralique méditerranéen, *J. Res. Oceanogr.*, 8, 15, 1983.
110. Bianchi, C.N., Tipologia ecologica delle lagune costiere italiane, in *Le lagune costiere: ricerca e gestione*, Carrada, G.C., Cicogna, F. and Fresi, E., Eds., CLEM, Massa Lubrense, Napoles, 1988, p. 57.
111. Carrada, G.C. and Fresi, E., Le lagune salmastre costiere. Alcune riflessioni sui problemi e sui metodi, in *Le lagune costiere: ricerca e gestione*, Carrada, G.C., Cicogna, F., and Fresi, E., Eds., CLEM, Massa Lubrense, Napoles, 1988, p. 35.
112. Pérez-Ruzafa, A. and Marcos, C., Colonization rates and dispersal as essential parameters in the confinement theory to explain the structure and horizontal zonation of lagoon benthic assemblages, *Rapp. Comm. Int. Mer Mediter.*, 33, 100, 1992.
113. Pérez-Ruzafa, A. and Marcos, C., La teoría del confinamiento como modelo para explicar la estructura y zonación horizontal de las comunidades bentónicas en las lagunas costeras, *Publ. Espec. Inst. Esp. Oceanogr.*, 11, 347, 1993.
114. Margalef-Mir, R., Distribución de los macrófitos de las aguas dulces y salobres del E y NE de España y dependencia de la composición química del medio, Fundación Juan March, Serie Universitaria, 1980.
115. Remmert, H., *Ecology*, Springer-Verlag, New York, 1980, p. 17.
116. Levinton, J.S., *Marine Ecology*, Prentice Hall, Englewood Cliffs, NJ, 1982, p. 74.
117. MacArthur, R.H. and Wilson, E.O., *The Theory of Island Biogeography*, Princeton University Press, Princeton, NJ, 1967.
118. Kjerfve, B. and Magill, K.E., Geographic and hydrographic characteristics of shallow coastal lagoons, *Mar. Geol.* 88, 187, 1989.
119. Kjerfve, B., Coastal lagoons, in *Coastal Lagoon Processes*, Kjerfve, B., Ed., Elsevier, Amsterdam, 1994, chap.1.
120. Pauly, D. and Yáñez-Arancibia, A., Fisheries in coastal lagoons, in *Coastal Lagoon Processes*, Kjerfve, B., Ed., Elsevier, Amsterdam, 1994, chap. 13.
121. Fonseca, L.C., Costa, A.M., and Bernardo, J.M., Seasonal variation of benthic and fish communities in shallow land-locked coastal lagoon (St. André, SW Portugal), *Sci. Mar.*, 53, 663, 1989.
122. González-Wangüemert, M., Variabilidad morfológica y del locus PGI de *Cardium glaucum* en el Mar Menor (SE de España) y su relación con las condiciones ambientales, M.Sc. thesis, University of Murcia, Murcia, Spain, 1997.
123. Krom, M.D. et al., Phytoplankton nutrient uptake dynamics in earthen marine fish ponds under winter and summer conditions, *Aquaculture*, 76, 237, 1989.
124. Carpenter, S.R. and Pace, M.L., Dystrophy and eutrophy in lake ecosystems: implications of fluctuating inputs, *Oikos*, 78, 3, 1997.
125. Entsua-Mensah, M., Ofori-Danson, P.K., and Koranteng, K.A., Management issues for the sustainable use of lagoon fish resources, in *Biodiversity and Sustainable Use of Fish in the Coastal Zone*, Abban, E.K., Casal, C.M.V., Falk, T.M., and Pullin, R.S.V., Eds., ICLARM Conference Proceedings, 63, 24, 2000.
126. Joyeux, J.C. and Ward, A.B., Constraints on coastal lagoon fisheries, *Adv. Mar. Biol.*, 34, 73, 1998.

127. Griffiths, S.P. and West, R.J., Preliminary assessment of shallow water fish in three small intermittently open estuaries in south-eastern Australia, *Fish. Manage. Ecol.,* 6, 311, 1999.

128. Repecka, R., Biology and resources of the main commercial fish species in the Lithuanian part of the Curonian Lagoon, in *Proc. Symp. Freshwater Fish and the Herring Clupea harengus Populations in the Coastal Lagoons Environment and Fisheries,* 1999.

129. Morais, A. and Carvalho, C., A pesca no Algarve: principais números, Actas do 7º Congresso do Algarve, Racal Clube, 1992.

130. Marcos, C., Planificación ecológica y ordenación del territorio en el litoral, Ph.D. thesis, University of Murcia, Murcia, Spain, 1991.

131. Clark, J.R., *Coastal Seas: The Conservation Challenge,* Blackwell Scientific, Oxford, U.K., 1998, p. 17.

132. Epstein, P.R., Sherman, B.H., Siegfried, E.S., Langston, A., Prasad, S., and Mckay, B., Eds., *Health Ecological and Economic Dimensions of Global Change (HEED). Marine Ecosystems: Emerging Diseases as Indicators of Change*, Center for Health and the Global Environment, Harvard Medical School, Boston, MA, 1998, chap. 3.

133. Folke, C. and Kautsky, N., The role of ecosystems for a sustainable development of aquaculture, *Ambio,* 18, 234, 1989.

134. Officer, C.B., Smayda, T.J., and Mann, R., Benthic filter feeding: a natural eutrophication control, *Mar. Ecol. Prog. Ser.,* 9, 203, 1982.

135. Takeda, S. and Kurihara, Y., Preliminary study of management of red tide water by the filter feeder *Mytilus edukis galloprovincialis, Mar. Poll. Bull.,* 28, 662, 1994.

136. Fresi, E. et al., G.D. Considerations on the relationship between confinement, community structure and trophic patterns in Mediterranean coastal lagoons, *Rapp. Comm. Int. Mer Mediter.,* 29, 75, 1985.

137. Huve, P. and Huve, H., Zonation superficielle des cotes rocheuses de l'etang de Berre et conparaison avec celles des cotes du golfe de Marseille (de Carry à Sausset), *Vie Milieu,* 5, 330, 1954.

138. Amanieu, M. and Lasserre, G., Organisation et évolution des peuplements lagunaires. *Oceanol. Acta,* N.SP., 201, 1982.

# 6 Modeling Concepts

*Boris Chubarenko, Vladimir G. Koutitonsky,*
*Ramiro Neves, and Georg Umgiesser*

## CONTENTS

1-56670-686-6/05/$0.00+$1.50
© 2005 by CRC Press

*Note:* The term *modeling* is used in this chapter in the sense of "numerical modeling." Physical modeling, conceptual modeling, or numerical modeling will only be used explicitly in relevant cases.

## 6.1  INTRODUCTION

In Chapter 3, the concept of transport equation was introduced, starting from the concepts of control volume and accumulation rate of a property inside this control volume. Diffusive and advective fluxes were also defined to account for exchanges between the control volume and its neighborhood, and the concept of evolution equation was introduced by adding sources and sinks to the transport equation. A "model" is built on the same concepts. Its implementation requires the definition of at least one control volume, the calculation of the fluxes across its boundary, and the calculation of the source and sinks using values of the state variables inside the volume. The number of dimensions of the model depends on the importance of relevant property gradients.

The simplest model is the "zero-dimensional" model. In this model, there is no spatial variability, and only one control volume needs to be considered. At the other extreme of complexity is the three-dimensional (3D) model, which is required when properties vary along the three spatial dimensions. Whatever the number of its dimensions, a model must include the following elements:

- Equations
- Numerical algorithm
- Computer code

The order of the items in this list can also be considered the order of their chronological development. Hydrodynamic equations are based on mass, momentum, and energy conservation principles, which were presented in Chapter 3. These have been known for more than 100 years. Actually, numerical algorithms used to solve hydrodynamic models were attempted even before the existence of computers. The analytical equations and the numerical algorithms developed before the existence of computers allowed the rapid development of modeling starting in the 1960s, when computers were made available to a small scientific community. Since that time, models and the modeling community have evolved exponentially. Modern integrated computer codes have done more for interdisciplinarity than 100 years of pure field and laboratory work.

The number of implementations of a model to solve various problems increases the knowledge of the range of validity of the model equations. The accuracy of the numerical algorithm is better known and confidence in the results increases. At that time, the major source of errors in the results is the existence of mistakes in the data files. Once the model equations, algorithms, and results are validated, the next priority is the development of a user-friendly graphical interface that simplifies the use of the model by nonspecialists. This reduces the errors of input files and simplifies the checking of those files. This chapter presents the concepts and methodologies used to build models and to understand their functioning.

## 6.2 NUMERICAL DISCRETIZATION TECHNIQUES

Computers can solve only algebraic equations. Analytic equations, integral or differential, must be discretized into algebraic forms. The procedure followed depends on the form of the analytical equation to be solved. The control volume approach is best for the integral form of evolution equations, while the Taylor series is best suited for differential equations.

### 6.2.1 COMPUTATIONAL GRID

The calculation of fluxes across a control volume surface is simpler if the scalar product of the velocity by the normal to each elementary area (face) composing that

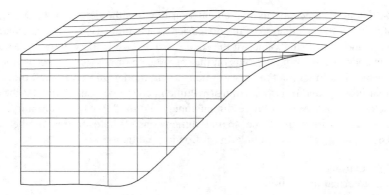

**FIGURE 6.1** Example of a grid for a three-dimensional (3D) computation. Two vertical domains are used. The upper domain uses a sigma coordinate. The lower one uses a Cartesian.

surface remains constant in each of them. The control volume that makes that calculation simpler must have faces perpendicular to the reference axis. If rectangular coordinates are used, the control volume generating the simpler discretization is a parallelepiped. In the case of a large oceanic model, a suitable control volume will have faces laying on meridians and parallels.

In depth-integrated models, also called two-dimensional or 2D horizontal models, the upper face of the control volume is the free surface and the lower face is the bottom. In three-dimensional or 3D models, a control volume occupies only part of the water column and its shape depends on the vertical coordinate used. In coastal lagoons, Cartesian and sigma-type coordinates (or a combination of both) are the most commonly used coordinates.

The ensemble of all control volumes forms the computational grid. In finite-difference-type grids, control volumes are organized along spatial axes and a structured grid is obtained. In contrast, typical finite-element grids are nonstructured. The latter are more difficult to define, but they are more flexible, thus allowing some variability in the spatial resolution. Figure 6.1 shows an example of a very general finite-difference-type grid using several discretizations in the vertical direction.

A system can be considered one-dimensional (1D) if properties change only along one physical dimension. In this case, control volumes can be aligned along the line of variation and one spatial coordinate is enough to describe their locations. Properties are considered as being constants across control volume faces perpendicular to that axis. Fluxes across the faces not perpendicular to that axis are null or have no net resultant.

### 6.2.2 CONTROL VOLUME APPROACH

Control volumes used in numerical models have the same meaning as the derivation of the evolution equation in Chapter 3. A discretization is adequate if it generates a simple calculation algorithm while maintaining the accuracy of the results. The

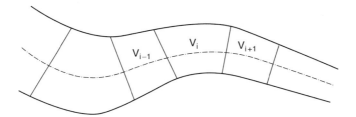

**FIGURE 6.2** Example of one-dimensional (1D) grid.

simpler calculation is obtained if properties can be considered as being constant inside the control volume and along parts of its surface. To make this possible without compromising accuracy, the control volume must be as small as possible; a fine-resolution grid is needed.

In a 1D model, properties can be stored into 1D arrays (vectors). Adjacent elements of a generic element $i$ are $i - 1$ on the left side and $i + 1$ on the right side (Figure 6.2). The length of a control volume must be small enough to allow properties in its interior to be represented by the value at its center. In that case, equations deduced in Section 3.2 apply and the rate of accumulation in volume $i$ will be given by

$$\text{Accumulation Rate} = \frac{(V_i C_i)^{t+\Delta t} - (V_i C_i)^t}{\Delta t}$$

where $\Delta t$ is the time step of the model. This equation is simplified if the volume remains constant in time. This is not the case in most coastal lagoons subjected to changing winds and it is certainly not the case in tidal lagoons.

Exchanges between $i$ volume and neighboring ones are accounted for by advective and diffusive fluxes. Their calculation requires some hypotheses. Let us consider Figure 6.2 and define the distances between the faces (spatial step) and the location points where other auxiliary variables are defined as shown in Figure 6.3. The net advective gain of matter to volume $i$ is given by

$$\left( Q_{i-\frac{1}{2}} C_{i-\frac{1}{2}} - Q_{i+\frac{1}{2}} C_{i+\frac{1}{2}} \right)^{t=t^*}$$

where $Q_{i-\frac{1}{2}} = u_{i-\frac{1}{2}} A_{i-\frac{1}{2}}$ while the diffusive flux, using the approach of Chapter 3, is given by

$$-\left( v_{i-\frac{1}{2}} \right) A_{i-\frac{1}{2}} \left( \frac{C_i - C_{i-1}}{\frac{1}{2}(\Delta x_i + \Delta x_{i-1})} \right)^{t=t^*} + \left( v_{i+\frac{1}{2}} \right) A_{i+\frac{1}{2}} \left( \frac{C_{i+1} - C_i}{\frac{1}{2}(\Delta x_i + \Delta x_{i+1})} \right)^{t=t^*}$$

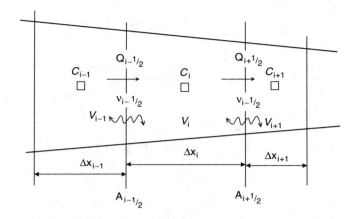

**FIGURE 6.3** Generic control volume in a 1D discretization.

In these equations, $t^*$ is a time interval between $t$ and $t + \Delta t$, to be defined according to criteria outlined in the next paragraph. $C_{i-\frac{1}{2}}$ is the concentration on the interface between elements $i$ and $i - 1$ and will be specified later. Combining the three equations, we obtain:

$$
\frac{(V_i C_i)^{t+\Delta t} - (V_i C_i)^t}{\Delta t} = \left( Q_{i-\frac{1}{2}} C_{i-\frac{1}{2}} - Q_{i+\frac{1}{2}} C_{i+\frac{1}{2}} \right)^{t=t^*}
$$

$$
- \left( v_{i-\frac{1}{2}} \right) A_{i-\frac{1}{2}} \left( \frac{C_i - C_{i-1}}{\frac{1}{2}(\Delta x_i + \Delta x_{i-1})} \right)^{t=t^*} + \left( v_{i+\frac{1}{2}} \right) A_{i+\frac{1}{2}} \left( \frac{C_{i+1} - C_i}{\frac{1}{2}(\Delta x_i + \Delta x_{i+1})} \right)^{t=t^*}
$$

(6.1)

In order to introduce the Taylor series discretization methods and to analyze stability and accuracy concepts, let us consider a simplified version of Equation (6.1). Consider the particular case of a channel with uniform and permanent geometry and regular discretization. The cross section ($A$), volume ($V$), and discharge are constant. Assume that diffusivity can be considered constant. Under these conditions, Equation (6.1) becomes

$$
\frac{C_i^{t+\Delta t} - C_i^t}{\Delta t} = \left( U \frac{C_{i-\frac{1}{2}} - C_{i+\frac{1}{2}}}{\Delta x} \right)^{t=t^*} + v \left( \frac{C_{i-1} - 2C_i + C_{i+1}}{\Delta x^2} \right)^{t=t^*}
$$

(6.2)

where $U$ is the constant cross-section average velocity and $\Delta x$ is the ratio between the volume and the average cross section. This is the most popular form of the transport equation but, as shown above, it is applicable only to particular conditions.

Additional approaches are required to calculate the advective flux, because the concentration is defined at the center of the control volumes and not at the faces. These approaches and their numerical consequences are described in the next sections.

### 6.2.3 Numerical Calculation of Advection

#### 6.2.3.1 Spatial Approach

Three common approaches are used to estimate concentration values at control volume faces:

- Linear approach
- Upstream stepwise approach
- Quadratic upwind approach (QUICK)

*6.2.3.1.1 Linear Approach*
In the linear approach it is assumed that:

$$C_{i-\frac{1}{2}} = \frac{C_i \Delta x_{i-1} + C_{i-1} \Delta x_i}{\Delta x_{i-1} + \Delta x_i}$$

Assuming a discretization where the grid size is uniform, it is easily seen that this approach generates central differences as obtained using the Taylor series (see Section 6.2.4).

*6.2.3.1.2 Upstream Stepwise Approach*
In this case, it is assumed that the concentration at the left face is

$$Q_i > 0 \Rightarrow \left( C_{i-\frac{1}{2}} = C_{i-1} \right)$$

$$Q_i < 0 \Rightarrow \left( C_{i-\frac{1}{2}} = C_i \right)$$

This discretization respects the transportivity property of advection. This property states that advection can transport properties only downstream or that information comes only from upstream. The linear approach does not respect this property because volume *i* will get information of downstream concentration through the average process. The violation of this property can generate instabilities and will create conditions to obtain negative values of the concentration. The upstream discretization avoids this limitation but, as shown in the following paragraphs, it can introduce unrealistic numerical diffusion.

*6.2.3.1.3 Quadratic Upwind Approach (QUICK)*
The quadratic upwind approach, or QUICK scheme, is an attempt at a compromise between respecting the transportivity property and keeping numerical diffusion at low values. In this case, it is assumed that the concentration distribution around a point follows a quadratic distribution centered on the upstream side of the face

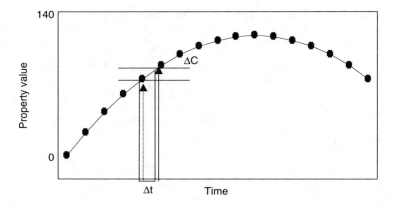

**FIGURE 6.4** Visualization of the consequences of temporal discretization. Property evolves within a time step, but values used to calculate flux do not.

being calculated. For the left face, we obtain

$$Q_i > 0 \Rightarrow \left( C_{i-\frac{1}{2}} = \tfrac{6}{8} C_{i-1} + \tfrac{3}{8} C_i - \tfrac{1}{8} C_{i-2} \right)$$

$$Q_i < 0 \Rightarrow \left( C_{i-\frac{1}{2}} = \tfrac{6}{8} C_i + \tfrac{3}{8} C_{i-1} - \tfrac{1}{8} C_{i+1} \right)$$

Using the Taylor series discretization described in the next paragraph, it can be seen that, in the case of a regular discretization, advection calculated using this approach is third-order accurate,[1] while pure upstream discretization is first-order accurate and the linear approach (central differences) is second-order accurate. The inconvenience of the QUICK discretization is that it requires additional approaches close to the boundaries. This is not a very limiting factor in 1D calculation but it is in 2D or 3D calculations, especially when the geometry is irregular.

### 6.2.3.2  Temporal Approach

In previous paragraphs, spatial discretization was analyzed. A solution was described for the diffusion term and three discretizations were suggested for the advection term but nothing was said about the time level at which the variables used to calculate advection or diffusion are evaluated. Figure 6.4 shows an example of a time evolution of a property $C$ at a point. The curved line shows the continuous evolution and filled circles show values at each time step. Vertical arrows show $C$ values at the beginning and end of a particular time step $\Delta t$. The flux in that time step is proportional to the product $\Delta C \Delta t$. Values at the beginning and end of a time step are shown, as well as concentration variation during that time step. The rate of accumulation at this point is proportional to the slope of this line. The slope of this line also gives an idea of the errors associated with the choice of $t^*$.

Models with **explicit** numerical schemes use $t^* = t$, while models with **implicit** schemes consider $t^* = t + \Delta t$. It can be seen from the figure that when the slope of the curve is positive, explicit models underestimate the advective fluxes,[†] while when the slope is negative, they overestimate them, introducing (at least) a phase error. Implicit schemes, on the other hand, underestimate or overestimate the fluxes by a value of the same order. The consideration of an intermediate value between $t$ and $t + \Delta t$ generates more accurate fluxes. The next subsection shows that $t^* = t + \tfrac{1}{2}\Delta t$ (semi-implicit method) gives the maximum accuracy. Values at $t^* = t + \tfrac{1}{2}\Delta t$ can be obtained by averaging the values of the properties calculated at time $t$ and time $t + \Delta t$. An increasing number of calculations to perform is the price to pay for accuracy improvement.

The next subsection shows that implicit methods have better stability properties than explicit methods. It can be shown that stability properties of the semi-implicit methods are similar to those of implicit methods. Because of their stability and accuracy properties, semi-implicit methods are the most efficient numerical methods.

### 6.2.4 TAYLOR SERIES APPROACH

Traditionally, discretized equations are obtained from partial differential equations by replacing derivatives with finite-differences obtained using the Taylor series. The Taylor series provides information on the truncation errors arising when replacing derivatives by finite-differences. In contrast, the control volume introduced in the previous subsection gives information about physical approaches used during discretization. When applied correctly, both methods must produce the same discretized equations.

In order to introduce the Taylor series discretization methods and to analyze stability and accuracy concepts, let us consider the differential equation corresponding to Equation (6.2):

$$\frac{\partial C}{\partial t} + U \frac{\partial C}{\partial x} = v \frac{\partial^2 C}{\partial x^2} \tag{6.3}$$

This equation describes the advection–diffusion transport in a channel with uniform velocity, a permanent geometry, and diffusivity.

#### 6.2.4.1 Time Discretization

The Taylor series relates the value of a property in a point (or time instant) with the values of the property in another point and the derivatives in the same point:

$$C_i^{t+\Delta t} = C_i^t + \left(\Delta t \frac{\partial C}{\partial t}\right)_i^t + \left(\frac{\Delta t^2}{2} \frac{\partial^2 C}{\partial t^2}\right)_i^t + \left(\frac{\Delta t^3}{3!} \frac{\partial^3 C}{\partial t^3}\right)_i^t + \cdots + \left(\frac{\Delta t^n}{n!} \frac{\partial^n C}{\partial t^n}\right)_i^t + 0(\Delta t)^{n+1}$$

---

[†] In explicit methods the flux during a time step is proportional to the area of the rectangle with side lengths $\Delta t$ and $C^t$, while in implicit methods it is proportional to $\Delta t$ and $C^{t+\Delta t}$.

Truncating this series at the first derivative, we obtain

$$\left(\frac{\partial C}{\partial t}\right)_i^t = \frac{C_i^{t+\Delta t} - C_i^t}{\Delta t} + O(\Delta t) \tag{6.4}$$

This equation states that the resolution of all the terms of the equation at time $t$ allows the calculation of the variable at time $t + \Delta t$ with first-order precision because the first missing term in the series is multiplied by $\Delta t$.

Similarly, we can relate the concentration at time $t$ with the concentration at time $t + \Delta t$:

$$C_i^t = C_i^{t+\Delta t} - \left(\Delta t \frac{\partial C}{\partial t}\right)_i^{t+\Delta t} + \left(\frac{\Delta t^2}{2}\frac{\partial^2 C}{\partial t^2}\right)_i^{t+\Delta t} - \left(\frac{\Delta t^3}{3!}\frac{\partial^3 C}{\partial t^3}\right)_i^{t+\Delta t}$$

$$+ \cdots + \left(\frac{\Delta t^n}{n!}\frac{\partial^n C}{\partial t^n}\right)_i^{t+\Delta t} + O(\Delta t)^{n+1}$$

Truncating this series after the first derivative as before, we obtain

$$\left(\frac{\partial C}{\partial t}\right)_i^{t+\Delta t} = \frac{C_i^{t+\Delta t} - C_i^t}{\Delta t} + O(\Delta t) \tag{6.5}$$

This equation shows that in implicit methods the truncation error is also of the first order, as in explicit methods, although processes are computed at time $t + \Delta t$. The difference between implicit and explicit methods is their stability properties, as described in the following.

From the above paragraph, it is expected that explicit and implicit methods should have the same truncation error and it is also expected that the calculation of the derivatives (or fluxes) at the center of the time step must have a smaller truncation error. To demonstrate this, let us use the Taylor series to relate properties at time $t$ and $t + \Delta t$ with variables at $t + \Delta t/2$.

$$C_i^{t+\Delta t} = C_i^{t+\Delta t/2} + \left(\frac{\Delta t}{2}\frac{\partial C}{\partial t}\right)_i^{t+\Delta t/2} + \left(\frac{\left(\frac{\Delta t}{2}\right)^2}{2}\frac{\partial^2 C}{\partial t^2}\right)_i^{t+\Delta t/2} + O\left(\frac{\Delta t}{2}\right)^3$$

$$\tag{6.6}$$

$$C_i^t = C_i^{t+\Delta t/2} - \left(\frac{\Delta t}{2}\frac{\partial C}{\partial t}\right)_i^{t+\Delta t/2} + \left(\frac{\left(\frac{\Delta t}{2}\right)^2}{2}\frac{\partial^2 C}{\partial t^2}\right)_i^{t+\Delta t/2} + O\left(\frac{\Delta t}{2}\right)^3$$

Subtracting the second equation from the first equation, we obtain

$$\left(\frac{\partial C}{\partial t}\right)_i^{t+\Delta t/2} = \frac{C_i^{t+\Delta t} - C_i^t}{\Delta t} + O\left(\frac{\Delta t}{2}\right)^2$$

This equation shows that semi-implicit methods are second-order accurate, and consequently allow for use of larger time step values. The implementation of these methods requires the computation of all derivatives and fluxes centered in time. Those values also can be computed with second-order accuracy, as the average between values at time $t$ and $t + \Delta t$, and can be demonstrated using expansions from Equation 6.6:

$$C_i^{t+\Delta t/2} = \frac{C_i^t + C_i^{t+\Delta t}}{2} + 0(\Delta t)^2$$

This temporal semi-implicit discretization is known as the Crank-Nicholson discretization. In this discretization we get

$$\frac{C_i^{t+\Delta t} - C_i^t}{\Delta t} = \frac{1}{2}\left(-U\frac{\partial C}{\partial x} + v\frac{\partial^2 C}{\partial x^2}\right)_i^{t+\Delta t} + \frac{1}{2}\left(-U\frac{\partial C}{\partial x} + v\frac{\partial^2 C}{\partial x^2}\right)_i^t + 0(\Delta t)^2$$

In order to solve this equation, the spatial derivatives have to be discretized.

### 6.2.4.2 Spatial Discretization

Spatial discretization using the Taylor series follows an approach similar to temporal discretization. Let us consider Taylor series developments for points on the left and on the right of point $i$, at a distance $\Delta x$ at an arbitrary time level:

$$C_{i+1}^* = C_i^* + \left(\Delta x \frac{\partial C}{\partial x}\right)_i^* + \left(\frac{\Delta x^2}{2}\frac{\partial^2 C}{\partial x^2}\right)_i^* + \left(\frac{\Delta x^3}{3!}\frac{\partial^3 C}{\partial x^3}\right)_i^* + 0(\Delta x)^3 \qquad (6.7)$$

$$C_{i-1}^* = C_i^* - \left(\Delta x \frac{\partial C}{\partial x}\right)_i^* + \left(\frac{\Delta x^2}{2}\frac{\partial^2 C}{\partial x^2}\right)_i^* - \left(\frac{\Delta x^3}{3!}\frac{\partial^3 C}{\partial x^3}\right)_i^* + 0(\Delta x)^3 \qquad (6.8)$$

Subtracting Equation (6.8) from Equation (6.7), we get the so-called central difference for the first-order spatial derivative of $C$:

$$\left(\frac{\partial C}{\partial x}\right)_i^* = \frac{C_{i+1}^* - C_{i-1}^*}{2\Delta x} + 0(\Delta x)^2 \qquad (6.9)$$

From Equation (6.7), we obtain an expression for a noncentered derivative (right side derivative), while from Equation (6.8), we obtain a left-side derivative, both with a first-order truncation error:

$$\left(\frac{\partial C}{\partial x}\right)_i^* = \frac{C_{i+1}^* - C_i^*}{\Delta x} + 0(\Delta x) \qquad (6.10)$$

$$\left(\frac{\partial C}{\partial x}\right)_i^* = \frac{C_i^* - C_{i-1}^*}{\Delta x} + 0(\Delta x) \qquad (6.11)$$

If Equation (6.10) is used when the velocity is negative and Equation (6.11) is used when the velocity is positive, the first derivative is computed using an "upstream method," since in both cases no downstream information is used.

Adding Equation (6.7) and Equation (6.8), we obtain

$$\left(\frac{\partial^2 C}{\partial x^2}\right)_i^* = \frac{C_{i+1}^* - 2C_i^* + C_{i-1}^*}{\Delta x^2} + 0(\Delta x)^2 \tag{6.12}$$

which is the finite-difference form of the second spatial derivative, discretized with a second-order truncation order.

In the next subsection, the stability criteria for some of these discretizations are analyzed. It will be shown that central differences for first-order derivatives generate unstable algorithms, and it will be shown that truncation error is not the unique aspect to take into account for estimating the accuracy of a numerical algorithm.

### 6.2.5 Stability and Accuracy

#### 6.2.5.1 Introductory Example

The exponential decay equation is considered first as an example because it illustrates the main features of stability without having to deal with spatial derivatives. This differential equation reads

$$\frac{\partial C}{\partial t} = -\alpha C \tag{6.13}$$

where $C$ is a generic concentration and $\alpha$ is a positive constant. The analytical solution to this problem is

$$C = C_0 \exp(-\alpha t)$$

where $C_0$ is the initial concentration at time $t = 0$. If the previous equation is discretized in time, we obtain

$$\frac{C^{t+\Delta t} - C^t}{\Delta t} = (-\alpha C)^{t^*}$$

As explained previously, we still must decide at which time level the term on the right-hand side has to be evaluated. Starting with an explicit approach, such that $t^* = t$, we can solve the equation directly for $C$ at the new time level:

$$C^{t+\Delta t} = C^t - \alpha \Delta t C^t = (1 - \alpha \Delta t)C^t$$

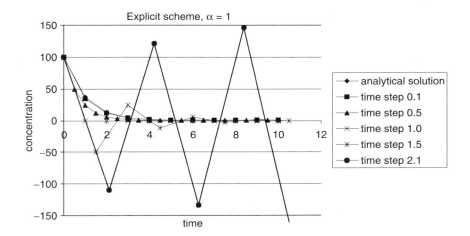

**FIGURE 6.5** Solution of the decay equation (Equation (6.13)) with the explicit scheme with different time steps.

As long as $\Delta t$ is small (more precisely $\alpha \Delta t < 1$), the solution is approximating the exponential decay. But once $\Delta t$ becomes equal to $1/\alpha$, the solution reads:

$$C_i^{t+\Delta t} = (0) \quad C_i^t = 0$$

and so, in the first time step, the value of the concentration drops to 0 and then stays there. Even worse, if $\Delta t > (1/\alpha)$ then $(1 - \alpha \Delta t) < 0$ and concentrations become negative, a completely nonphysical behavior.

However, even with these negative values, the solution of the decay equation is still stable because the oscillations generated are slowly decaying. However, if $\Delta t$ has been chosen to be $\Delta t > (2/\alpha)$, then $(1 - \alpha \Delta t) < -1$, and the oscillations start to amplify instead of decaying. There is no mechanism to dampen these oscillations and so they will amplify to reach arbitrary large (positive and negative) values. The solution has become unstable.

This behavior is shown in Figure 6.5 where the solution to the decay equation with $\alpha = 1$ has been plotted. As can be seen, all solutions with a time step of less than 1 are stable and are not undershooting. The solution with $\Delta t = 1$ drops to 0 in the first time step, whereas for $\Delta t = 1.5$ the solution produces negative values, but the solution is still stable. Finally, for $\Delta t = 2.1$ the solution becomes unstable.

The situation changes completely when the implicit approach is used. Now the discretized equation reads

$$C^{t+\Delta t} = C^t - \alpha \Delta t C^{t+\Delta t}$$

or, after solving for the concentration on the new time level,

$$C^{t+\Delta t} = \frac{1}{1 + \alpha \Delta t} C^t$$

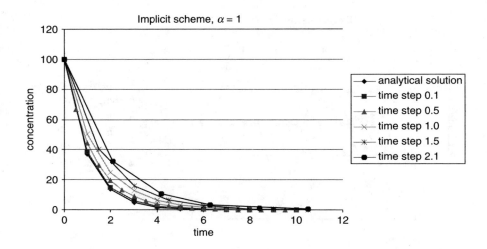

**FIGURE 6.6** Solution of the decay equation (Equation (6.13)) with the implicit scheme with different time steps.

As can be seen, this solution does not become unstable for any time step. The concentrations will always remain positive and no undershoots will occur. This is the desired property for the solution to the decay equation. Please note that the implicit solutions all have higher values than the analytical solution, whereas the stable and physical meaningful explicit solutions are all smaller than the analytical one. The solutions for the implicit scheme can be seen in Figure 6.6.

If the growth equation is considered instead of the decay equation, all arguments change. The growth equation reads

$$\frac{\partial C}{\partial t} = \beta C$$

Clearly this equation can be reproduced by the decay equation just by setting $\alpha$ to a negative value.

As can be seen easily, the growth equation remains stable if an explicit scheme is used. However, if an implicit scheme is used, the solution will be stable only if $\beta$ satisfies the stability criterion derived for $\alpha$.

In summary, it seems clear that for the decay equation, we should always use an implicit scheme in order to have a situation where solutions are stable for every time step used. On the other hand, if the growth equation is to be solved, an explicit scheme is better for the stability of the model.

The stability and accuracy associated with different options for temporal and spatial discretizations of the advection and diffusion equations (Equation (6.2)) can be examined by considering central explicit differences in the particular case of no

**TABLE 6.1**
**Example of a Time Evolution in a 1D Channel Computed Using Explicit Central Differences, a Unitary Courant Number, and No Diffusion**

| Time Step | Grid Point Number | | | | | | | Total Amount |
|---|---|---|---|---|---|---|---|---|
| | $i - 3$ | $i - 2$ | $i - 1$ | $i$ | $i + 1$ | $i + 2$ | $i + 3$ | |
| 0 | 0 | 0 | 0 | 1 | 0 | 0 | 0 | 1 |
| 1 | 0 | 0.00 | −0.50 | 1.00 | 0.50 | 0.00 | 0 | 1 |
| 2 | 0 | 0.25 | −1.00 | 0.5 | 1.00 | 0.25 | 0 | 1 |
| 3 | 0 | 0.75 | −1.13 | −0.50 | 1.13 | 0.75 | 0 | 1 |
| 4 | 0 | 1.31 | −0.50 | −1.63 | 0.50 | 1.31 | 0 | 1 |
| 5 | 0 | 1.56 | 0.97 | −2.13 | −0.97 | 1.56 | 0 | 1 |
| 6 | 0 | 1.08 | 2.81 | −1.16 | −2.81 | 1.08 | 0 | 1 |
| 7 | 0 | −0.33 | 3.93 | 1.66 | −3.93 | −0.33 | 0 | 1 |
| 8 | 0 | −2.29 | 2.94 | 5.59 | −2.94 | −2.29 | 0 | 1 |
| 9 | 0 | −3.76 | −1.00 | 8.52 | 1.00 | −3.76 | 0 | 1 |
| 10 | 0 | −3.26 | −7.14 | 7.52 | 7.14 | −3.26 | 0 | 1 |
| 11 | 0 | 0.31 | −12.54 | 0.38 | 12.54 | 0.31 | 0 | 1 |

diffusion. In that case Equation (6.2) becomes

$$\frac{C_i^{t+\Delta t} - C_i^t}{\Delta t} = \left( U \frac{C_{i-1} - C_{i+1}}{2\Delta x} \right)^t$$

$$C_i^{t+\Delta t} = \frac{1}{2} \frac{U\Delta t}{\Delta x} C_{i-1}^t + C_i^t - \frac{1}{2} \frac{U\Delta t}{\Delta x} C_{i-1}^t \qquad (6.14)$$

where $\frac{U\Delta t}{\Delta x} = C_r$ is the Courant number representing the ratio between the path length of a particle during a time step and the grid size. This is a critical parameter for most discretizations. Let us consider the case of a channel where initial conditions are zero everywhere except in a generic point $i$. Table 6.1 shows the temporal evolution along 11 time steps (0 to 11) for the case of a unitary Courant number ($C_r = 1$) and Table 6.2 shows the corresponding solution for the case of $C_r = 2$.

In both tables, columns $i - 3$ and $i + 3$ represent the boundary conditions (zero outside of the modeling area) and total amount stands for the total amount of matter inside the channel. Both solutions are unrealistic.

In such conditions, one would expect the contaminated water to move forward and, after a certain time, the entire channel should have a concentration equal to zero because the water entering the model area has concentration zero. The value of the total amount of matter inside the channel should remain constant until the matter reaches the outflow boundary, and then drop to zero while it leaves the domain.

### 6.2.5.2 Stability

A model is said to be unstable if errors generated inside the modeling area are amplified. This is what has happened in both the calculations. As time increased,

**TABLE 6.2**

**Example of a Time Evolution in a 1D Channel Computed Using Explicit Central Differences, $C_r = 2$, and No Diffusion**

| Time Step | $i - 3$ | $i - 2$ | $i - 1$ | $i$ | $i + 1$ | $i + 2$ | $i + 3$ | Total Amount |
|---|---|---|---|---|---|---|---|---|
| 0 | 0 | 0 | 0 | 1 | 0 | 0 | 0 | 1 |
| 1 | 0 | 0.00 | −1.00 | 1.00 | 1.00 | 0.00 | 0 | 1 |
| 2 | 0 | 1.00 | −2.00 | −1.00 | 2.00 | 1.00 | 0 | 1 |
| 3 | 0 | 3.00 | 0.00 | −5.00 | 0.00 | 3.00 | 0 | 1 |
| 4 | 0 | 3.00 | 8.00 | −5.00 | −8.00 | 3.00 | 0 | 1 |
| 5 | 0 | −5.00 | 16.00 | 11.00 | −16.00 | −5.00 | 0 | 1 |
| 6 | 0 | −21.00 | 0.00 | 43.00 | 0.00 | −21.00 | 0 | 1 |
| 7 | 0 | −21.00 | −64.00 | 43.00 | 64.00 | −21.00 | 0 | 1 |
| 8 | 0 | 43.00 | −128.00 | −85.00 | 128.00 | 43.00 | 0 | 1 |
| 9 | 0 | 171.00 | 0.00 | −341.00 | 0.00 | 171.00 | 0 | 1 |
| 10 | 0 | 171.00 | 512.00 | −341.00 | −512.00 | 171.00 | 0 | 1 |
| 11 | 0 | −341.00 | 1024.00 | 683.00 | −1024.0 | −341.00 | 0 | 1 |

the errors have increased. The error growth rate has been higher at a higher Courant number. To understand the reasons for such instability, we can use the following principle:

*"The influence of a point on its neighbors through advection or diffusion cannot be negative."*

This means that the consequence of increasing the concentration in one point can never be a reduction in any of its neighboring points. In order to guarantee the respect of this principle, no coefficient of the grid point values in Equation (6.14) can be negative. If a coefficient is null, there is no influence. In Equation (6.14), the coefficient of $C_{i+1}$ is negative whatever the Courant number. As a consequence, the higher the concentration in that point, the smaller the concentration in point $i$.

This method can be stabilized by adding diffusion. For example, if diffusion is considered, Equation (6.14) becomes

$$\frac{C_i^{t+\Delta t} - C_i^t}{\Delta t} = \left( U \frac{C_{i-1} - C_{i+1}}{\Delta x} \right)^t + v \left( \frac{C_{i-1} - 2C_i + C_{i+1}}{\Delta x^2} \right)^t$$

$$C_i^{t+\Delta t} = \left( \frac{1}{2} \frac{U\Delta t}{\Delta x} + \frac{v\Delta t}{\Delta x^2} \right) C_{i-1}{}^t + \left( 1 - 2 \frac{v\Delta t}{\Delta x^2} \right) C_i{}^t + \left( -\frac{1}{2} \frac{U\Delta t}{\Delta x} + \frac{v\Delta t}{\Delta x^2} \right) C_{i+1}^t$$

$$(6.15)$$

where $\frac{v\Delta t}{\Delta x^2} = d$ is called the diffusion number. In this case, positiveness of the coefficients is assured if

$$Re_g = \frac{U\Delta x}{v} \le 2$$

$$d = \frac{v\Delta t}{\Delta x^2} \le \frac{1}{2}$$

(6.16)

with $Re_g$ being the grid Reynolds number. The consideration of advection alone is equivalent to the consideration of an infinite Reynolds number and, consequently, whatever the time step (or $C_r$), central differences are always unstable. If on the one hand it is important for stability that $v$ is high enough (Equation (6.16a)), on the other hand it is limited by Equation (6.16b) and may not exceed a critical value given by $d \le 1/2$.

The consideration of diffusion does not always increase the stability properties of numerical models. Why did it in this case? Central differences do not respect the transportive property of advection. Physically, advection can only propagate information in the direction of the velocity. The analysis of Table 6.1 and Table 6.2 shows that information has also been propagated backward. This was a consequence of the use of a downstream value ($C_{i+1}$) to calculate the spatial derivative. Physically, diffusion propagates the information in any direction (according to the local gradients). In the case of Table 6.1 and Table 6.2, information diffusion transports matter upstream, making it available to be transported by advection.

When the advective flux is calculated using downstream information, one can remove matter from a control volume that is not to be removed. This is the mechanism that generates negative concentrations. The method is unstable because those errors are amplified in time. The consideration of (enough) diffusion makes the method stable but does not avoid the generation of negative concentrations. The upstream discretization was proposed first to avoid this problem. Consider now upstream explicit differences and again the particular case of no diffusion. In this case, Equation (6.2) becomes

$$C_i^{t+\Delta t} = \frac{U\Delta t}{\Delta x} C_{i-1}^t + \left(1 - \frac{U\Delta t}{\Delta x}\right) C_i^t \quad (U > 0)$$

(6.17)

It is easy to verify that the method is stable if the Courant number is not greater than 1. Table 6.3 shows results for $C_r = 1$ and Table 6.4 shows results for $C_r = 0.5$. $C_r > 1$ would generate an unstable model, which could not be solved adding diffusion. In fact, if diffusion were considered, the stability criteria would be $(C_r + 2d) \le 1$.

Table 6.3 shows that explicit upstream differences with $C_r = 1$ give the exact result. The concentration remains constant and travels at the exact speed of 1 cell per iteration. When the Courant number is reduced to 0.5 (Table 6.4) the solution is however degradated through the introduction of numerical diffusion. The method remains stable because the errors are reduced in time.

The results obtained in the above four examples show that small truncation errors as given by the Taylor series are not enough to guarantee accurate results. The upstream results also show that the reduction of the time step does not guarantee an improvement of the results.

**TABLE 6.3**

**Example of a Time Evolution in a 1D Channel Computed Using Explicit Upstream Differences, $C_r = 1.0$, and No Diffusion**

| Time Step | $i-3$ | $i-2$ | $i-1$ | $i$ | $i+1$ | $i+2$ | $i+3$ | Total Amount |
|---|---|---|---|---|---|---|---|---|
| | | | Grid Point Number | | | | | |
| 0 | 0 | 0 | 0 | 1 | 0 | 0 | 0 | 1 |
| 1 | 0 | 0 | 0 | 0 | 1 | 0 | 0 | 1 |
| 2 | 0 | 0 | 0 | 0 | 0 | 1 | 0 | 1 |
| 3 | 0 | 0 | 0 | 0 | 0 | 0 | 0 | 0 |
| 4 | 0 | 0 | 0 | 0 | 0 | 0 | 0 | 0 |
| 5 | 0 | 0 | 0 | 0 | 0 | 0 | 0 | 0 |
| 6 | 0 | 0 | 0 | 0 | 0 | 0 | 0 | 0 |
| 7 | 0 | 0 | 0 | 0 | 0 | 0 | 0 | 0 |
| 8 | 0 | 0 | 0 | 0 | 0 | 0 | 0 | 0 |
| 9 | 0 | 0 | 0 | 0 | 0 | 0 | 0 | 0 |
| 10 | 0 | 0 | 0 | 0 | 0 | 0 | 0 | 0 |

### 6.2.5.3 The Need for a Fine Resolution Grid

The reason why the upstream scheme with $C_r = 0.5$ gives such poor results is the coarse discretization used. In this case, matter travels only half of the grid size and consequently the matter contained in cell $i$ at $t = 0$ is distributed between two computing cells at time $t = 1$. Because the concentration is computed as the mass divided by the volume, its value is reduced to $1/2$. This result is obtained because the initial hypothesis that "the grid cell is small enough to allow the concentration

**TABLE 6.4**

**Example of a Time Evolution in a 1D Channel Computed Using Explicit Upstream Differences, $C_r = 0.5$, and No Diffusion**

| Time Step | $i-3$ | $i-2$ | $i-1$ | $i$ | $i+1$ | $i+2$ | $i+3$ | Total Amount |
|---|---|---|---|---|---|---|---|---|
| | | | Grid Point Number | | | | | |
| 0 | 0 | 0 | 0 | 1 | 0 | 0 | 0 | 1 |
| 1 | 0 | 0.00 | 0.00 | 0.50 | 0.50 | 0.00 | 0 | 1 |
| 2 | 0 | 0.00 | 0.00 | 0.25 | 0.50 | 0.25 | 0 | 1 |
| 3 | 0 | 0.00 | 0.00 | 0.13 | 0.38 | 0.38 | 0 | 0.88 |
| 4 | 0 | 0.00 | 0.00 | 0.06 | 0.25 | 0.38 | 0 | 0.69 |
| 5 | 0 | 0.00 | 0.00 | 0.03 | 0.16 | 0.31 | 0 | 0.5 |
| 6 | 0 | 0.00 | 0.00 | 0.02 | 0.09 | 0.23 | 0 | 0.34 |
| 7 | 0 | 0.00 | 0.00 | 0.01 | 0.05 | 0.16 | 0 | 0.23 |
| 8 | 0 | 0.00 | 0.00 | 0.00 | 0.03 | 0.11 | 0 | 0.14 |
| 9 | 0 | 0.00 | 0.00 | 0.00 | 0.02 | 0.07 | 0 | 0.09 |
| 10 | 0 | 0.00 | 0.00 | 0.00 | 0.01 | 0.04 | 0 | 0.05 |

to be uniform in its interior" is violated. This does not happen when half of the cell has matter and the other half does not. If the plume were contained inside many cells the problem would still exist but only in the plume limits and hence would not deteriorate the solution.

## 6.3  PRE-MODELING ANALYSIS AND MODEL SELECTION

### 6.3.1  HYDROGRAPHIC CLASSIFICATION

Characteristics of lagoons around the world are very different. Geomorphological characteristics depend on the type of shore, while hydrological characteristics are determined by marine influence and hydrological balance for the lagoon drainage basin. Lagoons with similar morphometry may exhibit completely different behavior in different ambient conditions. A careful classification of the lagoon type according to its geomorphology, hydrology, and mixing processes is a desirable first step toward the choice of the most appropriate physics to be included in the numerical model. The proper identification of a lagoon type allows the user to find a similar lagoon in another part of the world and benefit from the previous knowledge available for that lagoon. At the same time, the hydrographic classification database will be supplemented with new information that can be used for future studies in similar lagoons. It is very tempting to classify a lagoon according to its hydrographic features, i.e., utilizing only basic information on its morphometry and hydrology, which is usually available without additional field studies.

A proper lagoon, such as an atoll lagoon or a coastal lagoon (enclosed and much more shallow than the adjacent marine area coastal water body and separated from the marine area by an accumulative barrier), is a pure type of coastal water body. The majority of coastal waters are a mixture of such pure types, open bay, proper lagoon, and fjord (all of them without river outfall), and rivers (Figure 6.7), and exhibit the features of estuaries, the most widely investigated and most popular coastal water bodies. The hydromorphometric tetrahedron (Figure 6.7) provides the conventional coordinate system where any coastal water body may be described as a combination of the above pure forms and its position is expressed through specific quantitative characteristics.

### 6.3.1.1  Morphometric Parameters

Lagoons around the world have various shapes and bottom relief configurations that can change in the short run with time under the influence of tides, floods, erosion/deposition, wind surges, and seasonal run-off. As a start, it is convenient to consider a lagoon in terms of the classification proposed by Kjerfve,[2] which may highlight some of its hydrographic features. According to this classification, lagoons are divided into three types: choked lagoons, restricted lagoons, and leaky lagoons. The type of lagoon is determined by the water exchanges with the adjacent coastal sea, in the presence of tides and wind-driven circulation.[3] Related geomorphic

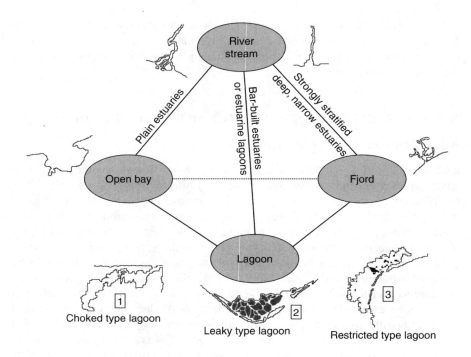

**FIGURE 6.7.** Hydromorphometric tetrahedron presents the concept of pure types of coastal water bodies and provides conventional coordinate systems and spaces where each point corresponds to a water pool with "mixed" properties. Examples that illustrate the main shape types of coastal lagoons[2,3] are (1) Darss-Zinst Bodden Chain Lagoon, Germany; (2) Ria Formosa Lagoon, Portugal; and (3) Venice Lagoon, Italy.

shapes of lagoons[2] are presented (Figure 6.7) and may be considered as qualitative features of these types, although, strictly speaking, the shape does not greatly influence lagoon hydrology.

A quantitative approach, based on some typical morphometric parameters, may provide a deeper understanding of the physical processes at work in the lagoon and highlight spatial scales of interest for the numerical model. For example, a lagoon can be considered an idealized rectangular basin (Figure 6.8) with a cross-shore length $a$, an along-shore length $b$, a volume $V$, and an average depth $H$. If the lagoon is round, it can still be considered as square, with equal sides $a$ and $b$. The lagoon entrance has a width $d$, a length $l$, and an average depth $h$ (Figure 6.8A,B).

This first-order approximation will yield the important spatial scales as well as some insight into the physical processes to be modeled. The length scales obtained will, in some cases, be comparable to those obtained by more elaborate methods that use the real topography of the lagoon. This morphometric approach is recommended

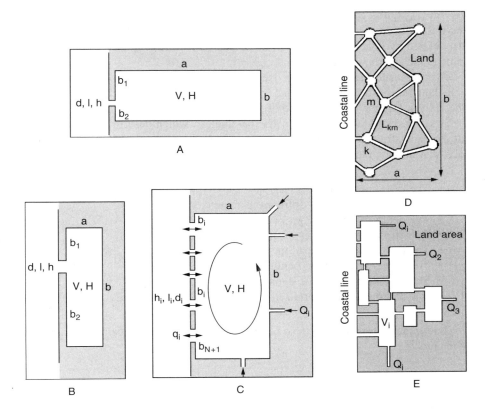

**FIGURE 6.8** Simple basic descriptions of lagoon shapes.

during the pre-modeling analysis of the lagoon. For example, the Vistula, Curonian, and Kara Bagaz Gol lagoons may be approximated by the rectangular shapes of types A and B in Figure 6.8.

Also, when the lagoon has several ($i = 1, N$) entrances (Figure 6.8C), each entrance can be described in terms of its own width, length, and depth ($d_i$, $l_i$, $h_i$). In such cases, barrier islands will have lengths ($b_i$). The number $i$ corresponding to each lagoon entrance and barrier island is set to increase in the counter-clockwise direction (as viewed from the top) in the northern hemisphere and in the clockwise direction in the southern hemisphere. As such, the influence of the Earth's rotation on the lagoon can be accounted for irrespective of the hemisphere. The Venice and Mar Menor lagoons can be represented by lagoons of type C (see Chapter 9.3 for details).

Other lagoons may feature a network of channels (Figure 6.8D), which become dry during hot seasons or during low tidal phases. These lagoons can be represented by a number of nodes ($m = \overline{1, M}$) connected by links. Each link has a length ($L_{km}$), a width ($D_{km}$), and a depth ($H_{km}$), $k$ and $m$ being the number of nodes connected by this link. The cross- and along-shore length scales of the total lagoon system are still defined by $a$ and $b$. The Ria Formosa Lagoon is an example of a lagoon made up of branched channels.

A complicated lagoon system occurs when different large basins, represented as rectangular basins, are connected through a network of channels (Figure 6.8E). The Dalyan Lagoon is an example of such a lagoon system (see case study).

A set of quantitative morphometric parameters, which describe the lagoon orientation and structure, its horizontal and vertical scales, and the potential sea influence, can now be introduced (Table 6.5):

- The restriction ratio ($p_r$)
- The orientation and anisotrophy parameter ($p_{or}$)
- The depth parameters ($p_{shell}$) and ($p_{deep}$)
- The openness parameter of potential sea influence ($p_{open}$)
- The three-component parameter of flow ($p_{resist}$)
- The shore development parameter ($p_{shore}$)
- The parameters of shore dynamics ($p_{er}$, $p_{acr}$, $p_{eq}$)
- The parameter of general sediment structure ($p_{sed}$)

Additional parameters can also be introduced for lagoons made up of a network of channels:

- The network "density" parameter ($p_{dens}$)
- The network "length" parameter ($p_{net}$)
- The network "multi-ways" parameters or entrance distance extremes parameters ($p_{short}$) and ($p_{long}$), characterizing the shortest and longest distances, respectively, between two marginal entrances.

Typical values of selected parameters for some lagoons are presented in Table 6.6. Although these geomorphic parameters alone can be helpful during the premodeling analysis, they are most effective when used in combination with the hydrological features of the lagoon.

## 6.3.1.2  Hydrological Parameters

Lagoons can also be described in terms of a set of hydrological parameters based on the water budget components: river water inflow ($Q_{riv}$), the atmospheric precipitation ($Q_{prc}$) and evaporation ($Q_{evp}$), the underground inflow ($Q_{grd}$), the marine water inflow ($Q_{inflow}$), and the outflow of the water from the lagoon to the adjacent open marine area ($Q_{outflow}$):

## TABLE 6.5
## Morphometric Parameters

| Parameter | Description |
|---|---|
| $p_r = \dfrac{d}{b}$, $p_r = \dfrac{\Sigma d_i}{b}$ | Restriction ratio, defined as the ratio between the total width of the lagoon entrances and the along-shore length, $p_r \in (0,1)$. |
| $p_{or} = \dfrac{b}{a}$ $p_{or} = \dfrac{b^2}{S_{lag}} = \dfrac{S_{lag}}{a^2}$ | Orientation or anisotrophy parameter. The lagoon has orthogonal dimensions of the same order if $p_{or} \approx 1$. It is more elongated in the parallel or perpendicular to shore directions if $p_{or} \geq 1$ or $p_{or} \leq 1$, respectively. In case of difficulties in explicit determination of cross-shore lagoon size (for example, the lagoon consists of series of connected elliptical cells), the transversal dimension together with lagoon surface area ($S_{lag}$) may be used for estimation of this parameter. |
| $p_{shall} \in \left( \dfrac{h_{avg}}{\max(a,b)}, \dfrac{h_{avg}}{\min(a,b)} \right)$ | Shallowness parameter, the range that characterizes the lagoon shallowness as a whole. This parameter is the inverse of the width-to-depth ratio usually applied to estuary classification.[4] |
| $p_{deep} = h_{max}/h_{avg}$ | Extreme depth parameter, which provides information on the deepest part of the lagoon and how it compares to the mean depth. |
| $p_{open} = \dfrac{\Sigma s_i^{in}}{S_{lag}}$, | Openness parameter, which characterizes the potential influence of the sea on lagoon general hydrology because flow velocities through the entrances are not included. Here, $s_i^{in}$ is the cross-sectional area of $i$th lagoon entrance for $i = 1$, $n$ entrances, and $S_{lag}$ is the area of the lagoon surface. |
| $\overline{p}_{resist} \in \left\{ \dfrac{s_{max}}{s_{min}}, \dfrac{s_{max}}{s_{inlet}}, \dfrac{s_{min}}{s_{inlet}} \right\}$ | A three-component parameter of flow, which illustrates the hydraulic resistance of the lagoon in different respects. Here, $s_{max}$ and $s_{min}$ are the maximum and minimum cross-sectional areas, respectively inside the lagoon and $s_{inlet}$ is the minimal cross-sectional area at the inlet. This set of components is valuable for pre-estimation of hydraulic resistance inside the lagoon. |
| $p_{net} \in \left( \dfrac{\Sigma L_{km}}{\max(a,b)}, \dfrac{\Sigma L_{km}}{\min(a,b)} \right)$ | Network parameter that characterizes the "length" of the channel network structure. Here, $L_{km}$ is the length of the link between nodes $k$ and $m$. |
| $p_{dens} = \dfrac{\Sigma L_{km} \cdot D_{km}}{a \cdot b}$ | Parameter that characterizes the "density" of the channel network structure. Here, $L_{km}$ and $D_{km}$ are the length and width of the link between nodes $k$ and $m$. |
| $p_{short} = L_{min}/\Sigma L_{km}$, $p_{long} = L_{max}/\Sigma L_{km}$ | Branching parameters for channel network structures. $L_{min}$ and $L_{max}$ are the minimum and maximum lengths of links between two remote marginal entrances. This parameter characterizes the "multi-variability" of ways through the lagoon system. |
| $p_{sed} = \Sigma d_i \cdot \dfrac{S_i}{S_{lag}}$ | Sediment structure parameter that characterizes the average diameter of sediment in the lagoon. It can be estimated as the spatial average between the diameters ($d_i$) of different sediment occupying the areas ($S_i$) in the lagoon. |
| $p_{shore} = 1 \cdot (4 \cdot \pi \cdot A)^{-0.5}$, | Shore development parameter, which is the ratio of the length of lagoon shore line ($l$) to the circumference of a circle whose area A is equivalent to that of the lagoon. |
| $p_{err}$, $p_{acr}$, $p_{eq}$ | Parameters that illustrate what fraction of the total lagoon coast line is under erosion ($p_{err}$), accretion ($p_{acr}$), or equilibrium ($p_{eq}$) conditions, and which are normalized as follows: $p_{err} + p_{acr} + p_{eq} = 1$. |

1. *The watershed parameter* $P_{wsh}$ showing the specific freshwater capacity of the lagoon watershed:

$$P_{wsh} = \frac{Q_{riv}}{S_{wsh}}$$

where $Q_{riv}$ is the freshwater river run-off [m³ a⁻¹] and $S_{wsh}$ is the catchment area of the lagoon [m²].

2. *The water budget components contribution,* $p_i^{WB}$. A comparison of absolute values of water budget components for different lagoons means nothing without comparing how each component influences the lagoon behavior. There are two approaches to derive corresponding specific parameters to evaluate the effect of these individual components: (1) by dividing each by the area of lagoon $S_{lag}$ and (2) by dividing each by the volume of the lagoon. The first set of parameters illustrates the effect of each component on the level variation:

$$p_i^{WB} = \frac{Q_i}{S_{lag}}$$

where $Q_i$ is the *i*th budget component. The dimension of these parameters is [m s⁻¹].

The second set of parameters characterizes the fraction that each component contributes to the lagoon volume, and in fact these are the inversed values of integral flushing time for each water budget component, which will be discussed in detail in Section 6.3.3.1.

It is always convenient to present the portrait of a lagoon water budget (regardless of the water budget component themselves or what specific parameters are considered) in the form of a rose diagram (Figure 6.14) for their absolute or relative magnitudes. This provides a means of comparing the hydrological features of different lagoons.

## TABLE 6.6
## Typical Values of Morphometric Parameters for Selected Lagoons

| Lagoon | $p_r$ | $p_{or}$ | $P_{shall}$ | $P_{deep}$ | $P_{open}$ [m²/km²] |
|---|---|---|---|---|---|
| Vistula Lagoon (Russia/Poland) | $4.5 \cdot 10^{-3}$ | 10 | $0.3 \div 2.9$ | 4.1 | 5.97 |
| Curonian Lagoon (Lithuania/Russia) | $8 \cdot 10^{-3}$ | 2.5–50 | $0.3 \div 16$ | 1.9 | 2.34 |
| Ria Formosa Lagoon (Portugal) | $45.9 \cdot 10^{-3}$ | 15.25 | $0.25 \div 3.7$ | 20 | 36.5 |
| Mar Menor (Spain) | $4.8 \cdot 10^{-3}$ | 2.6 | $3.3 \div 8$ | 1.5 | 1.8 |
| Grande-Entrée Lagoon (Canada) | $20.7 \cdot 10^{-3}$ | 6.4 | $1.5 \div 10$ | 2.7 | 51 |
| Odra Lagoon (Poland/Germany) | $11.9 \cdot 10^{-3}$ | 1.7 | $1.1 \div 1.9$ | $1.84 \div 2.9$ | 3.2 |

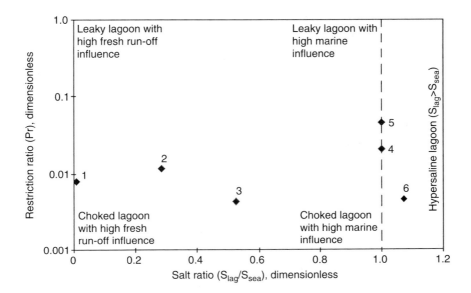

**FIGURE 6.9** Location of some selected lagoons (Table 6.7) on the morphometric-hydrological diagram.

The next step in classifying a lagoon based on its hydrological features is to position it on a morphometric-hydrological diagram (Figure 6.9), where the controlling parameters are the salt ratio ($s_{\text{lag}}^{(\text{avg})}/s_{\text{sea}}^{(\text{avg})}$) and the lagoon restriction parameter $p_r$ (Table 6.7). The salt ratio relates $s_{\text{lag}}^{(\text{avg})}$ and $s_{\text{sea}}^{(\text{avg})}$, which are the annual average salinity inside the lagoon and in the adjacent marine area, respectively. For example, this diagram (Figure 6.9) shows that the Curonian, the Odra, and the Vistula lagoons belong to the geomorphic class of choked lagoons; that the Vistula Lagoon is significantly influenced by the adjacent sea; and that the Curonian Lagoon is completely

**TABLE 6.7**
**Mean Annual Values of the Salt Ratio and Parameter of Lagoon Restriction for Some Selected Lagoons**

| N | Lagoon Name | Salt Ratio ($s_{\text{lag}}/s_{\text{sea}}$) | Restriction Ratio ($p_r$) |
|---|---|---|---|
| 1 | Curonian Lagoon | 0.007 | 0.008 |
| 2 | Odra Lagoon | 0.286 | 0.012 |
| 3 | Vistula Lagoon | 0.529 | 0.004 |
| 4 | Grande-Entrée Lagoon | 1 | 0.021 |
| 5 | Ria Formosa Lagoon | 1 | 0.046 |
| 6 | Mar Menor Lagoon | 1.081 | 0.005 |

influenced by river run-off. The Odra Lagoon is in an intermediate position. Both the Grande-Entrée and the Ria Formosa lagoons are totally under marine influence, regardless of the fact that they are significantly "restricted," as could be assumed from their shape. The Mar Menor Lagoon is restricted as well but it is also a hypersaline lagoon because of significant solar evaporation.

Salt ratios can be used as an average characteristic of long-term variations (years and decades). However, a more precise parameter is needed to evaluate the influence of the fresh- and saltwater budgets on the lagoon behavior. Such a parameter should provide answers to the following important questions: To what extent are lagoon waters under the influence of the sea? Or, conversely, do lagoon waters have an impact on adjacent seawater? For example, the lagoon salinity dynamics can also be described by a salting factor defined as

$$F_s = Q_{\text{inflow}}/(Q_{\text{inflow}} + Q_{fr})$$                                     (6.18)

where                $$Q_{fr} = Q_{\text{riv}} + Q_{\text{prc}} + Q_{\text{grd}} - Q_{\text{evp}}$$

and where the values $Q_{\text{inflow}}$, $Q_{\text{riv}}$, $Q_{\text{prc}}$, $Q_{\text{grd}}$, and $Q_{\text{evp}}$ are not negative.

It should be noted that although the absolute value of the sea inflow is important, it is the ratio of the sea inflow to the river freshwater inflow that actually determines the hydrological behavior of a lagoon. Strictly speaking, the salting factor reflects, at a certain point in time, the actual tendency of the hydrological changes in the lagoon. More precisely, it sets a limit to the salt ratio ($S_{\text{sea}}/S_{\text{lag}}$) to which the lagoon salinity is converging at this point in time. The lagoon tends to be less salty than the adjacent open marine or ocean water if current $F_s < 0.5$, and saltier if current $F_s > 0.5$. The salting factor averaged over 1 year shows the general relationship between water budget components and the lagoon hydrology. An example is shown in Figure 6.10. The mean annual salting factor $F_s < 0.5$ indicates that the freshwater influx is greater than the seawater flux and that the lagoon is under greater terrestrial influence. A mean annual salting factor between $0.5 < F_s < 1$ indicates that the lagoon is predominantly influenced by the sea.

The Curonian and Odra lagoons are examples of freshwater lagoons influenced predominantly by their catchments area ($F_s < 0.5$). The Vistula Lagoon, with relatively low salinity waters, is predominantly influenced by the Baltic Sea.

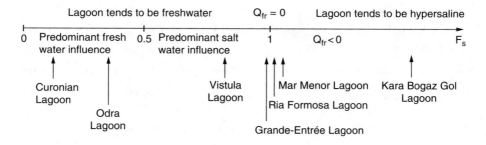

**FIGURE 6.10** Main gradations of salting factor and its annual average values for some selected lagoons.

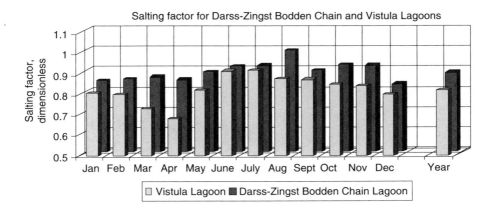

**FIGURE 6.11** Seasonal variations of monthly average salting factor for two Baltic lagoons: the Vistula Lagoon and Darss-Zingst Bodden Chain Lagoon.

The Grande-Entrée Lagoon can barely be distinguished from the open sea because it has no river inflow and only receives a small freshwater volume through precipitation. Both the Ria Formosa and the Mar Menor lagoons are examples of weak hypersaline lagoons where evaporation exceeds both precipitation and river inflow. The Kara Bogaz Gol Lagoon is a strong hypersaline lagoon where evaporation is extremely high.

The analysis of the temporal variation of the salting factor can be useful for lagoons where freshwater inflow plays a significant role. For example, monthly salting factors in the Vistula Lagoon over a year are presented in Figure 6.11. The average annual salinity (2 ÷ 5 psu) of this lagoon is less than that of the Baltic Sea waters (6 ÷ 8 psu). Even then, the salting factor indicates that the Vistula Lagoon remains more influenced by the marine water than the land influx during the whole year ($0.5 < F_s < 1$). The salting factor for the Vistula Lagoon is characterized by one spring minimum (maximum of river run-off, $F_s = 0.68$) and one summer maximum (before the rainy season, $F_s = 0.92$). The salting proceeds very actively from May to July. Smooth desalinisation then starts until an equilibrium is achieved between the salting and refreshing processes in winter.

The next desalinization of the lagoon coincides with the beginning of the fresh river run-off in spring. The temporal variation of the salting factor provides indications when two main processes, salting and desalinisation, are taking place.

Lagoons located in the same geographical conditions may express different behavior in terms of these two processes. This is seen, for example, in Figure 6.11 showing the salting factor for two Baltic region lagoons: the period of salting in the Darss-Zingst Bodden Chain is longer than that in the Vistula Lagoon and there is no period of equilibrium between the salting and desalination factors. The annual course for the salting factor is also characterized by a minimum and a maximum in the Darss-Zingst Bodden Chain Lagoon, but they are not as sharp as the corresponding extremes in the Vistula Lagoon and they do not coincide with the peak in the river run-off.

The lagoon hydrological annual cycle can also be characterized by using a temperature–salinity (T–S) diagram evolving with time (Figure 6.12A). This is useful

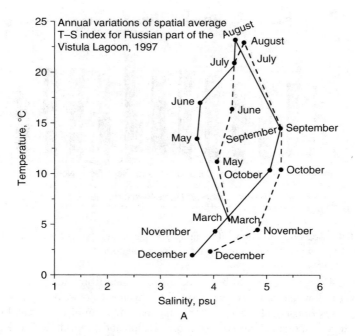

A

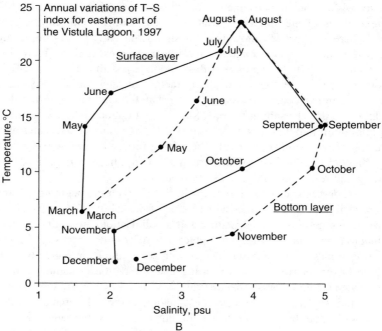

B

**FIGURE 6.12** Annual course of T–S index: (A) in the eastern freshwater corner and (B) near the entrance.

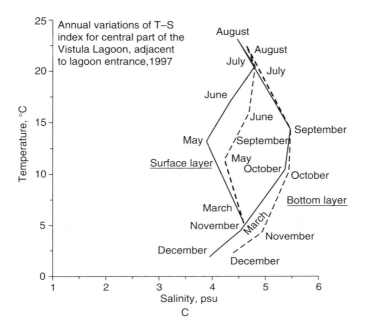

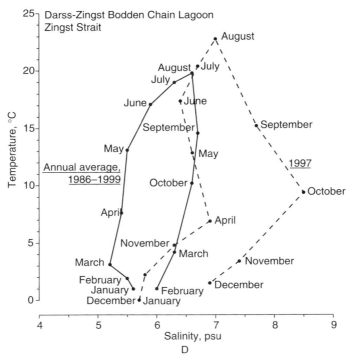

**FIGURE 6.12 (Continued)** Annual course of T–S index: (C) averaged over the northern part of the Vistula Lagoon, and (D) in the Darss-Zingst Bodden Chain Lagoon.

for model calibration by field data. Presenting both the upper and bottom layer T–S data, as in Figure 6.12B, reveals periods of vertical stratification and how they compare with the seasonal T–S variations. For example, it may not be appropriate to use a vertically integrated model when the vertical T–S variations are comparable to the seasonal T–S variations in both the upper and bottom layers (Figure 6.12C). Furthermore, since the temporal evolution of the T–S curve in these layers is significantly different, this may point to some independent forcing factors in each layer. This can be noticed in highly stratified estuarine lagoons that could express high variability from year to year (Figure 6.12.D).

The quantitative parameters introduced in this section define a multidimensional coordinate system from which lagoon types can be objectively defined using probability functions derived from similar or different lagoon combinations around the world.

In conclusion, the morphometric and hydrological features presented above may not quantitatively classify a lagoon as one type or another, and in that respect, Kjerfve's classification remains the most useful one in a qualitative sense. However, a combination of selected quantitative parameters may provide insights into the behavior of a lagoon, or in the absence of data, it may allow us to apply findings about other lagoons with similar dimensions to the lagoon under study and, eventually, to choose an appropriate model.

## 6.3.2 DESCRIPTION OF FORCING FACTORS

The state of any lagoon can be described at any point in time ($t$) by a number of variables. Some of these variables apply to the lagoon as a whole, such as its water volume $V_{lag}$, its average depth $H_{avg}$, and its free surface area $S_{lag}$. They are time-dependent functions of the lagoon water-level variations in time. Other variables, such as the depth $H(x, y, t)$ and the free surface variations $h(x, y, t,)$, have two-dimensional spatial and time-dependent distributions, while others have three-dimensional space and time-dependent distributions, such as the three velocity components $U_i(x, y, z, t)$ and various dissolved and suspended substances concentrations $C_k(x, y, z, t)$ including temperature $T(x, y, z, t)$ which is considered as a separate admixture in general terms. The relationships between these model variables are prescribed by the governing model equations, whereas all external driving forces causing the temporal and spatial variability of these variables are described as boundary conditions.

### 6.3.2.1 General Hierarchy of Driving Forces

Conceptually, a lagoon can be considered as a system acted upon by external forces and internal factors (Figure 6.13). This introduces the concept of different types of boundaries for the lagoon system. These are *physical boundaries* of the lagoon such as coastal line, bottom, and free surface; *conventional physical processes boundaries* where, for example, air–water or water–sediment interactions occur; and *conventional internal boundaries* for internal exchange parameterization (e.g., biological sink/source of mass and energy, or subgrid dissipation). As such, the lagoon becomes a separate system environment driven by external driving forces.

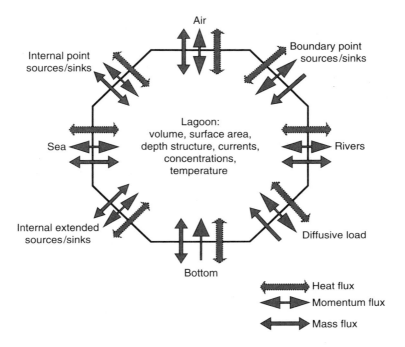

**FIGURE 6.13** Principal model boundaries with mass, momentum, and heat fluxes. Arrows show possible directions of fluxes.

Following this concept, the relationships between the lagoon and its surroundings are presented in Figure 6.13. The boundaries are the air, the sea or other adjacent basins, the bottom, the lagoon coastal zone including diffusive sources, rivers, and internal natural extended point sources/sinks as well as artificial internal and boundary influences.

The relationships between the lagoon and its external influences are prescribed by exchanges of mass, momentum, and heat specified by corresponding fluxes through all the boundaries. The fluxes are considered positive if they supply mass, momentum, or heat to the lagoon, and negative otherwise.

The driving factors defining the exchange processes across the lagoon boundaries can be subdivided into three groups:

- Processes responsible for mass exchange: These are all water balance terms, which also define the chemical and sediment balances in the lagoon, the coastal and the bottom erosion and sediment deposition. For example for salinity, the positive flux (input of the salt into the lagoon) is supplied by marine water inflow, by evaporation, and sometimes by groundwater infiltration if a salt-containing soil surrounds the lagoon. The total water outflow toward the adjacent sea, precipitation, and freshwater inflow (including fresh groundwater) will define the negative salt flux. Gain and loss for constituents are considered as loads.[4a] They arise from different sources and sinks: for example, sewage outlets (point sources), non-point sources or diffusive pollution, withdrawals, groundwater seapage (usually non-point and extended in space), etc.

- Processes responsible for heat exchange: These are direct solar radiation, direct turbulent heat fluxes through the air–water boundary caused by temperature gradients, and indirect heat fluxes due to evaporation. Advective transports of water from the sea, rivers, and the atmosphere also contribute to heat gains and losses in the lagoon due to temperature differences in lagoon and incoming waters.
- Processes responsible for mechanical momentum transport: Wind and tidal variations are the main factors here. Wind acts directly on the upper water mass, setting up a water level surge that in turn defines a pressure gradient that is responsible for the movement of the total water body. Any inflow water flux, irrespective of its origin (wind or tidally induced flow, river, or artificial source) provides a positive gain of momentum in the lagoon system. Negative inputs are provided by bottom and internal frictions.

The hierarchy of these forces is as follows: mass and heat fluxes at the boundaries define the state and time evolution of the lagoon system whereas momentum fluxes are responsible for internal transformations of mass and heat, their redistribution, and their exchanges between different parts of the lagoon system. A precise identification of all mass and heat sources/sinks for a lagoon under study requires a good description of the external driving forces. More specifically, the following parameters must be identified:

- The number of sources/sinks
- The name of the sources/sinks
- The position of sources/sinks in geographical coordinates (latitude, longitude) and in the computational grid coordinates
- The average water volume discharge [$m^3$ $s^{-1}$] in each source/sink
- The average concentrations of chemical substances under study [kg $m^{-3}$] at each source
- The average discharges of all studied chemical substances [kg $s^{-1}$ or t year$^{-1}$]

Additional parameters on air–water interactions and lagoon–ocean exchanges, on sediment erosion and deposition and on human-made changes must also be properly identified.

A good starting point is to estimate mean annual characteristics of all forcing factors. This provides information on the nature (sink or source) and range of the factors involved and first-order estimates of water, heat, sediment, and chemical budgets for the lagoon system.

The absence of information for some parameters will prompt additional field measurements. If these measurements cannot be performed, the estimations should be obtained through a numerical modeling approach. For example, the magnitude of the sea–lagoon water exchange and its temporal (seasonal or synoptic) variations, which are difficult to measure directly, could be defined as fluxes throughout the entrance during salinity model simulations. These simulations can be calibrated with annual salinity averages for the lagoon or with real salinity annual monitoring at some lagoon points.

## 6.3.2.2 Water Budget Components

The water budget is the fundamental knowledge that should be obtained before any hydraulic and water quality studies. The water balance of certain accuracy is helpful to ensure that all sources and sinks of water of appropriate capacity are known. The natural water budget in a lagoon is made up of contributions from surface evaporation and precipitation, from river inflow and from exchanges at the ocean–lagoon boundaries. In some lagoons, bottom seepage may also be significant.

One can propose a convenient presentation of water budget components in the form of a "rose" (Figure 6.14). According to the problem under consideration water budget components may be estimated in absolute values, either in [$m^3 s^{-1}$] or, more practically, in [$km^3 yr^{-1}$], as well as in specific values; for example, for the ratio of an absolute value of each budget component to the open lagoon surface dimensions are [$m s^{-1}$] or [$m yr^{-1}$]. Such a specific value describes the relative contribution of each water budget component to variations in the lagoon volume. The ratios of the average lagoon volume to the absolute value of each budget component (expressed in seconds or years) provide clear characteristics of the significance of water budget components for flushing the lagoon volume (see Section 6.3.3.1).

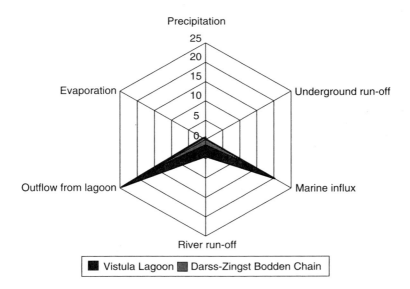

**FIGURE 6.14** Lagoon water budget rose. Budget components are represented in specific values (m·year$^{-1}$) as a ratio of corresponding component absolute value to the lagoon free surface to show the relative contribution of each water budget component to the lagoon volume.

Again, the parameters required for estimating the water budget are the river water inflow ($Q_{riv}$), the atmospheric precipitation ($Q_{prc}$) and evaporation ($Q_{evp}$), the underground inflow ($Q_{grd}$), and the marine water inflow ($Q_{inflow}$). The ($Q_{riv}$) and ($Q_{prc}$) are usually measured directly by standard hydrometeorological monitoring programs, and some empirical approaches are used to estimate the remaining terms. Some practical recommendations are provided below.

### 6.3.2.2.1  Surface Evaporation Budget

The evaporation from a lagoon surface area can be estimated as follows.[5,6] Under equilibrium conditions, when the difference between water and air temperatures ($\Delta t_1$) is 2 to 4°C, the evaporation in mm day$^{-1}$ is

$$Q_{evp} = 0.14 \cdot (1 - 0.72 \cdot u_2) \cdot (e_0 - e_2) \qquad (6.19a)$$

where $u_2$ [m s$^{-1}$] is the average wind velocity at a 2-m height, $e_0$ [mm Hg] is the absolute humidity for surface water temperature, and $e_2$ [mm Hg] is the absolute humidity for air temperature at a 2-m height above the water surface.

Under nonequilibrium conditions, the evaporation is

$$Q_{evp} = 0.104 \cdot (k_0 + u_2) \cdot (e_0 - e_2),$$

$$k_0 = 1.378 + 0.195 \cdot \Delta t_1 - 0.003 \cdot \Delta t_1^2 \qquad (6.19b)$$

### 6.3.2.2.2 Ocean–Lagoon Exchange Budget

Tides and wind surges are the main driving forces at the ocean–lagoon boundaries and water level variations can be used to calculate the corresponding ocean–lagoon fluxes. As a first approximation, the time-dependent, spatially averaged lagoon levels can be used. In this case, any influence of surface inclinations inside the lagoon on the exchange budget is filtered out and only seawater flux that penetrates deeply into the lagoon will affect the exchange. The remaining sea influx, which gives rise to short period inflow and outflow movements near the entrance, is not considered in this approach.

The spatially averaged and time-dependent lagoon levels $h_{avg}^{(t)}$ are normally obtained from averaging all levels measured at stations around the lagoon. These stations should be selected to represent the lagoon area proportionally, depending on its shape.

A more rigorous way to calculate a net water exchange through the ocean–lagoon boundary is to subtract, at each monitoring time step $\Delta t$, the other local components of the water budget such as the river inflow $Q_{riv}$, the precipitation $Q_{prc}$, and the evaporation $Q_{evp}$. First, the level variation time series $\Delta h_{avg}^{(t+0.5\Delta t)} = h_{avg}^{(t+\Delta t)} - h_{avg}^{(t)}$ are to be constructed from the spatially averaged level time series $h_{avg}^{(t)}$ over the entire period $T_{\text{total}} = N \cdot \Delta t$ ($\Delta t$ being the sampling interval). The terms of the ocean–lagoon exchange time series during any $n$th time step ($n = \overline{1,N}$) can then be obtained as

$$\Delta h^n_{\text{inflow}} = \Delta h^n_{\text{avg}} - \Delta t \cdot \left[ (Q_{\text{riv}} + Q_{\text{prc}} - Q_{\text{evp}}) / S^{\text{avg}}_{\text{lag}} \right]^n, \qquad (6.20)$$

where $S^{\text{avg}}_{\text{lag}}$ is the average lagoon surface area, which is a time-dependent function of the level variation $S^{\text{avg}}_{\text{lag}}(h(t))$. The volume inflow and flux are then easily calculated from $\Delta h^n_{\text{inflow}}$ by summation of positive terms only as:

$$V_{\text{inflow}} = S^{\text{avg}}_{\text{lag}} \cdot \sum \Delta h^n_{\text{inflow}}(+), \quad n = \overline{1,N} \qquad (6.21)$$

and

$$Q_{\text{inflow}} = V_{\text{inflow}} / T_{\text{total}}$$

Ocean–lagoon water exchange estimates and dominant periods of oscillation can also be obtained through spectral analysis. At first, the time-dependent level variation function $\Delta h_{\text{inflow}}(t)$ is decomposed into harmonic series, after subtracting the linear trend. A trend in local water-level measurements may be the result of a water-level trend in the adjacent ocean, of a geological subsidence of the coastline over yearly time scales, or simply an erroneous functioning of the tide gauge.

Once the local trend is removed from the series, the water-level time series can be reduced to its Fourier harmonic components as

$$\Delta h_{\text{inflow}}(t) = \Delta h_k \cdot \sin(\omega_k \cdot t + \varphi_k), \quad k = 1, K$$

where $\Delta h_k$, $\omega_k = 2 \cdot \pi / T_k$, and $T_k$ are the amplitude, the frequency, and the period of the $k$th harmonic, respectively. In this case, the harmonic frequencies are limited by the Nyquist frequency (or minimum period, $T_K$), given by $w_K = (2 \, \Delta t \,)^{-1}$, where $\Delta t$ is the sampling interval.

Assuming that the wet surface area change with water level $S^{\text{avg}}_{\text{lag}}(h)$ is not significant, the volume corresponding to one period of ocean–lagoon inflow $\Delta V^k_{\text{inflow}}$ and its flux $q^k_{\text{inflow}}$ due to the $k$th harmonic term at period $T_k$ is given by

$$\Delta V^k_{\text{inflow}} = 2 \cdot S^{\text{avg}}_{\text{lag}} \cdot \Delta h_k$$

$$q^k_{\text{inflow}} = \Delta V^k_{\text{inflow}} / T_k \qquad (6.22)$$

where the multiplication by 2 takes into account the fact that an integral over half of a period of harmonic function sin equals 2.

Finally, the total average seawater inflow flux during the period $T_{\text{total}}$ is

$$Q_{\text{inflow}} = \sum q^k_{\text{inflow}}, \quad k = \overline{1, K} \qquad (6.23)$$

The comparison of the magnitudes of $q^k_{\text{inflow}}$ provides information on which oscillation period constitutes the main input into ocean–lagoon water exchange.

This may be important for nontidal lagoons where random wind surge is the leading forcing factor for marine water influx. In turn, this information may also identify the physical forcing factors responsible for dominant ventilation of the lagoon by ocean waters. For lagoons where tides are the major forcing function, spectral analysis is not so relevant because the tidal harmonic frequencies are well known and are best resolved using harmonic analysis.

### 6.3.2.3  Heat Budget

The thermal structure of a lagoon water body and its temporal variations play an important role in the dynamics, mixing, water quality, and biotic conditions of a lagoon. In the case of a stratified lagoon, where temperature has a strong influence on water density, a three-dimensional numerical approach is required. For a nonstratified or a weakly stratified lagoon, the evolution of temperature in time and space may be described by a two-dimensional approach. This is expressed by two-dimensional partial differential equations in Eulerian coordinates (Equation (6.24)) where the left side of the temperature evolution equation is a total derivative of the temperature in the control volume and the right side is the net sum of heat fluxes, sources, and sinks:

$$\frac{\partial(VT)}{\partial t} + u_1 \cdot \frac{\partial(VT)}{\partial x_1} + u_2 \cdot \frac{\partial(VT)}{\partial x_2} = \frac{1}{C_p \cdot \rho}(A_s \cdot H_\Sigma + H_s), \qquad (6.24)$$

where $H_\Sigma$ is the net thermal-energy flux (Wm$^{-2}$), $H_s$ is the sum of heat internal sources and sinks $(W)$, $V$ is the control volume (m$^3$), $T$ is the temperature (°C), $t$ is the time (s), $u_1$ and $u_2$ are the depth-averaged water velocity components (m s$^{-1}$), $A_s$ is the surface area (m$^2$), $\rho$ is the density of water (e.g., 997 kg m$^{-3}$ at 25°C), and $C_p$ is its specific heat (e.g., 4179 J kg$^{-1}$ °C$^{-1}$ at 25°C). The advective transport of the heat into or out of the control volume is described by nonlinear advective terms on the left side of the equation and is not included in the net thermal-energy flux.

The net thermal-energy flux is a heat flux through the lagoon surface. A typical example of the internal or boundary heat source would be the outlet of a power plant cooling system. The advective heat flows caused by river run-off are usually considered boundary heat sources or sinks. The marine water influx is usually treated through the boundary condition for Equation (6.24).

The net thermal energy flux $H_\Sigma$ at the lagoon surface includes solar radiation in short and dispersed waves, back radiation from the water surface, direct heat exchange with the atmosphere (heat conduction), and latent heat exchange with the atmosphere in terms of evaporation heat loss or condensation heat gain. A detailed description of all components of the net thermal energy flux as well as all necessary formulas for their calculation or estimation can be found, for example, in Martin and McCutcheon:[4b]

$$H_\Sigma = (H_{SW} - H_{BSW}) + (H_H - H_{BH}) - H_B \pm H_L \pm H_S \qquad (6.25)$$

where $H_{SW}$ is the absorbed short-wave radiation (range of 50–500 Wm$^{-2}$); $H_H$ is the long-wave back radiation from atmospheric constituents or dispersed solar radiation (range of 30–450 Wm$^{-2}$); $H_{BSW}$ and $H_{BH}$ are the short-wave solar radiation (range of 5–30 Wm$^{-2}$) and dispersed solar radiation (range of 10–15 Wm$^{-2}$), respectively, reflected from the water surface; $H_B$ is the back long-wave radiation (range of 300–500 Wm$^{-2}$); $H_L$ is a latent heat exchange (the energy loss due to evaporation is within the range of order 100–600 Wm$^{-2}$); and $H_S$ is the net heat flux due to conduction or sensible heat transfer (with a range on the order of 50–500 Wm$^{-2}$).

*Net short-wave solar radiation* (the first two terms of Equation (6.25)) varies daily with the altitude of the sun, whose maximum depends on season, on dampening by radiation scattering and absorption in the atmosphere, and on reflection from the water surface:

$$H_{SW} = H_0 \cdot a_t \cdot (1 - R_S) \cdot C_a \qquad (6.26)$$

where $H_0$ is the extraterrestrial radiation reaching the Earth's outer atmosphere. $H_{SW}$ depends on latitude of location and time. Date and time of day determine the sun altitude, the sun-day duration, the standard times of sunrise and sunset, and the relative distance between the earth and sun. $a_t$ is the fraction of the extraterrestrial radiation reaching the water surface after reduction by scattering and absorption. It depends on the dust coefficient, reflectivity of the ground, the moisture content, and the optical air mass. $R_s$ is the albedo or the reflection coefficient, which depends on the solar altitude and cloud cover, and $C_a$ is the fraction of solar radiation not absorbed by clouds, which depends on the fraction of the sky covered by them.

*Long wave radiation* occurs when the atmosphere and clouds absorb part of the solar radiation coming at the top of atmosphere ($H_0$), become heated, and radiate heat at longer wavelengths. The magnitude of long-wave radiation is computed using the Stefan-Boltzman law modified for the emissivity of the air. It directly depends on air temperature and atmospheric moisture, and is influenced by chemical constituents of the atmosphere. The net long-wave radiation can be estimated from the following empirical relation cited in Martin and McCutcheon[4b] with the references of Swinbank[7] and Wunderlich:[8]

$$H_H = \alpha_0 \cdot 0.97 \cdot \sigma \cdot (T_a + 273.16)^6 \cdot (1 + 0.17 \cdot C_t) \qquad (6.27)$$

where $\alpha_0$ is a proportionally constant with a value of $0.937 \cdot 10^{-5}$, $\sigma$ is the Stefan-Boltzman constant of $5.67 \cdot 10^{-8}$ W m$^{-2}$ (K)$^{-4}$, $C_t$ is the fraction of the sky covered by clouds ($0 < C_t < 1$), and $T_a$ (°C) is the air temperature measured at a height of 2 m above the water surface. Reflectance from the water surface is generally assumed to be 3%.

*Back radiation* from the water is a black-body radiation type described using the Stefan-Boltzman law, considering that water emissivity $\varepsilon_w$ is approximately 0.97:

$$H_B = \varepsilon_w \cdot \sigma \cdot (T_S + 273.16)^4 \qquad (6.28)$$

where $H_B$ is the radiance or heat loss rate per unit surface area (Wm$^{-2}$) and $T_s$ is the water surface temperature (°C).

*Evaporative heat loss* ($H_L$, Wm$^{-2}$) depends on the water density ($\rho$), on the specific evaporative heat or latent heat of the water ($L_w$ is the heat energy required to evaporate a given mass of water, Jkg$^{-1}$), and on a rate of evaporation (E, m s$^{-1}$), which in turn depends on the wind and water vapor pressure gradient between the water and atmosphere. It is usually estimated by the following formula:

$$H_L = \rho \cdot L_w \cdot (a + b \cdot u_w) \cdot (e_s - e_a) \qquad (6.29)$$

where $a$ is an empirical coefficient representing the effect of vertical convection, which occurs even in the absence of the horizontal wind velocity, $e_s$ is the saturated vapor pressure at the water surface temperature (mb), and $e_a$ is the vapor pressure at the air temperature (mb).

*Sensible heat transfer*, or the transport of heat due to convection and conduction, can be estimated by an approach based on the Bowen ratio, which has been observed to be valid over an extended range of conditions. Practical means of estimating the sensible heat flux $H_s$(Wm$^{-2}$) is

$$H_S = \rho \cdot L_W \cdot (a + b \cdot u_w) \cdot C_b \cdot \frac{P_a}{P} \cdot (T_s - T_a) \qquad (6.30)$$

where $C_b$ is a coefficient (0.61 mb °C$^{-1}$), $P_a$ is the atmospheric pressure (mb), $P$ is a reference pressure at the mean sea level, $T_s$ is the water surface temperature, and $T_a$ is the air temperature. See Martin and McCutcheon[4b] for a more detailed description of the heat budget terms.

### 6.3.3 PRE-ESTIMATION OF SPATIAL AND TEMPORAL SCALES

It should be emphasized that although hydrodynamic numerical, physical, and other types of models can expand our knowledge about the hydrodynamics of a study area, the dominant physical processes must be known beforehand, i.e., before the model is applied. A numerical model, which incorporates the physics of these processes, can then be chosen to provide a quantitative description of these processes. Knowledge of the following temporal and spatial scales can contribute to the choice of a hydrodynamic model.

#### 6.3.3.1 Flushing Time

Without loss of generalities, the flushing time is the time required for a measurable volume of the water in a lagoon to be replaced by waters from river run-off, precipitation, and water exchange with the adjacent coastal marine waters. In a case where the trial volume of water equals the total lagoon volume, it will be an integral flushing time, which is also a measure of the self-cleaning capability of a polluted lagoon. This messure is commonly used in the assessment of rehabilitation schemes for lagoon ecosystems that are under stress from pollution. However, even when all pollutants in the lagoon are flushed, it must be remembered that they are still present elsewhere in the marine coastal zone.

### 6.3.3.1.1 Integral Flushing Time

The minimum value of the integral flushing time (i.e., the quickest flushing of the lagoon) required for a complete reduction of pollution in a lagoon can be estimated using the concept of well-mixed reactors, provided the lagoon is well mixed. If we consider the flushing time to be the characteristic time required for an $e$-fold renewal of water ($e$ is a Eulerian number approximately equal to 2.71828) or reduction of a pollutant concentration in the lagoon to 37% of its initial value when it is flushed by nonpolluted water, this concentration as a function of time is given by

$$C(t) = C_0 \cdot \exp(-t/\tau_{\text{flush}}) \tag{6.31}$$

$$\tau_{\text{flush}} = V_{\text{avg}}/q_{\text{flush}} \tag{6.32}$$

where $t$ is time, $C_0$ is the initial concentration of pollutant in the lagoon, $\tau_{\text{flush}}$ is the flushing time (or ventilation, or hydraulic replacement according to Kjerfve[3]), $V_{\text{avg}}$ is the average lagoon volume, and $q_{\text{flush}}$ is the discharge rate of the ventilation stream. The flushing time is the time it takes for the initial concentration to be diluted to a concentration $C = \frac{C_0}{e}$ (where $e = 2.71828$). A total of 93.4 and 95% dilution are attained in times equivalent to $\tau_{90\%} = e \cdot \tau_{\text{flush}}$ and $\tau_{95\%} = 3 \cdot \tau_{\text{flush}}$, respectively.

When a lagoon has several input sources of water (including inlets) with discharge $q_i$ and concentration $c_i$, the evolution of the concentration in the lagoon given by a well-mixed solution is

$$C(t) = C_0 \cdot \exp(-t/\tau_{\text{flush}}) + \frac{\tau_{\text{flush}}}{V_{\text{avg}}}\left(\sum q_i \cdot c_i\right)\exp(-t/\tau_{\text{flush}}) \tag{6.33}$$

$$\tau_{\text{flush}} = \frac{V_{\text{avg}}}{\sum q_i} \tag{6.34}$$

$$\frac{1}{\tau^{\text{flush}}} = \sum \frac{1}{\tau_i^{\text{flush}}} = \frac{\prod \tau_i^{\text{flush}}}{\sum \tau_i^{\text{flush}}}, \quad \text{where } \tau_i^{\text{flush}} = \frac{V_{\text{avg}}}{Q_i} \tag{6.35}$$

It is important to note that the flushing time depends only on the hydrographic characteristics of a lagoon and not on the absolute values of the initial and input pollutant concentrations. Equation (6.31) and Equation (6.32) describe the system adaptation to a new level of pollution load, and $\tau_{\text{flush}}$ (as well as $\tau_{90\%}$ and $\tau_{95\%}$) is the characteristic time scale for the system to adapt to a new equilibrium concentration.

In lake or water reservoir problems very often the *retention or residence* time, which is also estimated[4b] by Equation (6.32) as the ratio of the water volume to the inflow discharge, is used. This time is usually interpreted as the "lifetime" the water from the inflow water or an incoming particle resides in the lake or reservoir.[4b]

It is also possible to use Equations (6.33)–(6.35) to estimate the flushing time from periodic and/or stochastic water exchanges with the adjacent sea resulting from tidal or meteorological forcing at the entrance. In this case, these exchanges and the corresponding flushing times take the form:

$$q_{\text{flush}} = \Sigma\, q_i, \qquad q_i = 2 \cdot S_{\text{avg}} \cdot \frac{h_i}{T_i}, \qquad \tau_i^{\text{flush}} = \frac{V_{\text{avg}}}{q_i} \qquad (6.36)$$

where $q_i$ is the contribution of the $i$th harmonic oscillation to the ventilation stream (see Section 6.3.2.2.); $S_{\text{avg}}$ and $H_{\text{avg}}$ are the average surface area and water depth of the lagoon, respectively; and $h_i$ and $T_i$ are the amplitude and period, respectively, of the $i$th harmonic of the water-level oscillations. The resulting flushing time becomes

$$\tau_{\text{flush}} = \frac{H_{\text{avg}}}{2} \cdot \frac{1}{\Sigma\, \dfrac{h_i}{T_i}} \qquad (6.37)$$

Harmonic analysis of the water-level variations caused by irregular winds often shows that it is the higher frequency terms that contribute significantly to the flushing marine water flux in Equation (6.36). These terms prevail due to their short period, despite their small amplitude in comparison with the significant harmonic terms, which constitute the main deviation of sea level from its statistical mean value. These short period and small amplitude level variations contribute to the main part of the marine water pumping into the lagoon area. This is why the accuracy of statistical determinations of the amplitudes and frequencies for such short period harmonics should be carefully established.

However, estimating the marine water flux in the lagoon from Equation (6.36) is not straightforward. We cannot just include in this equation all harmonic terms obtained from the harmonic analysis. This may lead to a nonconvergent series of $(1/n)$ type because of the $n$-fold frequencies in the Fourier analysis and because of the possibility that the $h_i$ value will not decrease monotonously with $n$. In other words, including more short period terms may not guarantee the convergence of the flushing time value.

As an example, let us consider the statistical analysis of a 3-year time series of water-level variations at the Vistula Lagoon entrance, sampled at a 12-h interval. A formal Fourier analysis of this time series yielded a Fourier series with 1092 harmonic terms. As the Nyquist frequency is $1/2dt$, which corresponds to a period of 1 day, terms with periods less than 3 Nyquist periods were excluded from the series to obtain more reliable results. This reduced the number of harmonic terms in the Fourier series from 1092 to 367. Furthermore, harmonic terms with amplitudes less

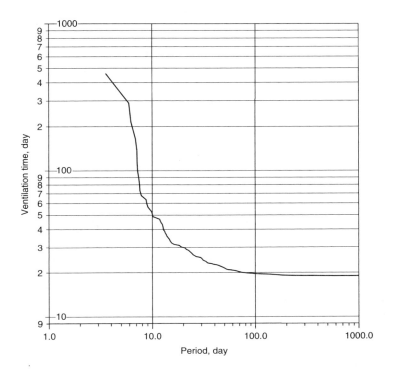

**FIGURE 6.15** Convergence of flushing times as longer period harmonics are added to Equation (6.37).

than the initial data random error (1 cm) were also excluded, leaving 286 terms in the Fourier series. These terms were finally used to calculate the marine water inflow (Equation (6.36)) and the corresponding ventilation time (Equation (6.37)).

Figure 6.15 shows results obtained from Equations (6.36) through (6.37) indicating that the flushing time for the Vistula Lagoon is $\tau_{flush} \approx 19$ days. This is the time necessary to reduce any conservative pollutant concentration to 37% of its initial value. This figure also shows that the initial concentration will diminish to 10% over $2.3 \cdot \tau_{flush} \approx 43–44$ days, and to 1% over $4.6 \cdot \tau_{flush} \approx 87–88$ days.

### 6.3.3.1.2 Local Flushing Time

It is instructive to distinguish between the *integral* flushing time and the *local* flushing time (Zimmerman[9]) for some applications, which means using the same approach but applied not only to total lagoon volume but also to any compartment of a lagoon, even to any grid cell.

Consider a lagoon of volume $V_0$, continuously flushed by a stationary volume flux $Q$ including a river discharge (if any) at the upstream boundary and tidal prism at the ocean boundary. The integral flushing time of the lagoon has been defined above as Equation (6.32). This expression implicitly assumes that the marine water added to the lagoon through its open boundaries is completely and instantaneously mixed with $V_0$ and that an equivalent mixed volume leaves without returning into the system.

In reality, mixing of incoming waters at either boundary of the lagoon will more likely occur with waters located near that boundary and the local flushing time will progressively increase as the distance away from this boundary increases. Hence, the integral and local flushing times are in fact functions of space. For nonstationary $Q$, it is a function of time as well. As such, we can also characterize a coastal system by its local flushing time distribution $\tau(x, y, z, t)$ where $x, y, z$ are the spatial coordinates and $t$ is time. More generally, the integral flushing time is simply the spatial integration of the local flushing time over the area of the system, under stationary conditions.

While the integral flushing time is useful for comparative coastal ecosystems studies, the local flushing time is useful for selecting the most appropriate sites for localized human interventions in a lagoon, such as an aquaculture site, an outfall diffuser site, etc.

Spatial distribution estimates of the local flushing time normally require the use of hydrodynamic-numerical models. In this case, two approaches can be used: the Lagrangian particle tracking approach or the Eulerian dissolved tracer advection–dispersion approach. In both methods, a hydrodynamic model forced by most representative forcing functions at its open boundaries is first developed for the lagoon.

Once the model has been calibrated and validated with measurements, coupled simulations are then carried out. For the Lagrangian method, a Lagrangian particle model is coupled to the hydrodynamic model and particles are introduced at each grid point in the lagoon at the beginning of the simulation. The simulation is then performed for a long period of time relative to the expected integral time and the time at which each particle leaves the lagoon is recorded. For the Eulerian method, an advection–dispersion model is coupled to the hydrodynamic model and a dissolved tracer concentration of 100% is set as an initial condition in the lagoon and 0% elsewhere. The coupled simulation is then performed for the same period as above and the time at which the initial concentration $C_0$ in each cell drops below $\dfrac{C_0}{e}$ is recorded.

In both methods, the result is a two-dimensional distribution of the local flushing time of the lagoon. A more formal discussion of the flushing time and related time scales can be found in Bolin and Rodhe,[10] Takeoka,[11] and Zimmerman.[9] An application of the tracer method to estimate local renewal times is found in Koutitonsky et al.[12]

### 6.3.3.2  Surface and Bottom Friction Layers

Both the maximum and the average wind-induced upper layer currents may be estimated from Ekman's (1905) theory.[13] A more recent development is presented by Madsen.[14] These theories describe currents in the upper layer of the open ocean, in equilibrium under the effect of shear stress, Coriolis force, and the local pressure gradient. Momentum is transmitted from the surface to deeper layers by vertical turbulent mixing. The vertical mixing is constant for Ekman's theory and variable for Madsen's theory. In both theories, nonlinear convective terms and horizontal turbulent mixing as well as temporal acceleration term are neglected. The resulting motion is in steady state in equilibrium and will remain so until the wind stops or changes direction. Actually, the currents do not reach this equilibrium state immediately but only after an adjustment time scale, as discussed in Section 6.3.3.3.

One can reasonably assume that, once the wind starts blowing and a pressure gradient is not yet established, currents in the lagoon will adjust toward the Ekman equilibrium, especially in regions far away from shorelines and over time scales much shorter than the relaxation time (see Section 6.3.3.3). At that moment, surface current estimates can be obtained from formulae derived for surface wind-induced currents by Ekman. They are directed at 45° to the right of the wind direction and their maximum speed is given by[15,16]

$$U_E = \frac{\tau}{\rho \cdot \sqrt{K_z \cdot f}} \quad \text{or} \quad U_E = \frac{\sqrt{2} \cdot \pi \cdot \tau}{\rho \cdot f \cdot h_E} \tag{6.38}$$

where $\tau$ is the wind stress, $K_z$ is the constant vertical turbulent coefficient, $\rho$ is the water density, $f = 2 \cdot \Omega \cdot \sin(\varphi)$ is the Coriolis parameter, $\Omega$ is the Earth's rotation speed ($\approx 7.2921 \cdot 10^{-5}\,\text{s}^{-1}$), $\varphi$ is the geographic latitude, and $h_E$ is arbitrarily taken as the effective depth of the wind-driven current (Equation 6.40). In the Vistula Lagoon, for example, $\varphi = 55°$ and $f = 11.94 \cdot 10^{-5}\,\text{s}^{-1}$, and current speeds estimated from Equation (6.38), Equation (6.42), and Equation (6.46) are 0.07, 0.20, and 0.46 m s$^{-1}$ for wind speeds of 5, 10, and 15 m s$^{-1}$, respectively.

As depth increases, horizontal velocities decrease exponentially and veer progressively toward the right, forming what is known as an Ekman spiral. The current distribution in this spiral, as a function of depth ($z$), is given by[15,16]

$$U = U_E \cdot \exp\left(-\frac{\pi \cdot z}{h_E}\right) \cdot \cos\left(45 - \frac{\pi \cdot z}{h_E}\right)$$

$$V = U_E \cdot \exp\left(-\frac{\pi \cdot z}{h_E}\right) \cdot \sin\left(45 - \frac{\pi \cdot z}{h_E}\right) \tag{6.39}$$

Wind-driven currents extend to a depth ($h_E$), called the Ekman depth, the thickness of which depends on the vertical turbulent coefficient ($K_z$) and on the geographic latitude through the Coriolis parameter $f$, as[15,16]

$$h_E = \pi \sqrt{\frac{2K_z}{f}} \tag{6.40}$$

At the depth $z = h_E$, the velocity is directed opposite to that at the surface and its speed is reduced by a factor of $\exp(-\pi)$ to about 4.3% of the surface speed. Current speeds and directions at various depths in the Ekman spiral are presented in Table 6.8. The wind-induced current may be estimated as the current prevailing at $z = 0.5 \cdot h_E$.

## TABLE 6.8
### Current Speed and Direction vs. Depth in the Ekman Layer

| Depth (z) | The Angle between the Ekman Surface Current and the Current at Depth (z) | Current Speed |
|---|---|---|
| 0 | 0° (the surface current direction is 45° to the right of the wind direction in the northern hemisphere) | $U_E$ |
| $0.25 \cdot h_E$ | 45° to the right | $0.456 \cdot U_E$ |
| $0.5 \cdot h_E$ | 90° to the right | $0.208 \cdot U_E$ |
| $0.75 \cdot h_E$ | 135° to the right | $0.095 \cdot U_E$ |
| $h_E$ | 180° (opposite to the surface current) | $0.043 \cdot U_E$ |

An opposite situation develops in the bottom boundary friction layer. Currents above this layer are assumed to be in geostrophic equilibrium, that is, the pressure gradient balances the Coriolis force. At a depth $h_E$ above the bottom, currents start feeling the effect of bottom friction and progressively veer to the left of the geostrophic current direction and vanish at the bottom. This forms the bottom Ekman spiral of depth $h_E$. Table 6.9 presents the vertical distribution of currents in the bottom boundary layer.

For typical turbulent vertical mixing coefficients between $10^{-4}$ and $10^{-3}$ m²/sec, the thickness of the friction layer ($h_E$) has typical values between 4 and 13 m at middle latitudes. The comparison of lagoon depths with $h_E$ provides information on the extent to which currents at various depths will be deflected by winds and surface currents. This thickness is also important for the evaluation of the vertical grid step for 3D models, which should not exceed 0.3–0.5 of $h_E$.

It should be noted once more that wind-induced Ekman currents (Equations (6.39)–(6.40)), which are directed at 45° to the right of the wind direction, only

## TABLE 6.9
### Current Speed and Direction in the Bottom Boundary Friction (or Ekman) Layer

| Elevation over the Bottom | The Angle between the Ekman Current and the Geostrophic Deep Water Velocity $V_0$ | Current Speed |
|---|---|---|
| 0 | 45° (the limit of current rotation to the left from the undisturbed velocity) | 0 |
| $0.001 \cdot h_E$ | 44.9° to the left | $0.004 \cdot V_0$ |
| $0.01 \cdot h_E$ | 36.5° to the left | $0.380 \cdot V_0$ |
| $0.25 \cdot h_E$ | 25.5° to the left | $0.750 \cdot V_0$ |
| $0.5 \cdot h_E$ | 11.7° to the left | $1.021 \cdot V_0$ |
| $0.75 \cdot h_E$ | 3.6° to the left | $1.069 \cdot V_0$ |
| $h_E$ | 0.0° | $1.043 \cdot V_0$ |
| $1.25 \cdot h_E$ | 0.8° to the right | $1.014 \cdot V_0$ |

characterize the initial phase of current adaptation. Once the pressure gradient is established, currents will adjust to a different equilibrium pattern under the influence of Coriolis acceleration, friction, and pressure gradients. In other words, pressure gradient currents will be added to Ekman currents.

### 6.3.3.3 Time Scales of Current Adaptation

In the absence of tides at the entrance, wind stress is the main force that generates or changes currents in a lagoon. Currents set up within a certain time interval after the wind starts to act on the surface because of the water mass inertia. In order to understand the time scale of the lagoon response to wind stress, it is necessary to estimate the current adjustment time scales. This time scale may be useful in determining the model simulation time step. For example, when studying currents generated by constant winds, it is desirable to have some 20 to 30 computational time steps during the adaptation time. For longer simulations under real wind conditions, at least 5 to 10 time steps are required to resolve the adjustment time.

#### 6.3.3.3.1 Wind-Driven Current

Once the wind stress starts acting on the surface of a lagoon, the upper layer currents respond with some lag because the Coriolis force does not act instantaneously. Kundu[17] showed that the influence of the Coriolis force becomes significant after $t > 0.25/f$. For the Vistula Lagoon (55° latitude), this adjustment time is about 30–40 min. As mentioned in Section 6.3.3.2, the current adjustment is a progressive process during which momentum is transmitted in the vertical direction by turbulence in the upper layer. The period of the current adjustment (the 63% development) under constant wind influence can be estimated as follows[16]

$$T = \frac{2 \cdot H}{\sqrt{2 \cdot f \cdot K_z}} \qquad (6.41)$$

where $H$ is the depth (m) at which the vertical mixing turbulence coefficient $K_z$ may be assumed to be zero or is the average depth of the lagoon, assuming it is completely mixed.

Since the turbulent coefficient is not known a priori, it may be evaluated as follows:

$$K_z = \frac{\gamma \cdot W_a \cdot h}{4 \cdot \rho \cdot k} \qquad (6.42)$$

where $\gamma = 3.25 \cdot 10^{-3}$ kg m$^{-3}$, $W_a$ is wind speed (m s$^{-1}$), $h$ is the depth (m) at which the turbulence coefficient may be assumed to be zero, $\rho$ is water density (kg m$^{-3}$), $k$ is the wind friction coefficient (~0.0125). When the mixing depth $H$ is not known a priori, it may be replaced by the lagoon depth.

The examples of current adjustment time for the Vistula Lagoon ($\varphi = 55°$, $H_{max} = 5$ m, $H_{avg} = 2.75$ m), calculated by Equation (6.41) and Equation (6.42), are presented in Table 6.10.

**TABLE 6.10**
**Friction Layer Current Adjustment Times (h) for the Vistula Lagoon Using Equations (6.41) and (6.42)**

| Depth of Turbulent Layer, m | Wind Velocity [m s⁻¹] | | |
|---|---|---|---|
| | 5 | 10 | 15 |
| 2 | 2.8 h | 2.0 h | 1.6 h |
| 3 | 3.5 h | 2.5 h | 2.0 h |
| 5 | 4.5 h | 3.2 h | 2.6 h |

*6.3.3.3.2 Equilibrium Current Structure*

Currents in a lagoon will adjust to winds in a time given by Equation (6.41) if the size of the lagoon is unlimited. In real cases, compensatory currents are set up through the continuity equation because of the enclosed nature of the lagoon.

In lagoons, as in any other closed water reservoirs, water levels under the influence of wind stress will rise at the downwind side of the lagoon and decrease at the upwind side. This hydraulic head sets up a longitudinal pressure gradient that modifies the current structure in the lagoon. Furthermore, when the depths near the sides of the lagoon are shallower, wind-driven coastal currents are set up in the direction of the wind, and an upwind return flow must be expected in the deeper central parts through continuity. This return flow can occur at a depth when the lagoon is slightly stratified in the vertical (e.g., in the presence of a thermocline and/or a halocline). See Chapter 9, Case Study 9.2, for an example of equilibrium current structure.

As a result, a rather complicated current pattern arises, which is not always possible to describe without a numerical model that takes into account the morphological characteristics, wind conditions, density stratification, and other parameters in the lagoon.

*6.3.3.3.3 Gradient Flow Development*

When the wind or the pressure gradient changes suddenly, the adjustment process of water fluxes under the influence of the Earth's rotation to its new equilibrium is accompanied by oscillations about this new equilibrium state. These oscillations, which are eventually damped out, are called inertial oscillations. They have a characteristic period called the inertial period, which depends on latitude as[18]

$$\tau_{\text{inert}} = \frac{2\pi}{f} = \frac{2\pi}{2 \cdot \Omega \cdot \sin(\varphi)} \tag{6.43}$$

For example, for the Vistula Lagoon (55° latitude), the inertial period is exactly 14.7 h, whereas for latitudes near 30°, this period is around 24 h. In the latter case, provisions must be made during the data analysis to distinguish between inertial oscillations and diurnal tidal oscillations when present. The rate at which inertial oscillations are damped depends on their period. For example,[18] it can be shown that, for a liquid on a rotating Earth, the time scale of the current adjustment to a new pressure gradient is $t \gg 1/f$. A good approximation is $t \sim 3$ to 5 inertial periods. A minimal estimate for the time of current adjustment can be half of a one-node seiche period (see Section 6.3.3.5) because level inclinations during a storm surge that generates a pressure gradient do not develop faster than this seiche period.

When the depth of a lagoon is deeper than $h_E$, a distinction must be made between two time scales of current adjustment in the upper layer. These time scales correspond to (1) Ekman adaptation due to the influence of the Coriolis force and internal frictional stresses in the upper Ekman layer, and (2) inertial motion due to the influence of the Coriolis force and pressure gradients. These two factors act independently on the currents. Their combination may produce one of the following results:

- A rapid formation (period less than the inertial period) of variable currents subjected to inertial oscillations adjusting to a constant Ekman solution over time scales $t \gg 1/f$
- A rapid dampening of inertial oscillations leading to Ekman currents within an Ekman adjustment time scale

A current adjustment time scale that combines the effects of both inertial acceleration and vertical turbulent diffusion (friction), and which gives a minimum estimate of the time during which 75% of the current changes following a wind change, is[19]

$$T = \frac{2 \cdot \pi}{f \cdot \sqrt{1 + 0.5 \cdot (0.5 h_E/H)^4}} = \tau_{\text{inert}} \cdot \left(1 + 0.5\left(\frac{0.5 \cdot h_E}{H}\right)^4\right)^{-0.5} \quad (6.44)$$

where $h_E$ (m) is the Ekman depth (Equation (6.40)) and $H$ is the depth (m) where the turbulence coefficient becomes zero, or the lagoon depth.

For example, the result of using both Equation (6.38) and Equation (6.44) for the Vistula Lagoon (Table 6.11) shows the current adjustment time corresponds closely to estimates in Table 6.10. Therefore, the current adjustment time in the

TABLE 6.11
Friction Layer Current Adjustment Times (*h*) for the Vistula Lagoon
Using Equations (6.40), (6.43), and (6.44)

| Depth of Turbulent Layer, m | Wind Velocity, m s⁻¹ | | |
|---|---|---|---|
| | 5 | 10 | 15 |
| 2 | 3.02 h | 1.54 h | 1.02 h |
| 3 | 4.4 h | 2.28 h | 1.54 h |
| 5 | 6.81 h | 3.72 h | 2.53 h |

Vistula Lagoon is primarily determined by the vertical shear stresses. The reason is that this lagoon is shallow and the frictional boundary layer occupies the whole lagoon volume.

### 6.3.3.4 Wind Surge

The local level inclination develops at all times when the wind acts on the water surface. Its upper bound value may be estimated from

$$\frac{d\varsigma}{dx} = \frac{\tau_w}{g \cdot \rho \cdot H} \tag{6.45}$$

where $\varsigma(x)$ is the water surface level, $x$ is the horizontal coordinate in the direction studied (not obligatory in the wind direction), $g$ is the gravity acceleration ($\approx 9.81$ m s$^{-2}$), $\rho$ is the water density (approximately $1.000$–$1.035$ kg m$^{-3}$), $H(x)$ is the depth, and $\tau_w$ is the projection of the wind stress vector along $x$.

The wind stress vector along the wind direction may be estimated as

$$\overline{\tau_w} = C \cdot \rho_a \cdot |W_a| \cdot \overline{W_a} \tag{6.46}$$

where $\rho_a$ and $W_a$ are the air density and wind velocity, respectively, and $C$ is the dimensionless wind drag coefficient. Several empirical values have been proposed for $C$ (as, for example, $0.0026$[20]). More elaborate expressions can be found, where $C$ increases with wind speed as[18,21]

$$C = 1.1 \cdot 10^{-3}, \quad \text{for } |W_a| < 6 \text{ m s}^{-1}$$
$$10^3 \cdot C = 0.61 + 0.063 \cdot |W_a| \quad \text{for } 6 \text{ m s}^{-1} < |W_a| < 22 \text{ m s}^{-1} \tag{6.46a}$$

Considering the standard of hydrometeorological observation when wind speed is measured at a height of 10 m ($W_{10}$), the value of the drag coefficient $C_{10}$ may be estimated according to the Garratt recommendation:[15]

$$C_{10} = 0.00075 + 0.000067 \cdot |W_{10}| \text{ for } 4 \text{ m s}^{-1} < |W_{10}| < 21 \text{ m s}^{-1} \tag{6.46b}$$

The wind-induced level set up on opposite sides of a lagoon ($\Delta h$) can be estimated by integration of Equation (6.45) from side to side along the length $L$ in the direction studied ($x$), considering the wind component along this direction and depth variations, or by using a simple formula with an average depth ($H_{avg}$):

$$\Delta h = \frac{\tau_w \cdot L}{g \cdot \rho \cdot H_{avg}}, \tag{6.47}$$

It follows from Equation (6.45) and Equation (6.47) that the level set up is proportional to the wind and inversely proportional to the lagoon depth, or to the lagoon shallowness ratio ($HL^{-1}$).

For example, let us consider two different water basins, the Vistula Lagoon (91 km long, 2.7 m average depth, and $HL^{-1} = 2.96 \cdot 10^{-5}$) and Lake Constance (Austria/Germany/Switzerland; 65 km long, 120 m average depth, and $HL^{-1} = 184 \cdot 10^{-5}$), and a wind stress of $3.5 \cdot 10^{-5}$ Pa corresponding to a wind speed of 15 m s$^{-1}$. The maximum wind-induced water level set-up ($\Delta h$) will be about 1.2 m for the Vistula Lagoon (which actually does occur during extreme wind surges) and only 0.02 m for Lake Constance.

Away from shore, the geostrophic current solution may be used for estimation of maximum currents or flows perpendicular to the line of wind-induced level set-up:

$$U_G = (g/f) \cdot \frac{d\varsigma}{dx} \qquad\qquad (6.48)$$

where $g \approx 9.81$ m/s$^{-2}$, $f$ is the Coriolis parameter, and $d\varsigma/dx$ is the surface inclination set up by the wind. The equation, however, leads to overestimation of speed due to the absence of bottom friction in the geostrophic solution. For example, the geostrophic current estimated for the Vistula Lagoon is about 1 m s$^{-1}$, whereas for Lake Constance, which is deep, it gives a more realistic value of 0.03 m s$^{-1}$.

The duration of wind action is of specific importance for the lagoon water dynamics as the wind stress (1) directly generates currents when the wind starts to blow and (2) establishes a level set-up that sustains a compensation near bottom flow within the time scale of long gravity wave propagation. For example, for the Vistula Lagoon (91 km × 9 km), this time scale is on the order of 5 h for alongshore winds and 0.5 to 1 h for transversal winds. In the Darss-Zingst Bodden Chain Lagoon (55 km × 3.6 km), there are several subbasins (3 to 10 × 3 km), and it will have a shorter alongshore length scale. The time for appearance of recurrent bottom flows in this case will be shorter: 0.5–1 h for alongshore winds and 12 to 30 min for transversal winds.

### 6.3.3.5  Seiches or Natural Oscillations of a Lagoon Basin

One more time scale that can be estimated for a lagoon is its natural (or eigen or fundamental) period of oscillation. It is the period of the free oscillation, which develops in the lagoon basin when, for example, the wind suddenly stops blowing. In the open ocean, inertial oscillations waves will develop. However, in confined areas such as a lagoon basin, these free waves, or inertial oscillations, combine to form standing waves whose characteristics depend on the size of the lagoon. These waves are called "seiches" and represent the eigenmodes of the particular basin under consideration. The only difference between these waves and their free open ocean counterparts is that their frequencies are determined by the size and the geometry of the basin.[22] Once the wind stops blowing, and in the absence of other forcing functions, these waves will control the currents in the lagoon.

Generally, this eigenvalue problem must be solved numerically for real lagoon basins. However, a preliminary estimate of the natural period of oscillation can be obtained for a constant-depth rectangular basin.[23] Equation (6.49) provides an estimate for the period of the complex seiche resulting from the superposition of the $N$-node longitudinal and the $M$-node transverse barotropic seiches:

$$T_{\text{barotrop}}^{(MN)} = \frac{2}{\sqrt{gH}} \left( \frac{M^2}{a^2} + \frac{N^2}{b^2} \right)^{-0.5} \tag{6.49}$$

where $g \approx 9.81$ [m s$^{-2}$], $H$ [m] is the depth of the basin, and $a$ and $b$ [m] are the width and length of the basin, respectively. The maximum period corresponding to a one-node longitudinal or transversal seiche is obtained by setting $M = 0$ or $N = 0$, respectively. The lowest nodes generally represent the seiche adequately.[22]

Seiches can exist as barotropic surface seiches, involving motions of the whole water masses that reach their maximum amplitude at or near the free surface, or as baroclinic internal seiches associated with the density stratification, with currents reaching their maximum amplitude at or near the pycnocline.[22]

For a two-layer basin, the period of an internal longitudinal one-node seiche may be estimated by[24]

$$T_{\text{barocl}}^{(1)} = \frac{2 \cdot b \cdot \sqrt{H_{\text{up}}^{-1} + H_{\text{down}}^{-1}}}{\sqrt{g \cdot (1 - \rho_{\text{up}}/\rho_{\text{down}})}} \tag{6.50}$$

where $H_{\text{up}}$, $H_{\text{down}}$, and $\rho_{\text{up}}$, $\rho_{\text{down}}$ are the thickness and density of the upper and the lower layer, respectively.

The phase speed of a barotropic seiche is $c = \sqrt{g \cdot H}$, where $g$ is gravity and $H$ is the water depth. For internal (baroclinic) seiches, the phase speed is reduced by one or two orders of magnitude because $g$ is replaced by the reduced gravity $g' = g \cdot (\Delta\rho/\rho)$, where $\rho$ is the average density and $\Delta\rho$ is the density difference between the upper and lower layers. For example, in elongated basins of 10 to 100 km in length, the period of oscillation of a barotropic seiche will be on the order of minutes or hours, while that of baroclinic seiches will be tens of hours or days. These different time scales are also associated with the effects of the Coriolis force, which modify the corresponding gravity waves into Kelvin or Poincaré waves, respectively.[22]

For example, in the Vistula Lagoon ($a = 91$ km, $b = 9$ km, $H = 2.7$ m) the natural period of oscillations are 9.8 and 1.0 h for the barotropic longitudinal and transversal seiches, respectively.

The simple seiche theory[23] does not provide information on the seiche amplitude. It totally depends on the amount of energy supplied to the basin by the wind action. However, if the maximum amplitude ($h$) is known, for example, from the measurements, the corresponding current at the seiche node points (where the amplitude is zero and the current is maximum) may be estimated by the formula

$$V = \frac{h}{4 \cdot H} \sqrt{g \cdot H} \tag{6.51}$$

A seiche will develop in a lagoon when the period over which the wind stops is smaller than half the period of its fundamental (or one-node) mode. In this case, the water surface relaxes from its previous wind-induced inclined position and starts to oscillate freely.

Using Equation (6.47) and Equation (6.51) for the Vistula Lagoon, a 10 m/s wind along the longitudinal axis will generate a corresponding level deviation of 25 cm in amplitude. Assuming this wind stops suddenly, the maximum nodal current associated with the pure seiche can be estimated to be 20 cm/s, according to Equation (6.51).

In contrast to a deep lake, it is rather difficult to observe clear seiches in shallow lagoons. Bottom friction tends to quickly damp the seiche oscillation before some periods of seiche oscillation are realized.

### 6.3.3.6  Wind Waves

Wind waves in lagoons are controlled by the barrier islands, the depth, and the limited fetch. As they progress from the sea into the lagoon through its entrance, sea waves are significantly deformed. High waves are damped due to bottom friction and wave spectra adjust to local bathymetry in the lagoon area adjacent to the barrier islands. This sudden constriction prevents sea waves from progressing deep inside the lagoon. The destruction of oceanic waves at barrier island boundaries results in a wave-pumping effect into the lagoon, which may be the main flushing mechanism in some coral reef lagoons.[15a]

Therefore, lagoon surface waves are generated mainly by local winds, and wave parameters quickly reach their upper limits for increasing wind speeds, wind duration, and fetch. The shallowness of the lagoon also increases the wave sharpness, and the ratio between the wave height and the wavelength can reach $\frac{1}{7}$, as, for example, in the Vistula Lagoon.[25] There are no residual long waves (or swell) in a lagoon after the wind relaxation. Much like seiches, they are rapidly damped out by bottom friction.

There are many empirical, semiempirical, or spectral methods for wind wave simulation in shallow waters.[15,26–29] These take into consideration all stages of wave development, wave refraction with depth, coastal configurations, and transformation into the surf zone.[28] A simple formula for the upper limits of the average wave height ($h_{avg}$ [m]) and the average wave period ($\tau_{avg}$ [s]) for a depth $H$ [m] and a wind speed $W_a$ [m s$^{-1}$] is[30]

$$h_{avg} = 0.062 \cdot W_a^{2/5} \cdot H^{4/5} \qquad (6.52)$$

$$\tau_{avg} = 1.46 \cdot H^{1/2} \qquad (6.53)$$

Although Equations (6.52) and (6.53) do not consider the wind fetch and only provide the upper limit of the average wave parameters at the local depth, they can be used in shallow waters where waves rapidly become depth limited. The relation between average wave length ($\lambda_{avg}$) and average wave period ($\tau_{avg}$) in shallow waters can be extracted from the following implicit relationship:[30]

$$\lambda_{avg} = \frac{g \cdot (\tau_{avg})^2}{2\pi} \tanh\left(\frac{2\pi \cdot H}{\lambda_{avg}}\right) \qquad (6.54)$$

Results of field measurements show that the average wave length can exceed five times the depth in shallow areas. The above relationship can then be further simplified for practical use, becoming similar to that for deep water:

$$\lambda_{avg} = \frac{g \cdot (\tau_{avg})^2}{2\pi} = 1.56 \cdot (\tau_{avg})^2 \tag{6.55}$$

since $(\tanh \frac{2\pi \cdot H}{\lambda_{avg}})$ is very close to 1 up to $H/\lambda = 1/5$.

The information about wind waves is important for estimating the depth of the upper mixing layer, where high turbulence coefficients are expected in the presence of waves. A reasonable estimation for the depth of this mixing is about one quarter of wave length. Actually, a wave becomes a shallow water wave when the water depth is less than half a wave length.

Suspension of bottom material by wind waves is another important parameter that can affect the water quality in a lagoon. Those areas affected by wave suspension under constant wind conditions can be delimited[31,32] on the basis of the simple criterion that waves are able to suspend fine bottom material if the water depth is less than one quarter of the wave length.[33]

We emphasize that all the relations developed in this section apply only to regular waves, using wave length as a key parameter. But, in reality, wave motion is highly irregular and various waves can be present at the same time. In this case, one should use instead the maximum wave heights and lengths, with low probability of occurrence (e.g., 1 or 0.1%). Relations for average waves can then be reformulated from known relations between maximum waves and average waves. These can be found in most wave calculation manuals.[29]

### 6.3.3.7 Coriolis Force Action

The Coriolis force will affect currents in lagoons as in other water bodies.[4b] The Coriolis force results from the rotation of the Earth, causing currents in the northern (southern) hemisphere to be deflected to the right (left). In stratified or estuarine lagoons, the effect is to move less-dense water to the right (left) looking seaward in the northern (southern) hemisphere. The interface between waters of different densities tends to be sloped as the pressure gradient force reaches equilibrium with the Coriolis force to achieve a geostrophic balance.[4c]

The Coriolis force starts to become significant for the dynamics of a lagoon when the width is greater than five $r_c$, $r_c$ being the radius of inertial circle, and dominates when the width is greater than $20 \, r_c$.[4c,34,35] In other words, the Coriolis force is considered important for a low Rossby number $R_\varphi$, $< 0.1$[4c,36] where

$$r_c = \frac{u}{f}, \quad \text{and} \quad R_\varphi = \frac{r_c}{L_{width}} \tag{6.56}$$

and $f$ is the Coriolis parameter (Equation (6.38)), $u$ is the mean water velocity (m s$^{-1}$), and $L_{width}$ is the characteristic lagoon width scale. For example, for the Vistula Lagoon,

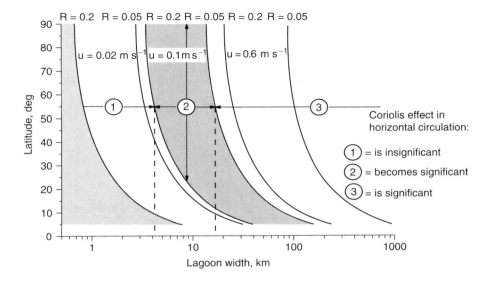

**FIGURE 6.16** Range of significant–insignificant Coriolis effects as a function of latitude and lagoon width for three fixed characteristic velocities.

where the latitude $\varphi$ is 55°, the Coriolis parameter $f$ is $11.94 \cdot 10^{-5}\ \mathrm{s}^{-1}$ and the characteristic velocity is within the range of 0.05–0.10 m s$^{-1}$, the radius of inertial cycle can be estimated as $r_c \approx 400\text{–}800$ m, and $R_\varphi \approx 0.044\text{–}0.088$. In this case, the Coriolis force is significant in the whole lagoon since the average width is 9 km.

Strictly speaking, the Coriolis force is at work in any basin at any movement. The period of the Coriolis cycle equals $2\ \pi f^{-1}$, which yields values of 23 h 56 min, 14 h 37 min, and 12 h and 44 min for latitudes of 30°, 55°, and 70°, respectively. This means that any advective movement of equivalent or longer time scales will be affected by the Coriolis force. Since the horizontal spatial scale in a lagoon is fixed, the slower the movement is in the lagoon, the greater the influence of Coriolis force on water dynamics and mixing processes. Although this may seem contrary to the idea that Coriolis acceleration is proportional to speed, this can be seen in Figure 6.16 and Figure 6.17.

Assuming that the wind-induced water velocity in a lagoon is 0.1 m s$^{-1}$, Figure 6.16 shows that the Coriolis effect at a fixed latitude is insignificant when a lagoon width $L_{\mathrm{width}}$ is less than $5 \cdot r_c$ or $R_\varphi > 0.2$ starts to be significant (shaded strip) when $5 \cdot r_c < L_{\mathrm{width}} < 20 \cdot r_c$, or $0.05 < R_\varphi < 0.2$, and dominates when $L_{\mathrm{width}} > 20 \cdot r_c$, or $R_\varphi < 0.05$. Similar arguments can be made for the same lagoon width at different latitudes. For example, the Coriolis effect is important for all lagoons with a width of 10 km only if they are located at latitudes greater than 45°. Furthermore, considering a velocity of 0.02 m s$^{-1}$ as a lower limit of natural lagoon velocities (as measured by most oceanographic instruments), Figure 6.16 indicates that the circulation in any lagoon with a length scale less than $L_{\mathrm{min}}^*$ will never be affected

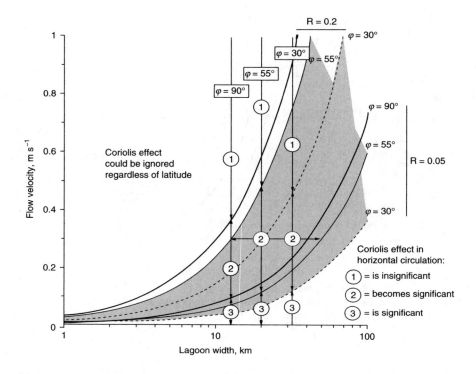

**FIGURE 6.17** Range of significant–insignificant Coriolis effects as a function of velocities and lagoon length scales (width) for three characteristic latitudes ($\varphi = 30°$, $\varphi = 55°$, $\varphi = 90°$).

by the Coriolis force, where $L^*_{min}$ is defined as

$$L^*_{min}[m] \approx \frac{686}{\sin(\varphi)} \qquad (6.57)$$

On the other hand, water movement in different parts of a lagoon may adopt different velocities such that the Coriolis force will only be important for movements of certain velocity. Figure 6.17 illustrates the ranges of velocities influenced by the Coriolis effect in lagoons of different scales. For lagoons of a given length scale, the velocities below the shaded zone will be significantly influenced by the Coriolis effect while the ones with velocity scales above the shaded zone are not influenced by it. The position of shaded zone selected for the intermediate velocity range depends on the latitude only. We can also see from Figure 6.17 that once a length scale $L_1$ is determined and the corresponding speed extremes $u_1$ and $u_2$ are determined for some latitude $\varphi_1$, the same $u_1$ and $u_2$ will correspond to another length scale $L_2$ at another latitude $\varphi_2$. In other words, at any other latitude $\varphi_2$, there will be a length scale $L_2$ associated with the same velocity interval where the Coriolis effect starts to be important. One can find this new length scale $L_2$ for which the Coriolis parameter will influence velocities in the range of $u_1$ and $u_2$ as

$$L_2 = L_1 \cdot \frac{f_1}{f_2} = L_1 \cdot \frac{\sin(\varphi_1)}{\sin(\varphi_2)} \qquad (6.58)$$

Finally, even if the Coriolis force might be important in lagoons when speeds are relatively low, it is important to remember that water motion in lagoons is governed not only by Coriolis accelerations but also by many terms in the equations of motion (see Chapter 3) and that it is the relative importance of these terms that determines which ones actually force the motion. Normally, for shallow water systems such as lagoons, we expect the pressure gradient terms to be balanced mainly by bottom friction and, to a lesser extent or in some tidal phase conditions, by nonlinear terms. Therefore, the Coriolis terms will influence the motion in lagoons only when they are on the same order of magnitude as the pressure gradient or friction terms.

### 6.3.4 OBJECTIVES OF MODELING

A modeling activity should start with a clearly defined objective. It is important for the user to clearly understand what the numerical model will be used for, i.e., what the problems in the area are and what kinds of answers are expected from the modeling activity. This is not always easy because a precise objective definition requires some skill and experience. For example, "to investigate a wind surge problem" or "to study heat transfer processes in the area" are not objectives as such but rather general directions of investigations.

The objectives should be formulated in such a way that they result in an expected model output, either in the form of justified recommendations, quantitative relationships between variables, or schemes and charts, etc. Such concrete objectives will optimize the time spent on solving the task and will keep the researcher on the right track from start to finish.[37]

From the outset, researchers should also make it clear what kind of results they intend to produce with the model. Is the goal maximum water levels and currents at certain points during a design event? Is it frequency distribution of the water levels for a certain period? Is it an impact study of the construction of an offshore breakwater on the eddy circulation in some approach channel, or some other clearly defined question?[37]

In some instances the modeling approaches or strategies may be changed during the course of investigation according to the results obtained. The objectives, however, must never change, unless the resources available become exhausted during the course of the study. Examples are the lack of time to finish the study, the lack of additional computer resources, or accidental loss of data. In such cases, conditions under which the original objectives may be attained must be stated.

There are as many objectives for a study as there are practical studies. One example of a short and precise objective could be the following:[37]

*The overall objective of the present study is to identify the best layout of the future cooling water system, on the following basis:*

- *compliance with the applicable environmental standard, both for the present power production and after planned capacity extension of the plant*
- *minimal recirculation of the cooling water, and minimal construction and operation costs*

*The planned target is a maximum excess temperature of 4°C at 500 m from the outlet. The future full capacity operation discharge has been specified at 151 m³/s with a temperature increase of 7.9°C, but higher and lower discharges should also be considered.*

The next example shows how it is possible to clarify a general objective statement such as *"to study the current structure around the island in the lagoon area"* into more precise statements of purpose:

- To establish a pattern of vertically integrated currents around the island in the lagoon area under typical wind conditions
- To estimate the characteristic time of current adaptation to wind variations
- To represent the results as both a set of current schemes for more statistically probable winds and plots for adaptation time against initial wind direction
- To calculate the barotropic steady-state fluxes through the channel between the island and shore for eight basic wind directions and represent these results in the form of a flux rose

It should be emphasized here that the formulation of precise objectives requires some advanced knowledge on the processes in the lagoon. This is the real reason so much attention must be paid to the description of preliminary hydrographic and processes, as outlined earlier in this chapter. Generally, the methods applied to undertake a given task do not depend on the objectives. It is understood that incorrect methods are not discussed here and that modeling is implied as the main approach. It should be clear, however, that modeling itself is not sufficient to solve most of the practical problems. It is a combination of modeling, data analysis, and expert assessment that will ultimately provide the required solution to the problem at hand.

In conclusion, some examples of typical lagoon hydrodynamic and transport modeling tasks are provided as guidelines for the formulation of modeling objectives.

*Analysis of typical situations.* In this case, either the current structure, the water exchange, the spatial distribution of chemical substances, or some other processes are simulated, using the most probable combination of driving forces (maximum/minimum of tidal level variations, typical wind for a given season, maximal river discharge, etc.). The simulation of extreme statistics for some parameters (e.g., water level) falls into this group. Usually, the driving forces of given statistical probability are determined first from data analysis. Then, these constant forces can be imposed on a nonstationary model.

*Environmental impact assessments.* Here, a set of scenarios is examined using some criterion in order to find the best solution to an environmental problem. Usually, time-dependent model solutions resulting from some forcing factor (e.g., new positions of wastewater inlets or some dredging work in the lagoon) are compared to a "basic state," which corresponds to some initial reference situation. These can be the situations of previous years, the absence of wastewater inlets, or other situations. Both the scenario and the basic solution usually represent the diurnal, synoptic, seasonal, annual, or inter-annual variability of the variables under study. These variables have to be identified in the formulation of the objectives.

*Long-term forecast modeling.* Here, the model solution is obtained from long-term forecasts of the driving forces. The accuracy and confidence of a forecast model will depend on both the accuracy of the driving forces forecast and on the model sensitivity to these forces.

*Operational modeling.* This type of modeling is used to provide real-time or short-term forecasts on the basis of real-time weather and boundary data measurements. For example, operational modeling can provide real-time information on daily currents and waves in a lagoon for safe navigation, fishing, or recreational activities. Operational modeling is also needed for contingency analysis (e.g., in the event of an oil spill), when the transport model solution must be available immediately. In this case, the constraints on the solution accuracy and the simulation time are very high because the model results are needed quickly for emergency actions, which are usually very expensive.

*Diagnostic analysis or analysis with data assimilation.* During this analysis the spatial distributions of some variables are obtained by imposing existing data on other variable distributions. One example is the study of the volume flux structure resulting from an imposed (constant or variable) density distribution to the model throughout the simulation.

*Basic studies of the hydrodynamic and transport processes in a lagoon.* In this case, all physical factors affecting these processes are studied. The researcher's task is to seek model solutions for some simple initial and boundary conditions (e.g., annual average wind and river discharge, on simple tidal mode) and to understand the lagoon response to the various factors. For example, the model can be used to compare current patterns caused only by river discharge, or by wind action, or by tidal inflow/outflow. Another example is the study of the influence of the Coriolis force on currents and transport in the lagoon.

The above list indicates some of the main types of environmental analysis tasks that can be solved by numerical modeling. Real practical applications, which are more diverse, usually will fall into one of the above types of modeling activities.

## 6.3.5 Recommendations for Model Selection

The information presented here should help the readers get their bearings in the choice of model. In the authors' opinion, it is not possible to make rigid recommendations for model selection that will be applicable to all modeling cases. The users should decide themselves what model is more appropriate to the task to be solved, what model is more accessible, etc. Therefore, direct recommendations on model choices will not be provided; instead, some information is presented to facilitate the model selection process. The approach proposed is that of *reductio ad absurdum*. The readers will find below a list of different opportunities and they could exclude everything inappropriate to reduce the scope of their choice. Some recommendations based on the authors' experience are also presented.

### 6.3.5.1 Selection Possibilities for Hydrodynamic and Transport Models

According to the spatial dimension of studied domain, we have the following spectrum of models:

- A zero-dimensional (0D), or one-box model, where the lagoon is represented as one box and all the variables depend only on time
- A one-dimensional (1D) model lagoon, considered as a transversally and vertically uniform lagoon, where all the variables are functions of longitudinal coordinates and time
- A two-dimensional (2D) model lagoon, considered as a vertically or a transversally uniform lagoon, where time-dependent variables are functions of plane $(x, y)$ or vertical slice $(x, z)$ coordinates, respectively
- A three-dimensional (3D) model lagoon, where at least one time-dependent variable is a function of all spatial coordinates

The model solutions could be subdivided according to *time resolution* into the following:

- A static solution (this solution is obtained from a governing equation that is made time independent by averaging in time or by assuming that the lagoon system has reached steady state and the variables can be considered as time independent)
- A dynamic solution (variables are considered as time dependent; driving forces are also time-dependent functions)
- A steady-state dynamic solution (time dependent but periodic in time steady solution during periodic or constant driving forces)

The following *simulated variables* may be included in the model:

- Level variations (in the 0D, 1D, and 2D models)
- Currents (in the 1D, 2D, and 3D models) or fluxes (in the 1D and 2D models)
- Wave parameters (2D surface field)
- Dissolved or suspended matter concentrations (in the 0D, 1D, 2D, and 3D models) and fluxes of them (in the 1D, 2D, and 3D models)
- Dissolved or particulate matter in terms of concentrations (in the 0D, 1D, 2D, and 3D models) and fluxes (in the 1D, 2D, and 3D models)

Including the following *effects* in the model defines its structure and complexity:

- Variations of variables in time
- Variations of variables in space
- Advection of momentum
- Advection of admixture
- Diffusion of momentum
- Dispersion of admixture
- Wind and bottom friction effect
- Wind waves and their interaction with general water movements
- Coriolis force
- Buoyancy effect due to stratification (i.e., vertical variations of water density)
- Gravity force influence due to spatial variation of water density or/and water level
- Eigen or induced lagoon behavior

In any case, the water movement in real lagoons is time dependent and three dimensional, and it develops under all forcing factors found in nature. We can use fewer spatial dimensions (geometry facilitation) or physically simplified models (some forcing factors or processes are reduced) only for estimation of the real situations with some uncertainty. In the most favorable cases, the residual error between the model solution and reality may be less than the accuracy required for the task, but that success depends on many circumstances.

### 6.3.5.2 Possible Simplifications in Spatial Dimensions

The real current structure in lagoons should be simulated only by the 3D model, but some simplification is possible because in many tasks the current solution itself is not needed. For instance, the fluxes between subbasins of a lagoon are the important variables needed for the ecological modules, so it is possible to reduce the spatial dimensions in the ecological modeling.

In an elongated lagoon, the solution given by horizontal 2D model (vertically integrated motion) for fluxes through a transversal slice is nearly the same as a 3D model would provide. Only in cases of the narrowing of the lagoon basin (e.g., by an island or a peninsula), when the cross section is reduced by, say, 80%, the difference between fluxes computed from 2D and 3D solutions becomes considerable when the winds are longitudinal. In the case of transversal winds, the 2D approach gives large errors in the flux solution.[38]

It could be possible to use a 1D model for advection–dispersion problems if the basin is essentially elongated: $p_{shell}^{max} \geq 10 \cdot p_{shell}^{min}$ or $\max(a, b) \geq 10 \cdot \min(a, b)$ (see the parameters in Table 6.1). In other situations it is more reasonable to use a 2D or 3D shallow water approach.

The 2D approach is useful for modeling of advection–dispersion transport and wind–wave resuspension if spatial mean water depth is less than wave penetration depth (see Section 6.3.3.6) for average wind, or if the water depth is much less than the depth of the friction layer (see Section 6.3.3.2).

For a leaky lagoon the correct inclusion of the inlets between the barrier islands in the computation defines the accuracy of the level variation and current predictions. Here a simple rule can be applied: The more leaky a lagoon, the bigger the adjacent sea area to be included in the computational grid.

Only a 3D approach is to be used if the local current or fluxes structure is studied in the vicinity of the considerable local depth variations.

It is possible to use the ratio between the time scale of advective transport and the vertical turbulent diffusion as one criterion for selecting the spatial dimension of the hydrodynamic model.[39] If $U$ and $L$ are typical velocity and length of advective admixture transport, then $T_{adv} = L/U$ is the characteristic time of advection. Similarly, if $H_0$ is the lagoon depth and $K_Z$ is the vertical turbulent diffusion coefficient, then the time $T_{diff} = H_0^2/K_Z$ characterizes the vertical diffusion processes. The dimensionless ratio $\Delta = T_{diff}/T_{adv}$ illustrates the relationship between the velocities of admixture spreading in depth and length. The use of a horizontal 2D model is reasonable if $\Delta << 1$. In all other cases, 3D models give more reliable results.

### 6.3.5.3  Possible Simplification in the Physical Approach

The spatial scale of lagoons influences physical processes and the general hydro-thermodynamic problem. For example, closed lagoon boundaries give rise to gradient forces due to wind surge level inclination and inflow of rivers. In the same way, the lagoon shallowness provides the basis for vertical mixing as turbulence generated in the upper and bottom friction layers occupies nearly all the water column. In this case, when freshwater inflow is low or wind–wave mixing is strong enough to overcome the density stratification, the general task may be reduced to a barotropic problem with density varying only in the horizontal direction.

It is recommended to include the Coriolis force in 3D or 2D models regardless of the ratio between spatial scale of the lagoon to the value of the Rossby radius (see Section 6.3.3.7). The reason is that the Coriolis term (similar to diffusion terms) inserts nonlinear cross-linkage between equations for three components of momentum (see Chapter 3) and the same one between water velocity components. Even if a scale analysis shows that the Coriolis term is smaller than other terms in the equation of motion, it provides a qualitative new linkage between velocity components and defines the vorticity of currents, especially at the areas with low motion (see Section 6.3.3.7).

The substitution of the real task by a steady-state or static problem where time dependence is omitted is reasonable only for barotropic solutions. However, it is not reliable for baroclinic situations where the complete time-dependent problem must be solved because even if the boundary conditions are time independent, the baroclinic currents never tend to steady-state solution. This is a fundamental property of fluid dynamics. Practically, in the presence of winds, tides, or varying freshwater input the water motion in a lagoon is never kept constant; it varies in time and space.

### 6.3.5.4  Possible Simplification According to the Task
### To Be Solved

The task of estimating the water balance for a lagoon, especially at seasonal or inter-annual time scales, may be solved using the 0D level. The accuracy of this solution will increase as the lagoon becomes more leaky.

For tasks involving lagoon water quality, 2D hydrodynamic and advection–dispersion models are usually enough to resolve the seasonal variations of simulated parameters. In that case, the model should be calibrated using some conservative tracer time series such as the seasonal salinity evolution. A 2D time-dependent hydrodynamic approach is also sufficient to simulate wind surges, tidal level, or current variations in a lagoon. The tidal water exchange between a lagoon and its adjacent coastal waters is also well simulated using a horizontally 2D approach. The same is true for nontidal lagoons where wind surge is the main external force.

However, when a two-layer circulation is suspected in the lagoon or at its leaky opening, the use of a 3D model is recommended. In some cases, local winds and bathymetry combine to produce a 3D current structure in the deeper basins of the lagoon.[40] This structure can only be simulated using a 3D hydrodynamic model (see Chapter 9.2, the Grande-Entrée Lagoon case study). Tasks involving short-term forecasts following accidental oil spills definitely require a 3D approach for the hydrodynamic and transport problems because of the possible different directions advective transports can have at different depths. These must be simulated precisely in order to predict the fate of the spill and optimize timely recovery operations. Finally, tasks involving the study of water exchange between the subbasins of a lagoon should be solved in 3D for precise short-term simulations and in 2D for simulations involving seasonal variations.

### 6.3.5.5  Computer, Data, and Human Resources

Recent developments in computer hardware have shown that processing speed and random access memories are increasing at a significant rate. Therefore, when a model is planned to be used for a long period of time, a modeler should not restrict the choice of models based on the model complexity (e.g., choosing 2D instead of 3D). Instead, the choice of the model complexity should depend on the data available to fit the model. Normally, a monitoring program should supply the data needed to calibrate and run the model regularly. The choice of model complexity should also depend on the qualification or the expertise of the model user. Recent models are very complicated tools and despite the easy-to-use graphical interface presented, the user must be able to interpret the modeling results. This requires some hydrodynamic expertise to interpret the results in the frame of the basic physical approach that was used. Therefore, computer hardware resources are not the major limitation for model usage. Although qualified personnel could be a temporal problem, the main difficulties lie in data availability, interpretation, and use within the model. This is a crucial point for model selection.

Models are developing in a way that considers as many significant processes as possible and formulates them as clearly as possible. Undoubtedly, the choice to use complex models (even in their simplest formulation) and to establish advanced data collecting and data supply programs (see Chapter 7) is in all cases more progressive than the use of a restricted simplified model based on the traditional monitoring data. Model simplicity is very attractive to administrators but modeling progress is in teaching environmental engineers to use more complex models instead of the "shortest" but unsustainable application of simple models to a wide range of different tasks.

## 6.4   MODEL IMPLEMENTATION

As described in Chapter 3, the implementation of a numerical model to a specific lagoon requires the specification of the boundary conditions and model parameters specific and suitable for the site and for the conditions envisaged in the simulations.

The bathymetry and the tidal characteristics at the inlet(s) are boundary conditions that can be considered as time independent or deterministic in most sites. On the other hand, river discharge, water column structure at the sea boundary (in the case of 3D simulations), and the atmospheric conditions are time-dependent boundary conditions.

Implementation of a model requires data to specify these boundary conditions, the initial conditions, and the various parameters for process simulation. Additional data are required to calibrate and validate the quality of model results.

### 6.4.1   BATHYMETRY AND THE COMPUTATIONAL GRID

Bathymetry describes the lagoon's geometry and is the basis of the whole modeling procedure. Bathymetry is generally measured by national hydrographic authorities using a fine resolution in order to provide data for navigation charts and to support various coastal engineering works. In terms of bathymetry and grid definition, laterally integrated models constitute a particular case, requiring much less information.

#### 6.4.1.1   Laterally Integrated Models

In the beginning of this chapter it was shown how to build a 1D model and the parameters characterizing the respective grid cross section and cell length. In that case, the volume of each cell was calculated as the product of the cross section by the cell length. Generally, the free surface level varies in time and the model calculates the value of the cross section as a function of that level. Under these conditions, the information required is the width of the cross section as a function of the level. This information has to be supplied at the two tops of each cell or at its center. Intermediate values can be calculated by interpolation.

In 2D laterally integrated models the procedure is similar to that in 1D models. In this case the difference is that a cross section is calculated for each layer, as a function of its thickness. In the case of sigma-type coordinates (see Section 6.4.1.2) the thickness is a function of the surface level for all the layers, while in Cartesian coordinates only the surface layer has a variable thickness.

In both types of models, the typical length of a cell is on the order of magnitude of the width. This does not mean that detailed bathymetric information is not required. In fact, in order to obtain realistic results, the geometry used in the model must represent the lagoon bathymetry. To accomplish that, every critical cross section must be defined as a cross section in the model (a section can be critical because it may be wider or narrower than the adjacent one).

### 6.4.1.2  Horizontal Resolution Models

Horizontal resolution models can be depth integrated (2D) or fully 3D. Here, the computational grid representing the modeled area is formed by rectangular cells in finite-difference methods and by triangles in finite-element methods. The depth is specified in the center of the grid or at the corners and is obtained by interpolation of hydrographic depth soundings or other sources of digitized bathymetry. The horizontal resolution in horizontally resolving models is much higher than that of laterally integrated models, and more detailed bathymetric information is required. In general, "grid-generating programs" create these computational grids. These programs require the supply of a detailed coastline and they compute the depth of each cell by averaging/interpolating a digitized bathymetry. When the information is scarce, special care has to be taken when verifying the depth generated for each model cell.

In the case of 3D models, once a horizontal grid is generated with a corresponding depth distribution, a vertical discretization has to be defined. Two discretizations can be easily defined: (1) sigma coordinates or (2) Cartesian coordinates. In Cartesian coordinates, the layers are horizontal and except for the surface layer, they maintain the same thickness in time. The surface layer thickness will depend on the free surface position. The bottom layers progressively disappear as the water becomes shallower. In the case of sigma coordinates, the water column is divided into layers of variable thickness (in space and time) in such a way that the ratio between the thickness of a layer and the local depth remains constant in space and time. In other words, in sigma-type models, the vertical resolution and the number of layers are independent of the local depth. This is convenient in systems where density effects play a secondary role compared to topographic effects. Special care must, however, be taken in intertidal areas, where the thickness of each layer approaches zero during the drying procedure. The consideration of two sigma domains in the water column can be a solution.

### 6.4.2  INITIAL CONDITIONS

Initial conditions must be supplied for each state variable (velocity and levels, temperature, salinity, nutrients, etc.). This information has to be obtained from field data measured synoptically. Unfortunately, information on most variables is unknown or available only for a few points, so that assumptions have to be made about initial conditions. For that purpose, properties can be grouped into rapidly dissipative and slowly (or non-) dissipative properties.

Dissipative properties rapidly "forget" the information related to the initial conditions and become dependent in time only on the boundary conditions. Hydrodynamic processes are usually driven by mechanical energy and are highly dissipative. In fact, errors in hydrodynamic initial conditions result in an artificial initial energy supply, which will be transformed into kinetic energy before being dissipated. Stability properties of the model are limited by the values of the velocity. Very unrealistic initial conditions can generate very high velocities and create conditions for numerical instability. For these reasons the most efficient way to initialize a hydrodynamic model is to start from null velocity and a horizontal free surface equal to the average level at the open boundaries.

Properties associated with the ecology of the system have to be initialized with realistic values. In fact, ecological systems are resilient and once they are submitted to a string of perturbations, they do not necessarily recover toward the state they assumed before that perturbation. Initializing ecological models with unrealistic values is equivalent to stating that the system has been highly disturbed. In fine-resolution models, a lot of information must be supplied by interpolating between points where information is available. Specific initialization software tools can simplify the initialization procedure.

Sediment transport models are also very dependent on the initial conditions. If initial suspended matter concentration is overestimated, settling will remove it. Conversely, if initial suspended matter is underestimated, erosion will supply the missing matter (if deposited matter is available). In this case the difficulty is not the initialization of the water column concentration, but rather the initialization of the erodable amount of sediment lying on the bottom. The user has to implement a procedure compatible with the information available and with the objectives of the simulation. If the information is scarce and the aim of the simulation is to obtain realistic values of the concentration in the water column, then a convenient procedure is to assume that there is no net erosion in any point of the lagoon. In this case the model is initialized assuming that there is a thin erodable layer everywhere in the lagoon, and it is run until an equilibrium solution is obtained. After this period, erosion and deposition areas are identified and the model is initialized assuming that there is no erodable sediment in the areas where erosion was identified. If the aim of the model is to simulate erosion processes in the lagoon (e.g., due to anthropogenic modifications of its morphology), a consolidation erosion module of the bottom has to be considered. In fact, after deposition, sediment are submitted to a consolidation process resulting in an increase in the critical shear stress for erosion. To simulate the erosion process an initial vertical profile of consolidation has to be specified.

## 6.4.3 BOUNDARY CONDITIONS

Boundary conditions can be specified in terms of a specified value, a specified flux, or a specified law of property variation at the boundary. The radiation boundary conditions used on ocean hydrodynamic models fit in the last group. A typical coastal lagoon has one or several sea inlets and receives land discharges through one or several rivers. In general, at sea inlets, the most convenient procedure is to specify the values of the properties or to estimate those values as a function of the values inside the modeling area. At the river boundary the flux is the river flow rate times the specific value of the property (concentration in case of a mass). In this case properties in the fresh water are the most convenient boundary condition. When multiplied by the river discharge the condition, in fact, becomes a flux condition.

### 6.4.4 INTERNAL COEFFICIENTS: CALIBRATION AND VALIDATION

Internal coefficients are used to parameterize empirical closures of the processes simulated by the model. These parameters have to be fitted using field data. This procedure is called calibration. The range of variation of each parameter is usually known from the literature. When the calibration procedure suggests that a parameter value is out of that range, a scientific explanation has to be found. In fact the most common reason for calibration values being out of range is the need to include additional processes in the model.

The calibration effort increases in a nonlinear way with the number of parameters (e.g., biological models). In the case of models with simple spatial grids the computational time is small and some automatic procedures can be established to select the best values of the parameters. In real systems horizontal transport simulation is required and a small number of trials can be done.

After calibration the model must be validated using a data set not used in the calibration process. This validation process will guarantee that the parameters are adequate for a range of conditions representative of those found in the system being studied.

## 6.5   MODEL ANALYSIS

### 6.5.1 MODEL RESTRICTIONS

Can natural processes be adequately modeled? This question could be argued at great length within the abstract concept of nature cognoscibility.[41] A practical answer, however, is that some natural processes can indeed be modeled adequately, but only within restricted limits and a predefined accuracy. This section discusses some of the restrictions that arise when attempting to model lagoon processes (Figure 6.18).

Physical restrictions arise when the equations used in the model are not complete, i.e., they do not include all factors influencing the process under investigation. For example, the exclusion of rotation effects of the Earth when modeling small-scale dynamics or the assumption that density is independent of depth when studying shallow-water motions.

Numerical restrictions, on the other hand, are introduced by the numerical schemes used to solve the equations. These schemes require spatial and temporal discretizations, which in turn impose restrictions on the computational grid size and the integration time step.

Mixed restrictions arise when the physics of a process is altered by computational constraints. An example is the phenomenon of subgrid processes. All processes occurring at spatial scales smaller than a computational grid size (or subgrid) are not simulated explicitly but are represented in the primitive momentum and transport equations by some empirical eddy diffusion/dispersion coefficients.

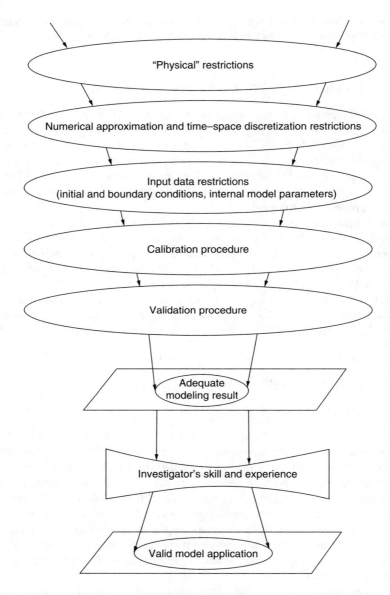

**FIGURE 6.18** Sequence of restrictions along the passage from modeling intentions to a valid model application. Subsequent restrictions narrow the sphere of model applicability. Only the investigator's experience and skills can expand the range of valid model applications.

Finally, some restrictions arising from data availability for initial conditions, boundary conditions, or internal parameter specification are discussed in some detail.

In summary, the solution accuracy should not exceed the range of the spatial–temporal variability of the simulated variables, and a model can only adequately simulate a particular process within specific temporal and spatial scales. Using the numerical methods terminology,[42] the terms neglected from the equations are the source of nonremovable errors in the solution, while the terms conserved are the source of numerical errors. These restrictions are now discussed in some detail.

### 6.5.1.1  Physical Restrictions

Physical equations constitute the heart of any numerical model. It is generally admitted that the physical background of most problems in marine sciences, including hydrology, hydrodynamics, oceanology, or climatology, are well investigated, especially when dealing with "averaged" dynamics and excluding turbulent or nonlinear processes. This physical background is well incorporated into the equations of matter, momentum, and energy conservation, and the nature of the driving forces is well understood.[41]

In some instances, the physics of a model is simplified by neglecting some terms that are thought to be "insignificant" in the equations. This is done in order to reduce the computational resources otherwise needed by a complex problem. However, in doing so, the question of whether the neglected terms are really "insignificant" remains unchecked.[41] Therefore, considering the capacity of modern computer resources and the availability of state-of-the-art models, it is now advisable to choose a numerical model that includes a complete physical background in order to solve any problem under consideration. It must be remembered that computer resources will only improve and very rapidly so.[41]

For example, a physical approach widely used for the study of lagoon hydrodynamics has been the shallow-water vertically integrated approach. The corresponding 2D numerical model has the following basic assumptions:

- The simulated variables (e.g., currents or density) have a homogenous vertical distribution in the water column
- Each point of the studied area has one water layer whose thickness is time dependent
- The flow is time dependent and vertically integrated through the whole layer
- The bathymetry varies slowly
- The water depth is small compared to the horizontal scale of the process under study (e.g., in comparison to the wavelength or the length of domain)

However, in some instances, this 2D approach may not be suitable for some lagoons. For example, when a wind blows along the major axis of a lagoon, the surface level at the downwind end is set up and an upwind return will occur near the bottom when the lagoon depth is on the order of a few meters. In this case, the numerical model should clearly incorporate the physics of 3D motion. In summary, a 3D numerical model is recommended for the study of lagoons where a deep basin (e.g., 5 m) can be found. For shallower lagoons, the 3D model can always be simplified to a 2D model during the simulation set-up by assigning only one layer in the vertical direction.

## 6.5.1.2  Numerical Restrictions

Numerical methods are used in numerical models in order to approximate the differential equations of motion and constituent transport into algebraic difference forms. These are then solved for unknown values at incremental finite points in space and time. However, the difference form approximations introduce numerical constraints and, occasionally, numerical errors.

For example, truncation errors arise when high-order terms are neglected in the approximations. These errors are directly proportional to the spatial grid size and the integration time step. On the other hand, attempts to reduce the time step or the grid size will introduce a number of calculations and thus round off errors. This error arises because the model computations are performed with a fixed number of decimal places, which is directly proportional to the number of calculations involved. A good numerical model should sum all numerical errors and provide the user with information on the space and/or time discretization limits that will minimize the errors for a given domain.

Numerical algorithms used in a model have convergence and sensitivity characteristics. Some widely used numerical schemes are known to converge toward the real solution. This is not the case for some nonlinear problems or some unconventional algorithms. Regarding sensitivity, it is suggested that the numerical methods used should not enhance the physical sensitivity of the problem under consideration, i.e., small variations in model parameters or boundary conditions should not give rise to solutions outside the physical range or to unbounded solutions.

## 6.5.1.3  Subgrid Processes Restrictions

Field measurements have shown that fluid motion in natural basins is mainly turbulent. In open oceans, the turbulent energy cascades down from large-scale gyres to small-scale turbulent eddies. In lagoons, however, only part of that turbulent spectrum will be present. The spatial spectral window is limited, on the one hand, by the size of the lagoon and, on the other hand, by molecular dissipation scales (~1 cm); however, two important processes contributing kinetic energy at intermediate scales can be found in lagoons.[43] These are the transformation of the potential energy into kinetic energy following baroclinic instability at internal Rossby scales (~10 km) and the transformation of the internal wave energy into turbulent kinetic energy due to the Kelvin–Helmholtz instability. In the latter case, horizontal and vertical scales are $L_H = 1$ m and $L_V = 20$ cm, respectively.

In addition, numerical models can describe only that part of the spectrum energy bounded in space at one extreme by the lagoon length scale and at the other extreme by the Nyquist wave number imposed by the grid size. Therefore, processes with wave numbers outside this range must be parameterized in the primitive equations of motion by turbulent diffusion, dissipation, and inter-scale energy exchange terms and are referred to as subgrid processes. In the ideal case, the parameterization of subgrid processes should be based on a detailed analysis of the subgrid vertex dynamics and its relationship with the large-scale processes directly resolved in the primitive models.

One way of reducing residual errors resulting from parameterization is to choose the spatial grid size such that the Nyquist wave number falls within a trough (minimum) of the energy cascade spectrum. This will ensure a distinct separation between motions determined deterministically and parametrically.[43]

Further constraints arise when assigning values to the diffusion coefficients. Usually, these values are not assigned from diffusion–dispersion field experiment results but rather from model numerical stability considerations or from model–data calibration tests.

### 6.5.1.4  Input Data Restrictions

A first type of input data restriction applies to *initial conditions*. How long does a solution "remember" its initial condition? The answer depends on the retention time of the system, which, in turn, depends on the boundary conditions. The retention time is the average time molecules of incoming waters remain in the system.[4b] Normally, a lagoon system will "forget" its initial conditions and will tend toward a state of equilibrium with the boundary conditions. The initial conditions influence the solution only during an initial period of about one to five times the retention time,[44] and then they are "flushed out" rather quickly by advection or mixing.

So, initial conditions can be specified from common sense (average values, for example) or simple balance estimations. Their influence on the solution lasts a limited period that can easily be determined. Therefore, field measurement requirements for initial conditions are minimal. It is sufficient to know an order of magnitude of the values of the required initial parameters.

In some instances, the initial conditions can be important, for example, during short-term emergency forecast simulations or in such specific cases when inflow and outflow are balanced and an initial error in the absolute water-level estimations is conserved during further studies of the system.[44] The initial conditions are also important in systems with long retention times as in some lakes or estuaries, especially if sediment dynamics is included in the simulation (the residence time of contaminants in sediment is extremely long).[44]

Finally, since a numerical model has a short "initial condition memory," initial values can be specified outside a critical range that could otherwise generate numerical instabilities. Practically speaking, after the corresponding initial estimations are "washed out,"[44] it is always possible to discard them from the solution.

A second type of input data restrictions is related to *boundary conditions*. Boundary conditions are the driving forces that cause circulation and water quality changes in a lagoon. Boundary conditions cannot be simulated within a model and must be specified. They include wind stresses and heat fluxes at the air–water interface and inflows, outflows, and difference in water–surface elevations across horizontal water boundaries, as a function of time. In some cases, a boundary condition can be a relationship between two dependent variables, such as flow vs. depth in a rating curve for an estuary.[4c]

Open boundaries of a model domain must not be located near regions of the grid where predictions are critically required.[4a] This is to avoid the propagation of boundary errors to that region of interest. For example, when attempting to model fluxes through a lagoon entrance, open boundaries must be located far offshore, into the sea.

Ideally, the best open boundary conditions for a model domain are field measurements of dependent parameters at these boundaries. An optimal choice can be
a combination of some "typical" and some "extreme" boundary conditions measurements. In the absence of field measurements, one must rely on dynamical boundary
conditions. Some examples are the (wave) radiation, the absorbent, the flow relaxation, the clamped, or the gradient open boundary conditions.[45]

Since the majority of lagoon problems are significantly boundary dependent, the
accuracy of the model solution will depend on the accuracy of the boundary conditions.
For example, the simulation of the water quality in a lagoon will be dependent on the
boundary nutrients loading. In some cases, one must also include nonpoint sources of
nutrient (diffusive) loading (see Chapter 9.4, the example of the Vistula Lagoon case
study). In summary, the accuracy of the model solution will depend, among other factors,
on the accuracy of the boundary condition measurements. However, when measurement
facilities are limited, the relationship between uncertainties in both boundary conditions
and model solutions must be studied using sensitivity or error analysis. This will dictate
the minimal measurement requirements for the desired solution accuracy.

A last type of input data restrictions relates to *internal parameter specification.*
The internal parameters are the independent variables that are normally required to
calculate the dependent variables. These parameters are the bottom topography, the
basin geometry and boundaries, and the flow resistance properties within the water
body, at the water surface and at the bottom.[4a]

The *bathymetry or bottom topography* and the *basin geometry* are conservative
independent variables that do not vary during a hydrodynamic simulation. Initially,
a few depth points may need to be changed in order to smooth out the bathymetry
during model calibration, when erroneous data are suspected, or in order to avoid
instability problems near open boundaries (e.g., lagoon or river entrances, dikes).
Bottom topography and basin geometry are the main factors that influence the spatial
structure of a numerical solution. As far as model computations are concerned, the
bottom structure and basin geometry should not be prescribed in more detail than
the computational grid resolution. However, practice shows that this information
should be as precise as possible.

*Resistance data* are also needed to estimate the frictional forces that dissipate kinetic
energy at the surface, at the bottom, and in the water body. Resistance data include drag
coefficients, bottom roughness coefficients, wind drag coefficients, and other parameters
related to turbulence production and dissipation. In addition, transport equations also
contain *dispersion parameters* as well as other *kinetic transformation constants.*

These internal parameters may be used for model calibration. However, considering their large number, and the probable uncertainty associated with some of them, it
is suggested that this number be reduced following a sensitivity analysis. Only those
parameters that significantly affect the solution may be kept as calibration parameters.

### 6.5.2 SENSITIVITY ANALYSIS

Sensitivity analysis is the process by which the response of different elements in
the model to external influences is investigated.[44] It is customary to compare the
model performance to some test cases that have analytical solutions.

During a sensitivity analysis, all the parameters and input data are varied individually, usually by a constant percentage, in order to determine which one causes the greatest change in the simulated solution.[4a] Results are presented as a fraction or percentage of the baseline conditions or as a ratio of input parameter change to the simulated dependent variable change. For example, a sensitivity analysis may show that a 1% variation of a selected parameter will produce a 3% variation in the steady-state solution variation over the total area.

Sensitivity analysis can be useful in two respects. First, it provides calibration guidelines by isolating the parameters that are best suited for calibration purposes. These parameters are those producing significant variations during the analysis. Should a model solution not be sensitive to some parameter, the value of this parameter can be taken out of the literature and need not be measured in the modeled domain.

Second, sensitivity analysis can define the error range in the initial and boundary conditions. This is called error analysis.[4a] It assigns estimates of uncertainty to all significant parameters, initial conditions, and boundary conditions and determines the combined uncertainty in the output or dependent variables. Monte Carlo methods can be used for such a purpose.[4a]

### 6.5.3  CALIBRATION

Unfortunately, there is no guarantee that, when provided with a carefully prepared bathymetry, well-planned boundary conditions, and other required input parameters, a model will give good results immediately. In fact, the opposite is often the case. In several instances, the model even becomes unstable during the first few simulations.[37]

This is why we must first calibrate a model with field observations of dependent variables measured inside the model domain. The purpose of calibration is to tune the model so that the differences between the computed and the measured values are reduced.[37] Some recommendations on data collection programs aimed to model calibration and validation are given in Chapter 7, Section 7.4.5 and Table 7.1.

The main parameters normally used to calibrate a hydrodynamic model are the bed resistance, the eddy viscosity, the bathymetry, the boundary conditions, and the wind friction. Some practical recommendations[37] on the use of these and other parameters during calibration are provided below.

The *bed resistance* can be used to calibrate or stabilize a model solution. This is especially true in shallow areas where, for example, bed resistance will change tidal amplitudes and phases significantly.[37]

The *eddy viscosity* is mainly used to stabilize a numerical solution. For example, increasing the eddy viscosity in an area will smooth out spurious results such as higher frequent oscillations in water levels or wiggles in the flow field—zigzagging current vectors—in that modeled area. The eddy viscosity can of course also be used to calibrate the model. Eddy viscosity changes will only change the tidal amplitude but not the tidal phase.

If the eddy viscosity is used for calibration in areas of variable water depths, a Smagorinsky formulation[46,47] relating the eddy viscosity to the (variable) local gradients in the velocity field should be chosen, because a constant eddy formulation will not give correct results in areas with such varying depth.

The *wind friction factor (drag coefficient)* is used as a calibration parameter when a simulation includes wind forcing, for example, surge level set-up.

The *bathymetry* is by far the most important calibration parameter. Initial calibration attempts must always focus on the model bathymetry before trying to vary other calibration parameters. In fact, several changes may have to be made in such areas where the bathymetry has been schematized in order to represent details that are finer than the grid dimensions.

The *boundary conditions* can also be used to calibrate the model. For example, the introduction of a sea-level slope along an open boundary to simulate geostrophy across that boundary may significantly affect the model results.

A *soft start* (or ramping time) can be applied at the beginning of the simulation to avoid initial numerical instabilities and to get rid of oscillations. If a simulation includes, for example, wave-generated radiation stresses along a shoreline, it is necessary to have a very long soft start to avoid shock waves generated by the instantaneous application of high long-shore current as normal boundary conditions.

A model should be calibrated for all different *situations* that are *relevant to the study*. In other words, calibration coefficients for simulations of seasonal processes could differ from those required for simulations over daily time scales. For instance, if tidal propagation in a region is expected to be highly variable over time and space, the calibration may be performed separately for neap and for spring tides. Or, if the hydrodynamic conditions vary with the seasons, e.g., due to monsoon influence, the calibration should be performed for each part of this seasonal cycle.

A basic rule to follow is that *all parameters should be tuned* during calibration, *but one at a time*. If more than one parameter is changed at one time, it may not be possible to clearly identify which parameter caused what changes.

Finally, we can ask what *level of accuracy* should be reached before the model calibration is completed? There is no clear answer to this question. It will depend on the type of study performed, the quality of the boundary data and measurements, the time allocated to the modeling investigation, and of course the required accuracy.

In any case, the *required accuracy must be established before the study* is started. An example of acceptance criteria for a 2D hydrodynamic model calibration and verification can be:[37]

- 5% on average (but not less than 0.1 m) for the tidal range
- 30% on average (but not less than 0.1 m/s) for current speeds
- 45° on average for current directions

However, the required accuracy of a solution *can never exceed the accuracy of the field measurements* with which it is compared. One also may ask how representative the model solution is of the field measurements. For example, a current velocity measured at a certain point above the bottom is not necessarily representative of the vertically averaged velocity simulated by a 2D model. In this case, some additional transformations are required. Assuming a logarithmic current profile in the vertical, currents measured at an arbitrary depth $y_0$ can be converted to a depth-averaged value by using the following relation:[37]

$$u = u_0 \bigg/ \left( 1 + (\ln \frac{y_0}{D} + 1) \cdot \frac{\sqrt{gD}}{k \cdot M \cdot D^{2/3}} \right) \qquad (6.59)$$

where $u$ is the mean velocity, $u_0$ is the measured velocity, $y_0$ is the measuring depth above seabed, $D$ is the instantaneous water depth, $g$ is the gravity, $k$ is Von Karman's constant ($\approx 0.4$), and $M$ is the Manning number.

The difficulty associated with a calibration procedure that utilizes a large number of parameters is that it is not clear which parameters are responsible for the change observed in independent variables. This can be avoided by limiting the number of parameters to those identified by a sensitivity analysis.

## 6.5.4 VALIDATION

Validation of a model is the comparison of its results to a set of independent observations not used for calibration purposes.[44] This means that two data sets are needed in the model domain: one for calibration purposes and the other for validation purposes. Good model validation results will convince potential users (managers, planners, scientists, or engineers) to use the validated model to solve a problem.[48] The criterion for a good model validation is that its numerical solution reproduces, within the prescribed accuracy, the system's variability during the simulation period. A first validation stage should reproduce normal conditions over a period covering at least one cycle of the system's natural variation. This cycle may be a day, a season, a year, a climatic period, or any other cycle of the problem under study.

A second validation stage should cover those extreme conditions that have not necessarily been encountered during the calibration stage. Those extreme conditions for a lagoon can be high river inputs, stormy winds, and large volume fluxes at the ocean–lagoon boundary. Field observations of these extreme conditions must be available for a good validation exercise.

The comparison of model results to observed data during the validation process provides one of the following outcomes:

- The simulation yields accurate results
- The simulation overestimates or underestimates the reality
- The simulation yields confusing results despite all efforts

When a model validation process is inconclusive, the calibration process may be repeated with parts of the data used for validation. It should be mentioned that it is very rare that a single data set of internal parameters will include normal and extreme situations. A reasonable model validation can be attempted with different sets of internal parameters for different situations or time scales.

Ultimately, a model simulation must be checked for volume, mass, and energy conservation. For example, the sum of incoming and outcoming volumes, heat, salt, or conservative pollutants must be equal through a cycle of variability.

Because the data requirement is crucial for model validation, available data must be representative of the natural conditions. For example, when winds from a coastal meteorological station are used to impose wind stress at the surface of a lagoon, do they really represent winds over the lagoon? Does the surrounding landscape relief influence it? In that case, what are the transfer functions for winds over the lagoon? Ideally, a set of experiments may be required to verify the validity of the available data for a specific task.

When existing data are not adequate or data are nonexistent for validation purposes, analytical solutions for simple geometry, but with realistic boundary conditions, may be used instead.[49] It should be noted, however, that even if tests with a simple geometry are successful, the model might still generate some errors or inconsistencies for the real conditions.

Finally, it is advisable to collect data for calibration and validation after a preliminary run of the numerical is made using approximate initial and boundary conditions obtained from the literature. This will point to those areas where hydrodynamic conditions are significantly different. Monitoring stations can then be positioned in those regions. Further recommendations can be found in the monitoring program design (see Chapter 7).

## ACKNOWLEDGMENTS

The authors deeply appreciate Dr. Irina Chubarenko's contribution to the planning and initial drafting of some of the sections. We are also thankful to colleagues from our respective institutions for sharing their ideas and personal results that led to the material presented in this chapter. Finally, we thank Ali Ertürk of Istanbul Technical University for his contribution during the editing stages of Chapter 6.

## REFERENCES

1. Leonard, B.P., A stable and accurate convective modelling procedure based on quadratic upstream interpolation, *Comput. Methods Appl. Mech. Eng.*, 19, 59, 1979.
2. Kjerfve, B., Comparative oceanography of coastal lagoons, in *Estuarine Variability*, Wolfe, D.A., Ed., Academic Press, New York, 1986, p. 63.
3. Kjerfve, B., Coastal lagoons, in *Coastal Lagoon Processes*, Elsevier Oceanography Series 60, Kjerfve, B., Ed., Elsevier, Amsterdam, 1994, chap. 1.
4. Martin, J.L. and McCutcheon, S.C., *Hydrodynamics and Transport for Water Quality Modelling*, Lewis Publishers, Boca Raton, FL, 1999; (a) part I, p. 62; (b) part III, p. 361; (c) part IV.
5. *Manual on Calculations of Surface Reservoir Evaporation*, Hydrometeoizdat, Leningrad, 1969, p. 83 [in Russian].
6. Babkin, V.I., *Water Surface Evaporation*, Gidrometeoizdat, Leningrad, 1984, p. 78 [in Russian].
7. Swinbank, W.C., Long-wave radiation from clear skies, *Quart. J. Royal Met. Soc.*, 89, 339, 1963.
8. Wunderlich, W.O., Heat and Mass Transfer between a Water Surface and the Atmosphere, International Memorandum, Tennessee Valley Authority Engineering Laboratory, Norris, TN, 1968.

9. Zimmerman, J., Estuarine residence times, in *Hydrodynamics of Estuaries, Vol. 1, Estuarine Physics*, Kjerfve, B., Ed., CRC Press, Boca Raton, FL, 1988, p. 75.

10. Bolin, B. and Rodhe, H., A note on the concept of age distribution and transit time in natural reservoirs, *Tellus*, 25, 58, 1973.

11. Takeoka, H., Fundamental concepts of exchange and transport time scales in a coastal sea, *Continental Shelf Research*, 3, 311, 1984.

12. Koutitonsky, V.G., Guyondet, K.T., St. Hillaire, A., Courtenay, S., and Bohgen, A., Water renewal estimates for aquaculture developments in the Richibucto estuary, Canada, *Estuaries*, 27(5), 839, 2004.

13. Ekman, V.W., On the influence of the Earth's rotation on the ocean currents, *Ark Math. Astr. Fyzik*, 2, 1, 1905.

14. Madsen, O.S., A realistic model of the wind-induced Ekman boundary layer, *J. Phys. Oceanogr.*, 7, 248, 1977.

15. Massel, S.R., *Fluid Mechanics for Marine Ecologists*, Springer-Verlag, Heidelberg, 1999, chap. 7.3; (a) chap 14.14.

16. Felzenbaum, A.I., *Theoretical Background and Methods for Steady Marine Currents Calculations*, USSR Academy of Sciences, Moscow, 1960 [in Russian].

17. Kundu, P.K., A numerical investigation of mixed layer dynamics, *J. Phys. Oceanogr.*, 10, 220, 1980.

18. Gill, A.E., *Atmosphere-Ocean Dynamics*, Academic Press, New York, 1982, chap. 6.

19. Dmitriev, N.V., *Mathematical Modelling of the Vertical Turbulence Mixing in the Upper Ocean Layer*, Novosibirsk, 1993, p. 47 [in Russian].

20. Connor, J.J. and Brebbia, C.A., *Finite Element Techniques for Fluid Flow*, Butterworth, London, 1976, p. 208.

21. Smith, S.D., Wind stress and heat flux over the ocean in gale force winds, *J. Phys. Oceanogr.*, 10, 709, 1980.

22. Hutter, K., Hydrodynamic modelling of lakes, *Encyclopedia of Fluid Mechanics*, Gulf Publishing Company, Houston, TX, 1986, chap. 22.

23. Proudman, J., *Dynamical Oceanography*, John Wiley & Sons, New York, 1953.

24. Watson, E.R., Movements of the water of Loch Ness, as indicated by temperature observations, *Geogr. J.*, 24. 430, 1904.

25. Lazarenko, N.N. and Majewski, A., Eds., *Hydrometeorological Regime of the Vistula Lagoon*, Hydrometeoizdat, Leningrad, 1971 [in Polish and Russian].

26. Bishop, C. and Donelan, M., Wave prediction models, in *Applications in Coastal Modelling*, Lakhan, V.C. and Trenhaile, A.S., Eds., Elsevier Oceanography Series 49, Elsevier, Amsterdam, 1989, p. 75.

27. U.S. Army Engineer Waterways Experimental Station, Coastal Engineering Research Center, Shore Protection Manual, 4th ed., 2 vols, U.S. Army Engineers Waterways Experiment Station, Vicksburg, MS, 1984.

28. Matushevskiy, G.V. and Kabattchenko, I.M., Complex parametrical integral model for wind wave simulation and its application, *Russian Meteorology and Hydrology c/c of Meteororlogiia i Gidrologiia)*, 15, 45, 1991 [in Russian].

29. U.S. Navy Hydrographic Office, Techniques for Forecasting Wind Waves and Swell, Publ. N 604, Washington, D.C., 1951.

30. Matushevskii, G.V., Ed., *Calculations of the Marine Wind Wave Regime: Manual No. 42*, Russian State Oceanographic Institute, Moscow, 1979, chap. 2 [in Russian].

31. Floderus, S., On the spatial distribution of wave impact at the Kattegat seabed, *Geogr. Ann. A.*, 70(3), 269, 1988.

32. Chubarenko, B.V., Lund-Hansen, L.Ch., and Beloshitski, A.A., Comparative analysis of potential wind-wave impact on bottom sediment in the Vistula and Curonian lagoons, in *Baltica: An International Year-Book on Geology, Geomorphology and Palaeogeography of the Baltic Sea*, 15, 31, 2002.

33. Sly, P.G., Sedimentary processes in lakes, in *Lakes: Chemistry, Geology, Physics*, Springer-Verlag, Heidelberg, 1978, p. 65.

34. Mortimer, C.H., Lake hydrodynamics, *Mitt. Internat. Verein Limnol.*, 20, 124, 1974.

35. Ford, D.E. and Johnson, M.C., An Assessment of Reservoir Mixing Processes, Technical Report E-86-7, U.S. Army Engineers Waterways Experiment Station, Vicksburg, MS, 1986.

36. Fisher, H.B., List, E.J., Koh, R.C.Y., Imberger, J., and Brooks, N.H., *Mixing in Inland and Coastal Waters*, Academic Press, New York, 1979.

37. *MIKE21 Training Guide*. Danish Hydraulic Institute, Hörsholm, Denmark, June 1993.

38. Chubarenko, B.V., Wang, Y., Chubarenko, I.P., and Hutter, K., Barotropic wind-driven circulation pattern in a closed rectangular basin of variable depth influenced by a peninsula or an island, *Ann. Geophysicae*, 18, 706, 2000.

39. Hutter, K., Ed., *Hydrodynamics of Lakes: CISM Lectures*, Springer-Verlag, New York, 1984, p. 19.

40. Chubarenko, B.V. and Chubarenko, I.P., The transport of Baltic water along the deep, channel in the Gulf of Kaliningrad and its influence on fields of salinity and suspended solids, in *ICES Cooperative Research Report*, N 257, Dahlin, H., Dybern, B., and Petersson, S., Eds., Copenhagen, 2003, p. 151.

41. Chubarenko, I.P., Adequate numerical modeling—the question of specifically orientated field experiment, Presentation at the Baltic Sea Science Conference, November 23–27, 1998, Rostok-Warnemuende.

42. Kahaner, D., Moler, C., and Nash, S., *Numerical Methods and Software*, Prentice Hall, New York, 1989.

43. Woods, J.D., Sub-grid movement parameterization, in *Modelling and Prediction of the Upper Layers of the Ocean*: *Proceedings of a NATO Advanced Study Institute*, Kraus, E.B., Pergamon Press, New York, 1976, p. 146.

44. Barretta-Bekker, H.J., Duursma, E.K., and Kuipers, B.R., Eds., *Encyclopaedia of Marine Sciences*, 2nd ed., Springer-Verlag, Heidelberg, 1998.

45. Roed, L.P. and Cooper, C.K., A study of various open boundary conditions for wind-forced barotropic numerical ocean models, in *Three-Dimensional Models of Marine and Estuarine Dynamics*, Nihoul, J. and Jamart, B., Eds., Elsevier Oceanographic Series 45, Elsevier, Amsterdam, 1987, p. 305.

46. Smagorinsky, J., General circulation experiment with the primitive equations, *Monthly Weather Review*, 91(3), 99, 1963.

47. Abbott, M.B. and Larsen, J., Modelling circulations in depth-integrated flows, *J. Hydraul. Res.*, 23, 309–326 and 397–420, 1985.

48. Wen-Sen, C., Remaining problems in the practical application of numerical models to coastal waters, in *Application in Coastal Modelling*, Lakhan, V.C. and Trenhaile, A.S., Eds., Elsevier Oceanography Series 49, Elsevier, Amsterdam, 1989, p. 355.

49. Ditmars, J.D., Adams, E.E., Bedford, K.W., and Ford, D.E., Performance evaluation of surface water transport and dispersion models, *J. Hydraul. Eng.*, ASCE, 113, 1961.

# 7 Monitoring Program Design

*Eugeniusz Andrulewicz and Boris Chubarenko*

## CONTENTS

## 7.1 INTRODUCTION

Monitoring is the application of fundamental scientific methods of observation of the environment. As a modern tool of water management, monitoring is deeply rooted in science. It is the assessment method of comprehensive determination of the current state of environmental conditions. Monitoring measures are for description rather than prediction; however, monitoring data are used for various purposes, including prediction scenarios/modeling.

1-56670-686-6/05/$0.00+$1.50
© 2005 by CRC Press

In contrast, modeling is a relatively new method rooted in engineering, especially its modification as computer modeling, which aims to simulate the behavior and response of water conditions to external and internal impacts. Monitoring is very useful for making an environmental assessment, while modeling is applied for an impact assessment. Modeling predicts trends and effects of future actions (see Chapter 6 for details).

This chapter first discusses what monitoring is and describes its various aspects. The relationships between monitoring and modeling as complementary tools for current water quality management are presented.

### 7.1.1 DEFINITION OF ENVIRONMENTAL MONITORING

Monitoring has been defined by the United Nations Environment Program (UNEP) as *"the process of repetitive observing for defined purposes, of one or more elements of the environment, according to prearranged schedules in space and in time and using comparable methodologies for environmental sensing and data collection."*[1] Implicit in this definition are a number of points:

- The purposes for undertaking monitoring vary, but it is understood that information is collected for a defined purpose, and not simply because it is available.
- Information gathering is undertaken following a prearranged schedule, which identifies frequency of sample collection, locations, and what information is collected.
- Monitoring involves repetitive, continuous sampling, resulting in a series of three-dimensional, cross-sectional, longitudinal, lateral, and temporal data.
- Sampling, storage, preservation, and analysis must be done systematically, utilizing compatible methodologies following rigorous procedures, to ensure that information is comparable.

Monitoring is distinguished from data collection by its long-term, continuous nature. Data collection efforts are sometimes referred to as short-term monitoring, but it is important to maintain a distinction from monitoring, because monitoring generally has different objectives than data collection.

Every environmental monitoring program should contain the following components:

- Monitoring guidelines (for sample collection, storage, preservation, and analysis)
- Quality assurance program (procedure of calibration and comparability of results)
- Data formats (for preparing data and relevant information for a data bank)
- Data bank (for storage and processing of data)

Monitoring is usually followed by environmental assessment, which is an indispensable step in decision making. Monitoring and research are very often treated as separate activities, but monitoring also should be regarded as a research activity. The basic difference between monitoring and research is already included in the definition of monitoring. Monitoring is a research activity that has three important features: *"prearranged schedule," "repetitive observing,"* and *"comparable methodologies."*

## 7.1.2 Objectives of Environmental Monitoring

Monitoring is not simply a scientific exercise—it is also a management tool. It is a crucial element in environmental decision support systems. Basically, the purpose of monitoring is to provide information that is needed by decision makers. The information desired by decision makers should be identified in the earlier stages of the decision support system, corresponding to the top box in Figure 7.1.

Monitoring usually serves the purpose of generating information needed to solve an environment-related problem. Furthermore, monitoring must be designed to fulfill the needs expressed in the lower portion of Figure 7.1; these needs relate to assessment of results and ultimate decision making. Further assessment of the effects of implementation of decisions forms a feedback loop, where improvements in monitoring programs are then identified. Information should be presented in such a way that it can be incorporated into decision making/implementation. Assessment must therefore reflect the ultimate needs of decision makers.

Decision-making requirements are the driving forces behind monitoring program design as explained in Chapter 8. These requirements may include one or more of the following: information on the state of the environment, natural and anthropogenic pressures, and trend analysis including possibly comparison with background values or other locations. It is therefore important that decision makers are involved in the monitoring program development process. The decision makers have the responsibility of defining clear, measurable goals and objectives.

Also, because long-term series of regular measurements are crucial for modeling (see Chapter 6 for details) and assessment, repetitive measurements of main parameters should be continued for a long period of time. Good decisions can only be made on the basis of long-term information.

## 7.1.3 Some Examples of Current Monitoring Programs

Perhaps the oldest marine monitoring program is related to biological resource assessment, including monitoring of commercial fish species in the North Atlantic and adjacent marine waters. Regular observations began under the International Council for the Exploration of the Sea (ICES) in the early 1900s and are still ongoing.

In the 1960s, when the effects of pollution started to become apparent, ICES expanded its efforts to advise on the development of marine environmental monitoring programs.[2] Advice has been utilized by commissions representing different water bodies (Baltic Sea, North Sea, Arctic seas, etc.).

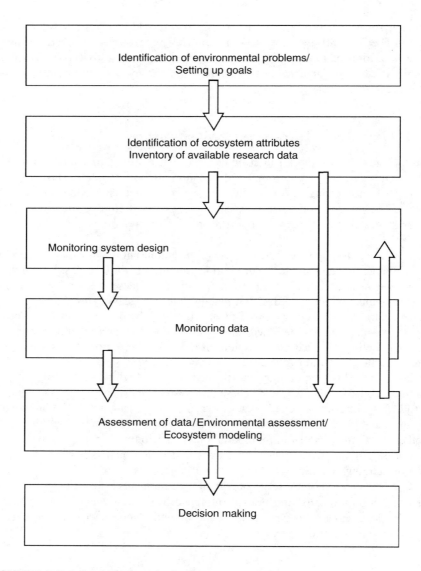

**FIGURE 7.1** Relationship of monitoring to the decision-making process.

Examples of current international monitoring programs related to the marine environment are the Joint Assessment and Monitoring Program (JAMP), established to monitor environmental quality throughout the North-East Atlantic; the Cooperative Baltic Monitoring Program (COMBINE); the Arctic Monitoring and Assessment Program (AMAP); and the Monitoring and Research Program of the Mediterranean Action Plan (MEDPOL). In addition to international monitoring programs, various national monitoring programs serve different purposes according to national needs and specific environmental problems.

During the past 25 years, monitoring has evolved from physico-chemical collection of information to monitoring of ecosystems and biological effects. Most of the present monitoring programs have become more integrated among disciplines (hydrology, chemistry, biology) and have expanded to cover effluents originating from within catchment areas.[3]

### 7.1.4 ISSUES SPECIFIC TO MONITORING OF LAGOONS

Lagoons are morphologically and ecologically complex, subject to constantly changing environmental conditions generally of much greater magnitude than is the case in the open sea (see Chapter 2 for details). For example, temperature may range from ice conditions to very warm waters; salinity may range from freshwater to hypersalinity; wave action usually reaches the bottom, causing dynamic conditions and high energy habitats; and current speed and direction may change frequently, particularly in inlets/outlets of lagoons and in their vicinity. Due to the transitional nature of lagoons, they usually display a number of specific features, which require development of monitoring methods and techniques specifically tailored to the ecosystem. In some cases, techniques utilized in freshwater and/or saltwater bodies may not be applicable or relevant.[4] These difficulties are compounded by the variable, dynamic nature of many lagoons.

Lagoons often contain a great variety of pelagic and benthic habitats (see Chapter 5 for details). For example, lagoons may include some or all of the following habitats: wetlands, marshes, sea grass meadows, intertidal flats, and upland areas, as well as others. Lagoons may have a variety of bottom sediment and sedimentation conditions.

Due to great variability of conditions, organisms usually live under a significant amount of natural stress; therefore, anthropogenic stress is particularly troublesome in such an environment.

There is no general scheme for monitoring of lagoons. Lagoon monitoring therefore needs to be designed with the specific water body in mind. A knowledge of the basic parameters of the given lagoon is essential, including trophic status, water exchange, morphology, salinity, annual variability, etc. Some aspects, which may be important for lagoon monitoring system design, are discussed in the following sections.

## 7.2 MONITORING SYSTEM DESIGN

There is no tradition of monitoring coastal areas as there is for monitoring open sea or freshwater areas. Monitoring system design for coastal zones is less advanced and in many cases needs to be developed from the beginning. Design efforts can borrow elements from the monitoring of marine waters and fresh waters, where monitoring has been under way for some time.[5,6]

As previously mentioned, the most important consideration regarding monitoring system design is the need to establish clear goals. These goals will then lead to determining what information is needed to fulfill the goals. However, this may be problematic due to differences in problem definition, understanding of cause/effect relationships, the interjurisdictional nature of problems, etc.

Monitoring should be designed to account for the unique characteristics of a lagoon ecosystem (see Chapters 2 and 5) and the specific environmental problems and the socio-economic systems (see Chapter 8) encountered in the lagoon watershed. There is also a large variety of different uses of lagoon ecosystems which should be considered for monitoring, such as tourism, fishery and mariculture, coastal technical developments, land reclamation, coastal defense, sand and gravel extraction, dredging and dumping, waste discharges, transportation, and other uses specific to the location. Uses within the drainage area also must be considered, e.g., agriculture (use of fertilizers and pesticides), industry (emissions to air and water), and settlement (domestic sewage).

Anthropogenic pressure is more evident on coastal lagoons than on open sea areas. It is generally agreed that the major environmental problems, including eutrophication, bacterial contamination, toxic compounds contamination, pressure on living resources and mineral resources, dumping of dredged materials, treated and untreated sewage discharges, and coastal erosion, are typical for many lagoons. Other problems, such as invasion of alien species, toxic species, and transport-related problems, may be specific to certain lagoons.

Developing a monitoring system involves trade-offs. The most obvious one is between "cost" and "power" of the information gathered. A greater frequency of observations will decrease the likelihood of erroneous results, all other things being equal. However, it must be kept in mind that costs of a monitoring program will increase as well. Consideration of costs is therefore critical during the monitoring program design phase. Another trade-off is between "power" and "time." A longer temporal string of measurements will likely flatten out abnormalities, but time is always limited for decision making.

Once the goals of the monitoring program are established, a number of "technical" issues remain to be addressed. These include:

- Spatial frequency of sampling—In a large-scale monitoring program, a sufficient number of stations must be selected to generate sufficient data for analysis. Depending on the goals of the program, stations may be randomly chosen or chosen based on hypotheses or results of preliminary modeling studies.
- Temporal frequency of sampling—In cases where biotopes are relatively unchanging, infrequent sampling (once every 5 years, e.g., for deep basin sediment) is usually adequate. If the ecosystem is very dynamic, more frequent sampling (at least 4 times per year, e.g., to characterize seasonal variations) is necessary.
- What is to be sampled—Information relevant to determining environmental quality of marine resources is manifested in the water column, in biota, and in sediment. A large-scale monitoring program should include all three to ensure that biological effects of anthropogenic activities are covered.

Monitoring programs should be regularly reviewed to ascertain that they have good quality control and are meeting established goals, basically providing the information needed for decision making. In addition, it may be wise to update a monitoring program based on availability of new technology or new information. Any changes in a monitoring program must take into account the importance of comparability of temporal data; thus, revisions that would result in breaks in temporal data should be limited to those that are absolutely necessary.

### 7.2.1 Monitoring for Meteorological and Hydrodynamic Parameters

Hydrodynamic models are the basis for ecosystem modeling. They include the background chemical and biological processes in the lagoon ecosystem: transport processes, internal water mixing and water exchange with adjacent open areas, vertical mixing, and interaction with the bottom.

Hydrodynamic modeling is the furthest developed type of modeling, but in the case of lagoons the implementation of models is rather difficult due to the technical difficulties in obtaining enough field data to calibrate the model for any type of application. The high variability of current patterns, its local peculiarities because of bathymetry variations, the complicated nature of water exchange between lagoons, and the adjacent open water body—all of these factors cause a dramatic increase in monitoring data needed for implementation of 2D and especially 3D hydrodynamic models as described in Chapter 6. An optimal monitoring strategy aimed at both future model applications and type of modeled processes is the key issue for reaching reasonable model precision at a reasonable price.

In lagoons, the major driving forces are usually wind stress, water level changes related to tides and wind action, density gradient of different origins, and direct atmospheric pressure (see Chapter 3 for details). Therefore, for hydrodynamic models, the following meteorological measurements are crucial:

- Wind speed
- Wind direction
- Barotropic pressure
- Precipitation and evaporation
- Solar radiation
- Cloudiness
- Air temperature
- Humidity

The list of monitored hydrodynamic parameters usually depends on the monitoring goals, but generally includes:

- Flows, salinity, and temperature through the lagoon entrance
- Discharges and water temperature from all rivers and artificial outlets
- Level variation at the open lagoon entrance
- Level variation at some points remote from lagoon entrances
- Current, salinity, and temperature vertical profiles at monitoring points inside the lagoon
- Spatial variation of salinity and temperature in the lagoon
- Wind wave height and spreading direction
- Parameters related to turbulent mixing
- Tidal characteristics of the adjacent marine area

The most critical hydrodynamic parameter is water exchange at entrances from the open sea. There are different types of water flows, including steady flows, pulsing flows, and backflows. More than one flow in different directions may occur in the water column at one time. These flows may be subject to considerable temporal variation and have a tremendous influence on lagoon hydrodynamics (e.g., for tidal lagoons).

The equipment needed includes standard meteorological and hydrological equipment; furthermore, a number of new technical developments should be applied to measuring hydrodynamic parameters, including the use of unattended equipment on buoys which relay continuous data records and remote sensing techniques. In fact, hydrodynamic modeling has been made possible due to technical development of measuring equipment.

Hydrodynamic modeling is a crucial component for the development of most types of emergency decision support systems. Currently, considerable effort is being devoted to developing hydrodynamic models for operational aspects in oceanography such as the Global Ocean Observing System (GOOS) and the High Resolution Operational Model of the Baltic Sea (HIROMB).

### 7.2.2 MONITORING FOR PHYSICAL PARAMETERS

Physical parameters are usually related to identification of three-dimensional properties of water masses. These properties usually include parameters defining water density structure (water salinity and temperature), which should be measured as vertical profiles at the lagoon entrances and at monitoring points given in the above section.

In addition, monitoring of the following optical parameters may be necessary for some modeling aspects at the entrances and inside the lagoon:

- Depth attenuation of solar radiation
- Secchi depth
- Water turbidity
- Inorganic suspended matter

Methodology for measuring these physical parameters is discussed in various guidelines for monitoring programs.[7-9]

Remote sensing is a useful tool applied for such parameters as water temperature, turbidity, surface water color, and ice cover.

### 7.2.3 MONITORING FOR CHEMICAL PARAMETERS

Chemical parameters given below relate to eutrophication and contamination of water masses and are usually recorded as:

- Oxygen
- Hydrogen sulphide (under anoxic conditions)
- pH
- Alkalinity
- Total inorganic carbon
- Nutrients (nitrogen and phosphorus compounds and sometimes silicates)
- Particulate and dissolved organic carbon

This is because most of the problems experienced in lagoons are related to eutrophication, nutrient enrichment/high primary productivity: unusually high organic carbon and nutrient concentrations, particularly after accumulation periods, high pH during productive season, large amount of suspended matter, and oversaturation with oxygen in productive layers. But due to the highly dynamic nature of some lagoons, it is very difficult to monitor nutrient fluxes. The effect of most important kinetic processes explained in Chapter 4 varies significantly between the intermediate mixing zone and the rest of the lagoon, and these variations create problems for effective monitoring.

Many examples show that morphology of the area and local hydrological conditions are important factors affecting behavior of nutrients. Inputs of nutrients can be severely influenced by tidal phenomena. Given the above considerations, it seems there is no universal, simple approach to monitoring of nutrients. It is unrealistic to develop a general monitoring strategy for nutrient exchange. In the case of riverine lagoons, gross nutrient flux from rivers should be determined. Satisfactory nutrient budget will require many cruises and sampling stations, which is often unrealistic in the long term. A short-term, intensive project is recommended; modeling will help plan the special scheme of measurements.

Special methodology is often necessary in the case of lagoons, due to their specific nature. For example, the analytical techniques utilized in measuring eutrophication parameters in marine areas or fresh waters are sometimes not applicable to lagoons, due to factors such as intermediate salinity, possibility of the presence of humid substances, differing water color, and presence of a large amount of suspended material. A further point to be considered is that lagoons are often remote, at a great distance from laboratories. Prior to analysis samples, which are usually gathered from a small research boat, must be preserved taking into consideration long distances and time.

Analyses of nutrients are based on spectrophotometric methods, so water color is an important consideration. Water color may vary in lagoons due to, for example, local events or the absence or presence of humic substances. Interpretation of results must therefore consider both temporal and spatial variability in water color. Results from spectrophotometric analysis will likewise be affected by the presence of suspended matter, which may be present in lagoons, but is seldom encountered in the water column in the open sea. Filtering or centrifugation (in the case of ammonia determination) is therefore necessary.

There is usually a need for immediate analysis, which could be problematic in the case of isolated lagoons. Strictly followed preservation methods are necessary for some chemical measures, although immediate analysis is definitely preferred. If this is not a possibility, samples should be kept cool or frozen. In the case of biogenic salts, samples can be stored safely up to 6 h at 0°C, and for a longer time if stored below –20°C. Preservation methods with addition of chemicals (chlorophorm, mercury, sulfuric acid), which have been used historically, are not recommended.

Automatic analytical methods may not be possible in a dynamic lagoon subject to varying water properties and concentrations, a typical feature of lagoons. Analysts must carefully consider calibration in such cases. Concentrations beyond the calibration range should be diluted. High levels of organic or particulate matter may introduce bias into results.

For contamination by harmful substances in water, biota, and sediment, the following parameters are usually determined:

- Trace metals (mercury, cadmium, copper, chromium, zinc)
- Pesticides, particularly chlorinated compounds such as dichlorodiphenylo-trichloroethane (DDT), hexachlorocyclohexanes (HCHs), and hexachloro-cydobenzene (HCB)
- Polychlorinated biphenyls (PCBs)
- Petroleum hydrocarbons (total hydrocarbons: PAHs)

Methodology is provided in various guidelines and specialized papers, although preference is given to ICES development and existing guidelines, e.g., the Helsinki Commission (Baltic Marine Environment Protection Commission), HELCOM[8] and/or OSPAR.[9] These guidelines have been developed for marine areas and include very useful precise measurement schemes, including sampling, sample preservation, sample pretreatment, and instrumental measurements. These guidelines can be used for measuring chemical parameters in lagoons, however, with some modifications because of specific conditions of the lagoon.

## 7.2.4 Monitoring for Biological Parameters

Biological parameters will depend on the specifics of the monitoring program. In the case of eutrophication, the following parameters are usually included:

- Primary production
- Chlorophyll *a*
- Phytoplankton (species composition and biomass)
- Zooplankton (species composition and biomass)
- Macrophytes (depth range, species composition, and biomass)
- Macrozoobenthos (species composition and biomass)
- Ichthyiofauna

Similar to other monitoring parameters, biological determinants should be especially well calibrated and agreed upon by participants. This is usually done through workshops where methodology and equipment are agreed upon.

Workshops on taxonomic determinations are under way in various international commissions. They are currently involved in unifying monitoring approaches. Guidelines are available for many monitoring programs, although none is comprehensive (e.g., HELCOM,[8] OSPAR,[9] JAMP, and COMBINE).

Various sampling equipment is adopted for monitoring phytoplankton, zooplankton, and zoobenthos. This is usually calibrated within one monitoring program, i.e., using the same mesh size, counting methods, etc. The intermingling of freshwater and marine organisms in samples originating from brackish lagoons may cause difficulties in laboratories, which may have capabilities in one or the other type of organism, but not both. Likewise, samples from hypersaline waters may contain organisms that are not easily identifiable by some laboratories. In general, lagoons have a large number of taxa, relatively high diversity, and a relatively high abundance of biomass. There is likewise a large amount of periphyton, which hampers analysis. In general, biological sampling in lagoons requires more time than required for the open sea and also requires a greater degree of knowledge about the ecosystem.

### 7.2.5 Monitoring of Impact of Different Uses of Lagoons

Coastal lagoons are characterized by intensive exploitation; therefore, they are subjected to various physical, biological, and social interactions (as explained in Chapter 8). In some lagoons, the presence of such activities may require the monitoring of additional parameters. These activities include tourism, sewage discharges, fisheries, mariculture, transport/shipping, coastal defense, dredging and dumping of dredged material, sand and gravel extraction, and other coastal engineering activities.

- Settlements and tourism—Human population tends to concentrate in coastal areas, particularly around lagoons. In addition, seasonal tourism may bring a growth in population amounting to a multiple of the permanent population as happens in Venice Lagoon (Italy) and in Mar Menor Lagoon (Spain). In many cases, this growth is beyond the carrying capacity of the environment as well as the capacity of local infrastructure. Consequently, such activity may create serious environmental damage, particularly since not many countries are able or willing to efficiently regulate tourism activities. Tourism tends to gravitate toward nature and landscape conservation areas as well as toward precious and pristine spots, consequently destroying the tourist amenities.

Environmental monitoring in such cases is related to the pressure of tourism and estimating the carrying capacity of the environment and the negative effects of tourism. Tourism and recreational activity usually destroy or make unintended changes in habitats, have effects on species diversity and rare species, cause changes in quality of bathing waters, etc. Modeling can be applied to demonstrate different scenarios of pressure effects of tourism on the lagoon ecosystem.

- Sewage discharges—The relationship between sewage discharges and microbiological pollution and the effects on sanitary conditions are evident. Species of pathogenic bacteria are found in the vicinity of sewage outflows. Monitoring is often undertaken to determine bathing water quality, utilizing established techniques based on concentrations of coliform bacteria. Distribution of microbiological pollution along the coast is often a subject of modeling activity.
- Fisheries—Lagoons are usually very productive and are often intensively exploited as fishery sites. They also are highly sensitive to overexploitation. The pelagic system as well as the bottom system may be adversely

affected by changes in the relative number of species, through removal
of commercial species, mortality of nontarget species, physical disturbance of the bottom, fish discards, and use of antifouling paints. It is
important to restrict fisheries to "safe biological limits" and to take
measures to eliminate and/or restrict potential negative effects of over-
harvest on the ecological community, which is what occurred in Dalyan
Lagoon, Turkey. Estimation of total allowable catches (TACs) and estimation of "safe biological limits" for fisheries are based on monitoring
of fish resources and environmental conditions for fish reproduction and
growth. Commercial fish species are well monitored to secure commercial catches for the future; however, monitoring of impacts on noncommercial fish species and on the environment is usually neglected and even
poorly understood.

- Mariculture—Lagoons are often utilized for mariculture, partly due to
their shallow nature and shelter from the open sea. Mariculture can have
a great effect on the ecosystem, due to excess of nutrient supply and
introduction of contaminants such as pesticides and antibiotics. For example, the Ria Formosa Lagoon, Portugal, is experiencing these effects.
Mariculture should be kept under control through management and regular
monitoring of its environmental effects.

- Transport/Shipping—Commercial shipping, including ferry boat traffic,
causes input of hazardous substances by cleaning tanks, illegal discharges
of fuel and bilge oil, burning of fossil fuels, discharges of waste water,
introduction and transfer of marine species (mostly by the discharge of
ballast water), use of antifouling paints, and loss of cargo and refuse
dumping. Thus, monitoring is required not only for organic chemicals but
also for organisms. Violation of the environment in coastal areas and
lagoons (e.g., illegal discharges) seems to be less frequent in lagoons than
in the open sea and off-shore areas. Such cases are most often detected
and punished. In some areas the use of antifouling paints, particularly
those containing organo-tin compounds, has caused serious biological
effects, including mortality of commercial and natural oyster beds and
some other mollusk species.

- Introduction of alien species—Although occasionally the result of natural
migration, introduction of new species into lagoons is mostly of anthropogenic origin, often the result of discharge of ballast water. The effects
in lagoons are often greatly magnified compared with similar events in
the open sea. For example, the effects of the spread of the zebra mussel
(*Dreissena polymorpha*) has been devastating in coastal and semienclosed
water bodies throughout the northern hemisphere. There are ecosystem
models for invasive species illustrating the ecosystem impact and trophic
cumulative effects of some introduced species in the Great Lakes and the
Mediterranean Sea.[10] The models aim at emphasizing effects other than
just prediction and competition for food, effects that are often less obvious
in the ecosystem, by incorporating quantitative data on abundances,
growth and uptake rates, etc.

- Coastal defense—Depending on the local situation, sea defense work is undertaken using rock armoring, construction of breakwaters, piers and jetties, or beach replenishment. In most cases these activities affect natural coastal dynamics and in some cases cause even more serious erosion problems in neighboring areas. Coastal defense requires monitoring and modeling activities as part of the planning process and optimizing the desired results and minimizing negative effects; unfortunately, monitoring the effects of such activities is often neglected.
- Dredging and dumping of dredged material—Dredging is often under- taken in lagoons to accommodate navigation and marine facilities as well as tourism/recreation. Dumping of dredged material is allowed according to the law of the sea. Decisions on dumping are based on the quality of the dredged material. Highly polluted material should be deposed of on land. Chemical-based assessment is generally not sufficient—biological analysis of the dumping site should be undertaken to determine the bio- logical fate of the dumped material at the site.
- Sand and gravel extraction—Some lagoons have mineral resources, which are subject to exploitation. Such exploitation may endanger the dynamics of the shore if it is undertaken in close vicinity. The mechanics of this exploitation result in mobility of locally inhabiting species and excessive turbidity, and adversely affect other species. Monitoring is usually related to habitat restoration and the possible effects of excess turbidity on marine plants.
- Beach nourishment—Beaches subjected to erosion are often regularly replenished with sand, and in some cases, beaches are built where they naturally would not exist, thus necessitating regular nourishment. Such intervention in natural dynamics causes adverse effects elsewhere, some- times increasing the rate of erosion. Monitoring and modeling should be incorporated into beach nourishment activity.
- Other coastal engineering activities, such as digging new channels, con- struction of bridges, construction of marinas, laying of pipelines and power transmission cables, building ports, and refinery terminals, etc., may cause serious environmental effects and should be preceded by mon- itoring of local conditions, environmental impact assessment, and mod- eling and monitoring of effects.

## 7.3  MONITORING-RELATED PROGRAMS

### 7.3.1  Monitoring Guidelines and Quality Assurance Program

In order to achieve appropriate and comparable results within the monitoring pro- gram, particularly in international and/or multilaboratory programs, there is a need for developing a manual for monitoring in which not only the methods and tech- niques are described but also general and specific guidelines on quality assurance are provided. There are good examples for these manuals, such as the OSPAR and HELCOM guidelines.[7–9]

Quality assurance, particularly in the case of international programs, is absolutely necessary for accuracy of data as well as for their comparability. It includes in-house quality assurance procedures (within a laboratory) as well as external procedures (between laboratories). There has been significant quality assurance procedure development activity within commissions such as ICES, HELCOM, and OSPAR as well as within the International Atomic Energy Agency (IAEA) and the EU (within the context of the Quality Assurance of Sample Handling [QUASH] program). Many programs have also been developed in the environmental protection agencies of the U.S. and Canada as well as on an international basis through regional commissions.

Only verified and quality-assured data should be stored in data banks, as explained below. Data obtained from field measurements should usually be supported by a number of additional/fixed parameters, which usually are necessary for proper interpretation of the indicated data.

## 7.3.2 DATA FORMATS AND DATA BANKING

Data formats are usually developed for data banking as well as to ensure that all important information to correctly interpret the data is noted. Apart from the desired data, additional information, e.g., on specific locations, meteorological conditions, etc., is needed for proper interpretation of data.

Generating data is an expensive and time-consuming procedure; therefore, particular attention should be paid to data collection and storage. Results of measurements carried out by different laboratories should be standardized according to previously agreed formats. Data banking should, in fact, only be undertaken in cases when comparability of data is assured. Ideally only certified laboratories should take part in a monitoring program.

The following additional information needs to be provided together with data:

- Coordinates of sampling place
- Type of sample
- Methods of sampling
- Methods of pretreatment
- Methods of preservation
- Methods of analyses (possible deviation from the manual)
- Limit of detection
- Quality assurance information (internal and external)
- Equipment
- Conditions during sampling and analyses
- Others

Data bank verification is necessary prior to acceptance of data, to ensure that proper procedures have been followed.

## 7.4  RELATIONSHIP BETWEEN MONITORING AND MODELING

Monitoring results are necessary for correct and productive ecosystem modeling (Figure 7.2). A well-designed model utilizing poor quality data will generate poor results. Many modeling efforts have been weakened because they were based on insufficient data sets. Existing results from many current national and international monitoring programs are not sufficient for ecosystem modeling, and these programs should be redesigned to accommodate the requirements of models.

For its part, modeling is helpful in designing monitoring programs and necessary for evaluation of monitoring results. It optimizes the development and implementation of monitoring programs and thereby may improve their efficiency and cost/benefit relationship.

Monitoring and modeling very often complement each other. For example, *feedback monitoring* is actually a loop in which both a short-term forecast and adaptive monitoring are incorporated. Such a complementary situation makes management efforts much more efficient.

### 7.4.1  PERSPECTIVE: MONITORING TO MODELING

Monitoring is undertaken according to a strictly prearranged spatio-temporal scheme (see definition at the beginning of this chapter). The preliminary determination of such a scheme to reach an optimization between the objectives and expenses is a key point of monitoring system design.

Modeling in conjunction with expert evaluation is usually used for preliminary monitoring design of an optimal scheme. For example, if the objective of monitoring is to determine a long-term trend of the system, preliminary hydrodynamic modeling could point out the areas in the basin where short-term variations of system parameters are minimal and therefore the errors of data collected are also minimal. Or vice versa, if the spatial scheme is previously fixed, analysis of the variations of the model solution at monitoring points can provide recommendations on the optimal time interval for monitoring sampling.

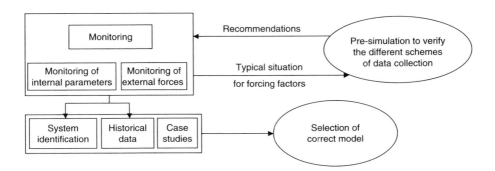

**FIGURE 7.2** Perspective: monitoring to modeling.

Thus, the preliminary verification of a monitoring program by modeling (Figure 7.2) could essentially increase the effectiveness of the design of the monitoring spatio-temporal scheme, which should be used for a long period without changes.

## 7.4.2 Perspective: Modeling to Monitoring

As discussed in Chapter 2, the value of the model for sustainable management as a forecasting tool directly depends on the amount and quality of data available. The more complex the model, the more parameters are needed to fit it; the smaller the amount of data used, the greater the uncertainty of the result. A general rule is that if the information base is not complete enough, an adequately simplified or "screening" model should be constructed.

The challenge is to fill the model with the sufficient data needed at any phase of model design, implementation, and use. Monitoring supplies the most important ingredients of the information needed during the phases of model design or model selection. At this phase any knowledge is useful to identify the system or lagoon type, and the more detailed the information is in space and time, the better. But the monitoring results are not completely exhaustive. Some historical information or results of case studies also are very valuable.

In the next phase, when the model is already chosen and its implementation is under way, the monitoring should be enlarged by studies dedicated to modeling. The results of monitoring (which is normally undertaken to reveal the long-term variations of the system) are usually not enough to calibrate a model, because modeling is not the purpose of standard monitoring. The data collected during monitoring are very often unacceptable or insufficient to completely fit a model; for example, not all the needed parameters have been measured, or these parameters have been measured at noncomparable periods or with noncomparable accuracy. Therefore, a special type of data collection program for model calibration and verification is essential.

Very often such projects are called "monitoring," but they are not actually monitoring because they do not satisfy the time duration requirement. Such programs, even being held according to a fixed prearranged spatio-temporal scheme as in standard monitoring, can be referred to as short-term data collection programs for model implementation to emphasize that the duration of the programs is short. In a practical sense it may vary from 2–3 months for calibration of the hydrodynamic module to 1–3 years for calibration of the advection–dispersion, water quality, or some biological modules.

Following the implementation, another type of modeling dedicated study— model accompanied (attendant) current data supply—has to be organized to provide the model with minimum data needed to run the model at any given time for impact assessment or prediction. These are primarily the data on driving forces for the lagoon system such as wind, water-level variations, river discharge, etc. (see Chapter 6). Some information on selected simulated variable or variables at least at one location in the lagoon is also desirable to have a reference point for current model simulations.

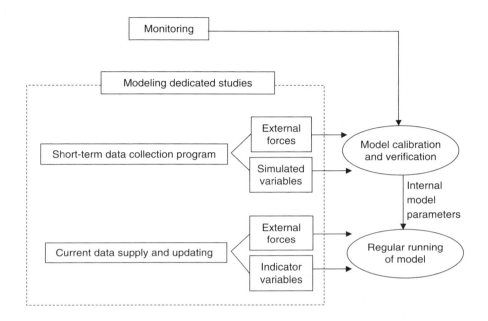

**FIGURE 7.3** Monitoring and modeling dedicated studies.

The modeling dedicated studies mentioned above are represented in Figure 7.3, where they are shown as rectangular boxes. The related types of modeling activity (for which the data are used) are noted by ellipses. Direct and feedback information streams are illustrated by arrows. These modeling dedicated studies will now be discussed, focusing on modeling-related problems.

### 7.4.3 SHORT-TERM DATA COLLECTION FOR MODEL IMPLEMENTATION

This special kind of study is executed during fixed periods only and is aimed at assisting in the model implementation process (see Chapter 6 for details). At this phase, precisely specified field data collecting programs are usually undertaken to determine the following: (1) initial values for all simulated variables; (2) time series for some simulated variables chosen as calibration variables at some calibration points of the studied area during this fixed period; and (3) corresponding time series for all boundary conditions. The series obtained are divided into both a calibration set used for model tuning and a validation set used for model verification.

As a typical example, Table 7.1 presents the specification on data collection aimed at hydrodynamic module calibration (usual monitoring period is 2–3 months) and calibration of an advection–dispersion module with salinity as the natural tracer (minimum period is 2 years; the first year is for model calibration, the second year is for model validation).

**TABLE 7.1**
**General Specification on Data Collection Aimed at the Calibration of Hydrodynamic (Currents, Level Variations, Waves) and Advection-Dispersion (Including Temperature Regime) Modules for a Two-Dimensional Model for a Nontidal Lagoon**

| No. | Parameter or Variable | Measurements Specification | Relation to Model |
|---|---|---|---|
| 1. | Water level variation at the open boundaries (lagoon entrances) | Persistent series or regular measurements with a frequency not less than four times per day | Boundary conditions |
| | Water level variation at some internal points remote from lagoon entrance (not less than two points) | Persistent series or regular measurements with a frequency not less than four times per day | Calibration series |
| 2. | Discharges and water temperature for all rivers and artificial outlets (95% of freshwater inflow) | Usually not more frequent than once a week or 10 days, but not less than once a month | Boundary conditions |
| 3. | Meteorological parameters over the lagoon (wind direction, wind speed, barotropic pressure, precipitation, solar radiation, cloudiness, air temperature, humidity) | Frequency is not less than four times per day (for calibration of seasonal variations) Frequency is not less than once per hour (for current or wave calibration) | Boundary conditions |
| 4. | Salinity and water temperature: • vertical profiles at the lagoon entrances | Frequency is not less than twice per day | Boundary conditions |
| | • vertical profiles at monitoring points inside the lagoon (usually more than three points) | Depends on requirements on model solution accuracy; for seasonal calibration it is usually 1–4 times per month | Calibration data |
| | • spatial variation in the lagoon area | The lagoon area should be surveyed as quick as possible and with regard to daily temperature variations | Initial conditions and calibration data |
| 5. | Wind wave height and spreading direction (minimum—at some points along the lagoon coast) | Persistent series to represent eight main wind directions; duration of series is not shorter than a few hours | Calibration series |
| 6. | Currents (vertical profile or upper and bottom layer values) at the entry and at several points in the lagoon (usually more than three points) | Persistent series with duration not shorter than several tidal periods or 10 days for nontidal lagoon, time step is as frequently as possible | Boundary conditions and calibration series |

## 7.4.4 MODEL-ACCOMPANIED CURRENT DATA SUPPLY

This special kind of regular data collecting is necessary to enable the running of the model at any time. For this purpose, in the case where the model is already calibrated, a limited measurement program is needed to supply the model with current data about initial and boundary conditions only (see Chapter 3 and Chapter 6 for details).

For example, the variations in levels at the open boundaries, discharges and concentrations of chemicals in rivers and other external and internal sources, wind, and initial distributions of modeled chemicals all offer possibilities to begin the model simulation.

This accompanied (attendant) current data supply is compulsory for support of operational modeling and simulation of transport phenomena in the case of an emergency. For all specific purposes, the specification on data collecting during accompanied monitoring is carefully designed according to the objectives of modeling and model requirements.

## 7.4.5 PRACTICAL RECOMMENDATIONS ON THE DESIGN OF SHORT-TERM DATA COLLECTION

The problem of data collection and sampling/measurements for model calibration, validation, and running is the key question in model implementation and in future operational model usage. Experience shows that standard monitoring programs executed by environmental authorities are often fragmented, both in time-space and in the parameters set; therefore, they are very often impractical because they do not satisfy the strict requirements on temporal and spatial compatibility for necessary specific calibration data (see Chapter 9, the Vistula Lagoon case study).

A special program of supplementary data collection aimed at model tuning (or current data supply) is the first step of model implementation. A typical general example of the specification for this type of monitoring is presented in Table 7.1. More detailed recommendations are not given here, because specification guidelines are to be based strictly on the specific model structure and must be prepared by an expert in modeling in each specific case. It should be emphasized that the program of data collecting dedicated to modeling is one of the most expensive tasks in the process of model implementation. Some recommendations are therefore useful to optimize it.

First, there is no way to avoid or to minimize the information about forcing factors and initial and boundary conditions (see Chapter 3 and Chapter 6 for details). The data must be collected in any case. The amount of data needed directly depends on the model used (zero-, one-, two-, or three-dimensional model).

Special field experiments that aim to define the internal modeling parameters (for example, for hydrodynamic models these are the surface and bottom drag coefficients, or turbulent diffusion or dispersion coefficients) are rather useful, but they are not as important as precise determination of lagoon bathymetry and geometry as well as geographical positions of sinks and sources, which have first priority together with initial and boundary conditions.

Precise parameterization of diffusion will not improve the model solution accuracy for prediction of distribution of chemical substances over the lagoon if the chemical loading from coastal sources (boundary conditions) is only roughly estimated. Internal model parameters preferably have to be free for model tuning, and field experiments on measurements have to determine only the range of its variations. The final list regarding what internal parameters are to be used for calibration should be defined by sensitivity analysis and/or for some specific practical reasons (see Chapter 6 for details).

How does one choose calibrated variables? In general, the model has to be calibrated against all the variables simulated, but this calibration should be closely connected with the objectives and spatio-temporal scale of expected model solutions. If the objective of the modeling is the prediction of the hot water plume mixing in the vicinity of a power plant inlet/outlet, it is senseless to expend great effort to achieve the same accuracy for model solution in the entire lagoon. In this case the calibration series for temperature (not for currents) has to be measured mostly in the focus area, but during a rather long period, for example, 1 year.

We now try to emphasize the essential minimum for calibrated variables that can ensure correct model tuning.

- Water level is an important variable both for water exchange and for currents and transport problems. Water level variations (Table 7.1) should be measured at a frequency according to tidal variations or once an hour for random wind surge, but in any case not fewer than four times a day. The tide gauges have to be sited at the lagoon entrance and at several points around the lagoon coast to reveal the local inter-lagoon level inclination and internal level variations. In the case of significant river inflow, level measurements at the river mouth are very important. The measurement technique is standard and well described in a number of national manuals (e.g., National Engineering Handbook[11]).
- Direct uninterrupted current measurements aimed at calibration are usually held during periods ranging from several tidal periods up to 2–3 months at several points in the lagoon simultaneously. The interval of measurements depends on equipment used and its positioning, but often it is 5–60 s if short-period pulsations are taken into consideration or 5–15 min if the general flow is rather stable. Generally speaking, the measurement interval should be selected according to the scale of time averaging used in the task considered. For example, Kundu[12] showed that the influence of the Coriolis force becomes significant after $t > 0.25f^{-1}$ (where $f$ is a Coriolis parameter). In this case, current adjustment time is in the range of $2^3/_4$ h–30 min for lagoons at latitudes in the range of 10–75°, respectively, and at least 5 to 10 measuring casts are required to resolve the characteristic time of current adjustment to wind variations (see Chapter 6).

Due to high variability of currents in the lagoon area it is very important to find the optimal positioning for current meters: on the one hand, their sites should correspond to both natural spatial current structures and the discreteness of the simulation grid, but on the other hand, their number is usually limited because this kind of field work is rather expensive. A formal technique for prior selection of current measurement points in the lagoon area, proposed in Chubarenko et al.,[13] is based on the minimum current direction variation approach. Points of measurements are chosen that represent the main structure of currents and that have the minimum current variations during wind changes, i.e., the currents at these points are sufficiently steady vs. the wind variations. If the model corresponds well with reality at such points it is a sufficient basic level of model calibration. If the model cannot reproduce the real currents at such points it must be reconstructed.

The following step-by-step procedure to choose such points for control calibration measurements is proposed:

1. Determine the duration of measurements and the limited numbers of points of measurements based on equipment and financial considerations.
2. Define the subareas with minimum range of current variations (absolute values of current also must be considered to exclude stagnation areas) that best represent the current spatial structure.
3. Make a final choice of limited measurement sites in these subareas with regard to both local depth field and coastal line configuration.

The first and third steps involve human expertise. The second and key step may be based on the formal mathematical procedure for preliminary simulations of steady-state currents for 8 or 16 basic wind directions and typical wind speed gradations. The current patterns obtained enable the total current structure in the lagoon area to be determined. The subareas of jet-type stream, which are characterized by the minimum range of current direction variation, must then be selected. The precise mathematical criteria to determine the jet-type current areas are defined in Chubarenko et al.[13] Jet-type current is defined as the minimum range of current vector variation that occurs, taking into account the ambient wind condition variation.

- The conservative tracer is used for calibration of the transport part of the model (advection–dispersion). In the majority of cases salinity is used for this purpose. It is a traditional measurement and equipment has been well developed. Measurements of salinity variations in time and space provide useful data for the water mixing and transport phenomena analysis and especially for transport model calibration for synoptic or seasonal variability in a lagoon.

The points for calibration measurements should be assigned to represent all important spatial differences in salinity in the lagoon area. First, the boundary values for salinity should be measured—salinity concentration in lagoon entrances and other inlet flows (rivers or artificial inlets). The frequency of measurements in the lagoon entrances (as well as in other places of salinity inflows) should be synchronized with level variation measurements or current measurements and should be measured not less than two times per day (Table 7.1). Second, the internal lagoon monitoring station should be (1) in areas under the influence of salinity inflows (e.g., near lagoon entrances) as well as areas under the influence of the main freshwater inlets; (2) in the remote semiclosed subareas important to modeling objectives; (3) in the transition zones of active water mixing, local hydrological fronts, etc. Vertical profiling of the salinity field is preferable at the points of monitoring measurements, or, as a last resort, sampling at the upper, intermediate, and bottom layers may be used.

If the implemented model aims at transport prediction in the scale of hours or days in the whole lagoon (not in the vicinity of the salinity front), the salinity data may not have enough contrasts, and in this case a special artificial conservative tracer should be used. The literature for recommendations on execution of tracer experiments includes Zac[14] and Martin et al.[15]

An artificial tracer is also very useful for obtaining calibration data on short-term water mixing and transport both in a hypersaline lagoon (where salinity is transformed not only by direct dissolving but by evaporation as well; for example, Kara Bogaz Gol Lagoon in the Caspian Sea, Turkmenistan), or in a freshwater lagoon, where salinity is usually found only in close vicinity to the lagoon entrance, for example, Curonian Lagoon in the Baltic Sea, Lithuania/Russia.

## 7.5  ASSESSMENT OF MONITORING RESULTS AND FORMS OF PRESENTATION

Monitoring is a costly, time-consuming endeavor. Therefore, its results should be well elaborated and made available not only to decision makers but also to other users, including the public. Thus, it may be necessary to develop more than one assessment document to suit different users and audiences.

For example, the following documents serve different but useful purposes:

- Background document (mainly for scientists, but also for decision makers)
- Executive summary (developed with decision makers in mind, and including references to the background document)
- Popular document (for media and public consumption)
- Indicator-based assessment (usually requested by the decision makers)

The form of presentation chosen will depend on the audience. Decision makers generally prefer summary documents, short on detail but including information useful for making environmental management decisions. People with a technical understanding of methodologies are interested in more detailed technical information. The public and mass media want general information, but access to more technical information if desired. It is useful to develop a communication strategy early in the monitoring program. Recently, an indicator-based document has been required by economists and decision makers. It is also desirable to produce demonstration programs/scenarios based on modeling results. This is particularly useful for decision makers, other potential users, and the public (see Chapter 8 for details).

## 7.6  FINAL REMARKS AND CONCLUSIONS

Monitoring and modeling are two modern tools for supporting decision making. Nowadays they complement each other in many ways and cannot be considered separately.

The quality of models and their output depends on high quality input data, generated by a different kind of monitoring and data collection programs. On the other hand the efficiency of monitoring programs directly depends on the modeling assessment of monitoring schemes and capability to use the monitoring results in model forecasts. The best example of such cooperation is feedback monitoring.

Some practical recommendations on the design of specific monitoring programs given in this chapter aim both at decision making and quantitative assessment model development. Implementation and use will help readers establish a link between these two powerful tools of environmental management.

Monitoring is undertaken in consideration of environmental objectives as well as the needs of the ultimate users of the information – decision makers.

Monitoring based on low frequency of measurements (e.g., on a periodic research cruises) is not very useful for modeling purposes. Coastal environments are extremely dynamic, therefore high frequency measurement programs are necessary. They may be built on an intensive/high frequency monitoring. New techniques, such as automated equipment, remote sensing, ships of opportunity, moored ships, and buoys, should be developed. They supply online or high frequency data necessary both for the understanding of natural variability and for environmental modeling. Selection of monitoring stations and parameters and frequency of measurements should take into account natural variability of the system.

Monitoring should be supported by suitable scientific programs. Assessment of the state of the environment should be based not only on monitoring results but should also include all the available scientific data.

Data originators should follow strict guidelines and implement a data quality system (in-house quality procedure and external intercalibrations). Data formats and reporting procedures are usually elaborated by a monitoring data center.

Occasional revisions of monitoring activities are necessary as a result of experience gained, improvements in monitoring methods, and the introduction of new techniques and parameters. Monitoring systems should thus be designed in a flexible manner to facilitate such revisions.

## REFERENCES

1.  ICES, Report of the Advisory Committee on the Marine Environment, ICES Coop. Res. Rep. No. 233, 1988.
2.  ICES, ICES Role in Environmental Monitoring, *ICES C.M.*1995/Gen.7, 1995.
3.  Andrulewicz, E., Developing monitoring and assessment strategy in the Baltic Sea, *ICES C.M.* 1992/E:46, Mimeo, 1992, 10 pp.
4.  Gürel, M., Nutrient dynamics in coastal lagoons: Dalyan Lagoon case study, Ph.D. thesis, Istanbul Technical University, Institute of Science and Technology, Istanbul, Turkey, 2000.

5. Andrulewicz, E., Proposals for the Baltic monitoring program (HELCOM BMP) and coastal monitoring program (HELCOM CMP) for the Polish marine areas of the Baltic Sea, *Oceanological Studies*, 1–2, 159, 1996.

6. Hameedi, M.J., Strategy for monitoring the environment in the coastal zone, in *Coastal Zone Management for Maritime Developing Nations*, Haq, B.U. et al., Eds., Kluwer, Dordrecht, the Netherlands, 1997, 111.

7. HELCOM, Guidelines for the Baltic Monitoring Program for the Third Stage, Baltic Sea Environment Proceedings, No. 27, 1988.

8. HELCOM, Manual for Marine Monitoring in the COMBINE Program of HELCOM, 1998.

9. OSPAR, Draft Guidelines on Contaminant-Specific Biological Effects Monitoring, Report of Ad Hoc Working Group on Monitoring, Oslo and Paris Commissions: Summary Record of the MON meeting, 1996.

10. ICES, Report of the Working Group on Introduction and Transfers of Marine Organisms, ICES, 1999.

11. National Engineering Handbook, U.S. Department of Agriculture, Soil Conservation Service, Washington, D.C., 1985.

12. Kundu, P.K., A numerical investigation of mixed layer dynamics, *J. Phys. Oceanogr.*, 10(2), 220–236, 1980.

13. Chubarenko, B.V., Chubarenko, I.P., and Sidorenko, A.V., Hydrodynamic modeling of the Vistula Lagoon and calibration currents measurements, information presented during 4th Workshop of NATO CCMS Pilot Study on Ecosystem Modeling of Coastal Lagoons for Sustainable Management, Gdynia, Poland, April 19–23, 1998.

14. Zac, V.I., Ed., *The Processes of Turbulent Diffusion in Seas*, Hydrometeoizdat, Leningrad, 1986, 208 pp.

15. Martin, J.L., McCutcheon, S.C., and Schottman, R.W., *Hydrodynamics and Transport for Water Quality Modeling*, Lewis Publishers, Boca Raton, FL, 1999, 794 pp.

# 8 Decision Making for Sustainable Use and Development

*Karen Terwilliger and John P. Wolflin*

## CONTENTS

1-56670-686-6/05/$0.00+$1.50
© 2005 by CRC Press

## 8.1   INTRODUCTION

### 8.1.1   PURPOSE AND SCOPE

The purpose of this chapter is to suggest a basic framework for making informed decisions and taking positive actions regarding the sustainable management of lagoon systems utilizing the information and tools described in this book. Previous chapters have presented the current status of available information on lagoon systems and models to describe the processes and mechanisms of the interrelationships and energy flow within a lagoon system. These data and models are useful for demonstrating the cause and effect relationship of changing input variables to predict the alternative future outputs for a lagoon ecosystem. They form the basis of a decision support system (DSS) that should be customized and enhanced on a continuing basis in order for sustainable management decisions to be effectively integrated into the socio-economic system (SES) influencing the natural lagoon system.

It must be recognized that many decisions that affect each lagoon will be made temporally (over a time scale) and spatially (across a wide geographic area and diverse societal infrastructure units and levels). It is critically important to provide the best available knowledge and information in a coordinated way. This will result in decisions that foster the sustainable management of these threatened coastal systems. It is the task of the decision maker to make choices that affect the lagoon system using the best available information and tools. These decisions inevitably center around finding the balance between the finite capacity of the lagoon system and the many demands placed upon it by the socio-economic system that depends upon it. It is a further task to establish a process or plan according to which informed decisions can be made over time about the future of the lagoon, with consistency and coordination by the multitude of "users" of the lagoon system.

In this chapter we suggest that a framework be established and used to guide the multidisciplinary decisions of the SES and diverse human communities of the lagoon watershed—an integrated lagoon sustainable management plan (ILSMP). The many sectors/disciplines/community functions and structures that make decisions that impact the lagoon should then incorporate the goals and principles of the overall lagoon plan into their respective arenas. Continued integration of this planning process with the best available knowledge and information into the many sectors of the socio-economic infrastructure of the lagoon area will promote informed decision making for sustainable management of the lagoon.

This chapter presents a stepwise process as a guide for decision making with the goal of sustainable management of a lagoon (Figure 8.1). It emphasizes the importance of developing and employing a good decision support system (DSS).

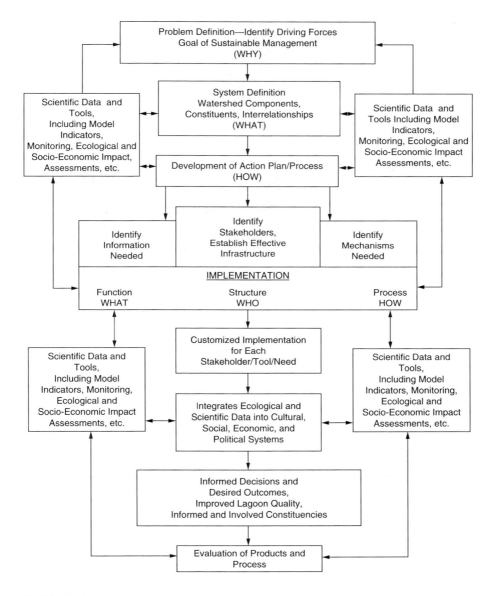

**FIGURE 8.1** Lagoon management decision-making process.

Such a DSS should include a wide range of ecological and socio-economic data along with the appropriate tools to collect, analyze, and evaluate this complex of information (Figure 8.2). This chapter provides examples and lists of the basic components of a DSS to assist in the decision-making process. This process attempts to answer the basic questions that are commonly asked in the course of lagoon management.

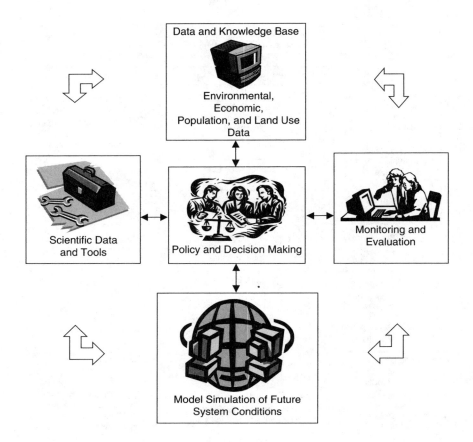

**FIGURE 8.2** Decision support system.

### 8.1.2 COMMON QUESTIONS AND ANSWERS ABOUT SUSTAINABLE LAGOON MANAGEMENT

- WHY—Why manage a lagoon for sustainability?

    The "health" of the SES is dependent upon the "health" of the lagoon ecosystem. Sustainable management means to meet the needs of the present without compromising the ability of future generations to meet their own needs.[1] Only through sustainable management will a healthy socio-economic and ecological system be maintained.

- WHAT—What lagoon system components and information do we need to make a decision about sustainable management?

    The best available socio-economic and ecological data are necessary along with a process for long-term integration of sustainable practices into the human infrastructure and footprint.

- HOW—How do we begin and continue to manage a lagoon for sustainability?

  Establishing an integrated lagoon sustainable management plan is a process for long-term integration into the temporal and spatial diversity of decisions made about the lagoon system.

- WHO—Who should be involved in the process?

  As many sectors/stakeholders as possible should be involved early on and continually throughout the process to maximize both short- and long-term successful implementation.

- WHERE—What area should be managed?

  The entire lagoon watershed and exchange area must be considered for sustainable management. This requires interjurisdictional coordination.

- WHEN—Over what time period is this necessary?

  Sustainability infers a long-term intergenerational timeframe and decision making will therefore need to be a continuing process that should employ the most current and best available data and knowledge. Regular evaluation intervals should be established to evaluate effectiveness of the management plan and its implementation for the benefit of this and future generations.

### 8.1.3 UNDERLYING PRINCIPLES AND ASSUMPTIONS

- Sustainable management is a conscious social decision that provides for the long-term health of both the ecological and economic systems of the lagoon area. The finite capacity of the lagoon's natural capital (NC) cannot meet the growing demands of the socio-economic system without a strategy of sustainable management.
- The use of the best available information, knowledge, and tools, infused throughout and an interactive process will result in improved, better-informed decisions.
- The use of a model as a tool in the decision-making process will enhance awareness of the interrelationships within the ecosystem, especially its input and output variables. This will further enhance accuracy of predictions for and awareness of the consequences of human actions and decisions concerning the lagoon system.
- Implementation and integration will need to occur at various national, regional, and state levels, but will be most effectively and ultimately accomplished at the local level.
- Public and stakeholder input and involvement into the process provides for increased acceptance of the plan and degree of implementation success.

### 8.1.4 WHAT IS DECISION MAKING?

Decision making means choosing between alternative courses of action when the consequences resulting from this choice are not always certain. Decision making involves information processing. Therefore, both the information and the process utilized are critical to effective decision making. The process of decision making calls for an assessment of existing and desired information and constraints as well as an analysis of the costs and benefits of each possible alternative choice. The accuracy of the analysis depends upon the accuracy of the information available for the analysis.

The information and tools available to the decision maker are referred to as the DSS (see Figure 8.2). The DSS therefore consists of a wide array of quantitative and qualitative data and tools, as well as the process and structure developed to ensure integrated and coordinated long-term decision making. A number of tools are available to the decision maker. This book focuses on providing these data and tools, including models, to be used as valuable means for predicting outcomes of alternative choices.

A conceptual example of a DSS for decision making in balancing socio-economic and NC development is presented in Vadineanu.[2,3] An example of a DSS used for management of a watershed is the Colorado River DSS by the Colorado Water Conservation Board and Division of Water Resources.[4] Other examples of applied decision-support systems and integrated modeling are the USDA's DSS for the Integrated Pest Management Program[5] and the Rousseau et al.[6] integrated model of land use and water management. These examples demonstrate model applications and scenario analyses based upon real data derived from the information collected as part of the project's DSS.

An underlying premise to any decision is the recognition of the interdependence between the lagoon ecosystem and the socio-economic infrastructure that directly and indirectly influences the ecosystem (i.e., catchment area, watershed, airshed). This important link is the basis for any DSS with the goal of developing a long-range integrated plan for sustainable management of lagoons. The area's socio-economic goals should reflect sustainability of the ecological life-support system of the lagoon. Both ecological and economic impacts should be quantified and analyzed as thoroughly as possible in any decision.

The following sections describe a process that facilitates decision making. Each section addresses an essential step that allows the decision maker to gather, process, and analyze information to produce effective decisions that result in sustainable management of a lagoon system. These sections further suggest the involvement of the public and stakeholders wherever possible in order to maximize information exchange and implementation success.

Decision making must be recognized as an ongoing process. Decisions that affect the lagoon area will be made continually. The important need and goal should be to develop a process or system that provides for informed decisions by the many different agencies and authorities that will be making decisions in this area. This is why an integrated, multidisciplinary plan is necessary. A plan or process that involves as many of the decision-making authorities as possible will increase effectiveness, consistency, and integration into the community. Decisions will continue to be made and the process should incorporate additional information and tools through regular evaluation of existing conditions.

## 8.1.5  What Is Sustainable Management?

As introduced in Chapters 1 and 2, sustainable management is managing to meet present needs as well as providing for future generations to meet their own needs.[1] Conceptually, it requires the awareness and consideration of the ecological system by the socio-economic system that has super-imposed itself upon the natural lagoon system. It requires measurements of and accountability for the values that the lagoon ecosystem provides to the SES that affects it.[2,7] Harris[8] has compiled a series of papers on rethinking sustainability in terms of institutional roles and many other societal considerations.

## 8.2 IDENTIFICATION OF THE PROBLEM AND NEED FOR SUSTAINABLE MANAGEMENT

Since early times, human settlement of coastal zones and utilization of these highly productive natural resource areas have created rural and urban landscapes reflecting cultures centered on trade and largely oriented toward the use of these special ecological systems. Agricultural development, urbanization, and associated industrial developments continue to modify and impact the coastal zones globally. It is no wonder that we find today that in most NATO coastal countries the vast majority of the population lives within a 50-km coastal band.[9] This has resulted in direct and indirect impacts that have considerably reduced the ability of these ecosystems to meet an ever-increasing demand for their use and development. Human population impacts have upset the delicate ecological balance and have resulted in the compromised health and productivity of both the ecological and economic systems no longer best serving the people of the area.

It is ironic that the values provided by the lagoon systems that have been the basis for human habitation and development are those values that have been most significantly impacted as a result of habitation. If coastal zones are going to continue to meet competing interests, an integrated, balanced management approach must be defined and implemented for the long term. The management approach must consider not only the broad interests for use and development (demand), but also the natural resource limits for delivery of goods and services (supply) and consequences of overutilization. Decisions about lagoon management should be based upon the best available scientific ecological and economic information and should be made with the best tools and processes available.

Just as the decade of the 1970s was considered the foundation of modern environmentalism, the 1980s were recognized as the emergence and framing of the concept of sustainable development. The World Conservation Strategy (WCS) launched by the IUCN in 1983 not only presented the popular definition of sustainable development, but also concluded that existing decision-making structures and institutional arrangements, both national and international, were inadequate to meet the demands of sustainable development.[1]

Numerous multilateral environmental agreements (MEAs) resulted, and the multilateral fund encouraged participation of developing countries.[10] The 1990s were then considered the decade for implementing sustainable development as the trend toward gobalization accelerated. The 1992 UNCED Rio Earth Summit produced major advancements in the implementation and application of sustainable development. Agenda 21 provided a blueprint for the environment and development into the 21st century and resulted in several major conventions and agreements. The Global Environmental Facility (GEF), created in 1991, and the Commission on Sustainable Development in 1992 facilitated the implementation of Rio agreements; however, the Rio + 5 report concluded that progress had been inadequate and too slow. The World Business Council for Sustainable Development (WBCSD) created in 1995 and the International Organization of Standardization 14000 mirrored sustainable development efforts in the private sector. Since the turn of the century, attempts to implement sustainable development have continued around the world, as more specific challenges and questions emerge with the additional integration into traditional approaches to development and planning.[10]

Below is a list of common problems and conditions associated with degradation of lagoons. The decision-making process must identify these conditions, prioritize these problems, and address them in an integrated, long-term plan in order to sustainably manage the lagoon.

Conditions that indicate a need for sustainable management of a lagoon are

- Eutrophication
- Contamination (by persistent and toxic substances)
- Oil pollution
- Presence of artificial radionucleides
- Exploitation of living (reduced shell and fin fisheries) and mineral resources
- Lack of sanitation of bathing waters
- Coastal degradation
- Threat to marine biodiversity

As described in Chapters 4 and 5, nutrient loading is a common problem throughout the coastal zone environment. Inadequate urban and industrial drainage and wastewater treatment facilities and run-off of nutrient-laden water (nitrogen, phosphorus) from agricultural areas in the catchment area are often responsible for nutrient loading in coastal zones. In addition, atmospheric deposition of nitrogen and other chemical constituents is significant in a broad geographic area and contributes to concentration of pollutants in coastal zones. Many other specific problems exist locally, including dumping of chemical wastes, weapons, industrial wastes, alien invasive species, and large-scale construction. These conditions are often the result of lack of land use planning and signal the need for sustainable management.

It is important to note that it is usually less costly and more efficient to detect and treat these conditions early in their development. These are some of the more visible reasons to sustainably manage a lagoon. They indicate that the health of the lagoon system is threatened and that the lagoon will not be able to continue to provide the life support systems for quality of life for all living plants and animals (including humans) in the lagoon watershed. This results in a compromised ecology and economy of the area.

Because coastal lagoon areas are highly sensitive and subject to overutilization and degradation, it is no surprise that more than 30% of the special protection areas designated under European Union directives for conservation are coastal, and that many NATO and partner countries have developed a considerable body of protective legislation in recognition of their value.

In order to arrive at the need for sustainable management, it is necessary to first identify the problems or issues driving the lagoon system and its management. The decision maker must assemble and utilize the best available data and the knowledge based on the lagoon system that can provide this information. This can be done by engaging the scientific community, regulatory authorities, nongovernment organizations (NGOs), and stakeholders with the data, expertise, and knowledge of the lagoon system. Compilation and analysis of the most current information and opinions will provide the decision maker with the best definition of the problems and driving forces of the lagoon system. Information must include data on both the lagoon's ecological and socio-economic values. Once the problem is recognized, the decision maker is faced with a multitude of choices on how to proceed toward a solution. The first step toward a solution, addressed in Section 8.3, focuses on identifying the critical driving forces and components of the lagoon system as well as the SES affecting it.

## 8.3 DEFINITION OF LAGOON SYSTEM AND CHARACTERISTICS USEFUL FOR THE DECISION SUPPORT SYSTEM

Decision makers may ask—What are the critical lagoon system components needed in decision making? In order to answer this question they must:

- Identify and incorporate the best available data and expertise on the ecological and socio-economic system (SES) of the lagoon area as well as the tools to assess them
- Inventory the watershed components, their ecological and economic value, and all existing and potential impacts that call for decision making
- Determine those critical driving forces and variables that are producing the problem

The first step in defining the lagoon system is to inventory available ecological and socio-economic data and expertise on that specific lagoon system as well as pooling data from other lagoon systems for a broader and more global perspective. Chapter 2 clearly introduced the lagoon system components of both the ecological and socio-economic systems. From this knowledge and these data, the critical driving forces or problems need to be determined. Not all the data are critical to a decision, and it is important to sort out which data are appropriate and needed for the decision. This can be done by locating and engaging those ecological and economic experts and stakeholders most familiar with the lagoon.

The lagoon ecosystem components identified and described in Chapter 2 are those components that we recommend be considered in the decision-making process. These ecological components and the hierarchical interrelationships between them are considered to be the basic elements used to define the natural system and identify the problems in the lagoon system. These components describe the "supply" side, or NC, provided by the lagoon system to the SES, or the "demand" side.[9]

Just as the ecological components were defined and inventoried (Chapters 2 through 5), the SES components should be, too. The basic units or elements of the SES and the hierarchical interrelationships within its infrastructure need to be identified and inventoried.[11,12] For example, the units and levels of agencies, private interest groups, industries, land use, education system, and information system that have an effect on the lagoon system should all be considered as basic conceptual and methodological units with which to approach sustainability. Ecological and economic variables representing both the potential capacity of the lagoon and the potential use (supply and demand) should be inventoried and viewed as important considerations to the decision maker.[12–15]

Once inventoried, the value of the basic components should be estimated. An example of an economic valuation or inventory is provided in Figure 8.3. It consists of two steps. The first step is to define the economic values derived from and intrinsic to the lagoon ecosystem, commonly referred to as the NC. *Capital* has been traditionally defined as the accumulated wealth in the form of investments, factories, equipment, etc. NC is similarly defined as the natural resources we use, both renewable and nonrenewable. More recently, this definition has been expanded to include not only the "goods," but also the myriad ecological "services" that the natural system provides (e.g., wetland filter pollutants from run-off water, buffer shorelines from erosion) and which are, in many cases, sinks to ground water.[16] For example, Costanza et al.,[17] Brouwer et al.,[18] and Wilson and Carpenter[19] describe the significant economic values of wetlands acknowledging their multifunctional resource role. The status of a methodology attempting to quantify these values is rapidly evolving and still considered to be inadequate in its early development stages.[13,20–23]

Defining the contribution of a lagoon to an area's economic system therefore consists of identifying the major economic and social values of the lagoon ecosystem. This can be done by categorizing the stocks that produce the wide range of ecological and economic goods and services used by the area's economy as NC or manmade capital.[17,24,25] Any analysis of NC should include both renewable and nonrenewable resources as well as the wide range of ecosystem processes that maintain and provide for the ecological life-support system of the area. Therefore, it is important to include in any analysis the many natural processes and functions that maintain the atmosphere, climate, hydrology, soils, biological fertility, and productivity as well as nutrient recycling, waste assimilation, and even the maintenance of genetic stocks (Figure 8.3). Unfortunately, estimation of these values is not a simple process and has only recently begun to be well quantified. Nevertheless, estimates of these values should be developed in order to comprehensively account for the value of the lagoon ecosystem to the area.

Additional anthropogenic values, including aesthetic and amenity values, have been placed on ecosystems.[14,15,17,19,24,26] These values should also be estimated in order to allow for a realistic cost-benefit analysis of the various alternatives for lagoon management. Section 8.7 presents a summary of some of the existing valuation tools and methods for quantifying the economic value of the ecological assets.

The second step in this inventory or SES definition is to identify those natural and human elements and actions (demands) that influence the lagoon ecosystem

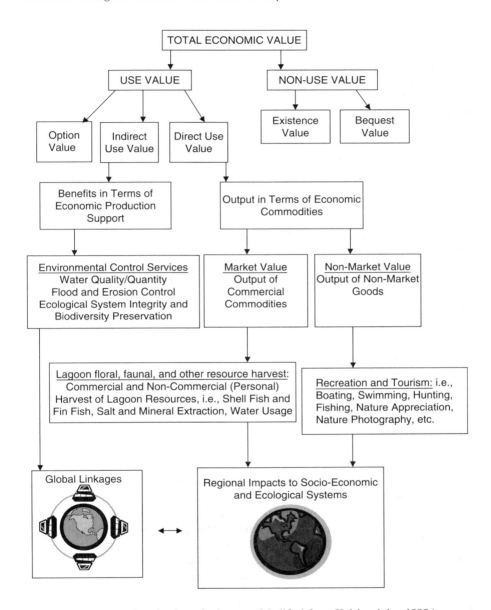

**FIGURE 8.3** Economic valuation of a lagoon. (Modified from Kulshreshtha, 1999.)

and therefore the economic values associated with it. In other words, it is first necessary to identify and inventory the NC provided by the lagoon ecosystem and estimate its economic value. It is then important to inventory all of the existing and potential socio-economic impacts, natural and anthropogenic, to the lagoon, in order to evaluate the cost-benefit of each impact on the NC. Here, the decision maker should gather the best data and expertise on the SES and assess each of the impacts in order to identify which are the key driving forces and variables in this system. An example might include the determination that the industrial sector is the key driving force in the point source contamination or pollution in the lagoon system, which has diminished the ability of lagoons to provide clean water for tourism, fisheries, etc.

Another example might be that of poor agricultural practices resulting in nonpoint source pollution and water quality degradation. It is often a combination of these variables that affect the lagoon ecosystem integrity, and it is the task of the decision maker to address each of these driving forces that are causing the problem in the most integrated, long-term method possible. This then calls for inventories of farms, agricultural co-ops and industries, agricultural agents, and the other existing socio-economic units that address the agriculture in the area. Further information and knowledge needed here might also include the banking and insurance structure, which provides resources to the farming community and affects their farming structure. It also includes an analysis for the farming community of the kinds and numbers of crops, farms, agricultural markets, equipment, financing, programs, and regulations. Oglethorpe and Sanderson[27] present an ecological economic model for agri-environmental policy analysis, which includes many of these factors.

Only through use of the most current data and knowledge of this system can the decision maker recognize and evaluate the various options to address the problem. A process for involving and engaging these agricultural and support socio-economic units needs to be developed as addressed in Section 8.4.7. Other SES infrastructures involve the tourism, fishing, or municipality planning sector. It is the task of the decision maker to identify which of these key SES units and structures are the driving forces behind both the problem and resolution for the lagoon conditions.

Decision makers should weigh the cost and benefits of their preferred alternative choices before making a final decision. Estimation of the economic value of each of the goods and services provided will supply critical data for a cost-benefit analysis.[28] The use of a model to predict outputs of the SES should provide the basis for estimating both the ecological and economic changes expected (see Chapter 6 for details). As a model predicts new outputs, these changes in the NC will need to be recorded as assets or liabilities for the affected economy. Each change in the model's input variables has a corresponding output change in the NC that needs to be documented in economic terms. Assigning monetary values to all goods and services is the critical link in providing more complete data for decision makers to determine the consequences of their actions in terms of benefit or loss to public or private financial interests (Figure 8.4).[29]

One scenario would be that of lagoon fisheries. If input variables such as toxics or nutrient loads increase, there could be a resulting decrease in the fisheries stock. The corresponding market values for the fish stock, the resulting decline in local employment, and other fisheries stock-related goods and services provided by the fisheries should be quantified in order to estimate the economic impacts of such a change. This is the additional data that the decision maker needs to analyze the cost and benefits of any decision that might influence this system.

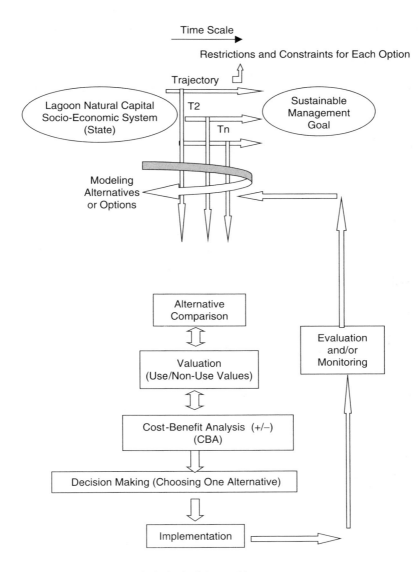

Time Scale

Restrictions and Constraints for Each Option

Trajectory

Lagoon Natural Capital
Socio-Economic System
(State)

T2

Tn

Sustainable
Management
Goal

Modeling
Alternatives
or Options

Alternative
Comparison

Evaluation
and/or
Monitoring

Valuation
(Use/Non-Use Values)

Cost-Benefit Analysis (+/−)
(CBA)

Decision Making (Choosing One Alternative)

Implementation

**FIGURE 8.4** Cost-benefit analysis in decision making.

This is an unfortunate but globally recurring scenario. Biological monitoring indicates that a decline in water quality (usually associated with an increase in nutrient loading and toxics) has shown corresponding declines in fisheries stock in the last half century. Use of those water quality indicators in a model would allow for prediction of the resulting estimated economic decline. Costs of clean-up or other water quality improvement efforts could then be factored into the equation to calculate both short and long-term net gains or losses. These are additional concerns and values that the decision maker must factor into the cost-benefit analysis.

Another example is that of tourism or other related outdoor recreational income attributable to the presence of "quality of life" and intrinsic contributions of a lagoon. An estimate of the income to the area from tourists, the associated employment, and income from other man-made or skilled products or services should be tallied. At the same time, however, the other side of the balance sheet, the impact cost to the NC by the increased use and degradation, must be quantified as well. Similarly, increased population growth and corresponding land or water use changes can be measured by changes in biological indicators. Chapter 7 deals with this concept in detail and provides examples and discussion of monitoring techniques.

## 8.4 TOOLS FOR DECISION MAKING

A wide assortment of implementation tools is available to those involved in the planning process and to decision makers. Many effective tools are already being used for improvements in lagoon water quality and the health of the socio-economic and ecological systems in the watershed. Involvement of more types of constituent groups and plan participants usually enriches the number and types of tools, options, and resources available. A sample list of tools now being used to assess and improve the health of these systems is provided below.

Technical (scientific ecological and socio-economic) tools available are

- Models—including modules of hydrodynamics, chemical processes, biological, ecological, land-use, and spatio-temporal factors that describe the lagoon system (Chapters 3–6, Section 8.4.1)
- Monitoring (Section 8.4.2 and Chapter 7)
- Indicators—biological, social, economic, etc. (i.e., the Organization for Economic Co-operation and Development (OECD) core set of indicators) (Section 8.4.3)
- Graphic user interface, including geographic information system (GIS) (Section 8.4.4)
- Economic valuation methods (Section 8.4.5)
- Environmental and social assessments (Section 8.4.6)
- Policy transformation and implementation (Section 8.4.7)
- Public input (Section 8.4.8)

Since most of these tools have been described in more detail in previous chapters, only a brief summary is provided here in the context of their use in decision making.

### 8.4.1 MODELING AS A DECISION-MAKING TOOL

A model is a useful tool for decision making because it provides for a better understanding of the elements, mechanisms, kinetic processes, and capabilities of the systems being modeled. Modeling allows for integrated interpretation of input scenarios by varying the existing parameters to the desired future conditions to produce numerical solutions for the differential equations describing the ecosystem's processes. The user is then able to model different alternative scenarios and analyze costs and benefits of each of the outcomes that has been input. A model provides the decision maker with a prediction, based upon the best scientific information available, of the outcome of changing input variables. This provides the necessary basis from which to analyze the costs and benefits of each alternative choice. A model, therefore, provides a predictive tool to the decision makers for use in their DSS, a tool that can predict the outcome of many different possible choices. Even though the results are predictive and not certain, they provide the decision maker a greater understanding of the ecosystem and the consequences of changing input variables. See Chapters 3 through 6 for examples of and references to watershed models.

Decision makers should weigh the cost and benefits of their preferred alternative choices before making a final decision. Estimation of the economic value of each of the goods and services provided will supply critical data for a cost-benefit analysis.[28] The use of a model to predict outputs of the SES should provide the basis for estimating both the ecological and economic changes expected. As a model predicts new outputs, these changes in the NC will need to be recorded as assets or liabilities for the affected economy. Each change in the model's input variables has a corresponding output change in the NC that needs to be documented in economic terms. Assigning monetary values to all goods and services is the critical link in providing more complete data for decision makers to determine the consequences of their actions in terms of benefit or loss to public or private financial interests (Figure 8.4).[29]

An example of how an integrated management of surface water model can be used in decision making can be found in Mailhot et al.[30] A number of land use scenarios such as timber harvest, agricultural practices, industrial, and urban land activities are modeled to show the varying outputs according to the change in land use or practice.

Another example of a watershed model is the Patuxent Watershed Case Study (PLM), which combines general models of ecosystem and economic site-specific processes with remote sensing and GIS data on land use changes.[25,31] This case study demonstrates simulation of detailed spatial dynamics of the watershed including the interaction of ecological and economic components and provides the link between science and policy. Another general model that might be useful in describing the ecological interrelationships can be found in the general ecosystem model (GEM). Chapters 2 through 7 provide a variety of other specific models that address and best model the units described in each chapter.

## 8.4.2 Monitoring as a Decision-Making Tool

Monitoring provides specific answers for specific questions posed about the specific aspects of a lagoon system; therefore, proper design of a monitoring program is essential for providing the right data to answer these specific questions. Monitoring collects selected data, with specific quality, format, collection, storage, and analysis guidelines, specifically designed to answer the questions desired to describe the status of the lagoon. The thoughtful and planned choice of variables to measure over time will provide decision makers and managers with the information needed to assess the status of a lagoon as well as to progress toward any goals and objectives set for the lagoon system plan. Therefore, informed decisions can best be made through a properly designed monitoring program and protocol including the selection of the most appropriate parameters. GEO 2000 recognizes effective monitoring as one of the most important needs in advancing sustainable development.[10] When a

sustainable management plan is developed for a lagoon system, certain goals and
objectives are set. If the correct variables are not identified and measured, then the
resulting information is of no use or is misleading to decision makers. It is essential
to select and agree upon the appropriate variables and how to measure them in order
to develop an effective monitoring program. Guidelines and more information on
the various aspects of monitoring are described in Chapter 7.

### 8.4.3 INDICATORS AS DECISION-MAKING TOOLS

Indicators have been used extensively as a tool for gauging the ecological status of an
ecosystem as well as the effectiveness of lagoon management efforts. However, little
progress has been made in the development of measures for sustainable development,
and there is no general agreement on what parameters should be used to measure
sustainability. Efforts to formulate a set of sustainability criteria that can be used to
indicate whether a path is sustainable have recently commenced.[32] Recent attention has
been focused on the need to develop indicators of sustainable development.[10,33] The
term *indicator* has been given various definitions, but it generally refers to a measure
of something. The OECD defines an indicator as "a parameter, or a value derived from
parameters, which points to, provides information about, or describes the state of a
phenomenon, environment, or area with a significance extending beyond that directly
associated with a parameter value."[34] Indicators include both measures of environmental
quality and anthropogenic pressures resulting from social and economic activity. Ecological indicators are also essential in determining what to monitor and how to interpret
what is found as well as assessing the effectiveness of management actions. The existence of indicators helps to facilitate and to stimulate long-term protection of the environment and to foster sound environmental decision making through credible science.

The OECD developed a systematic framework for environmental indicators commonly referred to as "pressure-state-response" or driving force-pressure-state/impact-
response (DPSIR),[34] which is based on the following causality chain: Human activities
(SES) exert pressures on the environment ("pressure") and change its quality and the
quantity of natural resources (NC)("state"). Society responds to these changes through
environmental, general economic, and sectoral policies (the "societal response"). This
format has also been adopted as a feedback loop by the Global International Water
Assessment Committee and European Environment Agency (EEA) to pressures resulting from human activities.[34]

Indicators of pressure (P) can also be called indicators of driving forces or
stressors. In the case of eutrophication or contamination, pressure on the lagoon is
caused by the direct input from point and diffuse sources of anthropogenic matter.
The pollution load can be regarded as a primary pressure indicator. However, the
original cause of pressure is sometimes created far from the lagoon. The identification of pressure requires knowledge of SESs within the catchment basin. The reason
for the original pressure might be poor governance or economic or social problems.
A causal chain analysis is needed to identify the original source of pressure. Therefore, we might have primary pressure indicators or secondary pressure indicators.
The following examples of the pressure indicators are related only to primary (direct)
pressure.

State (S) indicators are needed to assess the state of the lagoon system properly. This knowledge comes through research and monitoring. This is why establishing proper monitoring and research programs is an important task.

Response (R) indicators should help the decision-making process, developing regulatory standards and identifying the actions needed. The response in most of the cases should be based on modeling scenarios showing effects of different decisions. However, modeling can produce a reliable answer only if based on well-based data sets obtained within scientific and research programs.

The following examples of general indicators are provided to illustrate some of the most common symptoms and environmental issues. These examples are provided to present the use of the "pressure-state-response" framework[35] and are by no means a complete identification of all related indicators.

*Issue: Eutrophication*
Eutrophication has been defined by the OECD as "over-nourishment of adequate plants."[34] This is a particularly important issue in semi-enclosed basins and lagoons and has caused significant adverse biological effects over the past few decades. The following pressure/state/response indicators have been used to assess and monitor the eutrophication of lagoons.

---

**Anthropogenic Pressure Indicators**

Discharge of nutrients from point and diffuse sources
Discharge of nutrients from untreated sewage
Airborne discharges of nitrogen and phosphorus
Agricultural run-off of fertilizers

---

---

**Environmental State Indicators**

Winter concentrations of nitrogen and phosphorus N/P ratio
Chlorophyll *a* concentration in surface waters
Increased frequency and presence of toxic algal species
Secchi-depth visibility
Depth range of macrophytes
Species type and distribution
Oxygen depletion in historically oxygenated areas
Increase in opportunistic algal types

---

---

**Government/Society Response Indicators**

Reduction of nutrient discharges from point and diffuse sources
Adoption of best/sustainable agricultural practices
Technical measures to prevent/treat eutrophication, e.g., sewage
and wastewater treatment, buffer strips, wetland restoration

---

*Issue: Chemical Contamination*

Presence of chemicals, which are toxic, persistent, and liable to bioaccumulation, including inorganic (heavy metals), organic (some biocides and industrial compounds, usually chlorinated, and some polycyclic aromatichydrocarbons), and metal-organic compounds (organic compounds of mercury and tin). The following indicators can be used to monitor and assess this aspect of lagoon health.

---

**Anthropogenic Pressure Indicators**

Discharges of toxic compounds from land-based sources
Deposition of toxic compounds via atmosphere
Deposition of toxic compounds from sea-based sources

---

---

**Environmental State Indicators**

Contamination levels/concentrations in sediments and biota
Bioaccumulation levels/rates in organisms
Long-term trends in concentration levels
Ecotoxicological effects, e.g., reproductive declines,
immunosuppression, carcinogenic effects, genotoxic effects

---

---

**Government/Society Response Indicators**

Improvement/construction of wastewater treatment facilities
Ban or significant reduction in production and/or
use of substances (e.g., DDT, PCBs)
Reduction of toxic emissions from industry

---

### 8.4.4 GRAPHICAL USER INTERFACE AS A DECISION-MAKING TOOL

Anything that helps the user prepare input data for a model and/or to analyze and visualize the outputs of a model can be thought of as a user interface. Data input forms, file format conversion programs, and graphical post-processing software can all be part of user interfaces.

Usually, the model itself is a stand-alone computer program written in a standard programming language, which can read or write in a predefined format. This software development strategy usually makes the model operation harder, but the modeling software can be ported to different computers or operating systems easily. In the 1960s only computer scientists were able to use computers efficiently. The type of input was not very important for them, and some could even give commands in binary code.

Today, scientists, engineers, and even managers (some of them with little computer knowledge) are using software for engineering and decision-making purposes. These users are specialists in hydrodynamics, ecosystem modeling, water quality management, decision making, environmental management, water resources, etc., but they are not (and they need not be) computer scientists. In this case, the user interface has the task of isolating the user from the details of information technologies and supplying the user with an environment, where the user needs to know only about the processes and not about the computational details such as how to format the input files for the models.

### 8.4.4.1  Types of User Interfaces

A general classification of user interfaces would be as follows.

- Graphical User Interfaces (GUIs): These user interfaces use all the advantages of window-based operating systems and environments such as Microsoft Windows or X-Window of UNIX. They have graphical elements such as data boxes and command buttons, etc. GUIs are easy to design but in many cases the coding process needs considerable time and manpower. Most commercial and some freeware modeling software packages are distributed with GUIs. Innovative technologies like GIS are integrated into GUIs. The latest version of EPA's Water Quality Model WASP (Version 6.x) has a GUI with GIS and database support.
- Geographic Information System (GIS): In the last few decades, GIS has become an important tool for spatial database compilation, environmental assessment, and presentation of modeling results. GIS is used extensively to store spatially distributed data (land use, demography, bathymetry, roughness, vegetation, sediment physical and chemical parameters, etc.). A number of coupled hydraulic and ecological models include GIS as a part of user interface to enter the data and visualize the modeling results (MIKE model family developed by DHI, SMS models, ASA models, and WES = CH3D developed by the USACOE are GIS-compatible formats). Geostatistics, including spatial interpolation techniques, are critical to create the statistically reliable spatial grids, which are necessary for both hydraulic and ecological models. Further modeling applications are commonly used to combine outputs of ecological models together with other relevant information (i.e., socio-economic maps. pollution levels) as well as perform additional spatial analysis for use in integrated coastal zone management. Finally, GIS is a very valuable tool to present modeling results as self-explanatory maps and graphical interfaces to the public and decision makers.
- Command Line Based User Interfaces: The user runs data input commands from a nongraphical command prompt. These may also be used for data file management or file format conversion. This type of a user interface can also be used as a preprocessor, which checks the validity of the input data. The CE-QUAL-W2 hydrodynamic/water quality model developed by U.S. Army Corps of Engineers has a utility program for this task.[36]

- Hybrid User Interfaces: These interfaces have graphical and command line–based parts. A hybrid user interface consists of many computer programs interacting with each other and working together. Usually the graphical part takes the management of the command line–based parts. The command line–based parts interact with the raw data and the graphical part is responsible for visualization of modeling data.

### 8.4.4.2  What Can a User Interface Do?

A user interface will help a user enter the correct data in the correct format. A situation may arise, when a user makes a mistake when entering data, which cannot be detected by the model. A user interface will detect the error and ask the user for a correct entry, before running the model with invalid input data. Another thing a user interface can do is to create a model input file from different data sources such as databases or GIS data. A new user of a model would find a GUI very useful, because he or she would not need to deal with data formats of a model and could concentrate on other things. Some modeling packages have grid generators, which use GIS data to create computational grids for the models.

Some user interfaces also have post-processing facilities, which are even more important in analyzing and visualizing the results after a model has been run. Some models may create output files, which might be several hundred pages long. A graphical post-processor, which is a part of the user interface can make the long output file more understandable and accessible. Some of these post-processors can create outputs for GIS, making it even easier for the user to interpret the results.

User interfaces can do a lot of work for the user, and they make the lives of scientists, engineers, and managers easier, but they cannot do all the work for the users. The data analysis, preparation of the input data with valid and correct parameters, and interpretation of the results are (and will be) three important tasks that should be accomplished by the user(s). The user must never forget that "garbage in" means "garbage out." There are many cases where impressive and "colorful" results with no physical meaning were obtained from simulations within a few hours of computational time. So the user interface only helps the user to get rid of the information technology details.

### 8.4.4.3  Design and Development of User Interfaces

Three important steps in a user interface development process are planning, visual design, and code writing.

In the case of GUI and hybrid user interface development, the visual design is very important. User interfaces may get data from different sources such as electronic spreadsheet and database files. Another important issue in user interface design is presentation of outputs. Because model outputs may be used in scientific or engineering reports, they should be converted into readable data with a desktop publishing software package used for report writing.

The most important concept in user interface design as a part of a modeling software package is feasibility. Many information technologies such as object

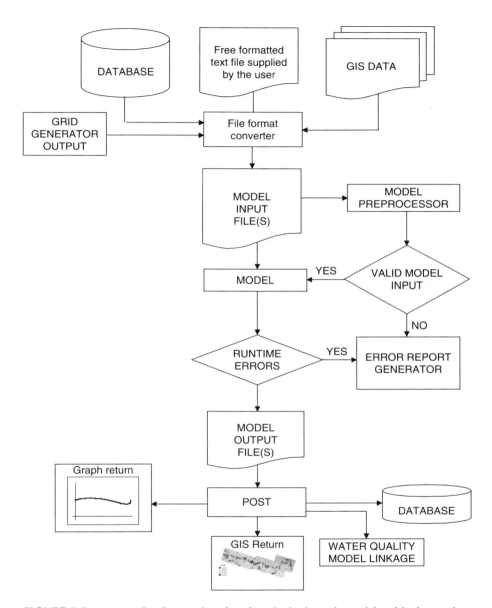

**FIGURE 8.5** An example of a user interface for a hydrodynamic model and its interaction.

oriented programming, GUIs, database engines, and artificial intelligence are available for user interface development. A good user interface is very difficult and in some cases expensive to develop. To design and develop a user interface that is better than needed is a waste of time and manpower. In some cases it is even unfeasible to develop a user interface. The friendlier a user interface gets, the more difficult its development. An example of a hydrodynamic model user interface and its interaction is illustrated in Figure 8.5.

## 8.4.5 Economic Valuation as a Decision-Making Tool

Section 8.2 provided the conceptual framework and need for economic valuation of the ecological lagoon system. Quantifying the various components of the lagoon system as accurately as possible will produce the best information for a cost-benefit analysis. Once an inventory of the goods, services, and other intrinsic values provided by the lagoon system has been conducted, a valuation of each component can be made.

A number of methods for evaluating environmental value in economic terms exist. However, due to the complexity and dynamics of the natural environment, the concern remains whether the existing technology and information adequately capture the comprehensive values of the system. Therefore, these methods are considered as tools to provide the best estimates available. There is a great deal of work in this emerging field of environmental economics and advances in environmental accounting are being made.[37] Significant information gaps and research needs remain.[17,19,28,38,39]

An example of an economic valuation or inventory is provided in Figure 8.3. It consists of two steps. The first step is to define the economic values derived from and intrinsic to the lagoon ecosystem, commonly referred to as the NC. "Capital" has been traditionally defined as the accumulated wealth in the form of investments, factories, equipment, etc. NC is similarly defined as the natural resources we use, both renewable and nonrenewable. More recently, this definition has been expanded to include not only the "goods," but also the myriad ecological "services" that the natural system provides (e.g., wetland filter pollutants from run-off water, buffer shorelines from erosion) that are, in many cases, sinks to ground water.[16] For example, Costanza et al.,[17] Brouwer et al.,[18] and Wilson and Carpenter[19] describe the significant economic values of wetlands and acknowledge their multifunctional resource role. The status of methodology attempting to quantify these values is rapidly evolving but is still considered to be inadequate in its early development stages.[20,21,40]

International attention has been focused on developing a better method of quantifying interrelationships between the natural environment and the economy.[41-43] The 1993 United Nations Handbook of the UN Statistics Division issued the integrated environmental and economic accounting system (SEEA). Environmental accounting was begun in the European countries by the Norwegian government in 1974, followed soon after by the French government. The Fifth Community Programme for the Environment and Sustainable Development[44] called for an environmental adjusted national account (to account for the natural resource stock of air, water, soil, landscape, heritage, etc.) to be formally adopted by the end of the decade.

A few examples of how countries are attempting to better account for natural resources can be found on the informative Internet sites.[9,45] Further examples, which discuss how these systems need to be improved, are contained in the Summit Conference on Sustainable Development.[46-48] These reports describe a national resource accounting framework that has been used to evaluate agricultural practices in the context of production systems including the cost of soil erosion, infrastructure costs, etc., and a service-level approach that has been proposed to value natural resource and man-made assets in making policy decisions. This approach relies on a set of performance indicators to measure user-base service levels. This provides decision makers data on regional and national performance of resources and assets.

Traditionally, the cost-benefit analysis (CBA) framework has been used extensively in program evaluation (Figure 8.4). CBA is an important informational tool available to policy and decision makers when evaluating a project. The goal of an economic CBA is to categorize and value all existing and future costs and benefits to society. These monetary values of cost and benefit flows over time are then discounted and compared, and a more informed decision can be made. An example of an environmental CBA performed on Albuquerque's engineered wetlands can be found in Holmes.[49]

CBA has been criticized because of its emphasis on short-term economic efficiency rather than considering the distribution of the welfare on present and future generations. Another approach to evaluation, adopted by the U.S. Water Resources Council and others, is the multiple accounts approach (MAF). The MAF recognizes the existence of several criteria or indicators that should be considered in evaluating a project, including monetary, social, and environmental data within the same framework. This provides the decision maker the flexibility to weigh the various indicators, but it also requires subjective judgments on the part of the decision maker. The MAF has been used to allocate forestland in British Columbia and for old growth timber conservation strategy evaluation.[30]

A list of variables considered in that study were the net value of commercial and noncommercial output and activities supported by the options, the environmental attributes affected by the options not amenable to quantification in the net benefit account, the direct employment generated by the options, and the government-related taxes and other revenues less the expenditures associated with the options on both local and regional levels.

In general, it has been easier to assign market values to NC stocks that are being used. These are referred to as use values for which a corresponding market value can be extrapolated. On the other hand, it has been more difficult to assign nonuse values to the wide array of environmental assets for which no market values can be extrapolated. It is critical to assess the value of all these assets in order to approximate the value of the system as a whole.

There are two main approaches, both direct and indirect, to reveal individual preferences for environmental assets, i.e., to estimate individuals' willingness to pay or willingness to accept. Indirect approaches seek to assign a value to environmental assets by observing people's behavior in related markets and by using date from these markets. The main methods in indirect valuation are the hedonic pricing (HP) and travel cost methods. The direct approach asks individuals directly about their willingness to pay (WTP) or willingness to accept (WTA). These approaches are commonly referred to as the contingent valuation method (CTV).[19,50] The many valuation methods available to price natural and environmental goods by means of individual preferences are summarized in Pearce and Turner.[50] One method estimates the costs necessary to keep the NC intact.[51,52] Turner identified wetland ecosystem services to include flood storage and protection, wildlife habitat, nutrient cycling and storage pollution control, landscape value, shoreline protection from storm damage, recreation, extended food web control, salinity balance mechanism, and commercial goods output.[53]

Brouwer et al.[18] performed a meta-analysis of the use and nonuse values generated by more than 30 wetland valuation studies across Europe and North America. Apart from providing an excellent overview of a qualitative descriptive analysis of wetland function, the study uses it as the basis for the subsequent quantitative meta-analysis of evaluations derived from a large number of CV wetland studies. These studies yielded over 100 value estimates of wetlands, which were considered in the meta-analysis. The analysis identified such values as flood control, water generation, and later quality attributes that exerted a stronger influence over the WTP than nonuse elements such as biodiversity of wetlands, which agree with the findings of other preference studies valuing nonmarket goods.

## 8.4.6  ENVIRONMENTAL AND SOCIAL IMPACT ASSESSMENT AS A DECISION-MAKING TOOL

### 8.4.6.1  Environmental Impact Assessment (EIA)

In many countries, the environmental impact assessment (EIA) process is the predominant methodology used to compile and integrate the array of environmental and socio-economic data, which provide decision makers information they can use to analyze and evaluate impacts.[54–56] It is important to integrate the economic, social, cultural, ecological, and other "larger picture" parameters that more fully represent the societal impacts through such an EIA process for use by decision makers. Such efforts have resulted in the development of such programs as the cumulative impact assessment (CIA), socio-economic assessment, life cycle analysis (LCA) links to EIA, and the strategic environmental assessment.[57]

The U.S. EIA process is framed and guided by the National Environmental Policy Act, 1969 (NEPA) procedures. This process has been widely utilized over the last 30 years; however, it still lacks a detailed socio-economic valuation component. A helpful summary presenting the fundamentals of an EIS has been developed by the U.S. EPA[58] and summarizes the principles of its environmental impact assessment program and supportive legislation in the form of an international training course.

Soon after, Canada established its environmental assessment and review process (EARP) where the responsibility for conducting EIA lies with the national government and ten provincial and two territorial jurisdictions. In a concise status summary of EIA used in over 30 developed and developing countries, Lemans and Porter[59] present the various methods used, problems encountered, and types of impacts experienced.

The proceedings of the European Commission (EC)-sponsored workshops on environmental impact assessment present valuable reviews on the status and examples of EU country EIA process and evaluation. Specifically, conclusions and recommendations of the discussion groups from EIA workshops outline the perceived deficiencies in methodologies of the member states' EIA.[20] Interestingly, one of the priority recommendations was to strengthen the role of the EIA in decision making. The EC provides a review of U.K. guidance on EIA, "Guidance for developers on the preparation of environmental statements."[20] Additional references provide significant guidance on EIA and its integration into planning.[60–63]

## 8.4.6.2 Social Impact Analysis (SIA)

Social impact analysis (SIA) is used interchangeably with socio-economic impact analysis.[64] SIA includes impacts to health and welfare, recreational and aesthetic values, land and housing values, job opportunities, community cohesion, life styles, governmental activities, physiological well-being, and behavioral response by individuals, groups, and communities. In SIA, advanced appraisal of the impacts of development projects or policy changes on individual and community quality of life are recommended.

Governmental and private programs, policies, and projects have the potential to cause significant changes in many aspects of the social environment.[65] These changes can be beneficial or detrimental. Accordingly, the SIA and EIA process should systematically be used to identify, quantify, and interpret the significance of the anticipated changes. SIA has been more recently included in the environmental impact assessment process.

The SIA process should:

- Involve all related social groups and individuals. Communication with the representatives of these social and economic groups should be established and maintained between the affected groups, the owner/developers of the project, and the decision makers.
- Identify impact equity. It is important to identify and separately assess the impacts to each different social group as well as to the general public as a whole. An impact may be determined to be beneficial, when assessed as a whole; however, it might adversely affect a number of social groups.
- Identify the focus of the study. It is often not possible or feasible to assess all social and economic indicators in an SIA; therefore, it is advisable to select indicators that best assess those socio-economic aspects of concern. Just as biological indicators can be used as a tool for the decision maker (Section 8.4.3), selection of the most relevant socio-economic indicators can produce more focused and effective results for management decisions.
- Identify the method. Assessment methods should be determined and their limitations and assumptions should be recognized. The data and predictions that cannot be adequately described should be documented.
- Inform project owners/applicants at various stages of the assessment. Proactive communication with project owners/applicants provides information on project costs and progress to all the parties involved.
- Involve social, economic, and environmental experts in order to provide accurate and balanced information.
- Develop a plan for monitoring and mitigating project impacts.

It should be recognized that social and economic impacts affect each other and are, therefore, clearly interrelated. Examples of categories of impacts that demonstrate this interrelationship are economic, demographic, social, fiscal, and quality of life. Modeling economic and demographic impacts provide basic information for addressing public service aspects such as education, health services, police and fire protection, utilities, and solid waste management. Social impacts include housing, transportation, urban land use, and land ownership. Fiscal impacts are dependent upon many of the public service and social impacts. Quality of life impacts represent a composite of public service, social, and fiscal impacts.

A suggested conceptual approach to SIA is similar to the traditional methods and process in EIA. The following steps represent such an approach:

1. Identification of potential socio-economic impacts
2. Preparation of description of existing socio-economic conditions
3. Procurement of relevant standards, criteria, or guidelines
4. Impact prediction for project action vs. no-action alternatives
5. Assessment of socio-economic impact significance
6. Identification of mitigation measures

Another basic approach to develop an SIA complies with the CEQ regulation inclusions and addresses the following issues of proposed project impacts:

1. Public involvement
2. Identification of alternatives
3. Baseline condition
4. Scoping
5. Projection of estimated effects
6. Predicting response to impacts
7. Indirect and cumulative impacts
8. Changes in alternatives
9. Mitigation
10. Monitoring

This process encourages the development of an effective public involvement plan to assess direct and cumulative impacts and the changes predicted from each alternative as well as to monitor the effects of each. This process also encourages the use of comparative methods to study the impacts in a community where an environmental impact has occurred and then extrapolate from that existing example of similar conditions what might be comparable and relevant in the proposed project.

Both of the above approaches to an SIA require four basic components, which are key to any SIA:

1. Identification of potential socio-economic impacts and the region of interest (scoping)

   Similar projects and SIA examples can be used for comparison purposes to help determine the potential factors to consider. Analyzing the results of the existing, relevant public involvement, and baseline condition studies can assist in determining potential impacts. The use of such tools as matrices, models, and listing is recommended.

   A general list of SIA factors has been developed by Canter[64] and includes demographic impacts, impacts on society, social conflicts between inhabitants and visitors, impacts on infrastructure, and the changes in individual and family life. The Canadian Environmental Impact

Regulations also list SIA impact factors to be considered and categorize them into economic and social impact categories.

Information on the area or region of interest must also be gathered. Data on the location, land and infrastructure requirement, project design and work power requirement, and the aim and scope of the project are essential to thorough impact evaluation.

2. Description of existing socio-economic conditions (baseline conditions)

Information on existing industry, housing, infrastructure, socio-demographics, culture, economics, employment, and workforce is necessary.[66] The existing local authority and political conditions should also be noted. The use of certain indicators or indices has been developed to help categorize and evaluate such factors. Canter[64] has developed indicators to define the existing socio-economic structure including life quality indices. The Asian Bank has developed the Human Improvement Index, which should also be included in an SIA and the UNEP-World Bank's Symposium on Environmental Accounting for Sustainable Development addresses important aspects to include in assessments.[67]

3. Impact prediction (projection of estimated effects)

Prediction of impacts can be estimated by the use of predictive tools such as models. Listing and matrices are other methods to anticipate impacts. Trend analysis, similarity comparison, and scenario development are also tools to consider when attempting to predict the potential impacts of a project.

4. Assessment of socio-economic impact significance (predicting response to impacts)

Several different methods have been used to assess impacts to an area. The Delphi method, rating scaling method, and the nominal group process have all been used. However, it is generally accepted that certain socio-economic factors are not adequately expressed in quantitative terms. Therefore, it is recommended that a qualitative assessment also be used to provide decision makers flexibility and a broader scope to the consequences of their choices.

### 8.4.7 Policy Transformation and Implementation

A wide assortment of policy implementation tools is available to those involved in the planning process and to decision makers. One of the most commonly used and effective tools to manage the growth and development of land areas is land use planning. It establishes and maintains the critical link between science, public policy, and regulation. Although directed by guidelines from a broader regional or federal level, it usually is implemented at the local level, thereby involving the community. Land use planning should be incorporated into the overall ILSMP and adopt its goals and objectives to drive the daily implementation of its own local processes. Each locality's municipal or land planning program offers the mechanism to implement the ILSMP at the local level. It is an existing process that the community has available

to incorporate the tasks and actions of the ILSMP for effective implementation at
the local level. It often involves a broad spectrum of the community including both
public and private interests. An example of watershed management with a more
detailed treatment of land use tools and their application can be found in the Saint
Alban's Bay in Vermont.[68] The European Union has additional references which
review and guide planning efforts in coordination with the environmental impact
assessment process. These are addressed and provided in Section 8.4.6. A sample
list of policy transformation and implementation tools is as follows:

*Regulatory Tools*
   Legislation
   Regulation
   Policy
   Zoning and ordinances
   Output standards, i.e., effluent, organic, and inorganic substance levels
   Environmental education policy and standards
   Public hearings, meetings
   Permitting process
   Guidance documents
   Environmental assessments
   Socio-economic assessments
   Land use regulations/policy
   Smart growth regulations and policy

*Nonregulatory Tools*
   Water quality improvement incentive programs
   Agriculture buffers and best management practices (BMP)
   Forestry buffers and BMPs
   Residential and commercial development guidelines and BMPs
   Conservation easements
   Transferable development rights
   Mitigation banking
   Land use planning options
   Cluster developments and greenspace options
   Resource use reduction incentives
   Adoption of environmental education curricula in all grades of education
   Public or stakeholder workshops, seminars
   Incorporation into public and private media and information delivery systems

## 8.4.8 Public Input as a Tool for Integration of the ILSMP

Public input, specifically for land and water use policy development, is an extremely
important element in implementing any plan in a community. It is an essential part
of the policy development process if long-term sustainable management of the
lagoon and watershed is to be achieved. The public, which is in fact composed of
many smaller "units" as stakeholders, can be represented by multiple entities, each
with a specific mission and goal to meet the needs of these stakeholder groups.

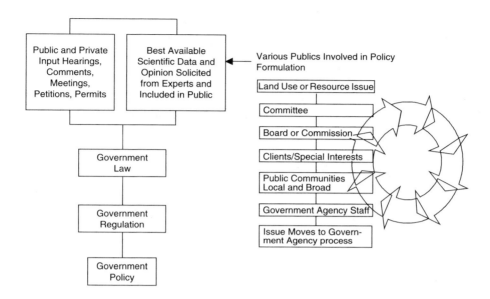

**FIGURE 8.6** Policy development and public use. (Modified from U.S. EPA, 1997.)

For example, planning commissions and wetland boards to which citizens are elected or appointed likely exist at the local level and address different aspects of land and water use. Citizen appointments at the federal, regional, state, and local levels can also represent the public. Public input should be built into the process of policy development at each level through a variety of methods (Figure 8.6).[69] The best available scientific data and opinion should be solicited from experts and included throughout the public input process.

One example of such a case study that incorporates public input into a lagoon-modeling project is the mediated modeling project in Ria Formosa, Portugal.[70,71] In that project, computer simulation modeling was used as an interactive tool in the facilitation of team learning by stakeholder groups in the lagoon area regarding the linkages between ecology and economics. Another project that utilized public input through surveys and questionnaires and collected socio-economic data is the Social Impact Assessment conducted in Turkey.[72] This project assessed the socio-economic conditions of the area and their relationship to the lagoon.

## 8.5 THE PROCESS OF DEVELOPING A PLAN TO SUPPORT DECISION MAKING FOR AN INTEGRATED LAGOON SUSTAINABLE MANAGEMENT

Decision makers may ask—How do I manage a lagoon for sustainability?

In answer to this question, the following steps are recommended:

- Identify the essential elements and process needed for a successful plan
- Develop a plan, which can be spatially and temporally integrated into the structure and function of the SES of a lagoon and catchment area

Decision making must be recognized as an ongoing process. Decisions that affect the lagoon area will be made continually. The important need and goal should be to develop a process and plan that provide for informed decisions by the many different socio-economic units that will be making decisions in this area. This is why an integrated, multidisciplinary plan is necessary. A plan that involves as many of the decision-making entities as possible will increase effectiveness, consistency, and integration into the community. Decisions will continue to be made and the process should incorporate additional information and tools through regular evaluation of the existing conditions.

Once the need for sustainable management has been identified and the lagoon system components defined, the next step is to develop an integrated lagoon management plan that will provide spatial and temporal guidance. *Sustainability* implies long-term commitment to the effort as well as providing a healthy system for future generations. *Integrated* means that the plan's goals, objectives, and tasks need to be incorporated into the many existing programs, structures, and entities that affect the lagoon system. The ideal result should be that each socio-economic entity, including individual citizens, industry, and other stakeholder groups, considers its impacts and demands on the lagoon and the limited capacity of the lagoon to provide for these demands on a daily operational basis.

Each government and nongovernment, public, or private entity or stakeholder in the lagoon system needs to adopt and integrate the ILSMP into its daily operations, administrative and financial accounting, and employment operations. All should be tasked with developing new approaches with creative incentives to make quantifiable progress toward the sustainable management of the system of which they are an integral part. The repetition of this integration process, recurring throughout the lagoon area and its daily functioning parts and participants, will eventually reach and permeate most aspects of the collective community to reinforce the philosophy of sustainability. Integrating the concept into the community's school system will provide for education of the next generation toward the adoption and maintenance of commitment over the long term.

The planning process can be simplified by describing it in several steps or components listed below and as depicted in Figure 8.1.

1. Set the goal—sustainable management of the lagoon
2. Define the problem—the problems/conditions that manifest the lagoon ecological and socio-economic systems imbalance (see Section 8.2)
3. Define the system—the lagoon's ecological and socio-economic structure and components (see Chapter 2)
4. Develop an effective DSS (see Section 8.1)
5. Use the best data and tools (see Section 8.4)
6. Develop and integrate this plan into infrastructure by function (see Sections 8.5 and 8.6)
7. Evaluate progress on a regular basis (see Section 8.10)

The planning process is used by most governmental and nongovernmental units as a means to set and accomplish goals. It results in a product that is only as effective as its accountability of implementation and evaluation. Recent programs, such as total quality management (TQM), demonstrate the integrated planning process with internal accountability.

## 8.6 PLAN DEVELOPMENT AND IMPLEMENTATION THROUGH INFRASTRUCTURE— THE INTEGRATION PROCESS

*Who should be involved in the development and implementation of a sustainable management plan? What are the main components of the socio-economic infrastructure that need to be involved?*

To answer this question, the decision maker should inventory and evaluate:

- Existing and potential organizational units of existing infrastructures
- Levels of government and community units
- Private and public entities and interrelationships
- Stakeholders' structure, interests, and resources
- Each unit's potential effect on the lagoon

A plan should be viewed as a product and a process, both leading to the agreed-upon goals and objectives. It is important to consider and involve the entities and stakeholders that can implement this plan and advance to sustainable management. One should begin with an inventory that includes all the local, regional, national, and international entities that have authority or impact or are impacted by the lagoon in any way. An inclusive approach, viewing all entities that affect the lagoon as "partners," provides for "common ground" and enlists more support and funding to accomplish the goal of sustainability. An inclusive format is important for effective information sharing and task assignments. The more input and partners, the more agreement ("buy-in") and informed consent for the common vision. Encouraging such groups as fisheries cooperatives, tourism associations, building associations, and agricultural cooperatives to reevaluate their organization's mission and strategic plan objectives to now include and integrate the philosophy of the ILSMP is necessary and helpful. This should be an objective of the planning process.

Accountability has traditionally been accomplished through regulations; however, a wide range of new, voluntary incentives should also be pursued to encourage new partners to contribute to the common cause of lagoon sustainability and see the economic benefits to them as well. Once a list of participants has been compiled, a strong effort to include their input should be made on a regular basis. Each stakeholder should be encouraged to develop a specific plan to integrate the overall goals and objectives of the plan into its own operations. Critical stakeholders, or those who have a significant effect on the lagoon system or area's economy, should have accountability or incentives built into the overall plan to ensure appropriate progress. In these cases, they should develop specific, quantifiable goals and objectives that will be evaluated on a regular basis as part of the overall plan.

Certain considerations and assumptions must be made in order to shape a workable model. A basic assumption for an ILSMP is that within it, each political entity will implement land and water use changes uniquely, according to local needs. Another assumption is that proactive planning is more effective than reactive land use policy and regulations. Additionally, regional planning accounting for structure, function, and process is effective in that it provides a framework for consistency on a broad scale.

Implementation is best accomplished at the local level because it can be customized to unique local conditions. Another important assumption is that input from stakeholders is preferable as early in the process as possible; this includes both public and private interests and scientific expertise and advice. The definition of *stakeholder* is a party that has a vested interest in the outcome of an issue. It is critical to inventory and includes as many stakeholders as possible or, at a minimum, a diversity of viewpoints and stakeholders.

Land and water use planning generally occurs at a minimum of three levels within a watershed. The broadest form is represented by national or regional watershed scale planning and is accomplished through the structure of a Regional Advisory Group. This entity establishes broad political commitment and accountability, consistency, goals, and recommendations for the entire watershed. It encompasses national, state, and local interests at the regional level. The second level is the state or provincial level, where the political administrations of the individual states or provinces establish their own regulations and policies for land use, using the regional information as guidelines. The third level is the local level, representing localities (counties and municipalities) that develop their own regulatory processes to implement regional and state/provincial guidelines and regulations. It is generally acknowledged that unique actions resulting in land and water use changes are best implemented through the positive engagement of the local entities.

*A potential list of stakeholders and structural levels to include in integrated lagoon sustainable management plan (includes both government (regulatory role) and NGO (nonregulatory, profit, and nonprofit roles)) would be the following:*

Local, regional, national and international levels
Planning boards and commissions
Environmental and wetland boards or commissions
Zoning and building agencies and organizations
Health and safety agencies and organizations
Agricultural and forestry agencies
Transportation agencies and organizations
Political and administrative branches
Natural resource conservation agencies and organizations
Soil and water, wildlife, and fisheries agencies and organizations
Agriculture, forest, fisheries cooperatives or associations
Agriculture, forest, fisheries private industries
Homeowner associations
Construction and builders associations and industries
Nature and environmental groups
Civic and religious organizations—youth and adult
Tourism agencies and organizations

Tourism facilities: hotel/restaurant associations
Finance and commerce agencies and organizations
Special areas commissions and boards
Engineering and surveying (coastal and other)
Land trusts and conservancies
School and school administrations
International conventions and treaty organizations

## 8.7   LINKING INFRASTRUCTURE AND FUNCTION

*Which infrastructure units have the most appropriate and needed roles or resources to accomplish the tasks identified in the plan? In order to answer this, a decision maker should:*

- Identify and inventory existing and potential mechanisms and tools needed to accomplish the goal
- Identify and select participants and infrastructure according to which functions are performed both by government at various levels and by stakeholders
- Identify and clearly communicate roles and responsibilities
- Emphasize continual communications among all groups
- Identify any gaps or future needs that may require additional structure to meet functional requirements

Lagoon management infrastructure is capable of enlisting significant financial and human resources and can, therefore, be a worthwhile organizational and planning effort. The structure of the program needed to implement the long-term sustainable management should directly address the tasks and needs identified in the ILSMP. Unless an entity's function matches a specific need and task, it may be superfluous and the process would thus be better served by a different or simpler infrastructure.

It is always important to recognize the power of incumbent and established structures in the implementation process. It is also, however, important to design specific integration and accountability tools because such established structures and entities are often the most entrenched and unlikely to accept change. Any structural entity and its role or function should be designed to specifically address a stated need in the plan. Accordingly, customized accountability and incentives need to be built into the process. This section describes how the process, structure, and function of the infrastructure that evolves to sustainably manage the lagoon ecosystem should lead to effective land use.

The value of including both government and NGOs (both private and public) is that it provides a system of checks and balances throughout the ILSMP process. Government plays the crucial role of developing policy, regulation, and legislation, while the NGOs and private interests provide for the implementation link into the community structure. Inclusion of a diversity of public and private entities maximizes integration and community participation.

The infrastructure created to implement lagoon management should be as simple or complex as the local situation and needs dictate. It could be that the existing

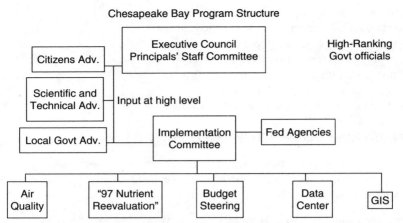

FIGURE 8.7 Example of lagoon implementation program structure. (Modified from U.S. EPA, 1997.)

public and private entities are sufficient to implement the new plan with little additional infrastructure necessary. On the other hand, the complexity of inter-jurisdictional boundaries and issues may dictate the creation of a new "coordinating entity" to direct and evaluate progress. The following examples are used to show the diversity of infrastructures dealing with lagoon management.

The function and structure of infrastructure committees and working groups are paramount to their effectiveness and therefore should be established and evaluated according to the depth and breadth to which they address their focus area. Each committee or workgroup should address a nonoverlapping task or issue and set goals, objectives, and specific action items with target dates. The committees are generally composed of working-level government staff and organizations or associations, which actually implement these tasks within the context of each respective program's responsibilities. Each of these committees develops, coordinates, and ensures the implementation of an important aspect of lagoon restoration.

An example of a structure established primarily for the administration and management of a lagoon system is shown in Figure 8.7. More complete descriptions of the Chesapeake Bay Program infrastructure, with committees and their functions, can be found in the EPA reference.[69]

## 8.8 THE IMPORTANCE OF EVALUATION IN THE DECISION-MAKING PROCESS

Evaluation is important because it:

- Is a means to measure progress and revise implementation strategies
- Recognizes both the tangible goals, i.e., pollution levels, fisheries catch, etc., and the intangible goals, e.g., partnerships, stewardship ethic

- Recognizes the value of both the process and the resulting products and outcomes
- Provides for buy-in and involvement
- Produces both biological and socio-economic outcomes
- Results in increased awareness of stakeholders, public input, and participation
- Results in an informed and educated public

Why evaluate?
To monitor the progress and evaluate the success of the project

When to evaluate?
At regular intervals and the end of the implementation period

What to evaluate?
Goals, assumptions, objectives against key indicators, progress

How to evaluate?
Through implementation teams or objective committees; regular stakeholder meetings, focus groups, steering committees

Figure 8.8 provides a schematic representation of the recommended steps of an effective evaluation process. Any evaluation method should account for the values of both the tangible and intangible results of such an endeavor. It is necessary to evaluate the impacts of the process as well as the resulting products. A common scenario is the development of a 5-year lagoon management plan. This plan would set goals for each year and have a time table to account for the progress of each task that has been prioritized in the plan. Evaluation should take place at the end of each year, at the target completion date for each task, and at the end of the 5-year period. At each of these evaluation periods, the goals, objectives, tasks and priorities, and schedule, as well as data and tools, should be reviewed. It should be recognized that with time, new information and tools become available because the uncertainties of limitations, restraints, and resources change with time. At each evaluation interval, these major components of the plan need to be evaluated in the context of available resources and constraints. The plan needs to be updated and corrected for the uncertainties that only time will manifest.

While evaluation of the plan itself and its tangible products and results is relatively straightforward and apparent, the results of the evaluation process are less evident. A process tends to affect every party participating in it to some degree. It is important to measure and evaluate these intangible effects of the process as well as the tangible ones. A likely scenario is that the stakeholders and participants, who have held monthly meetings for 5 years, have a greater awareness and knowledge of the lagoon components, as well as a greater understanding of the viewpoints, nature, and resources of the other stakeholders. A more educated and aware public and stakeholder coalition should be recognized as a result of the process, as should any negative or divisive impacts from the process. However, if evaluation is being conducted regularly and correctly, any negative or divisive tendencies should be addressed and effectively redirected into positive learning experiences. More often than not, the coalition, partnerships, and networks created from the process become

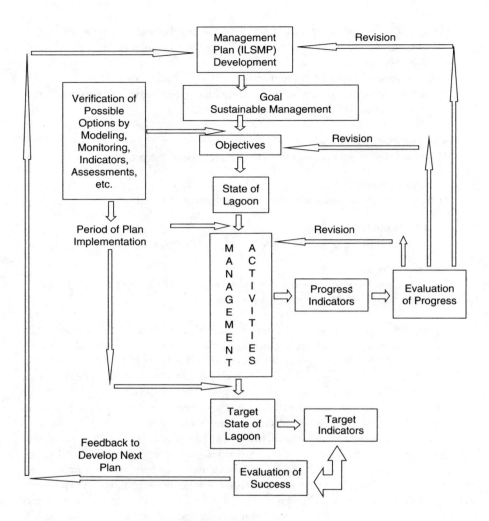

**FIGURE 8.8** Evaluation process for the ILSMP.

extremely valuable tools for future information dissemination, implementation, and lagoon management as the ILSMP now becomes integrated into the various aspects of the social and economic systems of the lagoon area.

## 8.9  CONCLUSION

This chapter has presented factors to consider and suggested a process to employ in the decision-making process of developing a DSS to use for sustainable lagoon management. The process is as important as the product of an ILSMP, and the intangible goals and results can be as beneficial as the tangible ones. The ILSMP represents a document that directs implementation, measures progress, and enforces accountability. It represents the collective thinking of a diverse group of participants who have committed time and resources to the common cause of a healthy lagoon and resulting SES. The plan and other products produced from this effort should have a built-in progress and accountability monitoring system. They should have a scheduled reevaluation regimen as old goals are reached and new ones are set. The long-term commitment to this effort should be emphasized at the onset, and measures to ensure its sustained viability need to be clearly stated in any documents produced.

The planning process itself, in addition to the products it creates, plays a major role in the integration of the information and philosophy of sustainable management of the system. The act of engaging existing and new constituents as participants and partners in the process raises the level of consciousness about the ecological and socio-economic link throughout the community. As a result, additional resources of expanded and nontraditional constituent groups promote new and creative approaches and solutions to old and new problems. Integrated planning and decision making are means to achieve sustainable management.

## ACKNOWLEDGEMENTS

The authors wish to acknowledge the sage advice and involvement of the team leader, Prof. I.E. Gönenç, and each of the other team members. Prof. Dr. A. Vadinaneau (Romania) provided invaluable conceptual input. Dr. E. Andrulewicz (Poland), Dr. Bade Cebeci (Turkey), Dr. A. Razinkovas (Lithuania), Dr. Chris Baker (U.K.), Dr. E. Tuska (Romania), and Ali Ertürk, M.Sc. (Turkey) all contributed to sections of this chapter.

## REFERENCES

1. WCED, World Commission on Environment and Development, The Brundtland Commission in Our Common Future, Oxford University Press, Oxford, 1987.
2. Vadinaneau, A., *Sustainable Development, Theory and Practice*, Bucharest University Press, Bucharest, 1998.
3. Vadinaneau, A., Decision-making and DSS for balancing socio-economic and natural capital development, in *Proceedings of International Conference on Globalization, Economy and Ecology—Bridging Worlds*, Tillburg, November 1999.
4. Colorado River DSS (CRDSS) by the Colorado Water Conservation Board and Division of Water Resources at http://cando.dwr.co.gov/overview/crdsscov.html.
5. USDA's DSS for the Integrated Pest Management Program (IPM), http://usda.dis.anl.gov/dsss.dsssywww.html. 2000.
6. Rousseau, A.N., Turcotte, R., Mailhot, A., Duchemin, M., Blanchette, C., Roux, M., Dupont, J., and Villeneuve, J.P., *GIBSI—An Integrated Modeling System for Sustainable Land Use and Water Management,* 1999.
7. Cairns, J., Jr., Sustainability, ecosystems services and health, *Int. J. Sustain. Dev. World Ecol.*, 4(3), 153, 1997.
8. Harris, J.M., *Rethinking Sustainability, Power, Knowledge, and Institutions*, University of Michigan Press, Ann Arbor, 2000.

9. World Resources Institute (WRI). A Guide to the Global Environment, Washington, D.C. http://www.wri.org/wri.org.wri/wr-96-97/ei_txt1.html. 1996–1997.

10. UNEP GEO-2000 Global Environmental Outlook, Chapters 1–5. http//www.grida. no.geo2000. 2000.

11. Odum, H.T., *Environmental Accounting: Energy and Environmental Decision-making*, John Wiley & Sons, New York, 1996.

12. Musters, M.J.C., De Graaf, J.H., and Keurs, J.W., Defining socio-environmental systems for sustainable development, *Ecol. Econ.* 26(3), 243, 1998.

13. De Groot, U., Problem in context: a framework for the analysis, expiation and solution of environmental problems, in *Environmental Management in Practice*, Vol. I, Nath, B., Hens, L., Campton, P., Devuyst, D., and Routled, G., Eds., Routledge, London, 1998, p. 22.

14. Brown, T.M. and Ulgiato, S., Energy evaluation of the biosphere and natural capital, *Ambio*, 28(6), 486, 1999.

15. Doherty, F.P. Jr., Marschall, A.E., and Grubb, C.T., Jr., Balancing conversation and economic gain: a dynamic programming approach, *Ecol. Econ.*, 29(3), 349, 1999.

16. Hinberberger, K., Luks, F., and Bleek–Schmidt, F., Material flows vs. "Natural Capital." What makes an economy sustainable? *Ecol. Econ.*, 23(1), 1, 1997.

17. Costanza, R., d'Arge, R., De Groot, R., Farber, S., Grosso, M., Hannon, B., Limburg, K., Naeem, S., O'Neil, R., Parnelo, J., Raskin, R.G., Sulton, P., and Van Den Belt, M., The value of the world's ecosystem services and natural capital, *Nature*, 387, 253, 1997.

18. Brouwer, R. et al., A meta-analysis of wetland contingent valuation studies, CSERGE Working Paper GEC 97-20, 1997.

19. Wilson A.M. and Carpenter R.S., Economic valuation of freshwater ecosystems services in the U.S.: 1971–1997, *Ecol. Appl.,* 9(3), 722, 1999.

20. European Commission, Environmental Impact Assessment Methodology and Research, *Proc. 3rd EU Workshop on Environmental Impact Assessment*, Delphi, 1994.

21. Diamond, P.A. and Hausman, A.J., Contingent valuation: Is some number better than no number? *J. Econ. Persp.*, 8(4), 45, 1994.

22. Ruijgrok, E., Vellinga, P., and Gossen, H., Dealing with nature, *Ecol. Econ.*, 28, 347, 1999.

23. Wackkernagel, M., Onisto, L., Bello, P., Callejas Linares, J., Falfan Lopez, S., Garcia, J., Mendez, I., and Suarez, M., National capital accounting with the ecological footprint concept, *Ecol. Econ.*, 29(3), 375,1999.

24. Costanza, R. and Daly, H.E., Natural capital and sustainable development, *Conserv. Bio.*, 6, 37, 1992.

25. Costanza, R., Economics growth, carrying capacity and the environment, *Ecol. Econ.*, 15(2), 89, 1995.

26. Costanza, R., Ecological economics, in *The Science and Management of Sustainability*, Columbia University Press, New York, 1991.

27. Oglethorpe, R.D. and Sanderson, A.R., An ecological economic model for agri-environmental policy analysis, *Ecol. Econ.*, 28(2), 245, 1999.

28. Pimental, D., Wilson, C., McCullen, C., Huang, R., Dwen, P., Flack, J., Tran, G., Saltman, T., and Cliff, B., Economic and environmental benefits of biodiversity, *BioScience*, 47, 747, 1997.

29. Van Der Straaten, J., *The Economic Value of Nature*, Katholike Univeristeit, Braband, 1998.

30. Mailhot, A., Rousseau, A.N., Massicotte, S., Dupont, J., and Villenuve, J.A., Watershed-based system for the integrated management of surface water quality: The GIBSI system, *Water Sci. Technol.*, 36(5), 381, 1997.

31. Voinov, A. et al., Watershed Management Over the Web, University of Maryland Institute for Ecological Economics, Solomons, MD, http://Kabir.Cbl.Umeces.Edu/ PLM/ Wbmod/Aconcepta.html 1998.

32. SEI, Creating an Agenda 21 for the Baltic Sea Region, Stockholm Environment Institute, Stockholm, Sweden, 1996.

33. Walpole, C.S. and Sinden, A.J., BCA and GIS: integration of economic and environmental indicators to aid land management decisions, *Ecol. Econ.*, 23(1), 25–44, 1997.

34. OECD, OEDC Core Set of Indicators for Environmental Performance Reviews. Environmental Monographs No. 83, Paris Synthesis Report OECD/GD (93), 179, 1993.

35. Andrulewicz, E., Development of Marine Environmental Quality Indicators, ICES Advisory Committee on Marine Environment, Doc 77.1, 1997.

36. Cole, T.M. and Wells, S.A., CE-QUAL-W2: a Two-Dimensional, Laterally Averaged, Hydrodynamic and Water Quality Model, Version 3.0, Instruction Report EL-2000, U.S. Army Engineering and Research Development Center, Vicksburg, MS, 2000.

37. World Bank, The World Bank Develops New System to Measure Wealth of Nations, Press Release, The World Bank, Washington, D.C., September 17, 1995.

38. Costanza, R., Segura O., and Martinez-Alier, J., Eds., *Getting Down to Earth: Practical Applications of Ecological Economics*, Island Press, Washington, D.C., 1996.

39. Harris, J.M. et al., Eds., A *Survey of Sustainable Development: Social and Economic Dimensions*, Island Press, Washington, D.C., 2001, p. 409.

40. De Groot, R.S., Environmental functions and the economic value of natural ecosystems, in *Investing in Natural Capital: The Ecological Economics Approach to Sustainability*, Jansson, A. et al., Eds., Island Press, Washington, D.C., 1994, p. 151.

41. UN Convention on Environmental Impact Assessment in a Transboundary Context. E/ECE/1250, UN, 1991. 20.

42. Krishnan, R., Harris, J.M., and Goodwin, N.R., Eds., A *Survey of Ecological Economics*, Island Press, Washington, D.C. 1995, p. 385.

43. UN Handbook of National Accounting: Integrated Environmental and Economic Accounting, Interim Version, UN (ST/ESA/STAT/SER.F/61), 1993.

44. Fifth Community Programme for the Environment and Sustainable Development, UN COM (92) 23, Vol. II, 27 March 1992, chap. 14.

45. IDRC, ECONOMICAL ISSUES from the RIO +5 Forum, 19 June 1997. http://www.idrc.ca/rio+5/index_e.html 1997

46. Bolivia, A., Presentation on integrated environmental and economic accounting, in *Proc. Ann. Conf. Society of Environmental Economics and Policy Studies*, 1997.

47. Ariyoshi, N.A., Complete system for integrated environmental and economic accounting, in *Proc. Annual Conf. of the Society of Environmental Economics and Policy Studies*, 1997.

48. Kulshreshtha analytical methods for integrating socio-economic and environmental effects of forest management strategies, in *Indicators of Sustainable Development Workshop*, Saskatoon, SK, Canada, http://mf.ncr.forestry.ca/conferrences/isd/ kulshreshthaeng, 1999.

49. Holmes, M., Cost benefit analysis: The effect of two variables on NPV on: http:// members.aol.com/econosite/graph.html, 1999.

50. Pearce, D.W. and Turner, K., *Economics of Natural Resources and the Environment*, John Hopkins University Press, Baltimore, MD, 1990.

51. Hausman, J.A., Ed., *Contingent Valuation, A Critical Assessment,* North-Holland, Amsterdam, 1993.

52. UNSO, SNA, *Handbook on Integrated Environmental and Economic Accounting*, Statistical Office of the United Nations, New York, 1990.

53. Turner, R.K., Valuation of wetland ecosystems, in *Persistent Pollutants, Economics and Policy*, Opschoor, H. and Pearce, D., Eds., Kluwer, Dordrecht, 1991.
54. Sadler, B., Environmental Impact Assessment in a Changing World: Evaluating Practice to Improve Performance, Final Report of the International Study of the Effectiveness of EIA Canadian Environmental Assessment Agency and the International Association for Impact Assessment, Ottawa, 1996.
55. Devuyst, D., Environmental impact assessment, in *Environmental Management in Practice* – Vol. I, Nath, B., Hens, L., Campton, P., Devuyst, D., and Routled, G., Eds., Routledge, London, 1998, p. 188.
56. Treweek, J., *Ecological Impact Assessment*, Blackwell Scientific, Oxford, 1999.
57. Verheem, R., Environmental Impact Assessment at the Strategic Levels in the Netherlands Project Appraisal, 7(3), 91, 1992.
58. USEPA, Principles of Environmental Impact Assessment, EPA315B98012, Washington, D.C., 20460, http://www.epa.gov.oeca/ofa, 1998.
59. Lemans, K.E. and Porter, A.J., A comparative study of impact assessment methods in developed and developing countries, *Impact Assess. Bull.*, 10(3), 57, 1992.
60. Department of the Environment, Guide on Preparing Environmental Statements for Planning Projects: Consultation Draft, London, 1994.
61. HMSO, *Evaluation of Environmental Information for Planning Projects: A Good Practice Guide*, London, 1994.
62. HMSO, *The Town and Country Planning (Assessment of Environmental Effects) Regulations*, London, 1988.
63. Harris, J.M., Ed., *Rethinking Sustainability: Power, Knowledge and Institutions*, University of Michigan Press, Ann Arbor, 2000, p. 295.
64. Canter, L.W. and Kamath, J., Questionnaire checklist of cumulative impacts, *Environmental Impact Assessment Review*, 15, 311, 1995.
65. Burdge, R.J., A brief history and major trends in the field of impact assessment, *Impact Assess. Bull.*, 9(4), 93, 1991.
66. Adenberg, A., Industrial metabolism and the linkages between economics, ethics and the environment, *Ecol. Econ.*, 24(23), 311, 1998.
67. Ahmad, Y.J., El Serafy, S., and Lutz, E., Environmental Accounting for Sustainable Development, A UNEP-World Bank Symposium. The World Bank, Washington, D.C., 1989.
68. Gale, J.A. et al., Evaluation of the Experimental Rural Clean Water Program, National Water Quality Evaluation Project, NCSU Water Quality Group, NCSU, Raleigh, NC, EPA-841-R-93-005, 1993.
69. U.S. EPA, The Chesapeake Bay Program, The Chesapeake Bay Agreement, Report to the Commission, Annapolis, MD, 1997.
70. Van Den Belt, M., Videira, N., Antunes, P., Santos, R., and Gamito, S., Mediated Modeling Project in Ria Formosa, Final Report, Portugal, 1999.
71. Commission of the European Communities, Note for the Inter-Service Working Group on "Environmental Statistics, Indicators and Green National Accounting," 1993.
72. Cebeci, B.A., New methodological approach to social impact assessment (SIA) studies, M.Sc. thesis, Institute of Science and Technology, Istanbul Technical University, Istanbul, Turkey, 1998.

# 9  Case Studies

*I. Ethem Gönenç, Vladimir G. Koutitonsky,
Angel Pérez-Ruzafa, Concepción Marcos Diego,
Javier Gilabert, Eugeniusz Andrulewicz,
Boris Chubarenko, Irina Chubarenko,
Melike Gürel, Aysegül Tanik, Ali Ertürk,
Ertugrul Dogan, Erdogan Okus,
Dursun Z. Seker, Alpaslan Ekdal,
Aylin Bederli Tümay, Kiziltan Yüceil,
Nusret Karakaya, and Bilsen Beler Baykal*

## CONTENTS

## 9.1  INTRODUCTION

### *I. Ethem Gönenç*

This chapter presents selected case studies from different areas of the world. These studies provide detailed information on how to apply the methodologies and practical approaches discussed in the other chapters of this book. These case studies are examples of how to integrate modeling into the decision-making process for sustainable management of lagoons. The brief summaries that follow outline the aim and scope of these case studies.

### 9.1.1  GRANDE-ENTRÉE LAGOON

The first case study area is the Grande-Entrée Lagoon in the Gulf of St. Lawrence, Québec, Canada. The aquaculture industries are focused on determining the carrying capacity of the lagoon for shellfish to maximize production. As defined in previous chapters, lagoons are shallow marine systems where tides and local and nonlocal winds play a major role in the dynamics of water motion. The management of shoreline ponds and/or caged/fenced areas for aquaculture depend on an understanding of water movement. Normally, when local winds blow along a lagoon axis, downwind drift currents (see Chapter 3 for details) develop in the shallow areas near both shores. Water pile-up at the lagoon end causes horizontal pressure gradients, which in turn force upwind gradient currents somewhere in the lagoon between the drift currents. The hypothesis put forth here is that gradient currents occur near the bottom in the deeper parts, away from the surface wind stress.

This hypothesis was tested in Grande-Entrée Lagoon, using current observations recorded at 1 m above the bottom. A three-dimensional (3D) numerical model and empirical orthogonal function (EOF) analysis of currents and winds show that bottom currents are negatively correlated with the wind directions. The numerical model is first used to simulate the horizontally induced two-dimensional (2D) wind-induced current fields. Results show that currents are quickly set up near both shores, and that a weak and sluggish return flow occurs in between those shores. The model is then used to simulate the 3D current structure under the same wind conditions. The hypothesis is verified: gradient upwind currents occur in the deep layers of the lagoon basin, while currents near the surface are oriented downwind. Such a 3D current structure can have significant effects on water renewal in the bottom layers and on the general lagoon ecosystem dynamics. It suggests that a 3D model should be considered when developing lagoon ecosystem numerical models and when selecting optimal sites for aquaculture development.

### 9.1.2  MAR MENOR LAGOON

The second case study area is the Mar Menor, a coastal lagoon in the Mediterranean Sea, Spain. As do many lagoons throughout the world, this area supports a wide range of beneficial uses and socio-economic interests. It is recognized as an emblematic environment of the Región de Murcia, on the southwestern Mediterranean coast of Spain. The area is a vital part of the regional development plans to provide high-quality tourist and recreational services.

The lagoon and surrounding coastal area maintain important fisheries, such as eel, grey mullet, gill-head bream, sea bass, striped bream, and crustaceans, particularly shrimp. The lagoon is, however, an object of social concern because of its rapid rate of development in the last decade and the detrimental impact of this development on the ecosystem, such as point and nonpoint pollution, shoreline and habitat destruction that result in degradation of the aquatic environment, and decreased fishery production. Some of these changes result from coastal work on tourism facilities, such as land reclamation; the opening, deepening, and extending of channels; urban development and associated waste; construction of harbors for sport; and creation of artificial beaches. Other factors include changes in agricultural practices in the watershed, such as the change from extensive dry crop farming to intensively irrigated crop farming and the increase in agricultural waste and nutrient input into the lagoon and coastal aquatic environment. These circumstances make the Mar Menor a useful example for analyzing the biological patterns and processes affected by changes in hydrogeographic conditions, nutrient inputs, and lagoon characteristics (see Chapter 5).

### 9.1.3 THE BALTIC SEA LAGOONS

The third case study area is the Baltic Sea lagoons. The Curonian, Odra, and Vistula lagoons are of great importance for the quality of coastal waters and open sea areas. These lagoons are natural filters for agricultural, industrial, and municipal waste loads. These anthropogenic pressures are particularly intense in the southern and southeastern parts of the Baltic catchment area. This region is densely populated. Industrial activity is intense, and a large proportion of land is used for agriculture.

The Vistula Lagoon experiences higher anthropogenic loading than the Odra or Curonian lagoons because of its relatively small water volume and very poor treatment facilities in its catchment area. The Vistula Lagoon was one of the first areas in the Baltic region where genuine strides in transboundary management have been attempted, including the implementation of modeling as a decision-making tool by both scientific institutions and national environmental authorities.

### 9.1.4 KOYCEGIZ–DALYAN LAGOON

The fourth case study area is Koycegiz–Dalyan Lagoon in the Mediterranean Sea in Turkey. This lagoon is widely regarded as a stellar example of ecosystem modeling for sustainable management in the context of the NATO-CCMS pilot study. A decision support system has been established, and monitoring and modeling are used to support decision making in the lagoon and watershed.

These four case studies are models that offer valuable insight into the decision-making process for ensuring thoughtful sustainable management of lagoons worldwide.

## 9.2   THREE-DIMENSIONAL STRUCTURE OF WIND-DRIVEN CURRENTS IN COASTAL LAGOONS

## Vladimir G. Koutitonsky

### 9.2.1   INTRODUCTION

A coastal lagoon is a shallow water body separated from the ocean by a barrier, usually parallel to the shore, and connected, at least intermittently, to the ocean by one or more inlets.[1] A lagoon can be choked, restricted, or leaky depending on the inlet configuration.[2] These geomorphologic features affect hydrodynamic processes inside the lagoon, namely the flushing time, the circulation, and the mixing of waters.[3,4] Depths of most lagoons seldom exceed a few meters so they respond quickly to forces acting at the air–sea interface and at the lagoon open boundaries. Forces acting at the lagoon lateral boundaries are tides, nonlocal forcing at the ocean boundary, and freshwater run-off, when present, at the upstream boundary. Forces acting at the air–sea interface include heat and water fluxes resulting from changes in air temperatures, precipitation and evaporation, and wind stress. These forces accelerate the motion and establish barotropic and baroclinic pressure gradients in the lagoon and across its inlets. They also modulate the turbulent mixing as well as the vertical and horizontal density gradients in the lagoon.[5] Bottom friction decelerates the motion and plays a significant role in the momentum balance.[6] This study focuses on the combined effects of wind stress, horizontal barotropic pressure gradients, and bottom friction in coastal lagoons.

Normally, winds blowing along the major axis of a lagoon set up downwind coastal currents in the shallower regions near both shores. The resulting water pile-up at the downwind end of the lagoon sets up horizontal (barotropic) pressure gradients that can induce upwind gradient currents elsewhere in the lagoon, between the coastal currents. This study suggests that wind-driven currents in lagoons are three-dimensional (3D) in space and that 2D models may not be adequate to describe them. Such a 3D structure may influence primary production,[7] nutrient fluxes,[8] and the transport of dissolved and particulate matter in the lagoon.[9,10] Modeling the response of ecosystems to the above physical processes is discussed in Hearn et al.[6] and in earlier chapters.

The objective of this study is to demonstrate that wind-induced currents can have a 3D structure in some coastal lagoons even when their depths are relatively shallow (~5 m). The hypothesis put forth is that return gradient currents occur in the deeper layers of the lagoon, away from the surface where motion is still responding to wind stress. This may have implications for water renewal in the bottom layer, for the transport of dissolved or particulate matter in the lagoon, and for biogeochemical fluxes at the sediment interface. The hypothesis is tested in Grande-Entrée Lagoon (Section 9.2.2) where predominant winds are oriented along the lagoon axis.[11] Low frequency current, sea levels, and wind time series obtained during a 1989 field experiment are analyzed in Section 9.2.3. Section 9.2.4 compares the results of 2D and 3D model simulations of these wind-driven currents. The numerical model used in this study is the MIKE3-HD model,[12,13] briefly described in Appendix 9.2.A. Conclusions are given in Section 9.2.5.

## 9.2.2 Grande-Entrée Lagoon

The Magdalen Islands are located in the middle of the Gulf of St. Lawrence in eastern Canada (Figure 9.2.1). They include several lagoons that have traditionally sustained aquaculture activities. Two of these lagoons, Grande-Entrée Lagoon (GEL) and Havre aux Maisons Lagoon (HML), are connected by a 60-m wide and 7-m deep pass under a bridge. The geometry of their entrance passes makes them leaky and restricted lagoons, respectively, and a tidal phase difference between these entrances generates significant tidal currents in some parts of the lagoons.[11] The focus of this study is GEL (Figure 9.2.1). Its bathymetry features a 7.5-m-deep navigation channel leading from its entrance to a harbor in its northern part. This channel separates a deeper basin (5–6 m) to the east from a shallower region (2–3 m) to the west leading to HML. The eastern deep basin sustains considerable mussel aquaculture activities. Tides in GEL are mixed and mainly diurnal with an average tidal range at the entrance of 0.58 m, reaching 0.95 m during spring tides.[11] The lagoon surface area is about 68 km$^2$ such that the tidal prism for normal tides is about $74.4 \times 10^6$ m$^3$. Assuming that a fraction $\beta$ of this volume remains inside the lagoon during each tidal cycle and is completely mixed with ambient waters,[14] a lower limit for the GEL flushing time of 6 days during strong wind conditions ($\beta = 1$) and 23 days during calm wind conditions ($\beta = 0.25$) is estimated. GEL has no freshwater river run-off and evaporation is negligible.

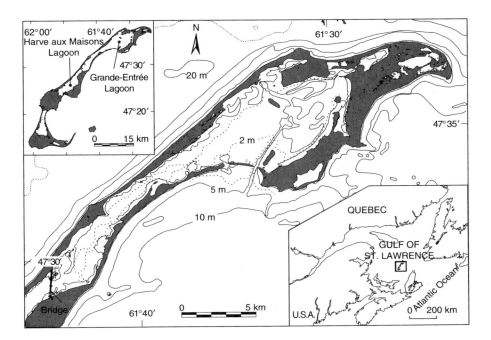

**FIGURE 9.2.1** Grande-Entrée Lagoon bathymetry in the Magdalen Islands (upper left inset), in the Gulf of St. Lawrence, Canada (lower right inset).

As a result, waters in the lagoon show little vertical density stratification except perhaps during the ice-melting period (April). Finally, winds are predominant from the southwest-northeast axis, that is, along the major axis of the lagoon.[11]

### 9.2.3 WIND-DRIVEN CURRENTS: OBSERVATIONS

A large-scale multidisciplinary field experiment was carried out in GEL during the summers of 1988 and 1989.[15,16] The objective was to study the impact of increasing mussel aquaculture on the biological production of the lagoon. Aanderaa current meters and tide gauges were moored exactly 1 m above the bottom from May 5 to May 20, 1989 at several stations in the lagoon, including C1, C6, C10, C12, and C11 in the deep basin (Figure 9.2.2). Mooring C1 outside the inlet was fitted with a second current meter near the surface that was left in the water for a longer period of time. Divers performed daily inspections of all moorings in order to remove possible rotor contamination by drifting algae. Conductivity and temperature profiles were obtained at several locations in the lagoon over neap and spring tidal cycles. Finally, winds and atmospheric pressure were measured at Grindstone (Figure 9.2.2). Analysis of the complete current and sea-level data set has been reported elsewhere.[11,16] It was shown that tidal currents reach speeds above 0.5 m/s at the lagoon entrance and in the shallow regions to the west, while in the deeper basin, to the right of the navigation channel,

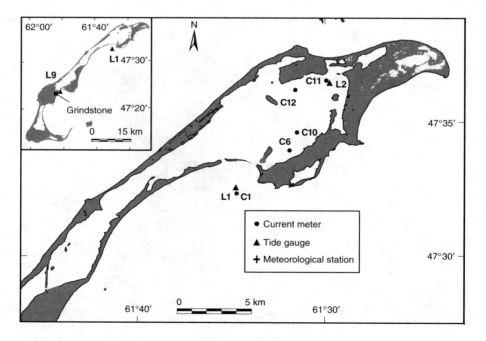

**FIGURE 9.2.2** Positions of sea level (L1) and current (Cx) recording stations during 1989 in Grande-Entrée Lagoon in the Magdalen Islands, Gulf of St. Lawrence.

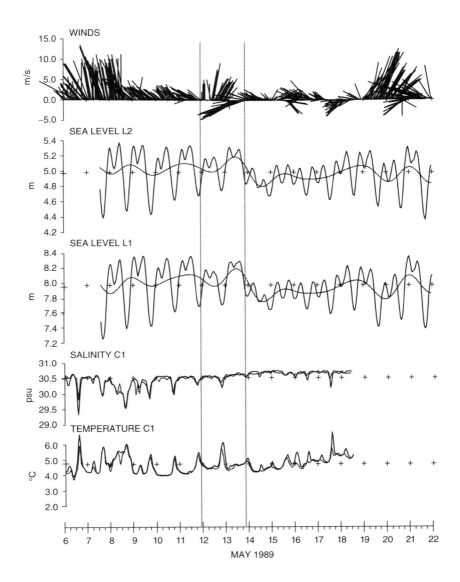

**FIGURE 9.2.3** Wind vectors and sea levels at stations L1 and L2 (low frequency, dotted line), salinity and temperatures at the surface (solid line) and at the bottom (dotted line) at station C1 from May 5 to 20, 1989.

they are almost nonexistent (ellipse major axis speeds ~ 0.01 m/s). The water renewal time in the deep basin was estimated to vary between 12 and 20 days. Since tidal currents are almost nonexistent, water exchange in this basin must be due solely to a combination of local winds and nonlocal forcing at the mouth. For completeness, local winds and the forcing functions at the mouth (station C1, L1) are presented in Figure 9.2.3 from May 5 to 20, 1989, a period during which current meter records are available inside the lagoon. Water salinity and temperature at the surface (solid line) and near the bottom (dotted line) do not show considerable differences. Hence, waters entering the lagoon are vertically mixed. Inside the lagoon, waters normally also remain well mixed[16] under the influence of omnipresent winds. Winds during the May 5–20 period were oriented along the lagoon axis only occasionally (arrows on top, Figure 9.2.3). This study focuses on the May 12–14 reversing wind event because current measurements inside the lagoon were available then. Winds on May 12 were southwesterly, along the lagoon axis, and they reversed on May 13 to become north-easterly. Low-frequency sea levels (dotted lines superimposed on sea levels in Figure 9.2.3) indicate that sea levels at the mouth (L1) and inside the lagoon (L2) were almost identical, as expected in a "leaky" lagoon.[11] Northeasterly (southwesterly) winds led sea-level rises (falls) at L1 and L2. It is not clear if this response was a local set-up or a gulf-wide response to large-scale winds. The application of a 3D hydrodynamic model to the Gulf of St. Lawrence[17] showed that, under prevailing southwesterly winds, sea levels rise in the northern gulf and decrease in the southern gulf and in the Magdalen Islands region. So it is possible that sea levels at the mouth (L1) and inside the lagoon (L2) respond to such nonlocal forcing as well as to local winds.

Winds, sea levels at L2, and wind-driven currents in the deep basin at C6, C10, C11, and C12 (see Figure 9.2.2) were then examined during the reversing wind event of May 12–14 to detect upwind return flows in the lower layers of the deep basin. The hourly time series were first low-pass filtered (cut-off frequency set at 34 h) in order to remove tidal oscillations and high-frequency noise.[18] These series are shown in Figure 9.2.4. The numerical simulations (Section 9.2.4) will focus on the same wind event. Several points are worth noting about the wind-driven current observations. Measured at 1 m above the bottom, the currents shown are located in the deeper layers of the basin. Their principal axes of variability (Table 9.2.1) indicate that they are oriented along axes that differ slightly (between 8 and 27°) from the lagoon longitudinal axis (40°). They also exhibit an out-of-phase relation with the wind direction. This out-of-phase relation can be objectively established from empirical orthogonal function (EOF) analysis[19] of the wind component resolved along the lagoon axis and the current vectors resolved along their principal axis of variability (Table 9.2.1). The correlation matrix for the resulting series at C6, C10, C12, and C11 and the longitudinal wind series is presented in Table 9.2.2a. Results from the EOF analysis of this matrix are presented in Table 9.2.2b in terms of the percentage of the total variance in the series explained by each empirical mode and the percentage of the variance in each series explained by each mode. Results suggest that all near-bottom currents resolved along their principal axis are negatively correlated with the longitudinal wind and that 75% of the total variance is explained by the first empirical mode. This mode explains 96% of the wind variance and the largest portion of the variance in each current series, except for the variance in currents at C6, which is partly explained by mode 2. In summary, these observations tend to support the hypothesis that the wind-driven circulation in a lagoon can be three-dimensional in space and that the return flow can occur at depth in an upwind direction.

Simple analytical reasoning provides insight into these findings.[20] Consider the steady-state motion in a lagoon of variable depth $H(x, y)$ resulting from horizontal pressure gradients and vertical shear friction, in the absence of Coriolis accelerations

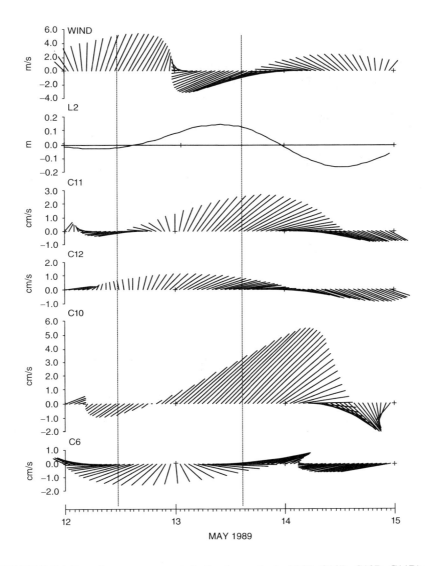

**FIGURE 9.2.4** Low-frequency currents in the deeper basin (C6B, C10B, C12B, C11B) and winds from May 10 to 15, 1989. Wind and current vectors are presented relative to the north pointing upward.

and density stratification. Assuming a wind-stress boundary condition at the surface, a no-slip friction condition at the bottom, and a constant vertical eddy viscosity $v$, the horizontal velocity vector $\vec{u}$ $(x, y)$ can be estimated:[20]

$$\vec{u} = \frac{\vec{\tau}H}{v}\left[\frac{\alpha}{2}\sigma^2 + \sigma(1-\alpha)\right]$$

**TABLE 9.2.1**

**Station, Major Axis Variance, and Percent of Total Variance Explained in Parentheses, Angle of Inclination of Major Principal Axis, and Series Length (hours) of Low-Pass Currents Series at Stations C6, C10, C12, and C11 in Grande-Entrée Lagoon**

| Station | Major Axis Variance/(percent variance explained) (m.s⁻¹) / (%) | Major Axis Inclination (degrees) | Number of Points (h) |
|---|---|---|---|
| C6  | 5.76 / (95.9%) | 13.0 | 238 |
| C10 | 7.28 / (82.0%) | 32.3 | 186 |
| C12 | 3.81 / (85%)   | 19.8 | 216 |
| C11 | 2.72 / (78.8%) | 26.6 | 188 |

**TABLE 9.2.2**

**Correlation (a) and Empirical Orthogonal Function (EOF) (b) Analyses of Principal Axis Low-Pass Currents at C6, C10, C12, and C11 and Winds along the Grande-Entrée Lagoon Longitudinal Axis**

(a) Correlation Analysis

| Station | C6 | C10 | C12 | C11 | Wind |
|---|---|---|---|---|---|
| C6   | 1.00  | 0.38  | 0.33  | 0.34  | −0.57 |
| C10  | 0.38  | 1.00  | 0.72  | 0.70  | −0.88 |
| C12  | 0.33  | 0.72  | 1.00  | 0.83  | −0.69 |
| C11  | 0.34  | 0.70  | 0.83  | 1.00  | −0.81 |
| Wind | −0.57 | −0.88 | −0.69 | −0.81 | 1.00  |

(b) EOF Analysis

| Mode | % Variance in All Series | % Variance in C6 | % Variance in C10 | % Variance in C12 | % Variance in C11 | % Variance in Wind |
|---|---|---|---|---|---|---|
| 1 | 75.52 | 44.5 | 78.1 | 56.0 | 68.0 | 96.5 |
| 2 | 16.67 | 55.0 | 8.7  | 7.8  | 9.0  | 1.2  |
| 3 | 4.49  | 0.3  | 2.5  | 29.0 | 14.8 | 1.2  |
| 4 | 2.68  | 0.2  | 9.8  | 5.1  | 5.2  | 0.9  |
| 5 | 0.64  | 0.0  | 0.9  | 2.1  | 3.0  | 0.2  |

In this equation, $\vec{\tau}\,(x,\ y)$ is the wind-stress vector normalized by a reference density, $\sigma$ is a dimensional depth $(0 \leq \sigma \equiv 1 + \frac{z}{H} \leq 1)$, and $\alpha$ is a dynamic ratio expressing the relative influences of horizontal pressure gradients $\vec{\nabla}_h \eta$ and the wind stress $\vec{\tau}(x,\ y)$ on $\vec{u}\,(x,\ y)$. This dynamic ratio is given by[20]

$$\alpha(x,\ y) = \frac{\text{Pressure gradient}}{\text{Wind friction}} \equiv \frac{\text{H g }\vec{\nabla}_h \eta \cdot \vec{e}_w}{\vec{\tau} \cdot \vec{e}_w}$$

where $\vec{\nabla}_h$ is the horizontal gradient operator and $\eta$ is the free-surface elevation above mean sea level $(z = 0)$. The resulting motion can be described as follows.[20] When $\alpha < 1$, the effect of the wind is dominant and the horizontal current flows downwind in the whole water column. This is the *wind drift current*, which usually occurs in the shallow areas (the value of H is small). On the other hand, when $\alpha > 2$, the effect of the pressure gradient is dominant, and the horizontal current flows upwind in the whole water column. This is the *gradient current*, which usually occurs in the deeper parts of the lagoon (the value of H is large). When $1 < \alpha < 2$, the horizontal currents reverse somewhere below the surface, from a predominant wind-drift current near the surface to a predominant gradient current near the bottom. Such analytical results, even if issued from simplifying assumptions about the constant eddy viscosity and the absence of nonlocal forcing at the mouth, form the basis for the potential of lagoons to adopt a 3D wind-induced current structure. More generally, lagoons have complex topographies, steady-state conditions seldom occur, and forcing at the mouth is significant for leaky and restricted inlets. In addition, waters can be vertically stratified and vertical eddy viscosities are not necessarily constant. Therefore, wind-driven currents in lagoons are best described using numerical 3D hydrodynamic models.

### 9.2.4 WIND-DRIVEN CURRENTS: NUMERICAL MODELING

The wind-driven circulation in GEL is next explored by using a 3D hydrodynamic model (MIKE3-HD, Appendix 9.2.A) to simulate first the vertically integrated or 2D current structure using one layer in the vertical. The model is then used to simulate the 3D current structure using the same initial and boundary conditions but imposing seven layers in the vertical, each 1 m in thickness. A comparison between the simulated 2D and 3D current structures in the deeper eastern basin should provide a test for the near-bottom upwind return flow hypothesis.

The computational domain includes the two-lagoon system shown in Figure 9.2.1 (GEL and HML, top left inset). The open boundaries were placed just outside the inlets connecting each lagoon to the Gulf of St. Lawrence. Since the inlets are not too distant (40 km, in the sense of free gravity wave propagation), the boundary conditions applied at both open boundaries were the same: the low-pass sea level fluctuations recorded at L1, outside GEL (Figure 9.2.2). Bathymetry data from the Canadian Hydrographic Service charts 4951, 4952, 4954, and 4955 were used to construct a rectangular grid with 360 cells along the $x$-axis (east) and 298 cells along the $y$-axis (north), each cell measuring 90 m $\times$ 90 m in size. The simulations in both the 2D and 3D model configurations were started at 12:00 h on May 10 and ended at 12:00 on May 15, 1989. The time step was set at 30 s, giving a Courant number of 3 for depths in the domain. The Smagorinsky eddy viscosity formulation (Appendix 9.2.A) was used for both 2D and 3D model configurations.

The vertically integrated or 2D currents in GEL under prevailing southwesterly and northeasterly winds are shown in Figure 9.2.5A and Figure 9.2.6A, respectively. Corresponding wind directions are also shown for clarity. Under southwesterly wind conditions (13:00 h, May 12, Figure 9.2.5A), currents are quickly established near both shores in the shallow areas. A weak return flow is seen in the basin, oriented toward the south. An almost opposite situation occurs for the vertically integrated coastal currents under northeasterly winds (16:00 h, May 13, Figure 9.2.6A). However, no return integrated flow is observed in the deep basin.

The 3D simulation was performed under homogeneous density conditions in order to avoid additional 3D baroclinic current structures. Horizontal currents at 0.75 and 4 m depths are presented in Figure 9.2.5B and Figure 9.2.5C, respectively, for prevailing northeasterly winds, and in Figure 9.2.5C and Figure 9.2.6C, respectively, for southeasterly winds. The major difference between the 2D and 3D current distributions can be seen in the deeper basin. In the 3D simulation, the wind-drift coastal currents near both shores are well defined in the surface layer, much as in the 2D case. However, the return flow or gradient currents from the 3D simulation occurs mainly at depth (Figure 9.2.5C and Figure 9.2.6C), in a direction opposite to that of the wind and the surface currents (Figure 9.2.5B and Figure 9.2.6B). When integrating the currents over the water column, opposing currents should cancel, as observed in the 2D case. In the shallower areas, to the west of the navigation channel, the drift current structures are almost identical in the 2D and 3D simulations.

### 9.2.5 CONCLUSION

The circulation and the transport of ecosystem variables in shallow lagoons are significantly affected by winds. Wind stress quickly establishes wind-driven currents in the lagoon shallow areas. Water accumulation at the downwind lagoon end then creates horizontal pressure gradients that generate a return flow in the form of upwind gradient currents. Wind-induced currents are further affected by friction at the bottom and, in wide subtropical lagoons, by Coriolis acceleration. In the presence of topographic variations, the resulting circulation becomes quite complex and a return flow can occur anywhere between the coastal currents. The hypothesis put forth in this study is that gradient currents will occur closer to the bottom in the deeper parts of the lagoon, away from the wind stress at the surface. This hypothesis was tested for GEL where a relatively deep (5–6 m) basin exists and where near-bottom current observations are available. An EOF analysis of low-frequency principal axis current fluctuations and the wind longitudinal component showed that near-bottom currents are negatively correlated with the wind direction. A numerical model was used to simulate vertically integrated currents under prevailing southwesterly wind conditions. Results show that coastal wind-driven currents are quickly established in the

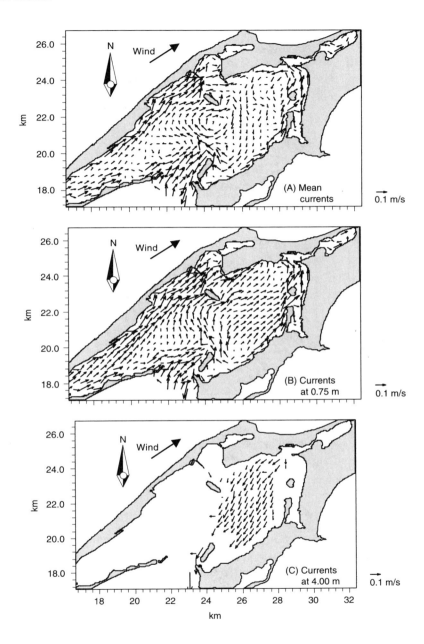

**FIGURE 9.2.5** (A) Simulated vertically averaged current, (B) Surface layer currents (at 0.75 m), and (C) near-bottom currents (at 4 m), under prevailing southwesterly wind conditions at 13:00 hrs, May 12, 1989.

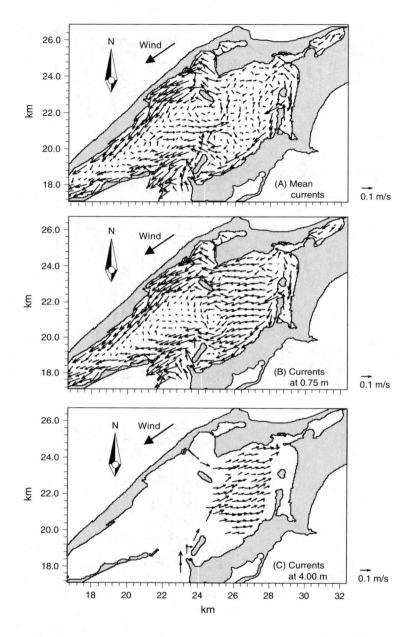

**FIGURE 9.2.6** (A) Simulated vertically averaged currents, (B) surface layer currents (at 0.75 m), and (C) near-bottom currents (at 4 m), under prevailing northeasterly wind conditions at 16:00 hrs, May 13, 1989.

lagoon near both shores but no obvious return flow can be detected against the wind in the deep basin. The model was then used to simulate the 3D structure of currents under the same wind conditions but with the water column divided in eight 1-m-thick layers. Results of the 3D simulation verified the initial hypothesis: the return flow does occur close to the bottom in the deeper basin, in a direction opposite to the winds. Surface currents in this case are still oriented downwind. These results should have significant implications for studies of lagoon ecosystem dynamics. For example, a stronger current near the bottom will enhance water and dissolved oxygen renewal for benthic species. It is suggested that, when attempting to model ecosystem dynamics in a coastal lagoon, currents used should be computed using a 3D modeling approach.

## APPENDIX 9.2.A:   THE MIKE3-HD NUMERICAL MODEL

As discussed in Chapter 3 and Chapter 6, the circulation and mixing processes in coastal lagoons are governed by the time-dependent, nonlinear equations of conservation of mass and momentum in three spatial dimensions. In some instances, 1D inlet-basin equations or 2D vertically integrated equations can be used. However, anticipating dynamic changes in the vertical dimension due to local wind stress, freshwater inflow from rivers, and mixing with denser oceanic waters, it is appropriate to start a lagoon hydrodynamics study by using the 3D governing equations of motion. Then, depending on local conditions (see Chapter 6), the equations can be simplified to one or two dimensions in space.

### 9.2.A.1  Governing Equations

The 3D governing equations of the MIKE3 numerical model are:[13]

1. The mass conservation equation:

$$\frac{1}{\rho\, c_s^2}\,\frac{\partial P}{\partial t} + \frac{\partial u_j}{\partial x_j} = S_{ss}$$

2. The momentum conservation equations, or the Reynolds-averaged Navier-Stokes equations in three dimensions, including the effect of turbulence and variable density:

$$\frac{\partial u_i}{\partial t} + \frac{\partial(u_i\, u_j)}{\partial x_j} + 2\Omega_{ij}\,u_j = -\frac{1}{\rho}\,\frac{\partial P}{\partial x_i} + g_i$$

$$+\frac{\partial}{\partial x_j}\left[\nu_t\left(\frac{\partial u_i}{\partial x_j} + \frac{\partial u_j}{\partial x_i}\right) - \frac{2}{3}\delta_{ij}k\right] + u_i\, S_{ss}$$

3. The salinity conservation equation:

$$\frac{\partial S}{\partial t} + \frac{\partial (S u_j)}{\partial x_j} = \frac{\partial}{\partial x_j}\left( D_S \frac{\partial S}{\partial x_j} \right) + S_{ss}$$

4. The temperature conservation equation:

$$\frac{\partial T}{\partial t} + \frac{\partial (T u_j)}{\partial x_j} = \frac{\partial}{\partial x_j}\left( D_T \frac{\partial T}{\partial x_j} \right) + S_{ss}$$

where $\rho$, $S$, and $T$ are the local density, salinity, and temperature of the fluid, respectively; $c_s$ is the speed of sound in sea water; $u_i$ is the velocity in the $x_i$ direction ($i, j = 1, 2, 3$); $\Omega_{ij}$ is the Coriolis tensor; $P$ is the pressure of the fluid; $g_i$ is the gravitational vector; $v_t$ is the turbulent eddy viscosity; $\delta$ is the Kronecker's delta; $k$ is the turbulent kinetic energy; $S$ and $T$ are the salinity and temperature; $D_S$ and $D_T$ are the associated dispersion coefficients; and $t$ denotes time. The $S_{ss}$ terms refer to the respective source-sink terms and thus differ from equation to equation. Water density $\rho$ is calculated from salinity, temperature, and pressure using the UNESCO equation of state.[21]

## 9.2.A.2 Wind Stress

The wind friction at the air–sea interface originates from the vertical shear term assuming a balance between the wind shear and the water shear at the surface. In a right-handed Cartesian coordinate system $(x, y, z)$, with corresponding current components $u, v, w$, the balance along the $x$-axis can be expressed as

$$v_t\, \frac{\partial u}{\partial z} = \frac{\tau_{xz}}{\rho} = \frac{\rho_{air}}{\rho}\, C_w\, W W_x$$

where $\rho_{air}$ is the density of air, $W$ is the wind speed with a component $W_x$ along the $x$-axis, and $C_w$ is the wind drag coefficient. The same formulation applies to the $y$-axis. The wind friction is thus a boundary condition to the vertical shear term. The wind drag coefficient can be calculated as:[2]

$$C_w = C_{w0} \qquad\qquad\qquad\qquad\qquad\qquad \text{for} \quad W < W_0$$

$$C_w = C_{w0} + (C_{w1} - C_{w0}) \cdot (W - W_0)/(W_1 - W_0) \qquad \text{for} \quad W_0 \le W \le W_1$$

$$C_w = C_{w1} \qquad\qquad\qquad\qquad\qquad\qquad \text{for} \quad W > W_1$$

where $C_{w0} = 0.0013$ for $W_0 = 0$ m/s and $C_{w1} = 0.0026$ for $W_1 = 24$ m/s.

### 9.2.A.3  Eddy Viscosity

The eddy viscosity coefficient $v_t$ in the water column accounts for the Reynolds stresses (turbulence) and other unresolved processes both in time and space (e.g., subgrid scale fluctuations). The eddy viscosity in the model can be specified in five ways: (1) a constant value; (2) a function of the Richardson number; (3) a time-varying function of the local gradients in the velocity field;[23] (4) from the solution of the turbulent kinetic energy equation, the so-called $k$-$\varepsilon$ model;[24] or (5) from the solution of both the turbulent kinetic energy and kinetic energy dissipation equations, the so-called $k$-model.[24] The simulations in this study are performed with the Smagorinsky formulation where the eddy viscosity is linked to the grid spacing and the velocity gradients of the resolved flow field as

$$v_t = l^2 \cdot \sqrt{S_{ij} \cdot S_{ji}} \qquad \text{where} \qquad S_{ij} = \frac{1}{2}\left( \frac{\partial u_i}{\partial x_j} + \frac{\partial u_j}{\partial x_i} \right)$$

Here, $l$ is a length scale replaced by the product of a constant $C_{sm}$ and the grid spacing $\Delta s$. Typical values for $C_{sm}$ in the vertical and horizontal directions are 0.176 and 0.088, respectively.

### 9.2.A.4  Bottom Stress

In shallow water depths, the bottom shear stress $\tau_{bottom}$ plays a major role in controlling the circulation and mixing of waters. It can be specified in terms of a drag coefficient formulation as

$$\tau_{bottom} = \rho \, C_D \, u^* |u^*|$$

where $C_D$ is a drag coefficient determined from the selected turbulence closure scheme and $u^*$ is the velocity at the top of the bottom boundary layer. When using the Smagorinsky eddy viscosity formulation, $C_D$ is given by

$$C_D = \left[ \frac{2\sqrt{2}}{3} \frac{D}{1} \left\{ 1 - \frac{z_m}{D} \right\}^{\frac{3}{2}} - \left\{ 1 - \frac{z}{D} \right\}^{\frac{3}{2}} + \frac{1}{\kappa}\log\left\{ \frac{z_m}{k_s/30} \right\} \right]^{-2}$$

In this equation, $z_m$ is the distance above the sea bed where the Smagorinsky profile matches the logarithmic velocity profile; $\kappa$ is the von Karman's constant; $z$ is the distance between the sea bed and the first computational mode; $k_s$ is the bed roughness length scale (range between 0.01 and 0.30 m); $l$ is the length scale given by the Smagorinsky eddy viscosity formulation; and $D$ is the actual depth.

## ACKNOWLEDGMENTS

This work is a Canadian contribution to the NATO Committee on the Challenges of Modern Society (CCMS) devoted to ecosystem modeling of coastal lagoons for sustainable management. It is funded by NATO-CCMS fellowship award

SA.8-3-01B (971261). Nestor Navarro contributed to the analysis of the 1989 data set as part of his M.Sc. thesis. The 1988–1989 Grande-Entrée lagoon project was funded partly by Fisheries and Oceans, Canada, and by an NSERC individual grant to the author.

## REFERENCES

1. Kjerfve, B., Coastal lagoons, in *Coastal Lagoon Processes*, Kjerfve, B., Ed., Elsevier Oceanographic Series 60, Elsevier, New York, 1994, 1.

2. Kjerfve, B., Comparative oceanography of coastal lagoons, in *Estuarine Variability*, Wolfe, D.A., Ed., Academic Press, New York, 1986, 63.

3. Smith, N., Water, salt and heat balances of coastal lagoons, in *Coastal Lagoon Processes*, Kjerfve, B., Ed., Elsevier Oceanographic Series 60, Elsevier, New York, 1994, 69.

4. Spaulding, M., Modeling of circulation and dispersion in coastal lagoons, in *Coastal Lagoon Processes*, Kjerfve, B., Ed., Elsevier Oceanographic Series 60, Elsevier, New York, 1994, 103.

5. Nunez Vaz, R.A., Periodic stratification in coastal waters, in *Modeling Marine Systems, Vol. II*, Davies, A., Ed., CRC Press, Boca Raton, FL, 1990, 69.

6. Hearn, C.J., Lukatelich, R., and McComb, A., Coastal lagoon ecosystem modeling, in *Coastal Lagoon Processes*, Kjerfve, B., Ed., Elsevier Oceanographic Series 60, Elsevier, New York, 1994, 471.

7. Knoppers, B., Aquatic primary production in coastal lagoons, in *Coastal Lagoon Processes*, Kjerfve, B., Ed., Elsevier Oceanographic Series 60, Elsevier, New York, 1994, 243.

8. Smith, S.V. and Atkinson, M.J., Mass balance of nutrient fluxes in coastal lagoons, in *Coastal Lagoon Processes*, Kjerfve, B., Ed., Elsevier Oceanographic Series 60, Elsevier, New York, 1994, 133.

9. Lacerada, L.D., Biogeochemistry of heavy metals in coastal lagoons, in *Coastal Lagoon Processes*, Kjerfve, B., Ed., Elsevier Oceanographic Series 60, Elsevier, New York, 1994, 221.

10. Nichols, M.M. and Boon, J.D., III, Sediment transport processes in coastal lagoons, in *Coastal Lagoon Processes*, Kjerfve, B., Ed., Elsevier Oceanographic Series 60, Elsevier, New York, 1994, 157.

11. Koutitonsky, V.G., Navarro, N., and Booth, D., Descriptive physical oceanography of Great-Entry Lagoon, Gulf of St. Laurence, *Estuarine, Coastal and Shelf Sci.*, 54(5), 833–847, 2002.

12. Rasmussen, E.B., Three dimensional hydrodynamic models, in *Coastal, Estuarine and Harbour Engineers Reference Book*, Abbott, M.B. and Price, N., Eds., Chapman & Hall, London, 1993.

13. DHI (Danish Hydraulic Institute), MIKE3–HD. Estuarine and Coastal Hydraulics and Oceanography—Hydrodynamic module, *Scientific Documentatio*, DHI Water and Environment Inc., Hørsholm, Denmark, 2002.

14. Booth, D., Tidal flushing of semi-enclosed bays, in *Mixing and Transport in the Environment* Beven, K., Chatwin, P., and Millbank, J., Eds., John Wiley & Sons, New York, 1994, 203.

15. Mayzaud, P., Koutitonsky, V.G., Souchu, P., Roy, S., Navarro, N., and Gomez-Reyez, E., L'impact de l'activié, mytilicole sur la capacité, de production du milieu lagunaire des Iles-de-la-Madeleine, Rapport INRS-Océanologie, Rimouski, Québec, 1991.

16. Navarro, N., Océanographie physique descriptive de la lagune de Grande-Entrée, Iles-de-la-Madeleine, M.Sc. thesis, Université du Québec à Rimouski, QC, Canada, 1991.

17. Koutitonsky, V.G. and Bugden, G.L., The physical oceanography of the Gulf of St. Lawrence: a review with emphasis on the synoptic variability of the motion, in *The Gulf of St. Lawrence: Small Ocean or Big Estuary?* Therriault. J.-C., Ed., *Can. Sp. Pub. Fish. Aquat. Sci.*, 113, 57, 1991.

18. Walters, R.A. and Heston, C., Removing tidal-period variations from time series data using low-pass digital filters, *J. Phys. Oceanogr.*, 12, 112, 1982.

19. Emery, W.J. and Thompson, R., *Data Analysis Methods in Physical Oceanography*, Pergamon Press, London, 1997.

20. Mathieu, P.-P., Deleersnijder, E., Cushman-Roisin, B., Beckers, J.-M., and Bolding, K., The role of topography in small well-mixed bays, with application to the lagoon of Mururoa, *Cont. Shelf Res.*, 22, 1379, 2002.

21. UNESCO, The practical salinity scale 1978 and the international equation of state of sea water 1980, UNESCO Technical Papers in Marine Science 36, Paris, 1981.

22. Smith, S.D. and Banke, E., Variation of the sea drag coefficient with wind speed, *Quart. J. Royal Met. Soc.*, 101, 665, 1975.

23. Smagorinsky, J., General circulation experiment with the primitive equations, *Month. Weather Rev.*, 91, 91, 1963.

24. Rodi, W., Examples of calculation methods for flow and mixing in stratified fluids, *J. Geophys. Res.*, 92(C5), 5305, 1987.

## 9.3 THE ECOLOGY OF THE MAR MENOR COASTAL LAGOON: A FAST-CHANGING ECOSYSTEM UNDER HUMAN PRESSURE

*Angel Pérez-Ruzafa, Concepción Marcos Diego, and Javier Gilabert*

### 9.3.1 FUNCTIONAL TYPOLOGY

#### 9.3.1.1 Location, Origin, Climate, and Hydrography

The Mar Menor is a hypersaline coastal lagoon, with a surface area of 135 km$^2$ and a perimeter of 59.51 km. It is located on the southwestern Mediterranean coastline (37°42′00″ N, 00°47′00″ W) with a mean depth of 3.6 m and a maximum depth of 6 m. La Manga, a sandy bar 22 km long and 100–900 m wide, acts as a barrier between the lagoon and the Mediterranean Sea. It is crossed by five more or less functional inlets called *golas*. Four are shallow (less than 1 m deep) and one of them, the El Estacio, was widened and dug to a 5-m depth to make it a navigational channel. Altogether the total width of lagoon entrances is about 645 m, giving Mar Menor a restriction ratio of 0.015. Mar Menor is therefore a restricted lagoon according to the classification proposed by Kjerfve[1] (see Chapter 6). There are two main islands and three other smaller islands, one of which is artificially connected to La Manga. Figure 9.3.1 and Figure 9.3.2 show the location of the Mar Menor Lagoon and its main physiographic characteristics.

The origin and evolution of the Mar Menor Lagoon have been greatly influenced by the changing levels of the sea since the Tortonian, the volcanic activity that occurred during the Pliocene and formed the small hills and islands in the Mar Menor basin, and the Quaternary compressive system that helped shift the sandy barrier that encloses the Mar Menor.[2,3]

At present, the main geomorphological elements that determine the lagoon dynamics are (1) the sandy barrier enclosing the Mar Menor; (2) the inlets or *golas* that determine the entrances from the Mediterranean Sea and its hydrography and confinement; (3) the islands and volcanic outcrops that constitute the only natural rocky substrates and generate environmental diversity for biological assemblages; (4) the gullies or *ramblas* that contribute waters and materials from agricultural run-off and mining mountains; and (5) marginal lagoons, now transformed into salt flats or salt mines.

The lagoon basin is located in a semi-arid region with low rainfall,[4] an annual mean of 300 mm, and high potential evapotranspiration (close to 900 mm) that results in a deficit of the net annual hydric balance that exceeds 600 mm/m$^2$ year (Figure 9.3.3). The orographic configuration of the basin, the scant vegetation, the impermeability of marly sectors, and the precipitation concentration all make torrential rainfall a characteristic of the area.[5] Winds show a well-defined and regular pattern during the year primarily from the east (*levantes*) followed by winds from the west and southwest (*lebeches*) (Figure 9.3.4). The annual mean velocities of the weakest winds (west and west-southwest) range from 9 to 12 km/h and the strongest

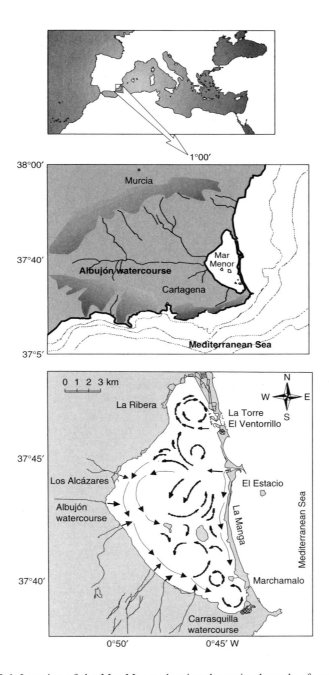

**FIGURE 9.3.1** Location of the Mar Menor showing the main channels of communication with the open sea, water courses, and current circulation diagram.

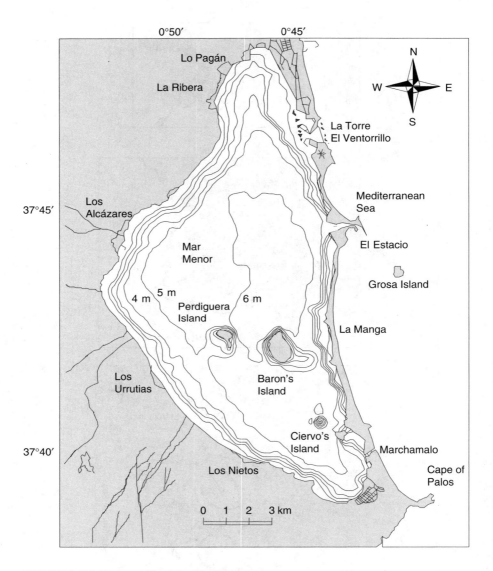

**FIGURE 9.3.2** Bathymetry of the Mar Menor.

(northeast, east-northeast, south-southwest, and southwest) from 18 to 26 km/h. The highest monthly mean velocities are usually in the range of 30 to 40 km/h.

Climatic and hydrologic features in this area of the Iberian southeast littoral (Table 9.3.1) combined with the lagoon's geomorphology cause it to behave like a concentration basin. Evaporation exceeds rainfall and run-off, and, until recent years, there was no permanent watercourse flowing into the lagoon. There are, however, more than 20 cataclinal watercourses on the watershed that collect rainfall water from the surrounding mountains. They get into the plains as real *wadis* but the waters

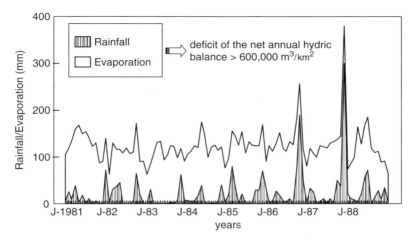

**FIGURE 9.3.3** Hydric balance in the Mar Menor area over a 10-year period. Rainfall is scarce and usually concentrated in brief, torrential downpours during the year. (Data from the San Javier meteorological station.)

evaporate and are lost through infiltration and do not reach the lagoon except in instances of strong torrential rainfall.[5] Most of the watercourses discharge in the southern half of the lagoon but that is determined by the sporadic and torrential rainfall. Among these, the Albujón watercourse, the main collector in the drainage basin, is an exception at present because it maintains a regular flux of water due to changes in agricultural practices, as described below.

Annual total inflow of fresh water through run-off and rainfall into the lagoon ranges from 27.9 to 122 Hm$^3$ while 155 to 205 Hm$^3$ evaporates, resulting in a hydric deficit ranging from 38 to 115 Hm$^3$ per year. The net loss of fresh water is compensated for by saltwater inputs from the adjacent sea,[6,7] and it is regulated by differences in the sea level between the lagoon and the Mediterranean Sea.[8] The lagoon water budget with the relative contribution of different components to the lagoon volume is shown in Figure 9.3.5.

### 9.3.1.2  Hydrodynamics

The lagoon hydrodynamics is mainly driven by winds and thermohaline circulation (Figure 9.3.1). The exchange of water between the lagoon and the adjacent Mediterranean Sea is mainly driven by differences in sea level phases,[8] with the El Estacio Channel playing the most significant role. The resulting exchange rates lead to residence times of the water bodies in the basin that change from year to year and are responsible for the development of the lagoon water characteristics and hence of the long-term ecological balance in the lagoon. The salinity of the lagoon waters ranges at present from 42 to 46 (with an annual mean value of 44.4) showing a north–south gradient (Figure 9.3.6).[9] Three main gyres can also be identified in a general circulatory pattern along this axis allowing us to differentiate three basins

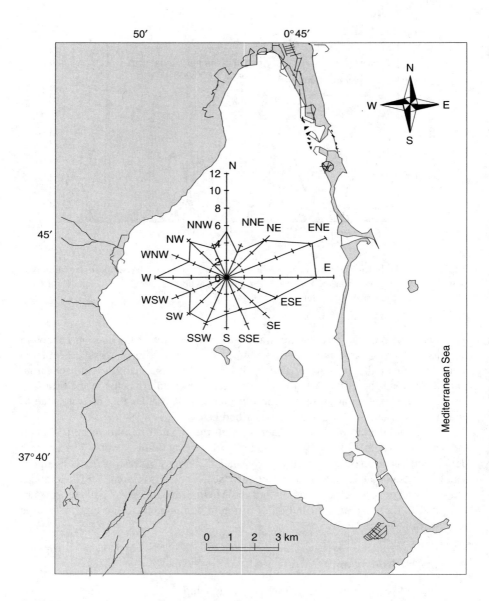

**FIGURE 9.3.4** Prevailing winds in the Mar Menor area during the 1980s. (Data from the San Javier meteorological station.)

at the Mar Menor: (1) the northern basin with the lower mean salinity values; (2) the southern basin with the most saline waters; and (3) the central basin with intermediate values and corresponding to the mixing area of Mediterranean and lagoon waters.

As in many other coastal lagoons, the water temperature is closely related to the atmospheric temperature.[10–12] The temperature distribution is relatively uniform over

## TABLE 9.3.1
## Minimum and Maximum Value of Climatic Variables in the Mar Menor Area (1981–1988)

|  |  | Maximum | Minimum |
|---|---|---|---|
| Monthly | Rainfall | 300 mm (November 1987) | 0 mm (usually in March and from June to August) |
|  | Evaporation | 165.5 mm (May 1981) | 50 mm (December 1987) |
|  | Solar radiation | 280.9 h of sun (July 1981) 623 cal/cm$^2$ * day (monthly mean, June 1987) | 110 h of sun (December 1987) 195 cal/cm$^2$ * day (monthly mean, January 1985) |
|  | Atmospheric temperature (absolute values) | 37°C (August 1986) | −3.4°C (January 1985) |
| Annual | Rainfall | 497.9 mm (1987) | 113.5 mm (1983) |
|  | Evaporation | 1442.9 mm (1981) | 115.2 mm (1988) |
|  | Solar radiation | 2344 h of sun (1985) | 2053.8 h of sun (1982) |

*Source:* Data from the San Javier Aerodrome Meteorological Station on the Mar Menor coast. See Reference 13.

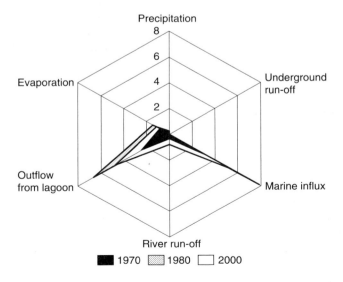

**FIGURE 9.3.5** Lagoon water budget rose (meters per year) for the Mar Menor. The diagram represents the relative contribution of each water budget component to the lagoon volume during the recent history of Mar Menor.

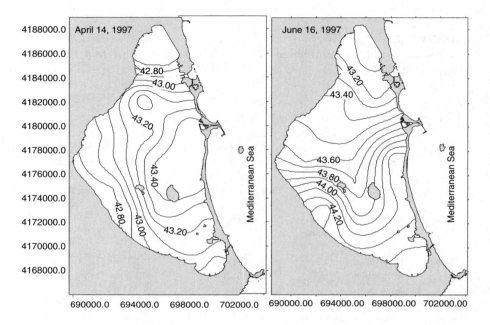

**FIGURE 9.3.6** Distribution of surface salinity at the Mar Menor lagoon in two hydrodynamic situations.

all the surface of the Mar Menor with some local differences, mainly related to the shallowest areas. Although the southern basin has warmer waters in summer and cooler waters in winter compared to the others, differences between them are usually less than 2°C at any time of the year.[13] Figure 9.3.7 shows the lagoon hydrological annual cycle using a temperature–salinity (T–S) diagram for mean monthly data.

Turbidity and suspended materials are highly variable depending on many topographical (distance to the coast, depth, nature, and slope of the bottom), biological (planktonic productivity), and climatic (wind and rainfall) variables. Values range from 2 mg/l of suspended solids in calm water conditions on rocky bottoms to 3.88 g/l in shallow waters on muddy or sandy bottoms under the action of the waves. It is possible, however, to distinguish two well-defined situations concerning water clarity: the first is clear waters associated with lower contents of nutrients and chlorophyll spanning along most of the year; the second is turbid water due to the increase of phytoplankton productivity mostly during the late summer.[14] Light is usually not a limiting factor for either phytoplanktonic or benthic biological productivity.

### 9.3.1.3  Sediment

According to their grain size composition, the bottom of the Mar Menor can be classified into two main sediment categories, muddy and sandy, with some areas of rocky bottoms.[9] On the one hand, muddy bottoms cover both the whole central area of the lagoon and those shallow bottoms that have lower hydrodynamism, at

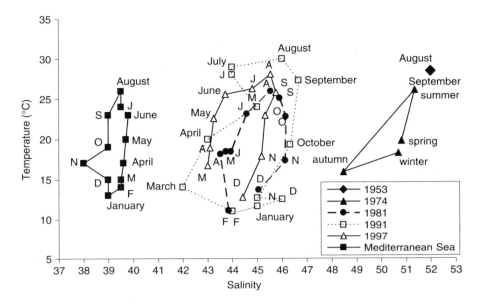

**FIGURE 9.3.7** Annual course of T–S index (monthly averaged values) for the Mar Menor during the last decades, compared with the Mediterranean shallow coastal waters of the Cape of Palos. Point at the right of the diagram corresponds to September and August monthly mean values at the Mar Menor according to data from Lozano.[40] The progressive "mediterranization" of the lagoon waters can be noted.

the same time covered by a dense meadow of the algae *Caulerpa prolifera* or patches of the seagrass *Ruppia cirrhosa*. On the other hand, sandy bottoms (with sand content up to 89%) are located on the margins of the basin and in the small bays surrounding the islands in which scant patches of the phanerogame *Cymodocea nodosa* grow.

The organic matter content in the sediment of the Mar Menor is highly variable, ranging from less than 0.34% in compacted red clays up to 8.6% in the *Caulerpa prolifera* areas. An increase in the organic matter content of sediment is observed seasonally, from autumn to winter, in both muddy and sandy bottoms. This increase is explained by the contribution of the fronds of the green algae *Caulerpa prolifera* and the phanerogame *Cymodocea nodosa*, respectively.

Dissolved oxygen values in sediment layers show a high range of variation oscillating between values in the range of 5 to 11 mg/l in surface waters (depending on wave action) to anoxic conditions and concentrations lower than 2 mg/l close to the bottom, in areas with dense meadows on muddy bottoms, high concentrations of organic matter, and low hydrodynamics. The water column shows homogeneous values, usually over saturation, with a small maximum just over the meadow (Table 9.3.2). Such a situation, with saturation levels in surface waters and anoxic conditions at the bottom, has been described previously in deeper lagoons.[12,15]

Submerged aquatic vegetation, rather than physical factors, seems to play a major role in the physical and chemical nature of sediment with only hydrodynamics

**TABLE 9.3.2**
**Oxygen Values (mg/l) in the Water Column on Distinct Kinds of Bottoms and Statement of the Macrophyte Meadows in a Location off the Los Urrutias 300 m from the Coast (July 1986)**

|  | Dense Meadow of Cymodocea nodosa–Caulerpa prolifera on Mud | Dense Meadow of Cymodocea nodosa–Caulerpa prolifera on Rock | Lax Meadow of Cymodocea nodosa on Mud |
| --- | --- | --- | --- |
| Surface water | 5.56 | 5.56 | 5.56 |
| 1 m | 5.64 | 5.64 | 5.64 |
| 2 m | 5.87 | 5.87 | 5.87 |
| 3 m | 5.87 | 5.87 | 5.87 |
| 3.2 m (bottom) | 1.76 | 4.12 | 4.04 |

*Note:* Profiles are less than 10 m from each other.

retrieving fine particles at the more exposed areas. Although sandy sediment usually show no vegetation coverage, *Cymodocea nodosa* meadows are also related to these bottoms determining slightly higher contents in the fine fraction and organic matter. *Caulerpa prolifera* or mixed *C. prolifera–Cymodocea nodosa* beds, in contrast, determine muddy bottoms with very high organic matter content, which also acts as a sediment trap thus favoring fine particles and organic matter accumulation (as stated in other places by Harlin et al.[16] or Genchi et al.[17]).

The contribution of the macrophytes to the organic content of sediment is a common feature in coastal lagoons, estimated by Mann[18] to be over 60% of the macrophytic production. In the Mar Menor the macrophytic production has been estimated as 165.6 g $C/m^2/year$[19] implying an input of detritic carbon to the lagoon bottoms of at least 13,400 mt C/year.

### 9.3.1.4 Biological Assemblages

Biota is characterized by eurihaline and euritherm species also present in the Mediterranean Sea, but they usually reach high densities in the lagoon. Many of them are generalist species (*r*-strategists) and their high density is the result of both their rapid growth rates and lack of competitors.

From an ecological perspective, the Mar Menor differs from other Mediterranean coastal lagoons in several aspects related to its environmental heterogeneity. Its size, depth, availability of rocky substrates related to the volcanic outcrop and docks, and the mediterranization process related to increased water interchanges account for much of the richness of species and actual bionomic diversity.

Phytoplankton assemblages follow a clearly defined seasonal succession in which four stages can be identified: (1) a winter period dominated by *Rhodomonas* and *Cryptomonas* with *Cyclotella* as the main diatom represented; (2) a spring phase where diatoms (mainly *Cyclotella*) are the dominant group with some monospecific blooms of other diatoms (mainly of *Chaetoceros* sp.); (3) a summer phase characterized by diatoms with blooms of *Niztschia closterium;* and (4) a post-summer and fall phase where diatoms still remain the major group but dinoflagellates increase in importance with peaks of *Ceratium furca*[20] with larger diatoms such as *Coscinodiscus* spp. and *Asterionella* spp. also present during the year.

Zooplankton at the nanoplankton (2–20 µm) level is dominated by flagellates eating bacteria. At the microplankton (20–200 µm) level ciliates, both oligotrichs and tintinnids, are well represented by a few species. At the mesoplankton level (>200 µm) copepods are the main group represented with smaller-sized species such as *Oithona nana* and larger ones such as *Centropages ponticus* and *Acartia* spp.[20] Appendicularians can also be found and gelatinous zooplankton is mainly characterized by the jellyfish *Aurelia aurita*. The importance of this trophic compartment has increased during the last few years due to massive proliferation of the jellyfishes *Rhizosthoma pulmo* and *Cotylorhiza tuberculata* in summer.

From a bionomic point of view, a number of benthic communities, depending on the type of substrata, wave exposition, and light, shows vertical zonation patterns that resemble the open sea communities but "miniaturized."[9,21,22]

Phytobenthos is represented by 33 species of Chlorophyceae, 20 species of Phaeophyceae, and 33 species of Rhodophyceae.[21] Soft bottom communities are mainly characterized by extensive meadows of the algae *Caulerpa prolifera*, with some areas of the phanerogam *Cymodocea nodosa* and small spots of *Ruppia cirrhosa* in very shallow areas.

Photophilic algae on hard substrates show different biocoenoses related to vertical zonation and the degree of confinement.[22,23] In low confinement conditions, close to the communication channels with the Mediterranean Sea, there is a narrow midlittoral fringe characterized by *Cladophora albida*, *C. coelothrix*, and *Enteromorpha clathrata*. In these areas, the infralittoral community is characterized mainly by *Jania rubens* and *Valonia aegagropila*.

In confined areas, the midlittoral is dominated by *Cladophora albida*, *Laurencia obtusa*, and *Cystoseira compressa* and the infralittoral by *Laurencia obtusa*, *Cystoseira compressa*, *Cystoseira schiffneri*, *Padina pavonica*, *Caulerpa prolifera*, and *Acetabularia acetabulum*.

Faunistic assemblages consist of up to 443 species, most of them benthic, included in 11 phyla:[9] Foraminifera (30), Porifera (21), Coelenterea (22), Nematoda (19), Anelida (100), Artropoda-Crustacea (48), Chelicerata (6), Unirramia (6), Molusca (106), Ectoprocta (7), Phoronida (1), Echinodermata (5), and Cordata (Tunicata (5) and Vertebrata-Osteichthyes (67)). Only certain species of each phylum dominate any one community with only rare occurrences of many of the species.[9] For example, the cnidarian *Bunodeopsis strumosa* and *Telmactis forskalii* reach densities of 2,000 individuals/m²; the polychaete *Filograna implexa* reaches a density of 2,500 individuals/m²; the amphipod *Caprella mitises* reaches a density of 36,700 individuals/m²; and the gastropod *Bittium reticulatumes* reaches a density of 39,800 individuals/m². Some of these densities are the highest reported for certain species, as in the case of the pycnogonid *Tanystylum conirostre*[24] with 3,600 individuals/m², or the ophiuroid *Amphipholis squamata*[25] with 475 individuals/m². The diversity of molluscs, taken as an indicator of the communities' structure, is rather low (0.5–2.2 bits/ind. on muddy bottoms and 1.7–2.8 bits/ind. on rocky bottoms).[9]

Fishes include mugilids, sparids, singnatids, gobids, and blennids. The benthic fish assemblage of the Mar Menor consists of 23 common species.[9,26,27] The dominant species are *Gobius cobitis, Lipophrys pavo,* and *Trypterigion tripteronotus* on shallow infralittoral rocks; *Pomatoschistus marmoratus, Solea vulgaris,* and *Solea impar* on sandy bottoms; and *Syngnathus abaster, Hippocampus ramulosus,* and *Gobius niger* on *Cymodocea nodosa–Caulerpa prolifera* mixed beds.

The lagoon also provides a habitat for many migratory seabirds. Its margins and islands attract nidificate species of international relevance such as *Himantopus himantopus, Recurvirostra avosetta, Charadrius alexandrinus,* and *Sterna albifron,* and others such as *Sterna hirundo, Tadorna tadorna, Anas platyrrhynchos, Burhinus oedicnemus, Larus ridibundus,* and *Larus cachinnans.* Marginal lagoons exploited for salt mining shelter stable colonies of flamingos (*Phoenicopterus ruber*), which reach concentrations of up to 1000 individuals.

### 9.3.2 RECENT HISTORY OF CHANGES IN THE LAGOON RESULTING FROM HUMAN ACTIVITIES

The Mar Menor has attracted humans since ancient times, and it has been the target of various types of aggression during its recent history. Since the early 1970s, tourist development has increased the demand for recreational facilities, resulting in the creation of new beaches, harbors, and channels. Mining, urban development, and changes in agricultural practices have increased the waste input into the lagoon. Many of these activities led to environmental changes that affected the biota and changed the configuration of the lagoon.

One of the first impacts on the lagoon environment caused by human activities resulted from the input of terrestrial materials, which increased the sedimentation rates from 30 mm/century to 30 cm/century as a consequence of the deforestation of the surrounding land for agricultural and pastural use during the 16th and 17th centuries.[28,29] Another ancient activity that developed in the Mar Menor area, beginning in 2000 B.C., was mining (argent, iron, lead, zinc, copper, and nickel among other minerals) in the southern mountains. The maximum extractive capacity occurred from 1960 to 1980 but mining continued until 1990. Waste from the barren mining lands was emptied into watercourses flowing into the south side of the Mar Menor until 1950. It was then diverted to silt Portman Bay (64 ha, 20 m depth) on the Mediterranean coastline facing the south. Although mining waste to the Mar Menor was stopped more than 50 years ago, the heavy metal concentration in sediment, specially in the southern part of the lagoon, still remains high and constant (mean values ≈ 2000 µg/g of lead and zinc)[28–30] (Figure 9.3.8).

Two marginal lagoons are at present used for salt mining. An additional four lagoons experienced either a natural process of filling up by sediment or were drained for agricultural use or urban development. This process has contributed to the reduction of the Mar Menor surface (from 185 km² in 1868, to 172 km² in 1927, 138 km² in 1947, to 135 km² in 1969 with a slight reduction since then (Figure 9.3.9) and a reduction in depth (Figure 9.3.10)).[5,29]

Records of the first tourist settlements date from the first half of the 19th century. Since then the lagoon has attracted an increasing seasonal population. The census of local population that lived year round in the Mar Menor for the year 2000 was 45,584 whereas the stable tourist population during the high season (July to September) was

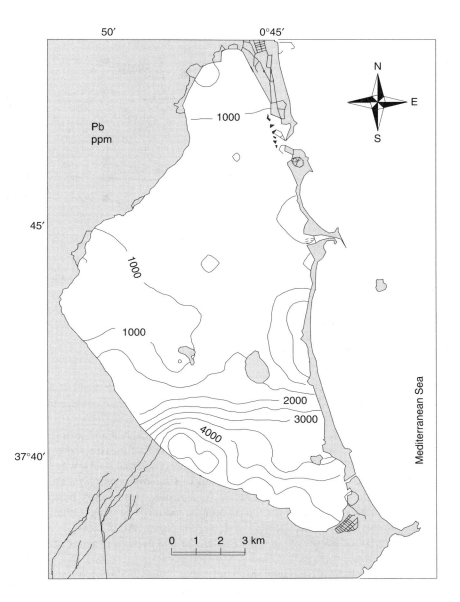

**FIGURE 9.3.8** Lead concentration in sediments of the Mar Menor Lagoon.[28]

estimated at about 450,000; together with the number of summer visitors approximately 748,211 people generated revenue of 198.4 million €.[31] The construction of tourist facilities has grown in parallel with the seasonal population. New roads, one of which connects Ciervo Island to La Manga, and highways surrounding the lagoon have been constructed recently. Between 1937 and 1976 the built-up lagoon perimeter increased from 12 to 54%, and to 56% in 1986 and 64% in 1994 (Figure 9.3.9). There

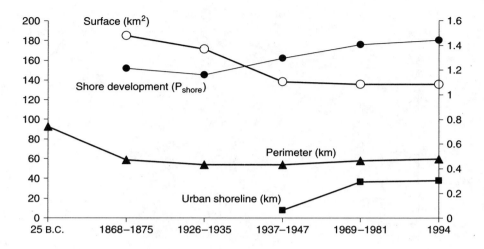

**FIGURE 9.3.9** Evolution of the perimeter and surface of the Mar Menor Lagoon from 25 B.C. to the present. Loss of surface and perimeter up to the 19th century probably was related to segmentation processes and fill-up of marginal embayments by erosion of surrounding lands. During the 20th century the surface still decreased but the perimeter increased as a consequence of land reclamation for different human uses. This led to an increase in the shore development index and a corresponding increase in the vulnerability of the system.

are nine yachting harbors, lodging 2784 boats, located along the lagoon's coastline, some of them less than 800 m away. At present, the major urban development takes place inland, perpendicular to the coastline.

Land has been reclaimed for construction of new beaches and promenades on the order of 270,275 m² with 640,456 m³ of sand, much of it extracted from inside the lagoon during 1987–1988. The resulting environmental stress was exploited by opportunist species such as floating masses of *Chaetomorpha linum* and meadows of *Caulerpa prolifera* in some areas. As a result, clean sandy bottoms were replaced by muddy ones with increased organic matter,[32] causing a change in the species composition of the benthic fish assemblages. At present, a new land reclamation plan of about 500,000 m² will create and expand beaches at the inner part of La Manga.

El Estacio is the only navigation channel that connects the Mediterranean Sea with the lagoon. It was artificially created in the early 1970s by dredging and widening one of the *golas* up to 30 m wide and 5 m deep at its minimum section. Opening this channel caused major changes in the lagoon dynamic. The modification of the renewal rate of the water changed the lagoon's physical and chemical properties, mainly salinity and extreme temperatures, which permitted access to new colonizer species, thus altering the lagoon's community structure with detrimental effects on fisheries.

On the other hand, agriculture in the watershed has experienced a great transformation in the last 15 years. Since 1986 surface waters diverted from the Tajo River, 400 km north, to the Segura River have changed extensive dry crop farming to intensive irrigated crop farming. This surface water for irrigation lessened the aquifer's overexploitation, thus raising the phreatic levels[33] and helping the main watercourse on the watershed maintain a continuous flow (about 24 1/s) fed by ground water with high nitrate levels to the lagoon. Due to overfertilization with nitrogen and pesticides used in agriculture, this flow is at present the main way nitrate enters the lagoon and pesticides enter the trophic food web.[34]

Not all human activity on the coastline has had negative effects on the biological assemblages. Some activities, such as the construction of small piers made of wood, sometimes on concrete pillars—a component of the traditional Mar Menor land-scape—have provided hard bottoms and shaded habitats that favor the settlement and development of sciaphilic assemblages, thus increasing the lagoon biodiversity. Such communities consist mainly of suspension feeders such as sponges, cnidarians, briozoans, and ascidians,[9] which actively contribute to maintaining the water quality.

### 9.3.3 Main Changes Affecting the Lagoon's Ecology

Two of the above-described human-induced changes have had, and continue to have, significant impact on the transformation of the lagoon dynamic. On the one hand, changes in hydrodynamics caused by the enlargement of the El Estacio Channel in 1972 produced an increase in the water renewal rates, decreasing salinity and lower extreme temperature, thus permitting access to new, mainly benthic and nectonic colonizers in the process of mediterranization of the lagoon (Figure 9.3.7, Table 9.3.3). Decreases in salinity values were observed from then until 1988 with an increase in colonizers such as the algae *Caulerpa prolifera*. On the other hand, changes in the nutrient input regimen are, at present, producing a chain of changes affecting mainly water quality, benthic vegetation, phytoplankton, and gelatinous plankton. A more detailed description of these changes is provided in the following section.

#### 9.3.3.1 Changes Induced by Water Renewal Rates

As mentioned previously, the hydrographic conditions of the Mar Menor have changed in the course of its geologic history because of the sea level fluctuations and the development of the sand barrier and communication channels with the open sea.

**TABLE 9.3.3**
**Influence of the Enlargement of the El Estacio Channel on Some Hydrographical Features of the Mar Menor[32]**

|                                                    | 1970        | 1980        | 1988        |
|----------------------------------------------------|-------------|-------------|-------------|
| Outflow of water to the Mediterranean ($m^3$)      | $3.6 \times 10^8$ | $6.1 \times 10^8$ | $6.4 \times 10^8$ |
| Inflow of water from the Mediterranean ($m^3$)     | $4.5 \times 10^8$ | $7.2 \times 10^8$ | $7.3 \times 10^8$ |
| Residence time (years)                             | 1.28        | 0.81        | 0.79        |
| Temperature range (°C)                             | 7.5–29      | 12–27.5     | 12–30.5     |
| Salinity range                                     | 48.5–53.4   | 43–46       | 42–45       |

The most recent event was the opening of the El Estacio Channel in 1972. Biological assemblages of the Mar Menor have been changing as a result of the degree of isolation and environmental conditions. Salinity increased after the last sea level regression in the Quaternary and the progressive isolation of the 18th century, reaching a maximum of 70 at the end of that century. After that period several sporadic storms broke the sandy bar, resulting in changes in salinity that allowed the colonization of several species, mainly fishes (striped sea bream, gilt-head, sea bream).[35,36] The last of these, which occurred in 1869 and was probably reinforced by the opening in 1898 of the Marchamalo *gola*, an artificial channel of communication with the Mediterranean to be used for fisheries, caused a significant decrease in salinity from 60–70 to 50–52. This decrease resulted in a marked change in the lagoon biology with the introduction of several species of submerged rooted vegetation—*Cymodocea* and *Zostera* throughout the basin and occasionally *Posidonia oceanica* in sandy areas of the south—and more than 30 new species of molluscs and fishes that became established in the lagoon.[29] A similar process took place after the opening of the El Estacio Channel. The above-mentioned increase in the renewal rate allowed the colonization of new marine species with a twofold increase in the number of molusc and fish species in the last 15 years.[9,13] As a consequence of the same process the allochthonous jellyfish species *Cotylorhiza tuberculata* and *Rhizostoma pulmo* entered the lagoon from the Mediterranean in the mid-1980s.[9] After an initial period of slow growth, their populations grew to massive proliferations. They also became pests as a consequence of the changes in the trophic status of the lagoon, resulting in a negative impact on tourism (see below for details).

In this way, an increase in the water renewal or interchange rates with the open sea is thus translated into an increase in the number of species inhabiting or visiting the lagoon (Figure 9.3.11). Related to this, changes in water renewal rates have also

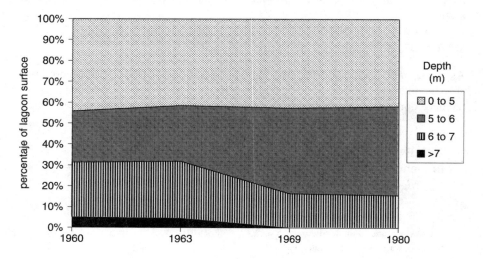

**FIGURE 9.3.10** Evolution of the depth distribution in the Mar Menor Lagoon. The maximum depth of 7 m disappeared after 1969 and areas with a depth greater than 6 m now constitute less than 20% of the bottom surface.

provided replacement of some facies in given communities. Assemblages of *Ceramium ciliatum* var. *robustum* and *Cladophora* sp. recorded previously[37] in photophilic community on the rocks were mostly replaced by facies of *Acetabularia acetabulum, Jania rubens, Padina pavonica* and, in some zones, by *Laurencia obtusa*.[21] In recent years, some of these assemblages have been replaced by *Caulerpa prolifera*.

Two main mechanisms operate in these processes: the increase in the colonization rates of marine species (both in larval or juvenile stages and by migration of adults) and a reduction in salinity and the mitigation of extreme temperatures, thus permitting the establishment of allochthonous species to the lagoon conditions[38,39] (see Chapter 5 for a more detailed discussion).

Some of the most important changes affecting the physiography and functioning of the Mar Menor took place at the benthic meadow level. *Cymodocea nodosa* (Ucria) Ascherson, *Zostera marina,* and *Z. nana* meadows dominated the Mar Menor bottoms before 1970[28,40] with scant mats of *Posidonia oceanica* (Linnaeus) Delile found in some small areas. In 1980, however, it was possible to find only a few mats of *Posidonia* close to some of the small islands to the south and close to the mouth of the El Estacio Channel.[9,29] At present, the benthic vegetation on the soft bottoms of the Mar Menor mainly consists of monospecific *Caulerpa prolifera* (Forskal) Lamouroux meadows on muddy and some rocky bottoms, covering more than 80% of the bottoms, favoring high levels of organic matter in sediment and low oxygen concentration. Scarce patches of *Cymodocea nodosa* are now restricted to shallow sandy bottoms and more or less dense spots of *Ruppia cirrhosa* (Petagna) Grande remain in the shallowest and calm zones.[41] (See Figure 9.3.13.)

Spreading of the *Caulerpa prolifera* at the expense of the *C. nodosa* monospecific meadows has been progressive since the opening of El Estacio Channel, starting in the north basin with intermediate states of mixed *Cymodocea–Caulerpa* meadows.[41] Two related processes seems to be involved in this change: (1) changes in environmental conditions of the lagoon, mainly moderation of salinity and extreme temperatures, permitting the entrance of the algae *Caulerpa prolifera* and (2) perhaps increasing stress in sediment and the increase of nutrients in water, in a second phase, thus giving competitive advantage to the algae over the seagrass *Cymodocea nodosa*.

Before the opening of El Estacio Channel, *C. nodosa* was reported at all depths in the Mar Menor.[28,29,40,41] In the 1990s it was present above the 3.3-m isobath estimated as the critical depth for this species in this ecosystem at that time.[42] After the opening of the channel, salinity dropped from 50–60 to 42–45 and the lowest temperatures were usually higher than 11°C. *Caulerpa prolifera* is sensitive to both salinity and temperature and cannot withstand temperatures below 10°C,[43] temperature readings frequently reached in winter before the enlargement of El Estacio. It has almost continuous growth throughout the year and a high capacity to generate vegetatively a new thallus from any fragment swept away by the water, resulting in a high colonization rate. It seems that most probably *C. prolifera* entered the lagoon after the opening of the El Estacio Channel, when salinity and temperature were no longer limiting factors, and settled progressively, mainly on muddy bottoms. Instability of sediment also provides macroalgae a competitive advantage over rooted vegetation. It seems that in the Mar Menor lagoon, both instability of sediment (dredging and pumping of sand, increase in sedimentation rates, increase in organic matter content) and

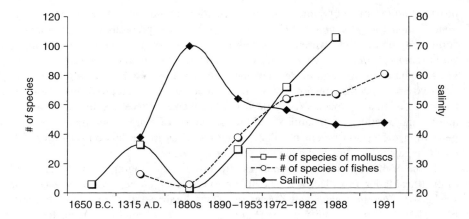

**FIGURE 9.3.11** Relation between the number of species of molluscs and fishes and the salinity of the waters in the Mar Menor Lagoon. Mollusc data before 1950 correspond to stratigraphical study of sediments by Simonneau.[28] Data for fishes for the same period come from works on fisheries[35,36] and probably underestimate benthic cryptic species, mainly for median age. Estimation of salinity in ancient time was based on historical data for the open conditions of the lagoon, which maintained a higher water surface level and active commercial navigation through its *golas*.[62]

overenrichment of nutrients in the water, in a second phase, acted in a synergistic way on the colonization rate of the *C. prolifera* up to the point that it now occupies most of the lagoon bottoms, including rocky substrates. Instability of sediment was mainly caused by dredging of parts of the lagoon bottoms, the fill-in of some beaches, and the input of terrigenous allochthonous materials from torrential rains. Nutrient enrichment mainly resulted from urban and agricultural wastewater.

Despite, or probably because of, the increased biodiversity in the lagoon there was a decrease in some fish production and catch. The colonization increase at the end of the 19th century resulted in a decrease in grey mullet. The enlargement of the El Estacio Channel in 1972 caused a decrease in Mugilidae and *Sparus aurata* (Figure 9.3.12). These changes have been related to the increase in interspecific competition and to changes in sediment properties and the bottom environment.[32,44]

### 9.3.3.2 Changes Related to Nutrient Inputs

Recent history of nutrient inputs into the lagoon has been closely related to urban, industrial, and agricultural development, either on the coastline or in the watershed, respectively. The eutrophication process, which has been described in many systems with observational rather than experimental data in coastal lagoons, starts with the increase in nutrients following a general trend in which seagrasses are replaced by macroalgae as the first step. Later small phytoplankton cells are replaced by larger ones that shade the bottom and hinder submerged vegetation growth by the decomposition of benthic organic matter and subsequent production of anoxia at the sediment level and in the water column afterward (see Chapter 5).

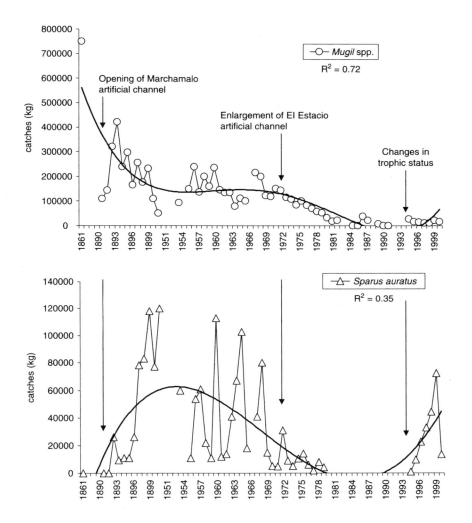

**FIGURE 9.3.12** Evolution of catches of *Mugil* spp. and *Sparus auratus* in the Mar Menor lagoon. Fisheries are artisanal and the fishing effort can be considered relatively constant over time.

Physical, chemical, and biological data recorded for many different Mar Menor programs over the years show some deviation from the rather classical pattern of the eutrophication process. As mentioned above, intensive urban development for tourism purposes started in the early 1970s, especially on La Manga. By that time, the El Estacio Channel had opened and the largest yachting harbor in the lagoon had been built. Simultaneously, summer residential areas were also built on the lagoon's western coastline. Wastewater treatment plants were installed in the main villages by the mid-1980s, but sewage overflow in many residential areas was, and still continues to be, filtered into the lagoon after primary treatment. Urban sewage is usually the main source of phosphorus in many Mediterranean coastal lagoons[45] and agriculture is the main source of nitrogen. By the time urban wastewater treatment plants were operational, agriculture started to change from dry crop farming with low amounts of nitrogen fertilizers to intensive crop irrigation with nitrogen overfertilization. During the dry agriculture period, nitrogen was the limiting nutrient for both benthic[42] and planktonic primary production in the lagoon; nitrogen entered mainly via run-off and phosphorus entered through urban sewage.[20]

In the 1970s, the Mar Menor was oligotrophic, and primary productivity was mainly benthic with the phanerogam *Cymodocea nodosa* as the main macrophyte. During the early 1980s, after the enlargement of El Estacio, the bottoms were covered by a mixed meadow of *Cymodocea nodosa–Caulerpa prolifera*, with a biomass of about 280 g dw/m² (Figure 9.3.13).[21,41,42] By the early 1990s a dense bed of the invasive macroalgae *Caulerpa prolifera* covered most of the bottom, restricting the seaweed *Cymodocea nodosa* to patches in the shallowest areas. The high benthic macrophyte biomass contrasted with the low phytoplanktonic density[46] and the oligotrophy of the waters.[20] Based on data from the mid- to late 1980s, it was estimated that 63.18% of the total primary production of the lagoon was due to *Caulerpa prolifera*, 0.42% to *Cymodocea nodosa*, and 0.24% to photophilic algae, with 11.62% due to microphytobenthos and 24.53% due to phytoplankton.[42]

Changes in the trophic status of the lagoon waters can be seen by comparing two extensive time series (weekly sampled), one for 1988 and the other for 1997. The time series of 1988 showed that nitrate concentrations were low throughout the year, in contrast to the higher phosphate values. It also showed the seasonal variation in the trophic state of the water due to different nutrient input regimes: nitrate mainly in winter from run-off and phosphorus mainly in summer from urban sewage. While nitrate concentration during 1988 (Figure 9.3.14A) was always under 1 μmol $NO_3^-$/l, much higher concentrations occurred in 1997 (Figure 9.3.14B), particularly during spring and summer (just at the harvest time when larger amounts of fertilizer are used in the lagoon's watershed) entering mainly through the major watercourse (El Albujón) due to the increase in the phreatic levels, as explained above. During 1997 higher nitrate concentrations were usually found on the west coast of the lagoon, close to the mouth of the main watercourses, while lower concentrations were found on the inner coast of La Manga and the El Estacio Channel influence area (Figure 9.3.15),[47] suggesting that nitrate input was related to the agricultural activity.

Drastic changes in phosphate levels were also found between 1988 (Figure 9.3.14C) and 1997 (Figure 9.3.14D). The seasonal distribution of phosphate found in 1988 was not evidenced in 1997, with much lower values probably due to the effect of wastewater treatment plants. The spatial distribution of phosphorus in surface waters, on the other hand, showed maximum concentrations close to the point sources (waste pipelines) of the main wastewater treatment plants in La Ribera and Los Alcázares (in this case through the Albujón watercourse) (Figure 9.3.16) probably due to treatment plant malfunctions. In 1997 the N:P ratio changed drastically as a consequence of higher nitrate load and phosphate removal, with phosphate becoming the limiting factor for planktonic productivity. As a consequence of changes in the nutrient input regimen, the water column in the lagoon changed from moderately oligotrophic to relatively

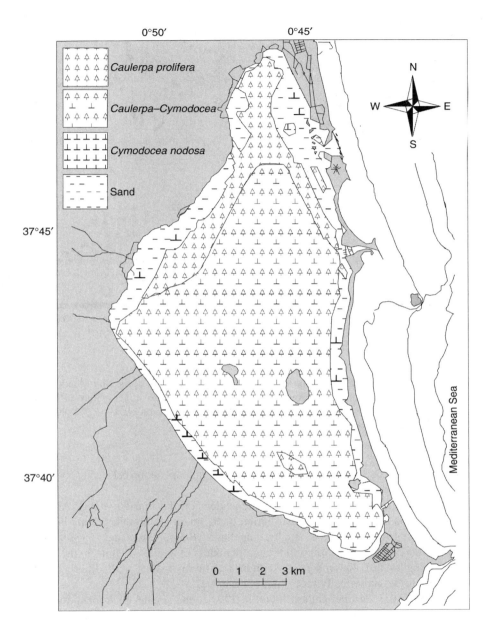

**FIGURE 9.3.13** Distribution of macrophyte beds at the Mar Menor Lagoon in 1982, showing the broad expansion of the algae *C. prolifera*.[9]

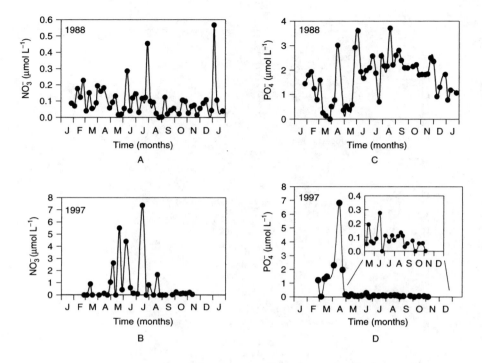

**FIGURE 9.3.14** Comparison of nutrient levels in the water column of the Mar Menor lagoon in 1988, under oligotrophic conditions, and in 1997, after changes in agriculture in the surrounding lands. (A) Nitrate concentration, 1988; (B) nitrate concentration, 1997; (C) phosphate levels, 1988; (D) phosphate levels, 1997.

eutrophic, providing conditions for growth of larger phytoplankton cells and subsequent changes in the trophic structure.

Planktonic communities are highly dependent on the nutrient status. As mentioned previously, the seasonal distribution of the main taxonomic groups in terms of density in 1988 reveals that smaller flagellates (*Rhodomonas* and *Cryptomonas*) were numerically the most important phytoplankters in winter, whereas in the period from spring to fall diatoms reached their maximum values with dinoflagellates appearing to be numerically more important in fall.

Although the seasonal pattern found in 1997 in the smaller phytoplankton fractions was relatively similar to the one found in 1988 there were differences in the large-sized phytoplankton cells compartment. Large diatoms such as *Coscinodiscus* spp. and *Asterionella* spp. were present throughout the year. These changes in cell size are in accordance with the ecophysiological theory based on the size-dependent nutrient uptake kinetics[48,49] and diffusion limitation of nutrient transport.[50,51] Similarities in the smaller phytoplankton fractions suggest that both the detritus pathway[52,53] and the microbial loop,[54] characteristic of shallow oligotrophic systems, still play an important role in transferring energy to larger organisms.

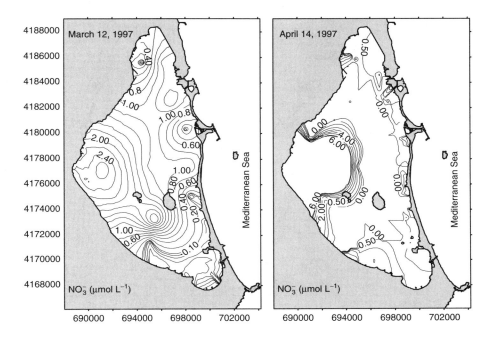

**FIGURE 9.3.15** Distribution of nitrate concentration in the Mar Menor surface waters in two situations with different discharge intensity through the Albujón watercourse mouth.

In the zooplankton compartment, small heterotrophic flagellates were numerically the major component in 1988. They showed a seasonal trend with higher values in summer and lower values in spring and fall. At the microplankton level, oligotrich ciliates appeared without exhibiting a clear seasonal trend throughout the year, whereas tintinnids showed increases mainly in summer. Eggs and veliger larvae, on the other hand, exhibited a strong seasonal trend whereas other larvae of benthic organisms (mainly Bippinnaria and Auricularia) occurred in lower numbers during periods of maximum temperature. Copepods were mainly represented by three species: *Oithona nana, Centropages ponticus,* and *Acartia* spp. (mainly *latisetosa*). They remain relatively constant throughout the year except for the warmer water period in which a steady decreasing trend is observed, which increases later up to the beginning of winter. Massive proliferation of copepods (>1000 indiv./l), mainly due to *O. nana*, may easily take place. Other large zooplankton, such as the appendicularian *Oikopleura dioica*, showed significant well-defined and well-characterized increases both in spring and at the beginning of winter. Again, at the micro- (20–200 μm size) and mesoplankton (>200 μm size) levels, seasonal trends in 1988 and 1997 showed no large differences in taxonomic composition except in the gelatinous compartment. During the summer of 1997 there was a bloom of the jellyfishes *Rhyzostoma pulmo* and *Cotylorhiza tuberculata*. Figure 9.3.17 shows the temporal distribution of jellyfish species during 1997. Each data point on the plot represents the mean density value estimated from 20 sampling stations spacially

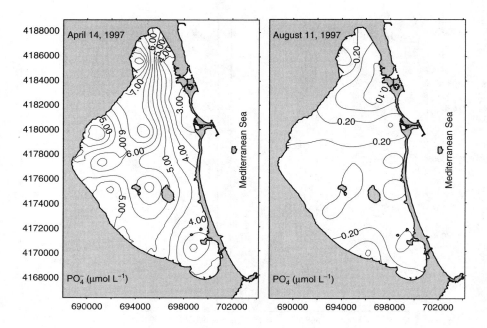

**FIGURE 9.3.16** Distribution of phosphate concentration in the Mar Menor surface waters in two situations with different discharge intensity through the urban waste pipelines.

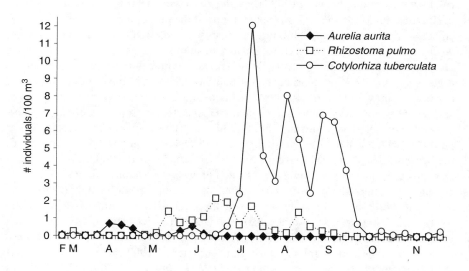

**FIGURE 9.3.17** Temporal distribution of jellyfish species in the Mar Menor Lagoon in 1997. F through N = February through November.

distributed. *Aurelia aurita*, the only autochthonous species, is the less abundant and has its maximum abundance in spring (April and May). *Rhyzostoma pulmo* started to increase in May, while *Cotylorhiza tuberculata* peaked in abundance in June and July, reaching more than 12 individuals per 100 m³. The total population of jellyfishes estimated in the lagoon by mid-summer of 1997 was on the order of 40 million.

Presence of large-celled phytoplankton in response to elevated nitrate concentrations has implications for the structure and function of the whole planktonic food web. The distribution of biomass in aquatic systems generally shows a regular decline in biomass with increasing organism size arranged in logarithmically equal size intervals.[55] The spectra slope has been related to the energy flow through the planktonic food web[56–58] and to the trophic state of the ecosystem.[59] Although the seasonal variation of the Mar Menor Lagoon's trophic state was also explained in the biomass size spectra study,[14] it was most interesting, nevertheless, to find that the interannual comparison of planktonic size distribution showed an almost invariable slope within a size range from 2- to 1000-μm equivalent spherical diameter.[47] The size range used for comparison excluded the jellyfish fraction (up to 40 cm in diameter). It can be seen, nevertheless, that this size fraction plays a major role in controlling the biomass spectra parameters. Jellyfish gut contents indicate clearly their preference for large diatoms (62–86%), tintinnids (3–33%), and copepods (1–3%).[47] High removal rates of larger plankton were expected in the Mar Menor Lagoon due to the large number of jellyfishes and their size-selective diet.

It is generally believed that nutrient loads stimulate primary production, thereby increasing planktonic biomass, but that was not completely true in our lagoon because of the top–down control of the planktonic food web by jellyfishes. It was paradoxical to find that chlorophyll *a* (Figure 9.3.18) and the total biovolume considered in the 2- to 1000-μm size range (Figure 9.3.19) at the same sampling station was always lower in 1997 (with higher nitrate loads, lower phosphate concentration, and very high densities of jellyfishes) compared to 1988 (with lower nitrate levels, higher phosphate concentrations, and where jellyfishes were not found) in the four spectra compared. The size spectra comparison suggests that jellyfishes can be an efficient top–down agent controlling the consequences of a eutrophication process.

In systems where nutrients are scarce, fast-growing small cells can provide available food for certain size ranges of grazers, resulting in relatively high densities of copepods. At higher concentrations of nutrients, large diatoms dominate, which copepods cannot eat. This means that the density of copepods is low to feed the largest zooplankton, such as fish larvae and jellyfish. As jellyfish gut contents indicate, high removal rates of larger plankton are expected in the Mar Menor Lagoon due to the large number of jellyfishes and their size-selective diet. While the origin of large diatoms in the water column can be explained as a direct consequence of nitrate loads, abundance of tintinnids, the second most numerically important of gut contents, feed mainly on bacteria, heterotrophic flagellates, and small phytoplankton cells. The effect of jellyfishes removing tintinnids can be seen as an indirect top–down control mechanism on small size fractions. By eating copepods, jellyfish also act indirectly on small phytoplankton, reducing the top–down control exerted by copepods on this fraction. Trade-offs between direct and indirect effects may explain why some eutrophic systems support viable populations of small-celled phytoplankton and large populations of

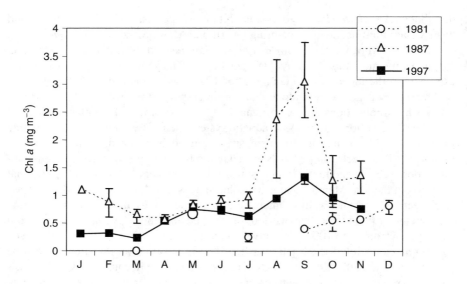

**FIGURE 9.3.18** Interannual monthly variation of chlorophyll *a* concentration in the Mar Menor waters in three different trophic states: 1981 oligotrophic state; 1987 oligotrophic state with increasing urban pressure in summer time without urban waste treatments and without jellyfishes; and 1997 eutrophic state with continuous flow through the Albujón watercourse, with urban waste treatments and with top–down jellyfishes control. J through D represent the months of the year.

large gelatinous zooplankton. On the other hand, removing large diatoms has a direct effect on nutrient loads as they uptake inorganic nutrients from the water column, but the simultaneous removal of grazers such as ciliates and copepods reduces the predation pressure on smaller phytoplankton. The feeding preference of jellyfish imposes a combination of direct and indirect effects on the planktonic structure at different size levels. The trade-off between competition for available resources (bottom–up) and predation (top–down) control mechanisms[60,61] results in a planktonic size structure different from that thought to occur under eutrophic conditions.

The presence of such direct and indirect mechanisms controlling the planktonic food web makes it necessary to know in depth the nature of those processes in order to establish causal relationships to be included in the eutrophication models. Knowledge about many of these relationships can only be acquired after implementation of long and detailed monitoring and scientific programs.

### 9.3.4 SUGGESTIONS FOR MONITORING AND MODELING PROGRAMS

Obviously, to control the processes that lead to the degradation of the Mar Menor or any other natural environment, it is necessary to have in-depth knowledge of the nature of such processes and the causal relationships between them and human activities that cause them. Such cause–effect relationships, in general, are unknown and to establish

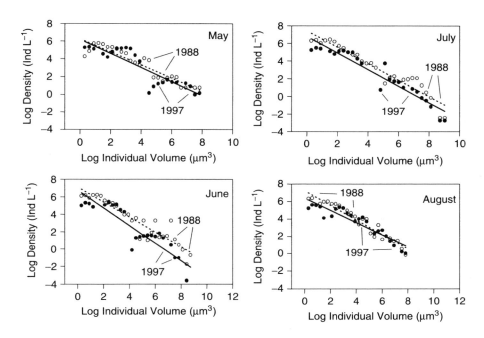

**FIGURE 9.3.19** Comparison of biomass size-spectra in the planktonic food web of the Mar Menor between 1988, an oligotrophic state without jellyfishes (open circles, dotted regression line) and 1997, a eutrophic state that jellyfishes control (filled circles, continuous regression line).

them demands detailed monitoring programs, which are sometimes incompatible with the accelerated evolution of the lagoon, forced by human-driven activities.

In the case of the Mar Menor, difficulties in the establishment of such cause–effect relationships have been compounded by the lack of detailed studies and data on the hydrographic conditions and biological assemblages before most human interventions. Furthermore, although in recent years the increasing demands on the Mar Menor coasts have led to the elaboration of *Indicative Plans of Uses of the Littoral* and distinct evaluations of the environmental impact of human activities in terms of physical, biological, and human environment, in most cases, only superficial analyses of the marine environment have been made, and they have focused on satisfying minimum administrative or legal requisites instead of generating the knowledge required for adequate management.

In any case, the large amount of information compiled during the last 20 years permits us to establish general guidelines for such cause–effect relationships in order to draw conceptual models to assign priorities for managerial actions and lines of research.

Figure 9.3.20 shows a diagram that relates three of the human activities in the Mar Menor discussed above: the enlargement of the El Estacio Channel, the building of artificial beaches, and the construction of sport harbors.[13] Many of the consequences of these activities involve potential risks to human use and the local economy. When the Figure 9.3.20 flowchart was created in 1994, the allochthonous

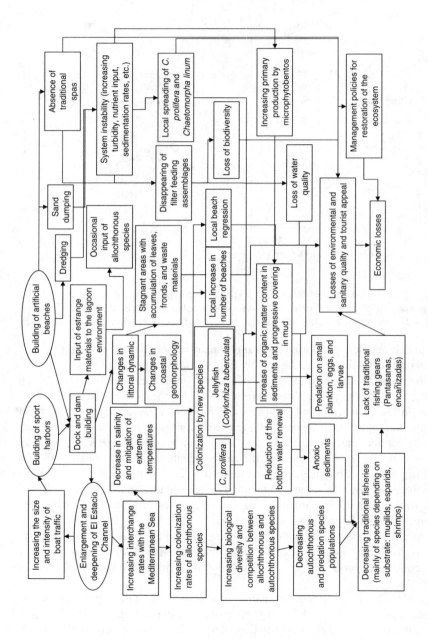

**FIGURE 9.3.20** Conceptual model of cause–effect interrelationships derived from three human activities in the Mar Menor Lagoon: (1) enlargement of the El Estacio Channel; (2) building of artificial beaches; and (3) construction of sport harbors.

jellyfish *C. tuberculata* was very scarce, but its very presence in the chart indicated that it could be a serious threat in the future if nutrient inputs became high enough.

On the other hand, legislation for protection and harmonizing of uses of the Mar Menor (*Protección y armonización de usos del Mar Menor*) or the establishment of specific statutes for protection, such as the one that exists for the salt mines area of San Pedro del Pinatar, will not be sufficient to stop the environmental damage of these singular spaces unless they are supported by a suitable management policy and a clear public awareness. Both must take into account that some urban-planning excesses, although attractive and with high short-term profitability, can shorten the life of lagoon ecosystems.

Any planning measure should define concise objectives for the use and exploitation of these environments; make an inventory of its natural resources; study the processes involved in the functioning of the ecosystem and of different subsystems (terrestrial, marine, interchanges with the open sea and surrounding systems); make an inventory of human activities and their influence on the ecological processes; classify the territory on the basis of previously defined objectives and its ability to sustain the different activities; establish the standards of environmental quality; and design monitoring plans and correction measures.

## REFERENCES

1. Kjerfve, B., Coastal lagoons, in *Coastal Lagoon Processes*, Kjerfve, B., Ed., Elsevier Oceanography Series 60, Elsevier, Amsterdam, 1994, 1.
2. Montenat, C., Les formations Neogenes et Quaternaires du Levant espagnol, Ph.D. thesis, University of Paris, Orsay, 1973.
3. Díaz del Río, V., Estudio ecológico del Mar Menor. Geología, report project no. 1005 Instituto Español de Oceanografía, 1990.
4. López Bermúdez, F., Ramírez, L., and Martín de Agar, P., Análisis integral del medio natural en la planificación territorial: el ejemplo del Mar Menor, *Murcia (VII)*, 18, 11, 1981.
5. Lillo, M.J., Geomorfología del Mar Menor, *Papeles del Departamento de Geografía* 8, 9, 1981.
6. Pickard, G.L. and Emery, W.J., *Descriptive Physical Oceanography: An Introduction*, 5th ed., Butterworth Heinemann, Oxford, U.K., 1990, chap. 5.
7. Hopkins, T.S., Physics of the sea, in *Western Mediterranean*, Margalef, R., Ed., Pergamon Press, Oxford, U.K., 1985, 100.
8. Arévalo, L., El Mar Menor como sistema forzado por el Mediterráneo. Control hidráulico y agentes fuerza, *Bol. Inst. Esp. Oceanogr.*, 5, 63, 1988.
9. Pérez-Ruzafa, A., Estudio ecológico y bionómico de los poblamientos bentónicos del Mar Menor (Murcia, SE de España), Ph.D. thesis, University of Murcia, 1989.
10. Fiala, M., Etudes physico-chimiques des eaux et sédiments de l'étang Bages-Sigean (Ande), *Vie Milieu*, 23B, 21, 1973.
11. Amanieu, M. et al., Etude biologique et hydrologique d'une crise dystrophique (malaigue) dans l'étang du Prévost a Palavas (Hérault), *Vie Milieu*, 23B, 175, 1975.
12. Baudin, J.P., Contribution a l'étude écologique des milieux saumatres méditerranéens. 1. Les principaux caractères physiques et chimiques des eaux de l'étang de Citis (B.-d.R.), *Vie Milieu*, 30, 121, 1980.

13. Pérez-Ruzafa, A., Les lagunes méditerranéennes, The Mar Menor, Spain, in *Management of Mediterranean Wetlands*, Morillo, C. and González, J.L., Eds., Ministerio de Medio Ambiente, Madrid, 1996, 133.
14. Gilabert, J., Short-term variability of the planktonic size structure in a Mediterranean coastal lagoon, *J. Plankton Res.*, 23, 219, 2001.
15. Dorey, A.E., Little, C., and Barnes, R.S.K., An ecological study of the Swanpool, Falmouth. II. Hydrography and its relation to animal distribution, *Estuarine and Coastal Marine Science*, 1, 153, 1973.
16. Harlin, M.M., Boyce, Th.M., and Boothroyd, J.C., Seagrass-sediment dynamics of a flood-tidal delta in Rhode Island, USA, *Aquatic Botany*, 14, 127, 1982.
17. Genchi, G. et al., Idrologia di una laguna costiera e caracterizzacione chimio-fisica dei sedimenti recenti in relazione alla distribuzione dei popolamenti vegetali sommersi (Lo Stagnone, Marsala), *Naturalista Sicil.*, 7, 149, 1983.
18. Mann, K.H., *Ecology of Coastal Waters*, 2nd ed., Blackwell Scientific, London, 2000.
19. Terrados, J., Pigmentos fotosintéticos y producción primaria de las comunidades macrofitobentónicas del Mar Menor, Murcia, M.Sc. thesis, University of Murcia, 1986.
20. Gilabert, J., Seasonal plankton dynamics in a Mediterranean hypersaline coastal lagoon: the Mar Menor, *J. Plankton Res.*, 23, 207, 2001.
21. Pérez-Ruzafa, I., Fitobentos de una laguna costera. El Mar Menor, Ph.D. thesis, University of Murcia, 1989.
22. Hegazi, M., Composición pigmentaria y propiedades ópticas de la vegetación submarina costera del Mediterráneo occidental, Ph.D. thesis, University of Murcia, 1999.
23. Pérez-Ruzafa, A. et al., Cartografía bionómica del poblamiento bentónico de las islas del Mar Menor, I: islas Perdiguera y del Barón, *Oecologia aquatica*, 9, 27, 1988.
24. Pérez-Ruzafa, A. and Munilla, T., Pycnogonid ecology in the Mar Menor (Murcia, SW Mediterranean), *Scientia Marina*, 56, 21, 1992.
25. Hendler, G. et al., *Echinoderms of Florida and the Caribbean. Sea Stars, Sea Urchins and Allies,* Smithsonian Institution Press, Washington, D.C., 1995, 163.
26. Ramos, A. and Pérez-Ruzafa, A., Contribución al conocimiento de la ictiofauna bentónica del Mar Menor (SE España) y su distribución bionómica, *Anales de Biología*, 4, 49, 1985.
27. Barcala, E., Estudio ecológico de la fauna ictiológica del Mar Menor, Ph.D. thesis, University of Murcia, 2000.
28. Simonneau, J., Mar Menor: évolution sédimentologique et géochimique récent en remplissage, Ph.D. thesis, University of Tolouse, 1973.
29. Pérez-Ruzafa, A. et al., Evolución de las características ambientales y de los poblamientos del Mar Menor (Murcia, SE España), *Anales de Biología*, 12, 53, 1987.
30. De Leon, A.R., Guerrero, J., and Faraco, F., Evolution of the pollution of the coastal lagoon of Mar Menor, *VI Journées Étud. Pollutions*, CIESM, 355, 1982.
31. Rosique, M.J., Recopilación y análisis de los trabajos existentes sobre el Mar Menor. Unpublished report, 2000.
32. Pérez-Ruzafa, A., Marcos, C., and Ros, J.D., Environmental and biological changes related to recent human activities in the Mar Menor (SE of Spain), *Mar. Pollut. Bull.*, 23, 747, 1991.
33. Pérez-Ruzafa, A. and Aragón, R., Implicaciones de la gestión y el uso de las aguas subterráneas en el funcionamiento de la red trófica de una laguna costera, in *Conflictos entre el desarrollo de las aguas subterráneas y la conservación de los humedales: litoral mediterráneo*, Fornés, J.M. and Llamas, M.R., Eds., Fundación Marcelino Botín-Mundiprensa, Madrid, 2003, 215.

34. Pérez-Ruzafa, A. et al., Presence of pesticides throughout trophic compartments of the food web in the Mar Menor lagoon (SE of Spain), *Mar. Pollut. Bull.*, 40, 140, 2000.
35. Butigieg, J., La despoblación del Mar Menor y sus causas, *Bol. de Pescas. Dir. Gral. Pesca del Ministerio de Marina. Inst. Esp. Oceanog.*, 133, 251, 1927.
36. Navarro, F. de P., Observaciones sobre el Mar Menor (Murcia). *Notas y Resúmenes Inst. Esp. Oceanog. ser. II*, 16, 63, 1927.
37. García, A.M., Contribución al conocimiento del bentos del Mar Menor: poblamientos bentónicos de las islas Perdiguera, Redonda y del Sujeto. Estudio descriptivo y cartografía biónomica, in *Actas I Simp. Ibérico de estudio del bentos marino*, San Sebastián, 1982.
38. Pérez-Ruzafa, A. and Marcos, C., Colonization rates and dispersal as essential parameters in the confinement theory to explain the structure and horizontal zonation of lagoon benthic assemblages, *Rapp. Comm. Int. Mer Medit.*, 33, 100, 1992.
39. Pérez-Ruzafa, A. and Marcos, C., La teoría del confinamiento como modelo para explicar la estructura y zonación horizontal de las comunidades bentónicas en las lagunas costeras, *Publ. Espec. Inst. Esp. Oceanogr.*, 11, 347, 1993.
40. Lozano, F., Una campaña de prospección pesquera en el Mar Menor (Murcia). *Bol. Inst. Esp. Oceanog.*, 66, 1, 1954.
41. Pérez-Ruzafa, A. et al., Distribution and biomass of the macrophyte beds in a hypersaline coastal lagoon (the Mar Menor, SE Spain), and its recent evolution following major environmental changes, in *International Workshop on Posidonia Beds, 2. GIS Posidonie*, Bouderesque, C.F. et al., Eds., Marseille, 1989, 49.
42. Terrados J. and Ros J.D., Production dynamics in a macrophyte-dominated ecosystem: the Mar Menor coastal lagoon (SE Spain), *Oecologia aquatica*, 10, 255, 1991.
43. Meinesz, A., Contribution à l'étude des Caulerpales (Chlorophytes), Ph.D. thesis, University of Nice, 1979.
44. Pérez-Ruzafa, A. and Marcos, C., Los sustratos arenosos y fangosos del Mar Menor (Murcia), su cubierta vegetal y su posible relación con la disminución del mújol en la laguna, *Cuadernos Marisqueros Publ. Téc.*, 11, 111, 1987.
45. Vaulot, D. and Frisoni, G.F., Phytoplanktonic productivity and nutrients in five Mediterranean lagoons, *Oceanol. Acta*, 9, 57, 1986.
46. Ros, M. and Miracle, M.R., Variación estacional del fitoplancton del Mar Menor y sus relaciones con la de un punto próximo en el Mediterráneo, *Limnetica*, 1, 32, 1984.
47. Pérez-Ruzafa, A. et al., Planktonic food web response to changes in nutrient inputs dynamics in the Mar Menor coastal lagoon, *Hydrobiologia*, 475, 359, 2002.
48. Malone, T.C., Algal size, in *The Physiological Ecology of Phytoplankton*, Morris, I., Ed., Blackwell Studies in Ecology University of California Press, 7, 1980, 433.
49. Hein, M., Pedersen, M.F., and Sand-Jensen, K., Size-dependent nitrogen uptake in micro- and macroalgae, *Mar. Ecol. Prog. Ser.*, 118, 247, 1995.
50. Chisholm, S.W., Phytoplankton size, in *Primary Productivity and Biogeochemical Cycles in the Sea*, Falkowski, P.G. and Woodhead, A.D., Eds., Plenum Press, New York, 1992, 213.
51. Thingstad, T.F. and Rassoulzadegan, F., Conceptual models for the biogeochemical role of the photic zone microbial food web, with particular reference to the Mediterranean Sea, *Prog. Oceanogr.*, 44, 271, 1999.
52. Newell, R.C., The energetics of detritus utilisation in coastal lagoons and nearshore waters, in *Coastal Lagoons*, Laserre, P. and H. Postma, Eds., Oceanologica Acta Special Publication, 1982, 347.

53. Newell, R.C., The biological role of detritus in the marine environment, in *Flows of Energy and Materials in Marine Ecosystems*, Fasham, M.R.J., Ed., Plenum Press, New York, 1984, 317.

54. Azam, F. et al., The ecological role of water column microbes in the sea, *Mar. Ecol. Prog. Ser.*, 10, 257, 1983.

55. Sheldon, R.W., Prakash, A., and Sutcliffe, W.H., Jr., The size distribution of particles in the ocean, *Limnol. Oceanogr.*, 17, 327, 1972.

56. Silvert, W. and Platt, T., Dynamic energy-flow model of the particle size distribution in pelagic ecosystems, in *Evolution and Ecology of Zooplankton Communities*, Kerfoot, W.C., Ed., University Press of New England, Hanover, NH, 1980, 754.

57. Rodríguez, J. et al., Planktonic biomass spectra dynamics during a winter production pulse in Mediterranean coastal waters, *J. Plankton Res.*, 9, 1183, 1987.

58. Gaedke, U., Ecosystem analysis based on biomass size distributions: a case study of a plankton community in a large lake, *Limnol. Oceanogr.*, 38, 112, 1993.

59. Sprules, W.G. and Munawar, M., Plankton size spectra in relation to ecosystem productivity, size, and perturbation, *Can. J. Fish. Aquat. Sci.*, 43, 1789, 1986.

60. Lehman, J.T., Interacting growth and loss rates: the balance of top-down and bottom-up controls in plankton communities, *Limnol. Oceanogr.*, 36, 1546, 1991.

61. Cottingham, K. L., Nutrients and zooplankton as multiple stressors of phytoplankton communities: evidence from size structure, *Limnol. Oceanogr.*, 44, 810, 1999.

62. Jiménez De Gregorio, F., *El Municipio de San Javier en la historia del Mar Menor y su ribera*, Ayuntamiento de San Javier, San Javier, 1957, chap. 3.

## 9.4  VISTULA LAGOON (POLAND/RUSSIA): A TRANSBOUNDARY MANAGEMENT PROBLEM AND AN EXAMPLE OF MODELING FOR DECISION MAKING

*Eugeniusz Andrulewicz, Boris Chubarenko, and Irina Chubarenko*

### 9.4.1  TRANSBOUNDARY MANAGEMENT PROBLEMS OF THE VISTULA LAGOON

The Vistula Lagoon experiences higher anthropogenic loading per cubic meter than other Baltic lagoons (e.g., the Odra or Couronian lagoons) due to its relatively small water volume and the very poor treatment facilities in its catchment area.[1-3]

Problems of transboundary environmental management are the subject of a special international agreement.[4] Recently, the European Union (EU) established a number of principles for planning and management actions.[5] HELCOM initiated the Baltic Sea Joint Comprehensive Environmental Action Programme,[6] Element 4 of which deals with the management of coastal lagoons and wetlands.[7]

The Vistula Lagoon is one of the first areas in the Baltic region where genuine steps in transboundary management have been attempted, including the implementation of modeling as a tool for decision making by both scientific institutions and national environmental authorities.[8-10]

#### 9.4.1.1  The Vistula Lagoon and Its Catchment Area

The Vistula Lagoon is separated from the Gulf of Gdansk by a narrow spit about 60 km long with a width varying from a few hundred meters to a few kilometers. The lagoon is connected to the Gulf of Gdansk by the narrow Baltijsk Strait (formerly known as the Pillau Strait; see Figure 9.4.1).

The Vistula Lagoon has a surface area of 838 km². It is shallow, with an average depth of 2.7 m. Its average salinity is low, in the range of 0.1–4.5 per mille (psu). The water volume of the lagoon is 2.3 km³ and the average retention time is about 40–50 days. About 56% of its total area is in Russia and 44% is in Poland. Water exchange with the Baltic Sea is very intense and is synoptically variable. This, combined with river inflows, results in considerable spatio-temporal variations in the salinity field. The Vistula Lagoon is very productive as a fishing ground. Freshwater species predominate, together with a few brackish-water ones. The lagoon serves as a filter for nutrients and contaminant excess from a large catchment area.[2]

The drainage area includes 23,871 km² within Poland and Russia, which is about 30 times larger than the surface area of the lagoon (Figure 9.4.2). The largest river in the drainage area is the Pregel. Of the more than 30 rivers in the region, the Pasleka and Prohladnaya are also relatively large. The largest city in the Vistula Lagoon drainage area is Kaliningrad, Russia, with a population of more than 470,000. It is located just above the mouth of the Pregel River. Other cities within the drainage area with populations exceeding 100,000 are Olsztyn and Elblag, both in Poland.

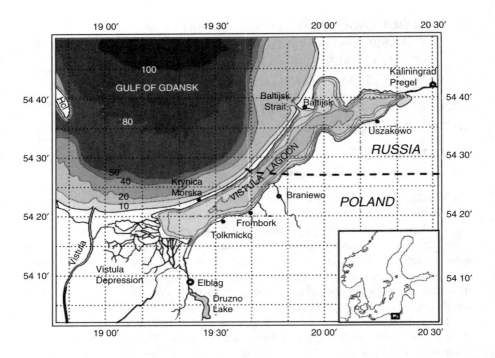

**FIGURE 9.4.1** Location of the Vistula Lagoon.

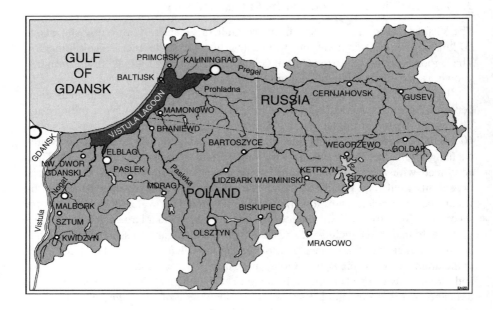

**FIGURE 9.4.2** Drainage area of the Vistula Lagoon including cities with populations exceeding 10,000.

Although fished only seasonally when it enters the lagoon for spawning, herring constitutes the most important catch in terms of biomass (1000 to 4000 t per year in the Polish part of the lagoon and about 8000 t per year in the Russian part).[11] Freshwater and migratory species are of great economic value. Pikeperch, bream, roach, and perch are the main freshwater species. Eel and sea trout are the main migrating species. The existence of some valuable species such as eel or pike is threatened by overfishing and the destruction of fishing grounds. Some species need regular restocking.[11]

### 9.4.1.2  Anthropogenic Pressure and Its Environmental Effects

According to the HELCOM Baltic Sea Joint Comprehensive Action Programme[6,7] (HELCOM 1992 and 1998), numerous "hot spots" are located near the Vistula Lagoon. The lagoon itself has been identified as a priority hot spot, which is in need of a comprehensive environmental management program. Environmental improvement investment is estimated to be 20 million EURO. The Kaliningrad area has been defined by HELCOM as one of the seven "multiple problem areas" in the Baltic Sea catchment area. These hot spots are by definition the most significant contributors to the pollution of the Baltic Sea. They include both point and nonpoint sources, as well as the above-mentioned hot spot of the lagoon itself.[12]

Based on available knowledge,[7,12] the most important environmental problems of the Vistula Lagoon are:

- Poor sanitary conditions in the southern part of the lagoon make the water quality so bad that no beaches along the southern lagoon shores are suitable for swimming, and this basin is not attractive for water sports.
- Eutrophication adversely affects the aesthetics of the water and may have negative biological effects.
- Treatment facilities in Kaliningrad and other coastal towns are poor, and there is direct run-off of a portion of untreated wastewater into the Pregel River and the Primorskaya Bight (the northern part of the Vistula Lagoon).
- Autumn wind-surge events in the lower Pregel River result in the movement of polluted water from the Kaliningrad harbor toward the drinking water inlets located upstream from Kaliningrad or the direct blocking of these inlets by saline lagoon water.[13]
- There is oxygen deficiency in the lower Pregel during warm summer periods.[14]
- The intensification of water exchange with the Baltic Sea due to continual dredging in the Baltijsk Strait because of shipping development leads to further increases in salinity of lagoon waters.[15,16]
- Use of portions of the Polish part of the Vistula Spit for recreational purposes during the summer season exceeds the carrying capacity of the resources (e.g., the Krynica Morska recreational area).
- Pressure on fishing resources is aggravated by legal and illegal fishing activities.

- There is an invasion of nonindigenous marine benthic organisms, and invading estuarine and marine species threaten the replacement of the original benthic population.[17]
- Degradation of landscape and natural habitats results from inappropriate construction of buildings and infrastructure.
- There is a potential danger of the flooding of low-lying areas due to the age and technical condition of antiflood and drainage infrastructure.[18]

During the last decade a sufficient number of sewage facilities have been constructed; however, sanitary conditions have not improved to the extent expected. This is most probably due not only to lack of efficient treatment facilities on the Russian side, but also to the re-emission of pollutants deposited in sediment. Further study of this issue is needed.

Information regarding the environmental effects of the pollution of the Vistula Lagoon is insufficient because of the lack of new studies on topics such as the loss of biodiversity due to eutrophication and levels of contamination by toxic substances and the related biological effects.

### 9.4.1.3 Economic Problems in the Catchment Area

The area encompassing the Vistula Lagoon drainage basin on the Polish side includes large population centers, scattered small towns and rural villages, and significant amounts of agricultural land (Figure 9.4.2). However, the farms are small in comparison to both western and eastern European norms. Industry is not concentrated in one place but scattered among the cities and smaller towns. A short list of economic problems affecting lagoon quality includes:

- High regional unemployment due to the deterioration of older economic structures, such as state farming, for which no viable alternatives are provided.
- Loss of tourism and the recreational potential of the Vistula Spit due to the poor water quality of the lagoon.
- Decline of commercial fishing activities, probably due to water quality problems and changes in composition of species, which is aggravated by eutrophication and possibly by overfishing.
- Loss of the city of Elblag's historical role as a port.
- Relative poverty of the region makes it difficult to place a high priority on environmental investments when other needs are more urgent. In recent years the municipality of Elblag has been successful in acquiring EU funds and has invested considerable sums in pro-ecological projects in the Polish part of the lagoon.
- Inefficient agriculture that generally does not generate enough profit to allow for the employment of the Best Agricultural Practices.

The problems facing the Russian side are similar. The most significant is the current lack of sufficient treatment facilities for industrial and domestic wastewater in the Kaliningrad area. A temporary economic slowdown and decrease in agricultural production and harbor activities have contributed to a provisional improvement in water quality on the Russian side of the lagoon and in downstream sections of the Pregel River. However, insufficient manmade and natural water treatment capacity will inevitably be the factor that limits future economic growth.

### 9.4.1.4 Issues Affecting Transboundary Management

Management difficulties are compounded by the fact that the Vistula Lagoon is transboundary in nature, i.e., it is shared by two countries with different approaches to water management.

Although the present political situation is much more favorable for cooperation on transboundary lagoon management, a number of issues hamper effective cooperation. These include several negative issues:

- Lack of tradition and experience—Historically, Baltic lagoons have not been considered fragile resources in need of careful management; therefore, there is a lack of experience in lagoon management. In addition, there is a tendency to manage resources on a sectoral basis, with little or no history of cross-ministerial cooperation or cooperation between national and local bodies. This is a common problem facing environmental management in central and eastern European countries, but it is particularly relevant to lagoon management because it entails the cooperation of a number of sectors, including environmental, economic, spatial planning, transportation, agriculture, and forestry.
- Differing economic interests and priorities—A relevant example is the use of the Vistula Lagoon inlet (the Baltijsk Strait), which is on the Russian side of the lagoon. The military use of the Baltijsk Strait, the only entrance from the lagoon to the open Baltic, restricts its use for commercial activity. This has led to discussions regarding the construction of a new channel across the southern part of the Vistula Spit within Polish territory.[16]
- Historical relationships and cultural differences between Poland and Russia—To account for these differences, different strategies may be needed on both sides of the Russian–Polish border to accomplish integrated management objectives.
- Different administrative and legal systems in neighboring countries—The bodies responsible for undertaking action and the legal and administrative tools to be utilized to promote integrated management objectives differ between the two countries. Local governments in Poland have been given a significant number of new responsibilities, while administrative bodies in Russia remain centralized. Responsibilities are diffuse and often unclear, leading to a lack of transparency. The lack of integrated environmental management experience is a shortcoming found on both sides of the border.
- Different mapping techniques and the use of different scales—Currently, the available data necessary for management are generally not comparable due to scalar differences, differences in procedures, etc.

Despite these problems, there are several positive aspects to the cooperative management of the Vistula Lagoon, including:

- Improved regional political situation—Since traditional post-war political divisions have faded, cooperation among the countries in the region has improved.
- Democratization—National governments in Poland and Russia are becoming more participatory and more responsive to the priorities of society. As a part of this process, responsibilities have devolved to more local governmental levels. Although there is a need to improve the capacity of local government to make informative decisions, the trend is encouraging.
- Constructive role of HELCOM—The new document[6] gave HELCOM the opportunity to develop an active role as a facilitator of regional cooperation on Baltic Sea–related matters. HELCOM PITF MLW's development of the Integrated Coastal Zone Management Plan is perhaps the foremost example of HELCOM's new role.
- Participation of "third parties"—The involvement of international financing institutions, the EU, outside countries, and other assisting organizations has benefited the region in more ways than financial. By prioritizing cross-border and regional projects, these entities have facilitated the sharing of knowledge and experience and laid the groundwork for future cooperation.

## 9.4.1.5  Efforts toward Transboundary Management of the Vistula Lagoon

The Vistula Lagoon requires an integrated, cross-sectoral management approach that transcends arbitrary political borders. Cross-border cooperation will facilitate integrated lagoon management. Already there is a spirit of cross-border cooperation in the region, mainly due to the impetus of local and regional contacts and joint scientific research activities, including modeling. Such regional efforts facilitate cooperation on a national scale with the additional benefit of improving regional security.

One successful result of this type of cooperation was the joint international project "Prioritizing Hot Spot Remediation in the Vistula Lagoon Catchment: Environmental Assessment and Planning for the Polish and Kaliningrad Parts of the Lagoon."[19,20] This project was conducted by different scientific institutions in Poland (GEOMOR Consult Ltd., Maritime Institute, Sea Fisheries Institute, Marine Branch of the Institute of Meteorology and Water Management, Marine Branch of the Geological Institute), Russia (Atlantic Branch of Shirshov's Institute of Oceanology, Center for Hydrometeorology and Environmental Monitoring), and Denmark (Danish Hydraulic Institute and Water Quality Institute) in close cooperation with local environmental authorities (Elblag and Kaliningrad Ecological Committees).

## 9.4.2 Ecological Modeling as a Tool for Transboundary Management

Taking into account the political processes in the Baltic area, it is obvious that as a border region shared by Poland and Russia, the Vistula Lagoon will remain under different management approaches and strategies in the near future. Given the different priorities declared by national authorities, the substandard water quality of the Vistula Lagoon is likely to degrade further.

The scientific community from both national regions in the catchment area of the Vistula Lagoon (northern Poland and the Kaliningrad Region of Russia) introduced the idea of developing a uniform tool to function as the basis of decision making for Polish and Russian regional environmental authorities. This joint tool had to be a numerical model, thereby ensuring a uniform quantitative approach to water management in the Vistula Lagoon.

This experience was unique in that the scientific community began to combine knowledge and efforts before the environmental authorities undertook cooperation on the political level.

### 9.4.2.1 Numerical Model of the Vistula Lagoon

The MIKE21 model system was used as the basic model for the Vistula Lagoon. This is a comprehensive modeling system for two-dimensional free surface flows where stratification can be omitted. A number of add-on modules describe physical, chemical, and biological processes related to environmental problems and water pollution.[21]

The following three modules of the MIKE21 model system were applied in the study conducted in 1994–1997 by experts[9,10,19,20,22,23] from all the above-mentioned Polish and Russian institutions and environmental authorities:

- MIKE21 HD is a hydrodynamic model that simulates water-level changes and flows in shallow areas and forms the basis for calculations in the other modules of the model system.
- MIKE21 AD is an advection-dispersion module which, coupled with the HD module, simulates processes of the spreading of dissolved and suspended substances under the influence of fluid transport and associated natural dispersion processes.
- MIKE21 EU is a eutrophication module that describes nutrient cycles, phytoplankton, and zooplankton growth as well as the general growth of benthic vegetation and oxygen conditions. This module is integrated with both previous modules, and calculations of hydrodynamics, transport, and nutrient cycles are carried out simultaneously. The results include phytoplankton carbon, nitrogen, and phosphorus; phytoplankton production; detritus carbon, nitrogen, and phosphorus; zooplankton carbon; inorganic and total nitrogen; inorganic and total phosphorus; benthic vegetation production; sediment oxygen demand; and Secchi depth.

### 9.4.2.2  Data Collection for Model Implementation

Despite declarations that lagoon waters had been "monitored" by both Polish and Russian environmental authorities for many years, the analysis of existing data revealed that this monitoring actually provided very poor data to fit the model. Parameters were neither measured regularly nor simultaneously. A typical situation occurred in the administrative approach to lagoon monitoring, namely that one parameter was measured one year and another parameter was measured the next year. The frequency of measurements was usually three to four times per year; these data can only be used to make rough evaluations of interannual dynamics.

These monitoring data combined with historical information[8] provided only a general understanding of the processes taking place in the lagoon. Therefore, a supplementary data collection program was undertaken.

All of the spatial characteristics of the lagoon were examined—configuration of the coastal line, bottom sediment structure, lagoon bathymetry, and root vegetation areas. Historical information on lagoon bathymetry was compared with current results, and it was determined that a general deepening of the Vistula Lagoon had taken place. Following detailed study, it was subsequently found that active wind resuspension processes promote the flushing out of sediment from the lagoon and that the lagoon may even have a negative sediment balance. The reason for this is the anthropogenic reduction of sediment loading, which is accomplished by routing Vistula River discharge directly into the Baltic Sea via other branches.[24]

A short-term (2.5-month) field program of current measurements at 16 stations in the lagoon together with four daily measurements of meteorological parameters at five stations around the lagoon was conducted. The lagoon entrance was carefully scrutinized, and variations in water level, salinity, and temperature were determined hourly during this period. The data obtained were used for calibrating the hydrodynamic module.

An 18-month-long data collection program was executed to gather data on salinity, nutrients, and suspended sediment dynamics. Cruises around 22 points in the lagoon (Figure 9.4.3) were undertaken by both Polish and Russian research teams monthly during previously agreed periods. Meteorological parameters, water level, and water temperature were measured from two to four times daily at four locations along the lagoon coast. In addition, salinity measurements were taken twice daily at the lagoon entrance.

Information on nutrient loading was collected during the same period for the main rivers and all wastewater discharge outlets. Loading estimations were done on the basis of field measurement data and using theoretical estimations, which took into consideration agricultural activity (a load of about 15 kg N/ha was estimated for farmland with medium and intensive use of fertilizers, and a load of 6.1 kg N/ha was estimated for minimum use), livestock, and population density.[20]

The comparison of theoretically estimated data with measurements and economic statistics from enterprises and environmental authorities revealed significant variations in loading values. Loading information is a crucial part of data collection, and errors in these data greatly exceed those in any physical data.

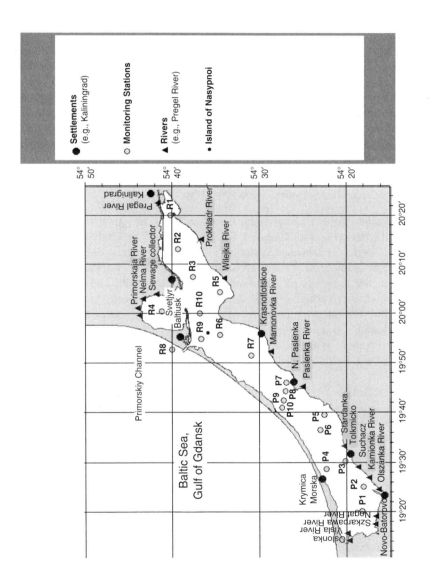

**FIGURE 9.4.3** Locations of monitoring stations, local settlements, and river mouths for model-dedicated short-term monitoring in the Vistula Lagoon in 1994.

The data on the vertical profiles of salinity at the monitoring stations obtained during 18 months of measurements were used to calibrate the advection–dispersion module. Analysis indicated that the artificial Kaliningrad Navigation Channel, which is twice as deep as the lagoon area, plays a key role in the water exchange processes between the Baltic Sea, the northern part of the Vistula Lagoon, and the Pregel River. This channel passes along the northern coast of the lagoon and is bounded by a set of artificial islands.

The inadequate description of the spatial distribution of salinity in the eastern part of the lagoon by the MIKE21 model was anticipated. This is because the model uses only a two-dimensional horizontal approach and does not describe the two-layer water dynamics that exists in the channel. It was decided to focus special attention on this feature during simulations. Specific values of calibration coefficients have been found that provided a simulation of the arrival of salt water in the eastern part of the lagoon, which is the same as the real water exchange, by the deep Navigational Channel.[25]

### 9.4.2.3  Hydrodynamic Modeling (MIKE21 HD)

The time step in the hydrodynamic simulations was 300 s, the grid step was 1 km per 1 km, and the period included all of 1994. The input data included wind speed and direction as well as water-level variation at the lagoon entrance. Furthermore, 28 sources were defined: 20 were rivers and eight were wastewater discharge outlets. The inflow from the Pregel River, the main river of the Vistula Lagoon, and the Pasleka River, the main river in the Polish part, were based on weekly measurements taken during 1994. For other rivers the mean annual discharge was estimated based on 1994 measurements and historical data. The seasonal variations of flows in these middle-sized rivers were assumed to be like those for the Prohladnaya River on the Russian side and the Pasleka River on the Polish side, where measurements were conducted.

Data were collected on the distribution of various kinds of bed sediment of the Vistula Lagoon (granulation of the lagoon bottom as well as rooted vegetation) for the lagoon bed resistance coefficient. However, after calculation using the HD model, it was clear that the water-level variations and currents in the lagoon are not sensitive to the variability of bed conditions, at least with the applied model resolution. For all simulations, a bed resistance of 32 $m^{1/3}$ $s^{-1}$ was applied.

Differing wind friction approaches were investigated. It was decided to apply detailed simulation (200-m grid step). In general, a wind friction coefficient of 0.0015 was used, and another of 0.0003 was used in the southwestern part of the lagoon for wind directions between 180 and 270° because of the influence of the nearby hills, which reduce wind stress. For simulation with a 1-km grid step used as a basic grid for eutrophication processes, a wind friction coefficient of 0.0017 was chosen.

Sensitivity analysis showed that the water levels were not sensitive to changes in eddy viscosity, whereas current structure is a function of eddy viscosity. In the case of currents, the best correspondence between simulations and measurements was obtained with an eddy viscosity of 20 $m^2$ $s^{-1}$.

### 9.4.2.4   Advection-Dispersion Modeling (MIKE21 AD)

The calibration of the transport module of MIKE21 was performed on the basis of 18 months of salinity measurements at 22 points in the lagoon. Good correspondence between measured and simulated salinity variations ensured accurate calculations of water exchange between the Baltic and the Vistula Lagoon. The diffusion coefficient obtained during calibration was used later in the eutrophication step. After several numerical experiments the specific values for calibration parameters were found, which provided acceptably accurate simulations for both the sea–lagoon water exchange and saltwater transport toward the remote eastern lagoon corner, where the Kaliningrad Channel meets the Pregel River. The best simulation salinity results were obtained after using a dispersion coefficient of 45 $m^2$/s.

### 9.4.2.5   Eutrophication Model (MIKE21 EU)

The input data for the eutrophication model covered the whole of 1994 and included:

- Initial concentrations of simulated chemicals in the Vistula Lagoon
- Concentrations of nitrogen and phosphorus in organic and inorganic forms and dissolved oxygen for all rivers and point sources
- Concentrations of nitrogen and phosphorus in organic and inorganic forms and dissolved oxygen at the open boundary (lagoon entrance)
- Water temperature
- Solar radiation

The eutrophication module describes the relation between available nutrients (nitrogen and phosphorus) and the algae growth in coastal and open waters. It simulates carbon, nitrogen, and phosphorus content in phytoplankton, zooplankton, detritus, and, with regard to nitrogen and phosphorus, the concentration of inorganic nitrogen and phosphorus in the water. It is assumed that dissolved carbon is present in excessive amounts and it is not included explicitly in the water.

The eutrophication module considers the chlorophyll $a$ concentration calculated from phytoplankton carbon (biomass). Chlorophyll $a$ concentration is included in the model because it is the most easily measured parameter representing phytoplankton biomass.

Nutrient loading was used for eutrophication module calibration along with results of direct measurements of nutrient and chlorophyll $a$ at monitoring stations in the Vistula Lagoon. Nearly 50 simulations with durations of 6 to 36 h were conducted during eutrophication module calibration. This resulted in identifying the set of internal parameters of the model that best fits the field data.[10]

Simulations were carried out with a 3-hour time step and grid spacing of 1 km × 1 km. Simultaneously, hydrodynamic and transport calculations were carried out with a 5-min time step and with the same grid spacing. A longer time step for eutrophication was used because of different time scales of the processes and also to minimize CPU time.

### 9.4.2.6  Scenario Assessment

Measurement results indicated a high rate of eutrophication in the Vistula Lagoon, although the situation differs in various regions of the lagoon. The challenge for the model was to estimate quantitatively the reduction of external nutrient loading, which would lead to the nutrient limitation of phytoplankton growth, improve water quality, and inhibit further eutrophication.

The simulated 1994 situation was chosen as a basic solution (to which all scenarios should refer). The spring and summer periods were used as an indicator period highlighting the differences between basic and scenario solutions.

According to Prognosis 2010, four nutrient-loading scenarios for the Polish part were applied:[10]

- The impact was estimated of the reduction in nutrient loading following the completion of all the treatment facilities constructed since 1993. The simulation showed a significant improvement in lagoon water quality.
- In the second scenario, the improvement of wastewater treatment technology was studied. It was assumed that mechanical-chemical-biological treatment would be applied in all cities with populations exceeding 10,000. In smaller towns, it was assumed that only mechanical-biological treatment would be applied. Further improvement of water quality was predicted, and the chlorophyll *a* concentration in spring and summer decreased by more than 25% in a large area of the Polish part of the lagoon.
- In the third scenario, the environmental effect of forecasted agricultural development and intensity of fertilizer use was evaluated. It was assumed that nitrogen and phosphorus fertilizer doses would be increased by 63 and 34%, respectively, and that nutrient loading from point sources would not differ from the previous scenario (improved treatment technology). These assumptions resulted in a 6% increase of the nitrogen load from the Polish part of the catchment area and in a 20% reduction of the phosphorus load. The concentrations of total nitrogen increased, especially in the central part of the lagoon. Chlorophyll concentration decreased only locally, which did not correspond to the previous scenario. On the whole, this scenario predicted a negative effect on lagoon water quality.
- In the last scenario, the resulting effect of improved wastewater treatment technologies (second scenario), the increased use of fertilizers (third scenario), and action undertaken to reduce the nonpoint run-off of nutrients to surface waters by 30% was evaluated. The resulting reduction of nitrogen and phosphorus loads from the Polish part of the catchment area was estimated to be 20 and 34%, respectively.

Two scenarios of bathymetric changes in the vicinity of the lagoon entrance were analyzed (one toward Kaliningrad and one toward the Polish ports). It was estimated that the assumed deepening of the two channels will not have negative consequences. Only local salinity increases in the vicinity of the lagoon entrance are expected.[15]

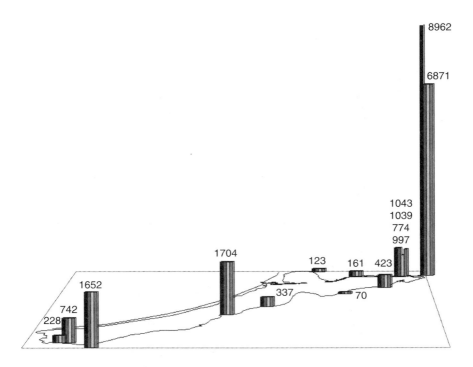

**FIGURE 9.4.4** Distribution of the nitrogen loading (t · year$^{-1}$) along the Vistula Lagoon coast based on four scenarios considered by the international Danish-Polish-Russian project "The Vistula Lagoon (1994–1997)."

The Kaliningrad hot spot can be divided into sources of untreated domestic and smaller-scale industrial sewage, and loading from direct discharges by the two existing paper mills into the Pregel River or those diverted to the existing outdated treatment facility. The four alternative remedial plans (see Figure 9.4.4. for nitrogen loading example) for reducing loading from hot spots in the Kaliningrad region result in the following scenarios:

1. Only the urban and industrial sewage presently discharged into the sewer systems is conveyed to the wastewater treatment facility for primary treatment, including chemical precipitation.
2. This scenario represents the hypothetical "present situation" as if all present urban and industrial sewage were collected and conveyed to the wastewater treatment facilities. The paper mills operate with outmoded technology, but the wastewater is conveyed to the wastewater treatment facilities. One of the mills operates at 40% capacity while the other runs at 80% capacity.
3. All urban and industrial sewage is collected and conveyed to the wastewater treatment facilities. One of the paper and pulp mills is completely renovated, introducing a new cleaner production technology resulting in a 90% reduction in load. The second mill is closed.

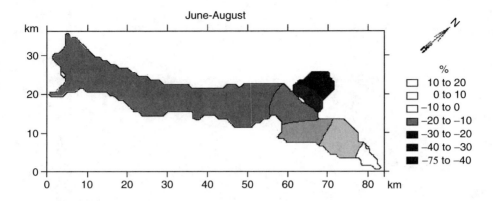

**FIGURE 9.4.5** Spatial distribution of change (in percent) of average summer concentration of chlorophyll *a* comparing scenario (3) to the basic solution for 1994. Negative percentages correspond to reduction of chlorophyll concentration.

4. All urban and industrial sewage is collected and conveyed to the wastes-water treatment facility. The first paper mill undergoes partial renovation. The second paper mill is not renovated.

In 1996, scenario (1) had an implementation target date of 1998, whereas scenarios (2), (3), and (4), which are extensions of the wastewater treatment in scenario (1), were planned to be fully implemented by the year 2010. In addition, different options for locations for the outlets from a new wastewater treatment facility were analyzed.

The results of scenario simulations were evaluated by comparing each with the reference basic solution for 1994. To condense the huge amount of information (a "basic" solution as well as eight scenario solutions were obtained for each parameter under study for each grid point in the lagoon for a 1-year period) the average seasonal maps of difference between scenario and "basic" solutions were developed. The example of such a seasonal map scheme is shown in Figure 9.4.5. Such a method revealed the benefits of different parts of the lagoon from different loading scenarios.

### 9.4.2.7 Summary and Management Recommendations

Integrated management of the Vistula Lagoon is particularly difficult because it is a transboundary area that faces multiple economic and environmental problems. In the last several years pronounced efforts have been made to improve wastewater treatment at the main point sources in the catchment area on the Polish side. The improvement of water quality on the Russian side of the lagoon depended directly on the reduction of urban sewage, including nutrients.

The need for a bilateral commission, with the possible participation of other partners, is crucial. For example, because the Polish and Russian project[19] partners often have their own points of view on critical issues, the responsibility of final conclusions was assigned to a third partner, an expert team from Denmark.[20]

The following final recommendations on Vistula Lagoon management were developed on the basis of modeling:[20]

- The study[19] revealed the vital necessity to focus attention on nonpoint pollution sources in the coming years. The sustainable development of agriculture in the catchment area will hopefully result in an improvement in the ecological state of the lagoon, especially in the Polish part.
- The water quality in both the Polish and Russian parts of the lagoon mostly depends on the local action taken to reduce nutrient loading. This conclusion strongly forced national authorities to apply local remediation measures instead of complaining about negative influence from the neighbor.
- The water quality in the southern Polish part of the lagoon is more sensitive to pollutants because the water retention time there is longer than in the northern Russian part. It is a clear example of difference in local flushing time (see Chapter 6.3.3.1.2) for various parts of a lagoon.
- It was concluded that limiting eutrophication processes in the Polish part of the Vistula Lagoon may be achieved by reducing the nitrogen and phosphorus loads from the Polish part of the catchment area. High loads of nutrients from the Russian part of the catchment area do not have a significant impact on the ecological state of the Polish part of the lagoon.
- It is presumed that the position of the planned outlet from the new Kaliningrad treatment facility to the Kaliningrad Navigational Channel that passes along the northern coast of the lagoon is not optimal since backward wind surge currents can transport the water toward populated areas near Kaliningrad.
- Due to complex water circulation between the Kaliningrad Navigational Channel and the Vistula Lagoon, it was recommended to establish a local fine-grid model. This model has to include the channel area to simulate the water quality in this part of lagoon and to identify the optimal situation between the costs involved and the environmental benefits achieved.
- The results of the scenario Prognosis 2010 and scenario (3) (which represents the best situation in terms of pollution load) indicate that the anticipated situation in the Vistula Lagoon ecosystem regarding eutrophication processes can be improved significantly.
- However, achieving such improvements requires effort and investment in the catchment area in order to achieve significant reductions in nonpoint nutrient run-off. Wetland restoration, reforestation, adequate regulation, and agricultural education should all be conducted. Simultaneously, point-source wastewater treatment technologies should be modernized.

Numerical modeling conducted during the project[19,20] has proven to be useful in the process of preparing impact assessments of different pollution sources located in the catchment area, as well as in evaluating the effects of alternative abatement actions. The conclusions drawn from the modeling study are valuable guidelines for the regional authorities responsible for environmental protection.

## ACKNOWLEDGMENTS

The work of the joint international modeling team (all persons cited in references related to the project[9,10,19,20] as well as Dr. A. Staskiewicz and Mr. O.K. Jensen) was organized by GEOMOR Consulting Ltd. in the form of periodic joint sessions. The results of the project as well as a dedicated version of the model for the Vistula Lagoon were transmitted to local Polish and Russian Environmental Authorities.

## REFERENCES

1. Andrulewicz, E., Chubarenko, B., and Zmudzinski, L., Vistula Lagoon—a troubled region with a great potential, *WWF Baltic Bulletin,* 1(94), 16, 1994.
2. Andrulewicz, E., An overview on lagoons in the Polish coastal area of the Baltic Sea, *Int. J. Salt Lake Res.*, 6, 121, 1997.
3. Chubarenko, B.V. and Valeyshina, E.V., Comparative analysis of the Baltic lagoons and bays on anthropogenic loading in 1970–1980, Report of the Institute of Complex Regional Studies, Kaliningrad State University, Kaliningrad, 1993 [in Russian].
4. UN ECE, Convention on the Protection and Use of Transboundary Watercourses and International Lakes, United Nations Economic Commission for Europe, 1992.
5. EU WFD Directive, European Water Framework Directive No. 41/1999, *Official Journal of the European Communities*, C 342/1, 1999.
6. The Baltic Sea Joint Comprehensive Environmental Action Programme (Preliminary Version), *HELCOM Conference Document No. 5/3*, Diplomatic Conference on the Protection of the Marine Environment of the Baltic Sea, April 9, 1992, Helsinki, Finland, 1992.
7. The Baltic Sea Joint Comprehensive Environmental Action Programme: Recommendations for Updating and Strengthening, *Baltic Sea Environmental Proceedings*, No. 72, 1998.
8. Lazarenko, N.N. and Majewski, A., Eds., *Hydrometeorological Regime of the Vistula Lagoon*, Hydrometeoizdat, Leningrad, 1971 [in Polish and Russian].
9. Kwiatkowski, J., Lewandowski, A., and Oldakowski, B., Water Quality Management for the Vistula Lagoon by the Application of a Mathematical Eutrophication Model, Report of GEOMOR Consulting Ltd., 1977.
10. Kwiatkowski, J., Rasmussen, E.K., Ezhova, E., and Chubarenko, B., The eutrophication model of the Vistula Lagoon, *Oceanological Studies*, 1, 5, 1977.
11. Netzel, J. and Feldman,V., personal communication, 2001.
12. Integrated Coastal Zone Management Plan (ICZM) for the Vistula Lagoon, HELCOM PITF MLW Report, Vol. 1, 1999.
13. Report on Environmental Situation in the City of Kaliningrad, Environmental Centre for Administration and Technology (ECAT-Kaliningrad), Kaliningrad, 1996.
14. Report on Environmental State in the City of Kaliningrad, Environmental Centre for Administration and Technology (ECAT-Kaliningrad), Kaliningrad, 2000.
15. Chubarenko, I.P., The influence of deepening works in Baltiysk Channel on the water salinity in Vistula Lagoon (numerical model MIKE21), in *Proc. 2nd Int. Workshop on Interdisciplinary Approaches in Ecology "System Research in Ecology: Linking Watershed, Riverine and Marine Processes,"* Lithuania, Klaipeda, August 25–27, 1997, *Monographs in System Ecology*, 2, 18, 1998.

16. Chubarenko, I.P. and Tchepikova, I.S., Numerical modelling analysis of artificial contribution to salinity increase into the Vistula Lagoon (Baltic Sea), *Int. J. Ecological Modelling and Systems Ecology*, 138, 87, 2001.

17. Ezhova, E.E. and Peretertova, O.V., Ecosystem effect of the successful invader *Marenzelleria viridus* in the Vistula lagoon, in *Proc. 16th Baltic Marine Biol. Symp. Functional Diversity and Ecosystem Dynamics of the Baltic Sea Alien Species in the Brackish Water Ecosystems*, Klaipeda, 52, 1999.

18. Szymkiewicz, R., A mathematical model of storm surge in the Vistula Lagoon, Poland, *Coastal Engineering*, 16, 181, 1992.

19. Lewandowski, A. and Malmgren-Hansen, H., Prioritising Hot Spot Remediation in the Vistula Lagoon Catchment: Environmental Assessment and Planning for the Polish and Kaliningrad Parts of the Lagoon, presentation at EUCC-WWF conference Coastal Conservation and Management in the Baltic Region, May 3–7, 1994, Riga-Klaipeda-Kaliningrad, 1994.

20. Rasmussen, E.K., Summary Report on International Project "Prioritising Hot Spot Remediation in the Vistula Lagoon Catchment: Environmental Assessment and Planning for the Polish and Kaliningrad Parts of the Lagoon," Water Quality Institute (DK); Danish Hydraulic Institute (DK); GEOMOR (Poland); P.P. Shirshov Institute of Oceanology, Atlantic Branch (Russia), 2nd ed., May, 1997.

21. *MIKE 21: User Guide and Reference Manual*, Danish Hydraulic Institute, Copenhagen, Denmark, 1994.

22. Chubarenko, I.P. and Kolosentseva, M.Y., Using of integrative model MIKE21 for hydraulic simulation of the Vistula lagoon, in *Ecological Problems of Kaliningrad Region*, Kaliningrad, 1997, 80 [in Russian].

23. Chubarenko, B.V., Chubarenko, I.P., and Ezhova, E.E., Integrated Numerical Model As an Instrument of Coastal Zone Management: Some Results of the Estimation on the Influence on the Vistula Lagoon System, in *Mankind and the Coastal Zone of the World Ocean*, Aibulatov, N.A., Ed., GEOS, Moscow, 2001, 438 [in Russian].

24. Chubarenko, B.V. and Chubarenko, I.P., New way of natural geomorphological evolution of the Vistula Lagoon due to crucial artificial influence, in *Geology of the Gdansk Basin, Baltic Sea*, Emeliayanov, E.M., Ed., Yantarny Skaz, Kaliningrad, 2001, 372.

25. Chubarenko, B.V. and Chubarenko, I.P., The transport of Baltic water along the deep channel in the Gulf of Kaliningrad and its influence on fields of salinity and suspended solids, in *ICES Cooperative Research Report*, No. 257, Dahlin, H., Dybern, B., and Petersson, S., Eds., Copenhagen, 2003, 151.

## 9.5 KOYCEGIZ–DALYAN LAGOON: A CASE STUDY FOR SUSTAINABLE USE AND DEVELOPMENT

*Melike Gürel, Aysegül Tanik, Ali Ertürk, Ertugrul Dogan, Erdogan Okus, Dursun Z. Seker, Alpaslan Ekdal, Kiziltan Yüceil, Aylin Bederli Tümay, Nusret Karakaya, Bilsen Beler Baykal, and I. Ethem Gönenç*

### 9.5.1 DECISION SUPPORT SYSTEM (DSS) DEVELOPMENT

#### 9.5.1.1 Identification of Environmental, Social, and Economical Characteristics of Koycegiz–Dalyan Lagoon System

*9.5.1.1.1 Environmental Characteristics*

*9.5.1.1.1.1 Geography*

The Koycegiz–Dalyan Lagoon, with an approximate watershed of 1200 km², is located in southwestern Turkey. It consists of two main drainage systems, Koycegiz Lake (1070 km²), and the Dalyan Channels and Lagoon Lakes (130 km²). The location and three-dimensional (3D) view of the watershed are shown in Figure 9.5.1.

The boundaries of the lagoon watershed are the high mountains (>2000 m), parallel to the coast, to the north; the Mediterranean coast to the south; and hills and ridges (500–1000 m), straight to the coast to the east and west. There are deltas

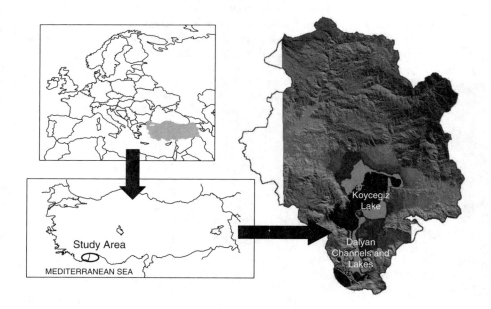

**FIGURE 9.5.1** Location and 3D view of the lagoon system study area.

formed by streams and plains that are located around Koycegiz Lake. The distances between the northwest boundary of the watershed and Koycegiz Lake are 22 and 27 km in west and north directions, respectively. The Dalyan Channel network connects Koycegiz Lake to the Mediterranean Sea with a delta formation downstream of the channel network.[1] Topography maps of 1:25,000 scale obtained from the Turkish Armed Forces General Command of Mapping (TAFGCM) are used to conduct topographical and geographical analyses.

*9.5.1.1.1.2 Climate*
A literature review, including information obtained from meteorology experts, indicates that the area is characterized by a Mediterranean climate with a hot, dry summer season and a warm, rainy winter season. Although the region is controlled by the terrestrial, marine or semimarine, and semiterrestrial low- and high-pressure systems, the high-pressure system is more effective. Precipitation usually occurs during the cold period and drought occurs during the hot period.

*9.5.1.1.1.3 Meteorology*
Although five meteorological stations exist within and in the vicinity of the watershed, only Koycegiz Station (+24 m), which is close to the northern shore of the lake, was relevant to represent the watershed for modeling purposes. The Dalaman (+13 m), Marmaris (+19 m), Fethiye (+3 m), and Muğla Stations (+646 m) are situated outside the watershed boundaries.

Long-term time series for meteorological data consisting of daily measurements of precipitation (47 years), pan evaporation (18 years), air temperature (25 years), humidity (28 years), wind (22 years), cloudiness (28 years), and solar radiation (6 years) were gathered from the Turkish State Meteorological Service. The data were reorganized and analyzed by a data analysis system developed via the integrated use of Microsoft® Excel* for analysis and Microsoft® Access† for database management. This system has the ability to create time series, to conduct queries, and to support statistical evaluations. The chart presented in Figure 9.5.2 is an example of the results of these statistical analyses and reflects the comparative analysis of precipitation, pan evaporation, and estimated potential evapotranspiration (PET) parameters as a basis for the water budget in the watershed.

General meteorological characteristics of the watershed were determined from these analyses. A comprehensive meteorological database to be used for further studies was also created.

*9.5.1.1.1.4 Air Quality*
No field study on the air quality of the watershed has been made. Coal with a high sulfur content is used for heating purposes. High $SO_2$ is the main cause of air pollution, which is greatly affected by the local topography and meteorological conditions. The impact of $SO_2$ on the biogeochemical cycles of the Koycegiz–Dalyan Lagoon is still unknown, but a future study is planned.

---

† Registered trademark of Microsoft Corporation, Seattle, WA, U.S.A.

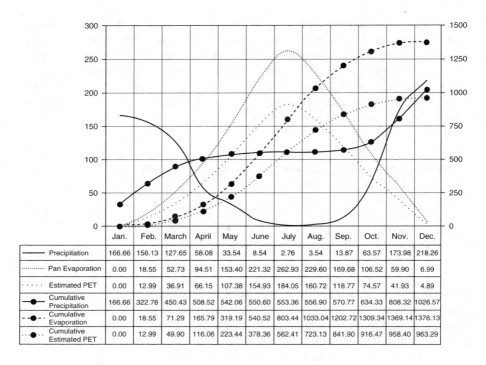

| | | Jan. | Feb. | March | April | May | June | July | Aug. | Sep. | Oct. | Nov. | Dec. |
|---|---|---|---|---|---|---|---|---|---|---|---|---|---|
| —— | Precipitation | 166.66 | 156.13 | 127.65 | 58.08 | 33.54 | 8.54 | 2.76 | 3.54 | 13.87 | 63.57 | 173.98 | 218.26 |
| ·········· | Pan Evaporation | 0.00 | 18.55 | 52.73 | 94.51 | 153.40 | 221.32 | 262.93 | 229.60 | 169.68 | 106.52 | 59.90 | 6.99 |
| · · · · · | Estimated PET | 0.00 | 12.99 | 36.91 | 66.15 | 107.38 | 154.93 | 184.05 | 160.72 | 118.77 | 74.57 | 41.93 | 4.89 |
| —●— | Cumulative Precipitation | 166.66 | 322.78 | 450.43 | 508.52 | 542.06 | 550.60 | 553.36 | 556.90 | 570.77 | 634.33 | 808.32 | 1026.57 |
| - -●- - | Cumulative Evaporation | 0.00 | 18.55 | 71.29 | 165.79 | 319.19 | 540.52 | 803.44 | 1033.04 | 1202.72 | 1309.34 | 1369.14 | 1376.13 |
| ···●·· | Cumulative Estimated PET | 0.00 | 12.99 | 49.90 | 116.06 | 223.44 | 378.36 | 562.41 | 723.13 | 841.90 | 916.47 | 958.40 | 963.29 |

**FIGURE 9.5.2** Annual fluctuations of meteorological parameters in the watershed.

*9.5.1.1.1.5   Geology and Hydrogeology*

Data on the geological and hydrogeological characteristics of the watershed were gathered from the General Directorate of Mineral Research and Exploration of the Turkish Republic and the General Directorate of Rural Affairs of the Turkish Republic (TRGDRA). A 1:10,0000-scaled geological map was used for geological considerations together with a literature review on the springs, hot springs, and groundwater resources of the area.

There are two major fault lines in the region. One lies to the south of Koycegiz Lake in a S–SW direction. Sultaniye Hot Spring, with a high level of radioactivity, is located on it. Velibey, Riza Cavus, and Kokargirme Springs and Hot Springs are located on another fault line and is separated from the major line. A second major fault line penetrates through Koycegiz Lake and lies in a NW–SE direction. At the north of the watershed is a fault line in the NW–SE direction. The locations of major springs and hot springs are shown in Figure 9.5.3.

The coastal geological structure of the system permits seawater flow into the Koycegiz–Dalyan Lagoon. Seasonal groundwater level varies between 0.05 and 6.55 m during May and November. Because of the carstic rock structure of the area, groundwater resources mainly nourish this lake. Efforts to obtain temporal and spatial high-resolution groundwater data are ongoing.

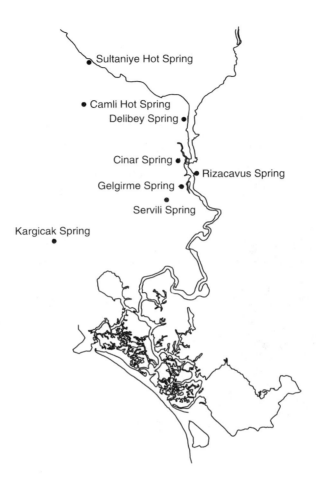

**FIGURE 9.5.3** Location of major springs and hot springs.

*9.5.1.1.1.6   Soil Characteristics*

An alternative approach for sustainable land-use planning is aimed at the assessment of not only the capability of land for various land-use purposes, but also of the suitability of land for specific land-use purposes.[2] Land capability classification, soil types, soil subgroups, land use, and other soil characteristics are studied and presented using different thematic maps obtained from the National Information Centre (NIC) of the TRGDRA. Concurrent with these studies, field soil surveys are carried out to investigate soil fertility, drainage and erosion conditions, and irrigation and fertilization requirements.

Figure 9.5.4A shows the eight major standard soil classes, ranked from "best" (I) to "worst" (VIII) in terms of specified categories of potential agricultural land use.[3] Figure 9.5.4B shows subsoil group distribution for the watershed.

Figure 9.5.4C shows other soil characteristics that were gathered separately. They indicate the drainage characteristics and fertility of different soil types. Since

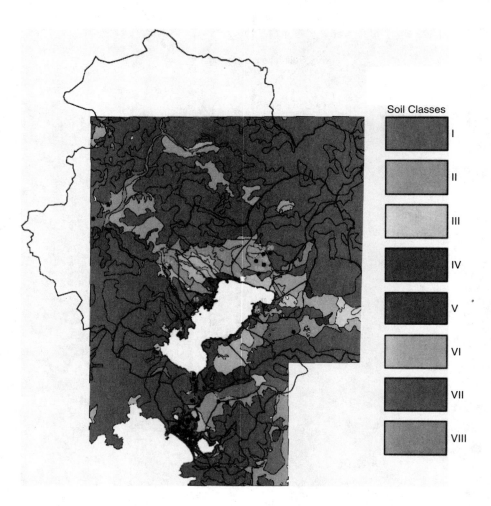

Soil Classes

I

II

III

IV

V

VI

VII

VIII

**FIGURE 9.5.4A** Land-use capability classification.

the land-based pollutant loads may drastically vary with respect to recent land-use distribution as well as soil type, a comparison of soil types and land use is made in the study by overlaying the two maps, as shown in Figure 9.5.4D.[4]

Lime-free brown soil, which is characterized by high clay content and sometimes stony clay texture with pebbles, is the dominant soil type in the area. The second major type of soil, the Mediterranean red-brown soil composed of hard limestone, granite, and rocks, is typically found in dry climate conditions.[4]

*9.5.1.1.1.7    Vegetation Cover*
The main vegetation cover in the watershed is:

- Rushes and reeds that cover the wetlands and marshes, in and around Koycegiz Lake and the Dalyan Lagoon delta

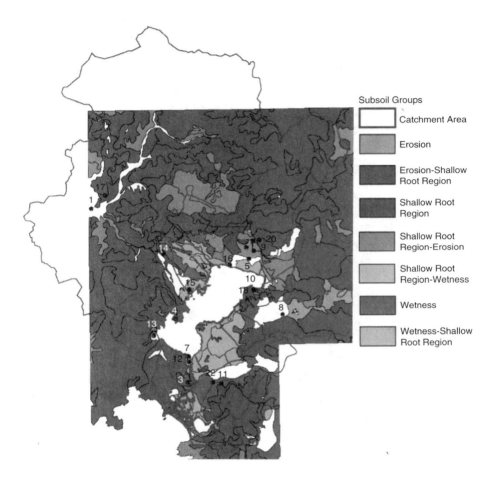

**FIGURE 9.5.4B** Subsoil groups of the watershed.

- Typical Mediterranean scrub vegetation
- Forests

The majority of wetlands are covered with rushes and reeds, whereas the hills are forest zones. *Liquidamber orientalis* forests, peculiar to southwest Anatolia and Rhodes Island, lie northwest of Koycegiz Lake.

Almost all the plain or close to plain land is devoted to agriculture. The main agricultural crops are citrus fruits, cotton, sesame, corn, and wheat. In higher regions olive groves enrich the vegetation cover. Detailed information on the vegetation cover obtained from the NIC of the TRGDRA and the reliability of most of the land-use information gained from different sources were verified via field observations and surveys. The region is predominantly undeveloped native landscape or agricultural land. Figure 9.5.4D shows the vegetative cover of the area.

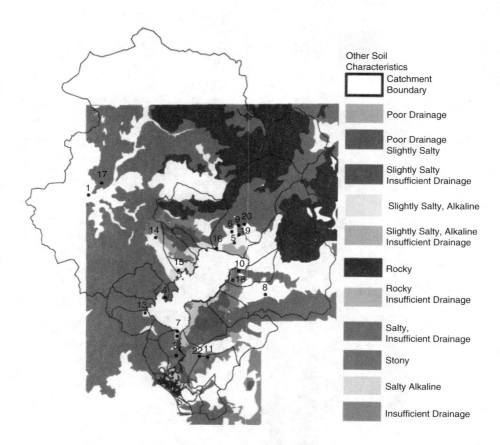

**FIGURE 9.5.4C** Other soil characteristics.

*9.5.1.1.1.8    Hydrological Characteristics*

As an initial step in hydrological studies, watershed delineation was made using conventional methods and watershed modeling system (WMS) 6.1[5] software. Figure 9.5.5 represents a sketch of watershed subbasins.

To estimate the annual water budget in the watershed, the rational method was preferred as a preliminary approach for computations, which was then improved by use of a more complex flow and hydrological models such as QUAL2E[6] and HSPF.[7] Application of these models to the Koycegiz–Dalyan Lagoon and its streams is explained in Section 9.5.1.2.2.

As a basis to assess acute diffuse pollution and flood risks due to catastrophic storms, an intensity-duration-frequency (IDF) analysis was completed. The results indicate that significant storm events might occur in December–February and that every 100 years a single storm event, where the total average monthly precipitation could fall in only 5 min, is likely. Under normal conditions, on the other hand, every 2 days during the rainy season a rainfall event of average intensity could be observed.

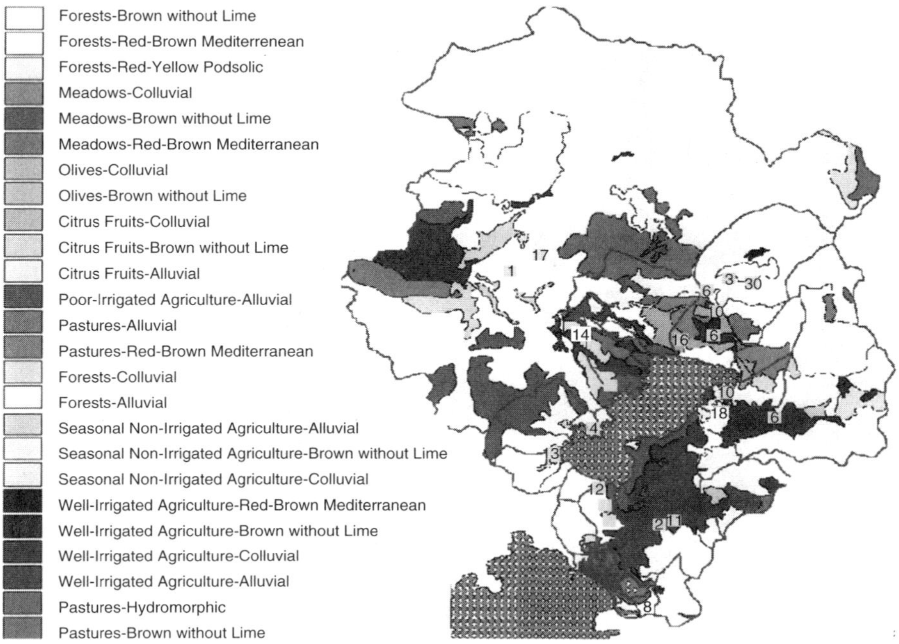
Forests-Brown without Lime
Forests-Red-Brown Mediterrenean
Forests-Red-Yellow Podsolic
Meadows-Colluvial
Meadows-Brown without Lime
Meadows-Red-Brown Mediterranean
Olives-Colluvial
Olives-Brown without Lime
Citrus Fruits-Colluvial
Citrus Fruits-Brown without Lime
Citrus Fruits-Alluvial
Poor-Irrigated Agriculture-Alluvial
Pastures-Alluvial
Pastures-Red-Brown Mediterranean
Forests-Colluvial
Forests-Alluvial
Seasonal Non-Irrigated Agriculture-Alluvial
Seasonal Non-Irrigated Agriculture-Brown without Lime
Seasonal Non-Irrigated Agriculture-Colluvial
Well-Irrigated Agriculture-Red-Brown Mediterranean
Well-Irrigated Agriculture-Brown without Lime
Well-Irrigated Agriculture-Colluvial
Well-Irrigated Agriculture-Alluvial
Pastures-Hydromorphic
Pastures-Brown without Lime

**FIGURE 9.5.4D** Overlaid maps of soil classes and land use.

The dry season (June–August), however, could be characterized by only one or two rainfall events of very short duration and relatively higher intensity.

*9.5.1.1.1.9 Hydrodynamic Characteristics*

An extensive study including both site and office work was undertaken to understand the hydraulic characteristics. Data analysis mainly focused on flows, salinity, and bathymetry data. The system was divided into six regions, as shown in Figure 9.5.6.

The northern part of the system is the moderate-sized lake called Koycegiz Lake. It is 13 km long, 2 to 6 km wide, and has an average depth of 25 m. It consists of two basins, as illustrated in Figure 9.5.7. The northern basin has a maximum depth of 28 m whereas the southern basin has a maximum depth of 32 m. The surface elevation of Koycegiz Lake is generally a few decimeters higher than the Mediterranean Sea except in the event of specific weather conditions. Several streams of various sizes at the surface and groundwater sources at the bottom feed the lake. Strong salinity and temperature stratification can be observed. The thickness of the epilimnion is about 10–13 m.

Long-term daily water surface elevation data gathered from the State Hydraulic Works of the Turkish Republic were analyzed. The results of these analyses showed that seasonal water level variations are 60–100 cm, with a few exceptionally high values like 150 cm.

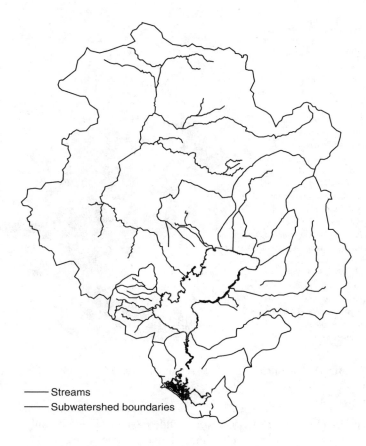

**FIGURE 9.5.5** Koycegiz–Dalyan watershed and its subbasins.

Alagol Lake, situated at the southwestern part of the Dalyan area, has a surface area of 0.55 km² with a round shape and an average depth of 3 m. In spring 1999, the average salinities in the upper and lower layers measured during 2 years of monitoring studies were 8 and 32 ppt, respectively, and salinity stratification was observed at approximately 1.4 m. Sulungur Lake, situated at the southeastern part of the Dalyan area, has a surface area of 3 km² with a maximum depth of 13 m. It showed a hydrodynamic behavior similar to Alagol Lake during the field study. The only difference was the depth of stratification, which was located at an average of 2.8 m below the surface. The upper layer had salinity of 3–28 ppt, and the lower layer had an average salinity of 35 ppt, which is very close to the salinity of the sea.

The Dalyan Channel is a narrow water body, 7 km long, 40–150 m wide, and 2–5 m deep. The vertical salinity profile of the Dalyan Channel varies both seasonally and spatially. The salinity in the upstream of the channel, where the Koycegiz Lake joins the Dalyan Channel network, is low compared to downstream and other regions in the system.

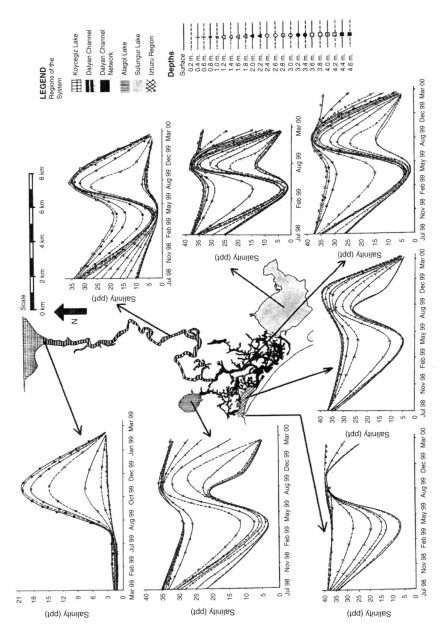

**FIGURE 9.5.6** Regions with different hydrodynamic properties in the system.[8]

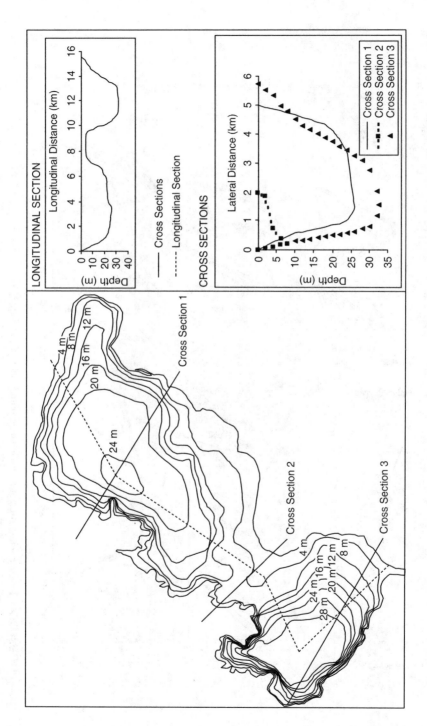

**FIGURE 9.5.7** The bathymetry of Koycegiz Lake.

**TABLE 9.5.1**
**Minimum and Maximum Concentrations of Various Parameters in Channel System for Surface Waters**

| Parameter | Cruise 2 April 1999 Min. | Max. | Cruise 3 August 1999 Min. | Max. | Cruise 4 November 1999 Min. | Max. | Cruise 5 March 2000 Min. | Max. |
|---|---|---|---|---|---|---|---|---|
| DO (mg l$^{-1}$) | 8.14 | 8.91 | 3.64 | 6.88 | 7.11 | 9.12 | 9.78 | 10.97 |
| NH$_4^+$–N (μM) | 2.59 | 6.63 | 0.45 | 1.42 | 0.92 | 1.47 | 1.17 | 6.66 |
| NO$_3^-$–N+ NO$_2^-$–N (μM) | 13.73 | 22.94 | 0.62 | 12.53 | 3.38 | 9.66 | 11.83 | 21.75 |
| DRP (μM) | 0.07 | 0.21 | 0.18 | 0.23 | 0.11 | 0.22 | 0.22 | 0.26 |
| Si (μM) | 299.73 | 512.82 | 41.48 | 100.72 | 120.26 | 209.39 | 104.74 | 299.44 |

*9.5.1.1.1.10 Water Quality*

Surface and bottom layer concentrations of water quality parameters were individually considered for further evaluation because of the dual layer flow in the lagoon. The ranges for various parameters observed along the channel are given for surface and bottom layers in Table 9.5.1 and Table 9.5.2, respectively.

Dissolved oxygen concentrations for both surface and bottom layers were generally found to be around their saturation values. In winter and spring, N and P loads transported via surface run-off and leaching increase due to high amounts of precipitation and snow melting.[9] NH$_4^+$–N oxidized nitrogen (NO$_2^-$–N + NO$_3^-$–N) concentrations were observed to be higher in spring than in summer months for both surface and bottom waters. Nitrate concentrations of the bottom water were lower than surface water concentrations. Freshwater input to the system increases in spring months so there may be an increase in nitrogen concentration carried to the water through run-off and groundwater. Fertilizers used in agricultural areas are among

**TABLE 9.5.2**
**Minimum and Maximum Concentrations of Various Parameters in Channel System for Bottom Waters**

| Parameter | Cruise 2 April 1999 Min. | Max. | Cruise 3 August 1999 Min. | Max. | Cruise 4 November 1999 Min. | Max. | Cruise 5 March 2000 Min. | Max. |
|---|---|---|---|---|---|---|---|---|
| DO (mg l$^{-1}$) | 7.30 | 9.29 | 4.29 | 6.73 | 5.49 | 7.3 | 7.38 | 10.37 |
| NH$_4^+$–N (μM) | 2.92 | 6.39 | 0.89 | 4.12 | 0.2 | 2.42 | 1.16 | 6.46 |
| NO$_3^-$–N+ NO$_2^-$–N (μM) | 3.53 | 28.43 | 0.49 | 3.09 | 0.76 | 4.21 | 10.12 | 20.52 |
| DRP (μM) | 0.13 | 0.28 | 0.11 | 0.22 | 0.08 | 0.21 | 0.21 | 0.39 |
| Si (μM) | 167.79 | 448.46 | 17.31 | 45.87 | 64.6 | 133.86 | 61.81 | 300.46 |

the agents that increase nitrate input, usually after precipitation. Groundwater level is high in Dalyan Lagoon, but there is a lack of information about nitrate input from groundwater. Since nitrite ($NO_2^-$–N) concentrations are low, the majority of oxidized nitrogen concentration consists of nitrate ($NO_3^-$–N). Total nitrogen (TN) concentrations were found to be high for surface and bottom waters, especially in winter due to terrestrial input. Variations in particulate organic nitrogen (PON) along the channel system were observed to be high, which might be the result of a settling process that varies spatially through the system. This is also valid for TSS and other particulate nutrients.[10] Dissolved reactive phosphorus (DRP) concentrations were in the range of about 0.1–0.23 μM, close to boundary concentrations for both surface and bottom waters during spring and summer. Compared with the Redfield ratio (N:P = 16:1), DRP concentrations in the system were low; therefore, it is assumed that phosphorus could be the limiting factor for plankton growth. However, this hypothesis must be verified with bioassay tests. Excessively high silicate concentrations (approximately 400 μM) were observed in spring. Silt comprising the main soil texture of the watershed causes high silicate concentrations; thus, silicate is not considered a limiting nutrient for algal growth.[10] Chlorophyll *a* average concentrations that can be used as an indicator for planktonic mass were generally lower than 4 μg/L.

This study compared chlorophyll *a* and nutrient concentrations in lagoon environments, as shown in Table 9.5.3, wherever reliable data are available. It may be

**TABLE 9.5.3**

**Comparison of Dalyan Lagoon Nutrient and Chlorophyll *a* Concentrations with Concentrations in Other Lagoons[11]**

| Lagoon | $NO_2^-$–N + $NO_3^-$–N | $NH_4^+$–N (μM) | DRP (μM) | Si (μM) | Chlorophyll *a* (mg.m$^{-3}$) | Reference |
|---|---|---|---|---|---|---|
| Celestun (Gulf of Mexico) | 0.9–13 (4.82) | 1–40 (7.82) | 0.3–3.2 (0.82) | 16–350 (54.4) | 0.5–28.5 (5.8) | Herrera-Silveria et al.[12] |
| Chelem (Gulf of Mexico) | 1–5 (1.89) | 1–15 (7.31) | 0.1–6 (0.41) | 11–39 (36.8) | 1.4–9 (3.8) | Herrera-Silveria et al.[12] |
| Dzilam (Gulf of Mexico) | 1–7.2 (4.07) | 1–35 (2.5) | 0.2–8.1 (1.45) | 20–300 (163) | 2–6 (2.7) | Herrera-Silveria et al.[12] |
| Rio Lagartos (Gulf of Mexico) | 0.2–5 (0.7) | 2–11 (8.5) | 0.3–8 (1.55) | 5–120 (24.7) | 2–10 (4.9) | Herrera-Silveria et al.[12] |
| Mar Menor (Spain) | 0–0.55 (0.1) | | 0–4 (2) | | 0.5–4 (1) | Gilabert et al.[13] |
| Dalyan Channel (Turkey) | | | | | | |
| Surface | 0.62–22.94 | 0.45–6.66 | 0.07–0.26 | 41.48–512.8 | 1.21–4.18 | Gürel[10] |
| Bottom | 0.49–28.43 | 0.89–6.46 | 0.08–0.39 | 17.31–448.5 | 0.02–7.47 | |

concluded that phytoplankton production in the Dalyan Lagoon is comparatively lower than in the other lagoons. It is important to note here that these lagoons have different hydrodynamic and ecological properties, which affect the response of primary production to nutrient concentrations.

During the evaluation of water quality, the difficulty usually faced is that there is no universal limiting nutrient criteria for lagoon systems. The response to nutrient enrichment and the related changes in ecosystem dynamics that influence the selection of assessment parameters vary considerably both regionally and among different types of coastal ecosystems. The U.S. EPA uses a specific regionalized waterbody type approach to develop nutrient criteria, whereas the European Union makes regional comparisons of eutrophication status on a common basis; however, these studies are still ongoing.

### 9.5.1.1.1.11 Ecological Characteristics

An intensive literature survey conducted to determine ecological characteristics of the watershed indicates the presence of some endangered and endemic species, which also has been verified by site visits.

*Iris xanthospuria*, an iris species peculiar to the region, is spreading in the aquatic zones, marshes, and *Liquidamber orientalis* forests. Red pine and *L. orientalis* are of great importance in the watershed as they are trees endemic to Turkey. There is an *L. orientalis* forest of 8 ha northeast of Koycegiz Lake, which is under protection.

During the long dry period, mosses found at the bottom parts of leafy trees and pines are important to keep soil moist. The flora of Koycegiz Lake–Dalyan Lagoon and its surroundings is still being studied. It is estimated that there are about 700 flower plants, needles, and fern species in the region; environmentally important flora species in the region are *Cistus cretius, Colchicum, Cyclamen trochopteranthum, Erica, Hyacinthus arientalis, Iris xanthospuria, L. orientalis, Lunipus angustifolius, Nerium oleander, Pinus brutia, Sternbergia eischeriana,* and *Typha* sp.[14]

To date, 180 bird species have been observed in the watershed.[15] *Holycon smymensis* and *Cryle rudis* are the two major species. *Holycon smymensis* is a rare species of bird and the distribution of *C. rudis* is limited in Turkey. Their population is about 100–150 and 250 couples, respectively. The number of species that breed or have the possibility to breed in the watershed is decreasing.

Iztuzu beach, the nesting and breeding area of *Caretta caretta* sea turtles, is among the 17 important sea turtle breeding regions of Turkey. Site surveys have identified 5 turtles, 2 frogs, 9 snakes, and 12 mammal species.[1]

Blue crab, which consume dead fish and similar wastes, thus preventing bad odor and decay in the environment, is an important species in this sensitive region.

## 9.5.1.1.2  Socio-Economic Characteristics

### 9.5.1.1.2.1  Demographical Characteristics

Data gathered from the State Institute of Statistics of the Turkish Republic (SIS) were used for demographical analyses. According to the 1997 census, the total population of the watershed area is 43,585, 74% of which lives in the Koycegiz

**TABLE 9.5.4**
**Current Land Use of the Dalyan Lagoon Watershed[9,17]**

| Land Use | Area (km²) | % Distribution |
|---|---|---|
| Forests | 80.93 | 62.55 |
| Agriculture | 29.70 | 22.95 |
| Wetlands | 6.56 | 5.07 |
| Sulungur Lake | 3.01 | 2.33 |
| Alagol Lake | 0.55 | 0.43 |
| Iztuzu Lake | 0.19 | 0.15 |
| Lagoon | 2.50 | 1.93 |
| Others (historical sites and springs) | 5.94 | 4.59 |
| Total | 129.38 | 100 |

Lake area and the rest in the Dalyan Channel network drainage area. The largest town in the lagoon drainage area is Dalyan, with a population of 3,357 according to the 1997 census. Koycegiz is the largest town in the lake drainage area, with a population of 7,526.[16]

The population increase in the lagoon area does not reflect the rapid and huge increase in other regions of Turkey where industrial activities are significant. As expected, there are no significant industrial activities in this area, which is an important factor promoting high population increase in Turkey. It may be concluded that the economy of the watershed area is mainly based on agriculture, tourism, fishery, and forestry.

*9.5.1.1.2.2   Land Use*

Detailed information on current land use was gathered from the NIC of the TRGDRA, as shown in Figure 9.5.4D. The data were validated by the site visits. The current land use of the area is given in Table 9.5.4.

Because towns and small communities of the lagoon drainage area are scattered in agricultural lands, they cannot be shown separately. Forests and agricultural land cover approximately 85% of the watershed.

*9.5.1.1.2.3   Transportation, Energy, and Communication Facilities*

Regional, national, and international transportation is facilitated by roads, marine transportation, and airlines; there is no railway. Koycegiz town center is 40 km from the international airport of Dalaman. Thus, it is easy to travel to Dalyan and other places with high tourism potential. Marine transportation to the coastline of Mugla province and its vicinity is provided by Turkish Maritime Lines. There is extensive bus transportation to the major cities of Turkey from Koycegiz town center.

*9.5.1.1.2.4   Infrastructure Facilities*

Sewer systems and common wastewater treatment systems for both Koycegiz and Dalyan were recently constructed within the context of the Special Protection Authority Project supported by the KfW Fund, Germany. Leaky/leak-proof cesspools have been used to solve the wastewater problem in rural areas. It was decided to treat these wastewaters at a common wastewater treatment plant. Treatment plant effluent is then discharged to the wetlands in Dalyan and to a drainage channel, which is 1 km away from Koycegiz Lake. Wastewater from the Dalyan and Koycegiz urban areas is collected by a separate pipe system. Sewer service will also be constructed in new development areas in the future. The project also deals with the renovation of the water supply and distribution systems of Koycegiz and Dalyan.

### 9.5.1.1.2.5    Social Facilities

A major social problem is that Koycegiz and Dalyan have been unable to develop a social structure and policy that correspond to their development level. Education level and social awareness have not kept pace with economic developments.[18] Education level and social awareness are more developed in Koycegiz than in Dalyan. There are two high schools and three vocational high schools in Koycegiz. Efforts are being made to change the existing vocational high school educational program to a 2-year undergraduate program in aquatic products at the nearby university. Koycegiz has a library with 4,772 patrons. Young people attend activities at the public education center.

According to the information gathered by SIS, the population of Koycegiz has a high percentage of young people between the ages of 10 and 34. The average household has three to four members. Although the literacy rate is high, most people are only primary school graduates. The number of men and women is almost the same but there are three times more literate men that women. University graduates account for 7% of the population. The number of young people who go on to university is low because most young people prefer to become involved in trade activities at an early age. There are two hospitals and six village clinics in the watershed.

### 9.5.1.1.2.6    Sectors and Their Characteristics

#### Agriculture

Agricultural activities are the major sources of economic income in the watershed. An area of 7111 ha of irrigated land or land receiving sufficient rainfall is devoted to agriculture in Koycegiz. The agricultural activities are similar to other Mediterranean countries based on polyculture, basically cotton, citrus fruits, wheat, corn, peas, and horticulture. Land distribution of crops is as follows: 2630 ha of citrus fruits, 550 ha of horticulture, 700 ha of greenhouse, 1580 ha of olive, 5 ha of vineyards, and 20 ha of poplar. The rest of the agricultural land is used for potato, melon, onion, and garlic production. About 3259 ha of land is used for agriculture in Dalyan, and product distribution is identical to Koycegiz. Some companies have tried ecological farming around Koycegiz, but despite their efforts, these attempts were unsuccessful. Especially in the rural regions, they could not convince farmers to use these new techniques. Floriculture has also developed in this region.

#### Tourism

Tourism is a major industry in Koycegiz and Dalyan. The ruins of the ancient city of Caunos are located in this area and the 4th century B.C. Lycian rock tombs are near the seaside. Iztuzu Beach on the Mediterranean coast is one of the most beautiful beaches in Turkey and the nesting and breeding ground of *Caretta caretta* sea turtles.

The number of beds in hotels and other accommodation facilities is low compared to the rest of Turkey. During July and August there is full capacity. Almost all accommodation facilities were established in the ancient city areas according to land and planning constraints. National and international environmental organizations and scientists studying the protection of *Caretta caretta* turtles have played a major role in the increase of tourism. Mud baths and hot springs are among the other major tourist attractions.

*Fishery*
Fishery activities are carried out in Koycegiz, Alagol, Suluklugol, and Sulungur Lakes, and along the Iztuzu coast by means of a cooperative (DALKO), which has 507 shareholders and 60 employees. Fish production fluctuates. Mullet, which has economic value, is found in the sea near the coast, but during the breeding period adult mullet travel to brackish streams and lagoons and their offspring grow here. One third of adults (300 t per year) are caught as they travel to brackish waters, and they are used in caviar production.

*Forestry*
The majority of the watershed is covered with forests, which are of major economic value. Between the years 1970–1991, 13,748 ha of land was forested. A total of 750–850 ha is planted annually.

*Public Services*
Public service is developed in Koycegiz because it is a district center. Almost 700 people are employed in the public service sector, mostly in municipal posts, forestry administration, education, and medical services.

*Industry and Trade*
There is no major industrial activity in the watershed. Cotton gins process cotton collected from the neighborhood villages. Almost all companies are small scale and located in the "Koycegiz small industry site." Koytas is a moderate-scale enterprise that produces equipment for agricultural activities and municipality services. A fish-processing factory has been established in Beyobasi. A citrus fruit processing company is important because it provides jobs during the winter and exports provide economic income. Trading activities that parallel tourism activities are developed in both Koycegiz and Dalyan.

*9.5.1.1.2.7    Income and Manpower Distribution*
A total of 78% of the population in Koycegiz is involved in agriculture; it is the major source of income in the rural regions. Tourism is the major source of income for Dalyan. Interestingly, the education level and social activities are more developed in Koycegiz than in Dalyan, and they constitute a positive contribution to the increase in income in Koycegiz.

*9.5.1.1.3   Pollution Sources and Waste Loads*
A detailed study including site visits, surveys, office work, and laboratory work was conducted to calculate nutrients and pesticide loads in the system. During on-site visits information was gathered from both the District Directorates of Agriculture of Koycegiz and Ortaca and the pesticide dealers. Laboratory studies for residual pesticide analysis

in water were carried out in cooperation with Mersin University, Department of Environmental Engineering, and the Middle East Technical University (METU) Institute of Marine Sciences.

### 9.5.1.1.3.1   Point Sources of Pollutants

Because no major industrial plants exist in the area, domestic wastewater is considered to be the only point source. The infrastructure of towns and small communities has been constructed but is not yet fully completed, so, at present, domestic wastewater is filtered through septic tanks. In May 2002, the common wastewater treatment plant for the area began operations; however, all the information gathered from the area relates to the period before 2002. Pollutant loads of domestic origin are calculated based on the assumption that approximately 30% treatment occurs in septic tanks prior to percolation and urban run-off and that no further treatment occurs until the loads reach the water environment. Unit loads in terms of Total-N and Total-P are therefore taken as 8 g/capita/day and 3 g/capita/day, respectively.[19]

### 9.5.1.1.3.2   Nonpoint Sources of Pollutants

Detailed studies on diffuse pollutants have been conducted only at the Dalyan Channel network drainage area. Each type of fertilizer was converted to active nitrogen and phosphorus values according to charts supplied from the manufacturers in order to calculate the consumption in terms of the two nutrients. Annual fertilizer consumption for the year 1998 was calculated as 146.8 kg/ha/year N, and 54.2 kg/ha/year P for the Dalyan Channel network drainage area, which are almost twice the country's averages. The major reaction occurring after fertilizer application on soil is crop uptake, which was calculated as 63.16 kg/ha/year N and 15.14 kg/ha/year P based on the crop types. Furthermore, these two nutrients undergo reactions such as denitrification, volatilization, adsorption, etc. on soil. It is the remaining part, called excess or surplus nutrients, that is transported with soil until it reaches the water environment through surface run-off and/or leaching. Nitrogen and phosphorus loads in excess were estimated as 83.66 kg/ha/year N and 39.06 kg/ha/year P. According to the literature, the optimum surplus N value is 50 kg/ha/year,[20] above which certain control measures are needed.

Calculations were made to estimate the amount of monthly surplus nutrients. The main nitrogen reactions were assumed to be denitrification, ammonia volatilization, surface run-off, and leaching, and were defined as certain percentages of the amounts remaining after crop uptake. The soil properties of the region were considered suitable for high denitrification ranging from 20 to 40%.[21,22] Thus, it was assumed that the amount of nitrogen that underwent detrification was 15% in February; 20% in March, April, October, and November; and 25% for the rest of the year. The amount was taken as zero for months when soil temperatures were around 10°C, as in January and December. The pH of the soil in the region is between 6.0 and 8.0, where ammonia volatilization may occur up to 20% depending again on the soil temperature.[21] It was assumed that 5% volatilization will occur in January, February, March, and December, and 10% in the rest of the months. Because all types of the remaining nitrogen are converted to nitrate form, most of the remaining amounts are transported to the lagoon through leaching. Adsorption is assumed to be the major phosphorus reaction on soil.[23–25] Depending on the soil properties, 65% of phosphorus was considered to be adsorbed in May, 10% in June, 64% in July, 90% in August and in October, and 27% in November. These assumptions take into account the monthly applications and crop uptake values. The leaching and/or surface run-off distributions were estimated based on the soil properties and net monthly precipitation and run-off values.

A detailed forest survey has not yet been carried out. Because the mechanisms are too complex in these systems, unit polluting loads were selected from the literature, representing and reflecting similar climatic conditions and forestry, as 2 kg/ha/year for nitrogen and 0.2 kg/ha/year for phosphorus.

Pesticides consumed in agricultural areas were also examined in the Dalyan Channel network drainage area because they form the basis of toxic pollution of the soil and aquatic environment. The impact of 41 pesticides on the soil system based on the main mechanisms like persistence and mobility were investigated, and findings were then used to determine the changes in pesticide amount being lost via run-off or by leaching with the aim of confirming their probable existence in the lagoon environment.

Pesticide use in the area is approximately 12 kg-1/ha, which is quite high compared to overall annual consumption value for all of Turkey (1.25 kg-1/ha) and for the metropolis Istanbul (3.5–4 kg-1/ha).[26] However, the consumption value, even in the Dalyan Channel network drainage area, is low compared to other European countries.[27] Still, some residual pesticides were observed in the aquatic environment because of misuse.

In order to predict how the pesticides in the soil move to the water environment, a rough estimate was made by running the EPIC model,[28] where meteorological and soil data together with intrinsic properties of the selected pesticides were used. The EPIC model was used to determine biological degradation and leaching amount of the six selected pesticides. Pesticide uptake by crops is assumed to be negligible.[29] The frequency and amount of pesticide application on a monthly basis over a 1-year period were considered to be part of the pesticide properties used in the rough estimation of excess loads. The net water distribution on the ground, the infiltration rate, and run-off rates required by the model were estimated. Table 9.5.5 shows two typical examples of excess pesticide calculations.

*9.5.1.1.3.3    Estimated Pollution Loads*
The distribution of estimated excess nitrogen and phosphorus loads according to current land use were estimated. Almost 87% of N and 59% of P were found to be due to nonpoint sources in spite of point-source loads due to untreated wastewater characteristics. It is important to note here that livestock (organic fertilizer) application is not the usual practice in the area, which may be another important source of diffuse pollution for many other lagoon watersheds. During the calculation of the estimated loads, only the net monthly precipitation values were considered for the transportation of water either vertically or horizontally. However, the two types of crops, cotton and citrus fruits, are irrigated during the dry seasons, which account for almost 5–6 months per year. Under such circumstances, the total annual load emitted to the environment will not change; only the monthly distribution trend will vary.[30]

## TABLE 9.5.5
### Typical Calculation of Excess Pesticides

| Pesticide | Dichlorvos<br>Pathway: Leaching | Endosulfan<br>Pathway: Surface Run-Off |
|---|---|---|
| Initial amount applied | 2.5 kg/ha/year | 3577 kg/year |
| Lost by degradation | 0.24 kg/ha/year (9.6%) | 50.1 kg/year (1.4%) |
| Amount remaining on ground | 2.5 − 0.24 = 2.26 kg/ha/year | 3577 − 50.1 = 3526.9 kg/year |
| Amount remaining on ground after leaching | $2.45 \times 10^{-8}$ kg/ha/year (almost none remains) | 387.96 kg/year (11%) in November[*]<br>2193.7 kg/year (62.2%) in December[*]<br>945.2 kg/year (26.8%) in February[*] |

[*]These 3 months have a high tendency toward intense surface run-off.

### 9.5.1.1.4 Administrative and Legal Structure

The Turkish government declared part of the area a special protection area in 1988 because of its unique exemplary structure as an important ecosystem with a high diversity of species and intense biological activity. In 1990 the Koycegiz and Ortaca districts were included in the boundaries of the special protection area, increasing the total area to 385 km$^2$.

The Environmental Protection Union (EPU), consisting of more than 50 settlement representatives responsible for the operation and maintenance of infrastructure facilities, was established recently. The EPU is increasing its responsibilities through regulation changes.

This special area is protected by the following international agreements and protocols: Bern Agreement, Barcelona Agreement, Genova Declaration, World Heritage Agreement, Biological Diversity Agreement, and Ramsar Agreement.

## 9.5.1.2  Decision Support System Tools

### 9.5.1.2.1  GIS

During the last 3 years of the Koycegiz–Dalyan Lagoon study, a GIS study was initiated to support the decision support system of the entire watershed. In the previous years of the GIS studies, various versions of ArcView® software from ESRI®[†] were used, but then it was decided that future GIS work would be conducted using the ArcGIS®[†] software package directly.

UTM was selected as the base coordinate system because most of the data gathered from various institutions were in the UTM coordinate system.

Spatial data such as land use, vegetation, topography, and soil characteristics explained in the previous sections are transferred into GIS as layers. Watershed boundaries are delineated and imported to GIS. In order to import the bathymetry

---

[†] Registered trademark of Environmental Systems Research Institute, ESRI, Canada.

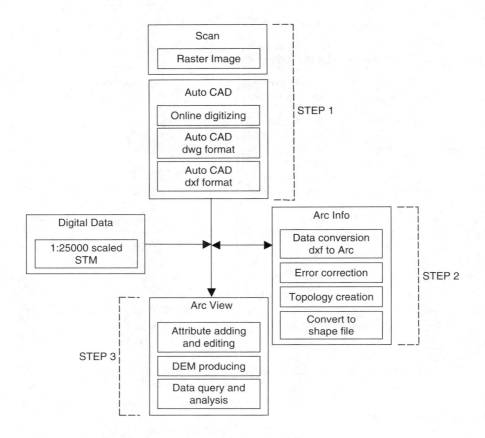

**FIGURE 9.5.8** Flow diagram of the GIS methodology.

of the lagoon system into GIS, the available bathymetry data are digitized and converted to the UTM coordinate system.

The overall flow diagram of the GIS methodology is given in Figure 9.5.8. Step 1 involves the integration, manipulation, and transformation of elevation data, sub-basin boundaries, and coastlines. ArcGIS was preferred for geographical data creation, management, integration, and analysis. Thus, in Step 2, the data developed in the preceding step are converted to ArcGIS file format. Finally, in order to enable queries via the GIS environment, attribute tables are prepared in Step 3 for all map layers.

Figures given in previous sections are good examples of using GIS as a tool for planning studies.

### 9.5.1.2.2 Models

#### 9.5.1.2.2.1 Rivers and Rural Area Run-Off Modeling

Namnam Creek and Yuvarlakcay Creek, the two most important streams, carry 85% of the water flow and most of the nutrient loads into Koycegiz Lake. Therefore, models that simulate the water quality of these streams are important components of a decision support system for the entire Koycegiz–Dalyan Lagoon system. According to the data analyses one-dimensional (longitudinal) stream water quality models were found to be appropriate for both streams.

Modeling of the streams was done in two stages. First, a preliminary modeling study, using QUAL2E and QUAL2E-UNCAS, was done in order to better understand the system behavior and the impact of different nutrient loads on the water quality of the streams. Uncertainty analyses were done with the Monte Carlo Simulation option in QUAL2E-UNCAS. To obtain the uncertainty parameters (type of distribution of parameters, mean value, and input variable variances), a simple spreadsheet-based database management system was developed using Microsoft® Excel. The details of the preliminary modeling study for rivers are given in Ertürk et al.[31] The results obtained from QUAL2E and QUAL2E-UNCAS applications were used as initial values for the dynamic simulation of rivers. These simulations have already begun. WASP 6.1,[32] developed by the U.S. EPA, was selected as the most appropriate model for modeling of both streams. The EUTRO module of WASP 6.1 was chosen as the water quality model for dissolved oxygen, BOD, and nutrient calculations of streams.

A parallel study for calculating the run-off flow rates and nonpoint source loadings into the streams and the lagoon is still ongoing. Despite its large data requirement, the Hydrological Simulation Program FORTRAN (HSPF) is preferred over various other hydrological models due to the wide range of output parameters it is capable of producing. This is a commonly cited U.S. EPA model that has been successfully applied to rural/agricultural zones for more than two decades.

The model requires various time series and input parameters, which differ according to the selection of modules to be run. To simplify the modular structure of the model the first step is to eliminate the impervious land (IMPLND) module, which simulates the run-off from impervious catchments because almost the entire basin is unpaved, i.e., pervious. The results of the meteorological analysis show that the dominant type of precipitation within the watershed is rainfall. Thus, rainfall was assumed to be the major source of run-off. Processes to melt snow packs at the highest zones of the watershed area would be utilized only if a justifiable water budget could not be attained.

As HSPF uses a standard time series format called watershed data management (WDM), datasets obtained from the State Meteorological Service of the Turkish Republic were converted to WDM format by a sequence of programming operations.

Subsequent to a successful run on a fictitious basin with input parameters gathered according to a literary study to set compliant values for the real basin, the watershed was then divided into model segments. The segmentation is based on discrete hydrological patterns and basin topography. Thus, the subwatershed boundaries determined by the watershed modeling system (WMS) 3D topographical model were further segmented regarding changes into soil structure, crop types, and agricultural practices.

Hydrological modeling studies to adjust a water budget for the basin are ongoing and the ultimate goal is to attain a reliable understanding of the significance of nonpoint loads on the aquatic environment of the entire watershed.

*9.5.1.2.2.2    Hydrodynamic and Water Quality Modeling*
*of the Lagoon System*

The ongoing modeling study for the Koycegiz–Dalyan Lagoon system has a seven-step approach. The first step includes seasonal data analysis. In the second step, the southern part of the system (Dalyan channels and lakes) was divided into computational boxes, which were later used as model boxes of finite difference computational elements. In the third step, the U.S. EPA's one-dimensional finite difference hydrodynamic model DYNHYD5[33] and the Lung O'Connor method[34] were used to estimate the flow distribution and velocities in the channels of the Dalyan Lagoon. In the fourth step, a mass balance approach was used to verify the results obtained in the previous step. A modified version of the U.S. EPA's WASP5[35] model was used. In the sixth step, simple hydraulic calculations were made for the northern part of the system (Koycegiz Lake) to estimate inflow and mixing effects. In the seventh step, the CE-QUAL-W2[36] model was applied to used Koycegiz Lake and Dalyan Channel and lakes for hydrodynamic simulations.

The data analysis step used salinity data obtained from field studies, long-term daily inflow data gathered from the State Hydraulic Works and State Electric Works, and water-level elevation data of the Koycegiz Lake gathered from the State Hydraulic Works. The field data indicated that the system is under a saline water influence that changes spatially and temporally. The hydrodynamic behavior of the lagoon system is explained in detail in Section 9.5.1.1.1.

As shown in Figure 9.5.9, following the data analysis of the Dalyan Channel network, the system was divided into 62 computational elements for preliminary modeling studies.

Data analysis results concluded that a two-dimensional (longitudinal and vertical), dynamic, laterally averaging, numerical model (x-z model) is required to simulate the flow in the Dalyan Channel networks. Instead of running a two-dimensional dynamic x-z model and conducting the whole calibration work for that model, a simpler technique was used. The U.S. EPA's one-dimensional hydrodynamic model DYNHYD5 was run for a computational network. DYNHYD5 calculates the flow rates and depths in a node-link computational network. The results were used to estimate the net flow (upper layer flow plus lower layer flow) carrying capacity of each channel. Upper and lower layer flows were estimated with a modified version of the Lung O'Connor method which was modified by Ertürk[37] to handle the nonlinear vertical salinity gradients.

The flows estimated were checked with a salinity mass balance. Mass balance and transport equations had to be solved for each control volume in the Dalyan Channel network. The one-dimensional computational network has 49 control volumes. As there is a two-layer flow, the number of control volumes increases to 98. The EUTRO module in the WASP5 model was used to solve the mass balance equations. The estimated flow rates were used as flow input and the control volumes were used as WASP junctions. The standard version of WASP5 allows the user to simulate 50 control volumes. Because of some other program limitations related to number of flows, the source code of WASP5 has been modified and recompiled. The verification was made with the field data from the August 1999 cruise, when a high influence of saline water was observed. The results for flow velocities calculated for summer are illustrated in Figure 9.5.10.

**FIGURE 9.5.9** Computational elements of the Dalyan Channel network.

The estimated flows and salinities from the mass balance approach were used for preliminary water quality modeling studies with WASP5 and as initial values for CE-QUAL-W2 hydrodynamic simulations. The CE-QUAL-W2 modeling study was begun for a simplified version of the model network. The results obtained from the CE-QUAL-W2 simulations are compatible with field data.

The modified densimetric Froude number was used to estimate the stratification potential of Koycegiz Lake. The values calculated for each month were between $10^{-4}$ and $10^{-3}$, which indicate a high stratification potential in the lake. Separation/plunge

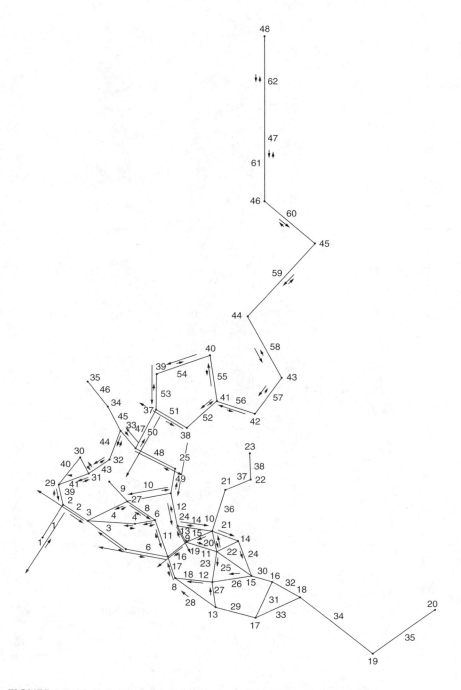

**FIGURE 9.5.10** Calculated velocities for the Dalyan Channel network.

point depths, distances of separation points from the coast, and overflow velocities for the five most important streams entering Koycegiz Lake were calculated each month for inflow mixing analysis. The calculations for average monthly flows indicated that the distances of the separation points for the streams have a range of 2–100 m. Another result obtained from these analyses was that there was no plunge point for five streams for each month of the year, because the lake water is always denser than the stream water. Internal wave heights calculation results were 10–46 cm for average winds, but they may increase to several meters for stronger winds observed in the area. In that case, strong mixing may occur and destroy the picnocline/halocline in the system. For preliminary studies, the CE-QUAL-W2 model was run for hydrodynamic simulations. The results obtained from these simulations correlate (depth of hypolimnion, salinities, and temperature) with field data.

Data analysis and verified flow estimation results indicate that the longitudinal (x) and vertical (z) changes are important for the Dalyan Channel network, therefore a two-dimensional (longitudinally and vertically), laterally averaging, hydrodynamic model (x-z model) such as CE-QUAL-W2 would be an appropriate choice. For the entire system, however, a three-dimensional, hydrodynamic model coupled with a water quality model would be appropriate. A parallel study for deciding on such a modeling system has already been initiated.

### 9.5.1.2.3  Monitoring

#### 9.5.1.2.3.1    Monitoring System Design

In order to understand the nutrient dynamics in the Dalyan Lagoon subsystem and to collect data for modeling studies, a monitoring system was designed. The locations of monitoring stations were determined according to the following factors.[10]

- Stations forming the boundary conditions for the system are to be selected.
- The Dalyan Lagoon system consists of channels and lakes. Monitoring stations are needed to observe the interaction between channels and lakes.
- Stations representing the lakes are necessary.
- Stations along the channels are needed to detect the spatial variations in the channel systems.
- Stations near the pollution sources are required.

Taking into account these factors, 16 monitoring stations were selected for the Dalyan Lagoon system, as shown in Figure 9.5.11. Station 0 and Station 14 represent the two boundary conditions. Station 3 and Station 4 were at the entrance and inside of Alagol Lake, respectively. Station 8 and Station 9 were selected as the entrance stations to Sulungur Lake. Station 10, Station 11, and Station 12 were inside Sulungur Lake. Other stations were the subsequent stations along the lagoon channel system.

Considering the financial and practical limitations it was decided that seasonal monitoring for 5 years would produce enough data for modeling and quality assessment objectives.

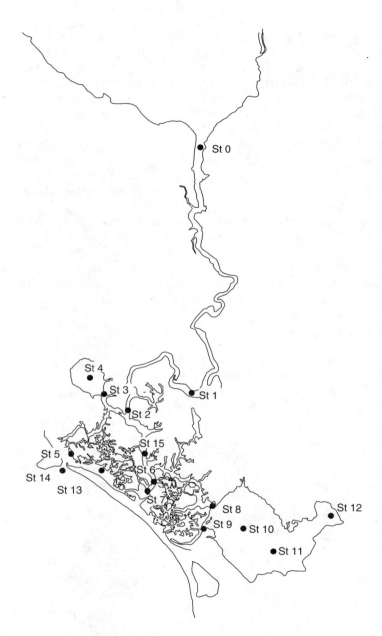

**FIGURE 9.5.11** Sampling stations along the lagoon system.

Five cruises were conducted to the area to collect water samples for the analyses of the monitoring parameters. Almost the entire first sampling was used to implement the sampling and experimentation methods and to overcome and minimize the difficulties and troubles encountered during experimentation. The rest of the experimentation values were considered during the discussion of results, which further led the scientists to achieve overall seasonal evaluations. Sampling periods were selected in order to make seasonal evaluations and comparisons.

As horizontal and vertical salinity gradients were observed in the system, surface and bottom water samples were collected at each station. Surface water samples were collected from an 0.5-m depth. Bottom water sampling depths were chosen according to the depth of the stations and vertical salinity gradients. Salinity measurements were carried out with the SBE25 Sealogger CTD probe at various depths.

The measured and calculated parameters from the five cruises are listed below.

*Measured Parameters*
Measured parameters include salinity, temperature, depth, Secchi depth, light, pH, total suspended solids (TSS), dissolved oxygen (DO), biochemical oxygen demand (BOD), alkalinity, $NO_3^- - N + NO_2^- - N$, $NO_2^- - N$, $NH_4^+ - N$, total nitrogen (TN), dissolved reactive phosphorus (DRP), total reactive phosphorus (TRP), total dissolved phosphorus (TDP), total phosphorus (TP), Si, chlorophyll *a*, and particulate organic carbon (POC).

*Calculated Parameters Using Measured Parameters*
Calculated parameters using measured parameters include $NO_3^- - N$, dissolved inorganic nitrogen (DIN), particulate organic nitrogen (PON), dissolved organic nitrogen (DON), particulate phosphorus (PP), particulate inorganic phosphorus (PIP), and particulate organic phosphorus (POP).

*9.5.1.2.3.2   Using Monitoring Results as a Tool*
After comprehensive data analyses, reliable data have been used for assessment of

- Hydraulic/hydrodynamic characteristics of the system
- Water quality and eutrophication characteristics of the system as well as model calibration and verification studies, which are discussed in detail in the relevant sections

*9.5.1.2.4   Indicators/Indexes*
Indicators and indexes are still being evaluated. A methodology for indication of indicators and indexes has been developed, but its assessment is still in progress.

*9.5.1.2.5   Economic Evaluations and Cost-Benefit Analysis*
A parallel study is ongoing for cost-benefit analysis. Results of the social impact assessment studies will provide data for the benefit analysis, whereas the results obtained from decision support system tools such as models or GIS will be used to support the cost analysis. The methodology and details of social impact assessment studies are given in Section 9.5.1.2.7.

### 9.5.1.2.6  Public Involvement

As mentioned in Section 9.5.1.1.4, the EPU was established, based on recommendations by the NATO-CCMS Pilot Study Group, to provide stakeholder involvement
in the decision-making process. The EPU consists of city governors, town mayors,
and governmental institutions, but representatives from public and nongovernmental
organizations cannot be members of the EPU under current Turkish legislation.
Therefore, the pilot study group advised the EPU to establish a legal entity to allow
public and nongovernmental organizations to be partners. The pilot study group is
still supervising this legal entity and related institutions to prepare an integrated
sustainable management plan for the watershed.

As part of the social impact assessment studies (see Section 9.5.1.2.7), questionnaires were prepared and distributed to the inhabitants to provide the opportunity
for additional public involvement. The questionnaire results satisfactorily show the
willingness of the inhabitants to support and participate in planning studies.

### 9.5.1.2.7  Social Impact Assessment (SIA)

A survey was conducted to evaluate the benefits to households from reduced pollution loads by operating common water and wastewater treatment systems in the
Koycegiz–Dalyan area. During the survey period the construction of wastewater
treatment facilities had just been completed. According to an agreement signed
between the Turkish Government and the German grant authority KfW, after completion of the system, the contractor would operate the plants for 1 year and the
inhabitants would not be charged during that period. After that period, the EPU
would be responsible for the operation of the system. Thus, the EPU attempted to
plan a billing process for operation of water and wastewater systems.

Since the contingent valuation method (CVM) is the only practical means of
estimating the environmental benefits and willingness to pay, this method was
applied in the Koycegiz–Dalyan area in June 2002 by the project team, who are
the authors of this chapter. CVM is a method that attempts to elicit information
about individual preferences for goods or services. Generally, if policy makers want
to know people's opinions regarding the quality of life of the area in which they
live, the CVM is the only available benefit estimation procedure.[38] Of the three
interview methods—by mail, telephone, or in person—for a CV study, personal
interviews were preferred. A contingent valuation (CV) survey has three basic parts.
First, a hypothetical description of the terms under which the good or service to
be offered is presented to the respondent. Second, the respondent is asked one or
more questions to try to determine how much he or she would value a good or
service if actually presented with the opportunity to obtain it under the specified
terms or conditions. Third, a CV survey includes a series of questions about the
socio-economic and demographic characteristics of the respondent and his or her
family.

The questionnaire was designed based on these three basic components of the CV
survey, and it contains 37 questions, 20 of which are related with first and second
components. A total of 17 questions are about the socio-economic and demographic
characteristics of the people. Households were asked about their willingness to pay

(WTP) for the wastewater treatment system that would improve future water quality in the area. In this survey, the evaluation question consists of two steps:

- The respondents were first asked whether they would vote in favor of making an extra payment in support of the operation of the wastewater treatment plant.
- Those in favor were asked whether they would accept the bid amount.

The common techniques for asking CV questions are iterative bidding, payment cards, and dichotomous choice. The single-bounded dichotomous choice approach was preferred. In this technique, the item being evaluated and the hypothetical market for this item are described to respondents, who are then asked to state whether they accept or reject a single "take-it-or-leave-it" offer for the item being evaluated. As the first step, a pretest was applied for each town on the first day of the CV survey. After the feedback from the pretest, the questions were revised. The opinions of the people in each town were better understood during the pretest—they were aware of the importance of pollution and environmental values. A total of 400 households from each town (800) replied to the questionnaire.

All the questionnaires were entered into a database using the SPSS software. After the completion of data transfer, as a first step the data were sorted by removing "protest responses" given by individuals, who for one reason or another rejected the hypothetical scenario and refused to give meaningful answers.[38] Afterward, the survey was analyzed according to the methods explained below.

This kind of survey is typically analyzed in three increasingly sophisticated ways. First, analysts examine the frequency of distribution of the responses to the evaluation questions. Second, analysts look at cross-tabulations between WTP responses and such variables as socio-economic characteristics of the respondent and attitudes toward the environment. Third, analysts use multivariate statistical techniques to estimate a valuation function that relates the respondent's answer to socio-economic characteristics of the respondent and attitudes toward the environment. The types of statistical procedures utilized depend on whether the respondent answered a direct, open-ended, valuation question (asking the respondent a direct question about the most he or she would be willing to pay for the good or service) or a yes/no question.

### 9.5.1.2.8  Discussions on Evaluation of Tools

The Koycegiz–Dalyan Lagoon system is a very complex and complicated system, composed of channels and lakes. Satisfactory evaluation of such a system for sustainable management studies necessitates a large amount of data. Therefore, the field data should be maintained for a long period and an ecological model should be used for better understanding of the nutrient dynamics in the system.

Hydrodynamic characterization of the lagoon systems has a vital importance in nutrient dynamic evaluations. Hydraulic retention times in channels are very low, i.e., 1 or 2 days. On the other hand, sampling in 16 stations takes more than a few days for each cruise. Therefore, in order to observe the spatial variations in the system, it could be even better to take samples simultaneously in the stations along the channels, or to continue sampling at the same station for a few more days. In practical terms, the application of such a monitoring strategy in the field is difficult. Because of the dynamic character of the system, continuous measurements of velocities and water surface elevations at the system boundaries should be conducted.

Because sediment can be an important internal nutrient source, especially in Sulungur Lake, sediment sampling should also be performed. Ions such as iron, calcium, and manganese that have roles in sediment-water column chemistry should be studied. In particular, fish production should be kept under control to prevent sedimentation.

There is no simple classification of brackish water and seawater because the species composition in these systems varies considerably. The tendency among ecologists is to determine the species one by one in the system. The trophic level of the system may be estimated by conducting ecological studies. Classifications, based on the existing species composition present, should be made for systems with such a dynamic character.

Groundwater is important in terms of the water budget as well as the transportation of pollutants, especially nitrate nitrogen. Information on groundwater flow and pollutant concentrations that lead to the Dalyan Lagoon system is not known, but such information on groundwater would be useful for calculating water and nutrient budgets.

Because the Koycegiz–Dalyan Lagoon system has a significant area covered with rooted aquatic plants their impact on the water column should also be investigated.

During the studies, two new wastewater treatment plants were completed. To evaluate the expected improvements in water quality, additional monitoring stations are recommended.

## 9.5.2 BASIS OF SUSTAINABLE MANAGEMENT PLAN

Characteristics of the watershed, the sensitive and specific ecosystem characteristics of various wetlands located in the watershed, the incomplete present administrative structure that cannot establish a direct relation with the public and integrate with the socio-economic system, and current developments that undervalue the ecosystem underscore the need for sustainable planning. Thus, as a result of the meetings with the stakeholders, development of an integrated sustainable watershed management plan (ISWMP) was decided.

The main components of this planning may be summarized as follows:

The target is sustainable management of the watershed as a whole, to improve the socio-economic system by using the potential of tourism as well as promoting natural, ecological, health, educational, and cultural assets of the area, ecological or organic farming, and aquaculture production.

Problems anticipated for implementation of the plan:

- The future of the watershed is threatened because of the disintegrating of the two main components of the ecosystem, namely natural capital and the socio-economic system.

- On the one hand, stakeholders in the socio-economic system are trying to use the entire potential provided by the natural capital, whereas on the other hand governmental administration implementations are not considering the needs of the socio-economic system for protecting the natural capital. This leads to conflicts between both sides, compounded by the implementations of the third parties that do not comply with the legislation.
- Problems arise from excessive use of the environment for tourism and fisheries, and the eutrophication indicators that are beginning to occur in certain coastal areas.

Decision-Making Aspects:

- Developing an effective decision support system using the information and knowledge produced by these studies
- Increasing the awareness of all individuals in the socio-economic system about the watershed ecosystem
- Developing regional land-use alternatives and preparing the final plans in cooperation with the representatives of the socio-economic system and stakeholders
- Application of plans as appropriate to the local scale
- Ensuring the participation of stakeholders in the management and decision-making process
- Implementing the plans by strengthening administration mechanisms, particularly the EPU structure and improvement of the plans by continuous assessment
- Extending the authority of the EPU in a progressive, stepwise manner
- Announcing the results obtained from the decision support system
- Effective use of water and wastewater income
- Obtaining income from other sources, e.g., fees from the tourism sector, international funds, etc.
- Establishing and improving the monitoring system
- Enforcing central government legislative changes for watershed management principles
- Encouraging and promoting investigators for ecological tourism, farming, and aquaculture

The studies are still ongoing. Modern tools such as GIS and models are being used to prepare alternative plans. Continuous evaluation systems will be set up.

## ACKNOWLEDGMENTS

Such a detailed case study on ecological modeling and management of a lagoon watershed could not have been initiated and conducted without the financial support and participation of various organizations. The authors would like to express their thanks to the NATO-CCMS, the ITU Research Fund, TUBITAK, NATO Collaborative Research, and Istanbul University Institute of Marine Sciences. Special appreciation goes to the local authorities of the towns of Koycegiz and Dalyan, the Regional Environmental Protection Agency, and the Fisheries Cooperative (DALKO) for their kind hospitality and help during field surveys conducted since 1998.

## REFERENCES

1. Gönenç, I.E., Ertürk, A., Ekdal, A., Tümay, A., Tanik, A., Beler Baykal, B., Gazioglu, C., Polat Beken, C., Seker, D.Z., Hepsag, E., Okus, E., Dogan, E., Altiok, H., Yüceil, K., Gürel, M., Karakaya, N., and Topcu, S., Modeling and Land Planning of Koycegiz–Dalyan Lagoon and Its Watershed, Vol. 1 and 2, Istanbul Technical University Research Fund, Istanbul, Turkey, 2002 [in Turkish].
2. ESCAP-UN, Guidelines and Manual on Land-use Planning and Practices in Watershed Management and Disaster Reduction, ST/ESCAP/1781, Economic and Social Commission for Asia and the Pacific, 1997.
3. Frevert, R., Schwab, G. Edminster, T., and Barnes, K., *Soil and Water Conservation Engineering*, 4th ed., Wiley Interscience, New York, 1993.
4. Tanik, A., Seker, D.Z., Gürel, M., Yüceil, K., Karagoz, I., Ertürk, A., and Ekdal, A, Towards integrating land based information for watershed modeling in a coastal area via GIS, in *Proc. 7th Int. Specialized Conference on Diffuse Pollution and Basin Management*, Bruen, M., Ed., Dublin, Ireland, 2003, 6–122.
5. Brigham Young University, Watershed Modeling System, WMS 6.1, Reference Manual, Environmental Modeling Research Laboratory, Provo, UT, 1999.
6. Brown, C.L. and Barnwell, T.O., The Enhanced Stream Water Quality Models QUAL2E and QUAL2E-UNCAS: Documentation and User Manual, U.S. Environmental Protection Agency, Environmental Research Laboratory, Athens, Georgia, EPA/600/3-87/007, 1987.
7. Bicknell, B.R., Imhoff, J.C., Kittle, J.L., Jr., Jobes, T.H., and Donigian, A.S., Jr., Hydrological Simulation Program—FORTRAN, HSPF, Version 12, User's Manual, U.S. Environmental Protection Agency, National Exposure Research Laboratory, Athens, Georgia, 2001.
8. Ertürk, A., Gürel, M., Koca, D., Ekdal, A., Tanik, A., Seker, D.Z., Kabdasli, S., and Gönenç, I.E., Determination of model dimensions for a complex lagoon system—a case study from Turkey, in *Proc. XXX IAHR Congress Water Engineering and Research in a Learning Society: Modern Developments and Traditional Concepts*, Ganoulis, J. and Prinos, P., Eds., Vol. A, Thessalonica, Greece, 2003, 53.
9. Karak, P., Investigation of nutrient behavior in land based sources of pollutants, M.Sc. thesis, Istanbul Technical University, Institute of Science and Technology, Istanbul, Turkey, 2000 [in Turkish].
10. Gürel, M., Nutrient dynamics in coastal lagoons: Dalyan lagoon case study, Ph.D. thesis, Istanbul Technical University, Institute of Science and Technology, Istanbul, 2000.
11. Gürel, M., Okus, E., Polat Beken, C., Tanik, A., and Gönenç, I.E., Dalyan Lagunu nutrient dinamigine bir yaklasim, *Turkiye'nin Kiyi ve Deniz Alanlari III. Ulusal Konferansi*, Turkiye Kiyilari 01 Bildiriler Kitabi, Yildiz Teknik Universitesi, Istanbul, 2001, 437 [in Turkish].
12. Herrera Silveira, J.A., Ramirez, R.J., and Zaldivar, J.A., Overview and characterization of the hydrology and primary producer communities of selected coastal lagoons of Yucatan, Mexico, *Aquatic Ecosystem Health and Management*, 1, 353, 1998.

13. Gilabert, J., Perez-Ruzafa, A., and Marcos-Diego, C., Food web structure in coastal lagoons, NATO-CCMS Project on Ecosystem modeling of Coastal Lagoons for Sustainable Management, 4th Workshop, Gdynia, Poland, 1998.
14. Secmen, O. and Leblebici, E., *Su Urunleri Avciligi ve Av Teknolojisi*, G.T.H.B., Su Urunleri Genel Mudurlugu Yayini, Ankara, Turkey, 1980 [in Turkish].
15. Buhan, E., *Koycegiz Lagun Sistemindeki Mevcut Durumun ve Kefal Populasyonlarinin Arastirilarak Lagun Isletmeciliginin Gelistirilmesi*, T.C. Tarim ve Koyisleri Bakanligi Su Urunleri Arastirma Enstitusu Bolge Mudurlugu, Bodrum, Turkey, Seri B, Yayin No. 3, 1998 [in Turkish].
16. DIE, State Institute of Statistics, Prime Ministry of Turkish Republic, Bulletin dated November 30, 1997, Item No. 48, Mugla, Turkey, 1997.
17. Guvensoy, G., Fate of pesticides on soil and their impact on water environment, M.Sc. thesis, Institute of Science and Technology, Istanbul Technical University, Istanbul, Turkey, 2000.
18. Hengirmen, M., *Kaunos, Dalyan, Koycegiz Tanitim Kitapcigi*, Engin Yayinevi, 2000 [in Turkish].
19. Metcalf & Eddy, *Wastewater Engineering, Treatment Disposal and Reuse*, 3rd ed., McGraw-Hill, New York, 1991.
20. Boers, P.C.M., Nutrient emissions from agriculture in the Netherlands, causes and remedies, *Water Sci. Technol.*, 33(4–5), 183, 1996.
21. Broadbent, F.E. and Reisenhauer, H.M., Fate of wastewater constituents in soil and groundwater: nitrogen and phosphorus, in *Irrigation with Reclaimed Municipal Wastewater—A Guidance Manual*, Pettygrove, G.S. and Asano, T., Eds., Lewis Publishers, Chelsea, MI, 1988, 12–1.
22. Burt, T. P., Heathwaite, A. L., and Trudgill, S.T., *Nitrate: Processes, Patterns and Management*, John Wiley & Sons, Chichester, U.K., 1993.
23. Tunney, H. and Carton, O.T., Phosphorous leaching under cut grassland, *Wat. Sci. Technol.*, 39(12), 63, 1997.
24. Johnes, P. J. and Hodgkinson, R.A., Phosphorus loss from agricultural catchments: pathways and implications for management, *Soil Use Man.*, 14, 175, 1998.
25. Holas, J., Holas, M., and Chous, V., Pollution by phosphorus and nitrogen in water treams feeding the Zelivka drinking water reservoir, *Wat. Sci. Technol.*, 39(12), 207, 1999.
26. Tanik, A., Beler Baykal, B., and Gönenç, I.E., The impact of agricultural pollutants in six drinking water reservoir, *Water Sci. Technol.*, 40(2), 11, 1999.
27. Ozturk, S., *Tarim Ilaclari*, extended 2nd ed., Ak Yayinevi, Istanbul, Turkey, 1997 [in Turkish].
28. Leonard, R.A., Knisel, W.G., and Still, D.A., Gleams: Groundwater loading effects of agricultural management systems, *Trans ASAE*, 30(5), 1403, 1987.
29. Tanik, A., Gürel, M., Guvensoy, G., and Gönenç, I.E., Pesticide distribution and dynamics in the catchment area of a coastal lagoon—A case study from Turkey, *5th Int. Symp. and Exhibition on Environmental Contamination in Central and Eastern Europe*, Prague, Czech Republic, 2000 (proceedings on CD-Rom).
30. Tanik, A., Gürel, M., and Gönenç, I.E., Effect of irrigation on the distribution of surplus fertilizer loads on land, in *Proc. 13th Int. Symp. of the International Scientific Centre of Fertilizers (CIEC), Fertilizers in Context with Resource Management in Agriculture*, Tokat, Turkey, 2002, 194.
31. Ertürk, A., Gürel, M., Ekdal, A., Yüceil, K., and Dertli, B., Application of mathematical modelling and uncertainty analysis to a stream for the estimation of nutrient loadings transported into a lake, in *Proc. Asian Waterquality 2003 IWA Asia-Pacific Regional Conference*, Bangkok, Thailand, 2003 [proceedings on CD].

32. Wool, T.A., Ambrose, R.B., Martin, J.L., and Comer, E.A., Water Quality Analysis Simulation Program WASP, Version 6.0.0.12, U.S. Environmental Protection Agency, Athens, GA, 2001.

33. Ambrose, R.B., Jr., Wool, T.A., and Martin, J.L., The Dynamic Estuary Model Hydrodynamics Program, DYNHYD5, Model Documentation and User Manual, U.S. Environmental Protection Agency, Environmental Research Laboratory, Athens, GA, 1993.

34. Lung, W., A Water Quality Model for the Patuxent Estuary, Final Report submitted to Maryland Department of Environment, University of Virginia, Charlottesville, 1992.

35. Ambrose, R.B., Wool, T.A., and Martin, J.L., The Water Quality Analysis Simulation Program, WASP5, Part A: Model Documentation, U.S. Environmental Protection Agency, Athens, GA, 1993.

36. Cole, T.M. and Wells, S.A., CE-QUAL-W2 A Two-Dimensional, Laterally Averaged, Hydrodynamic and Water Quality Model, Version 3.1, Instruction Report EL-2002-1, U.S. Army Engineering and Research Development Center, Vicksburg, MS, 2002.

37. Ertürk, A., Hydraulic modeling of Koycegiz Dalyan lagoon system, M.Sc. thesis, Institute of Science and Technology, Istanbul Technical University, Istanbul, 2002.

38. Pearce, D., Whittington, D., Georgious, S., and James, D., Project and Policy Appraisal: Integrating Economics and Environment, Paris, OECD Documents, 1994.

# Index

## A

*Acartia,* 401, 413
Accuracy, 242–249, *243–246,* 302
*Acetabularia,* 401, 407
Acidity, denitrification, 85
Administrative structure, 459
Adriatic Sea, 72, 211
Adsorption, 159
Advection and advective flux
  basics, 44–45
  numerical discretization techniques, 237–239
  physical processes, 47–48
Advection-dispersion modeling, 429, 433
Aegean Sea, 211
Agriculture, 23
Agriculture sector, 455
Air quality, 441
Alagol Lake, 448, 456, 465
Aland Sea (Baltic), *114*
Alien species, 318
*Alosa fallax,* 220
AMAP, *see* Arctic Monitoring and Assessment Program (AMAP)
Ambrose studies, 165, 179
Ammonia, 89, 155
Ammonia nitrogen (benthic), 116–118, *117–119*
Ammonium, 87–88
Ammonium nitrogen, 110–111, *112–113*
*Amphipholis,* 401
Andersen studies, *123*
Andrulewicz studies, 307–329
ANNs, *see* Artificial neural networks (ANNs)
Anthropogenic pressure
  decision support system, 340
  indicators, decision-making tool, 347–348
  monitoring programs, 312
  Vistula Lagoon, 423, 425–426
AQUATOX
  dissolved oxygen, 133
  nitrogen cycle, 106
  phosphorus cycle, 121
Arcachon Basin, France, 83, 87
Arcachon Bay, *124*
ArcGIS software, 459–460

Arctic Monitoring and Assessment Program (AMAP), 310
Arctic seas, 309
ArcView software, 459
Army Corps of Engineers (U.S.), 349
Arrhenius and Arrhenius-type equations
  nitrification, 84
  organic matter oxidation, 102
  organic nitrogen mineralization, 88
  oxidation, 180
  reaeration, 100
  sediment oxygen demand, 104
*Artemia,* 217
Artificial neural networks (ANNs), 35
ASA model, 349
Asian Bank, 357
Asko Area (Baltic), *114*
Assawoman Bay, Maryland, *119, 131*
Assumptions, sustainable use and development, 335
*Asterina,* 219
*Asterionella,* 401, 412
*Aurelia,* 401, 413, 415
Azam studies, 207

## B

Bacteria, boundary processes, 74
Baltic Marine Environment Protection Commission, 316
Baltic Sea and lagoons, *see also* specific lagoons
  anthropogenic pressure, 425
  basics, 375
  catchment area, 423
  fish assemblages, 211
  hydrological parameters, 256–257
  monitoring programs, 309
  short-term data collection, 328
  silicon cycle, 96
Baltic Sea Joint Comprehensive Environmental Action Programme, 423, 425
Baltijsk Strait, 423, 425, 427, *see also* Vistula Lagoon
Bank End, U.K., *115*
Bansal studies, *112*

Wind
  drift current, 383
  friction factor, 302
  stress, 388
  surge, 278–279
  waves, 281–282
Wind-driven currents
  basics, 376, 384–387
  Grande-Entrée Lagoon, *377,* 377–378
  numerical modeling, *377–378,* 383–384,
    *385–386*
  observations, *378–379,* 378–383, *381–382*
  pre-modeling, 275
Wishart, Knowles and, studies, *114*
WMS, *see* Watershed modeling system (WMS)
Wolfe studies, 163
Wolflin studies, 1–4, 331–367
Wollast studies, *132*
Wool studies, 133
World Business Council for Sustainable
    Development (WBCSD), 338
World Conservation Strategy (WCS), 337
WTA, *see* Willingness to accept (WTA)
WTP, *see* Willingness to pay (WTP)
Wunderlich studies, 267

**Y**

Yaquina River Estuary, Oregon, *147*
Yoon and Benner studies, *124*
Ythan Estuary, Scotland, 206
Yüceil studies, 440–471
Yuvarlakcay Creek, 460

**Z**

Zac studies, 328
Zebra mussel, 318, *see also* Mussels
Zimmerman studies, 271
Zooplankton
  biological assemblages, 401
  nutrient inputs, 413, 416
  oligotrophic state, 201–204
*Zostera*
  eutrophication process, 198
  leaky lagoons, 219
  Mar Menor Lagoon, 406–407
  nitrogen uptake, 83